$150

NEURAL NETWORK MODELING AND IDENTIFICATION OF DYNAMICAL SYSTEMS

NEURAL NETWORK MODELING AND IDENTIFICATION OF DYNAMICAL SYSTEMS

YURY V. TIUMENTSEV

Moscow Aviation Institute

Department of Flight Dynamics and Control

Moscow, Russia

MIKHAIL V. EGORCHEV

Moscow Aviation Institute

Department of Computational Mathematics and Programming

Moscow, Russia

ELSEVIER

ACADEMIC PRESS

An imprint of Elsevier

Academic Press is an imprint of Elsevier
125 London Wall, London EC2Y 5AS, United Kingdom
525 B Street, Suite 1650, San Diego, CA 92101, United States
50 Hampshire Street, 5th Floor, Cambridge, MA 02139, United States
The Boulevard, Langford Lane, Kidlington, Oxford OX5 1GB, United Kingdom

Library of Congress Cataloging-in-Publication Data
A catalog record for this book is available from the Library of Congress

British Library Cataloguing-in-Publication Data
A catalogue record for this book is available from the British Library

ISBN: 978-0-12-815254-6

For information on all Academic Press publications
visit our website at https://www.elsevier.com/books-and-journals

Publisher: Mara Conner
Acquisition Editor: Chris Katsaropoulos
Editorial Project Manager: Leticia M. Lima
Production Project Manager: Anitha Sivaraj
Designer: Christian J. Bilbow

Typeset by VTeX

Contents

About the Authors vii
Preface ix
Acknowledgment xiii
List of Acronyms xv
List of Notations xvii

Introduction 1

1. The Modeling Problem for Controlled Motion of Nonlinear Dynamical Systems

1.1 The Dynamical System as an Object of Study 7
1.1.1. The General Concept of a Dynamical System 7
1.1.2. Classes of Dynamical Systems 11
1.1.3. Classes of Environments 14
1.1.4. Interaction Between Systems and Environment 15
1.1.5. Formalization of the Dynamical System Concept 16
1.1.6. Behavior and Activity of Systems 20
1.2 Dynamical Systems and the Problem of Adaptability 21
1.2.1. Types of Adaptation 22
1.2.2. General Characteristics of the Adaptive Control Problem 23
1.2.3. Basic Structural Variants of Adaptive Systems 24
1.2.4. The Role of Models in the Problem of Adaptive Control 27
1.3 A General Approach to Dynamical System Modeling 28
1.3.1. A Scheme of the Modeling Process for Dynamical Systems 28
1.3.2. The Main Problems That Need to Be Solved During Design of a Model for a Dynamical System 32
References 33

2. Dynamic Neural Networks: Structures and Training Methods

2.1 Artificial Neural Network Structures 35
2.1.1. Generative Approach to Artificial Neural Network Design 35
2.1.2. Layered Structure of Neural Network Models 39
2.1.3. Neurons as Elements From Which the ANN Is Formed 47
2.1.4. Structural Organization of a Neuron 49
2.2 Artificial Neural Network Training Methods 51
2.2.1. Overview of the Neural Network Training Framework 52
2.2.2. Static Neural Network Training 58
2.2.3. Dynamic Neural Network Training 62
2.3 Dynamic Neural Network Adaptation Methods 67
2.3.1. Extended Kalman Filter 67
2.3.2. ANN Models With Interneurons 69
2.3.3. Incremental Formation of ANN Models 72
2.4 Training Set Acquisition Problem for Dynamic Neural Networks 73
2.4.1. Specifics of the Process of Forming Data Sets Required for Training Dynamic Neural Networks 73
2.4.2. Direct Approach to the Process of Forming Data Sets Required for Training Dynamic Neural Networks 73
2.4.3. Indirect Approach to the Acquisition of Training Data Sets for Dynamic Neural Networks 80
References 88

3. Neural Network Black Box Approach to the Modeling and Control of Dynamical Systems

3.1 Typical Problems Associated With Development and Maintenance of Dynamical Systems 93
3.2 Neural Network Black Box Approach to Solving Problems Associated With Dynamical Systems 94

3.2.1. Main Types of Models 94
3.2.2. Approaches to Consideration of Disturbances Acting on a Dynamical System 95
3.3 ANN-Based Modeling and Identification of Dynamical Systems 99
3.3.1. Feedforward Neural Networks for Modeling of Dynamical Systems 99
3.3.2. Recurrent Neural Networks for Modeling of Dynamical Systems 101
3.4 ANN-Based Control of Dynamical Systems 102
3.4.1. Adjustment of Dynamic Properties of a Controlled Object Using Artificial Neural Networks 102
3.4.2. Synthesis of an Optimal Ensemble of Neural Controllers for a Multimode Aircraft 112
References 126

4. Neural Network Black Box Modeling of Nonlinear Dynamical Systems: Aircraft Controlled Motion

4.1 ANN Model of Aircraft Motion Based on a Multilayer Neural Network 131
4.1.1. The General Structure of the ANN Model of Aircraft Motion Based on a Multilayer Neural Network 131
4.1.2. Learning of the Neural Network Model of Aircraft Motion in Batch Mode 133
4.1.3. Learning of the Neural Network Model of Aircraft Motion in Real-Time Mode 133
4.2 Performance Evaluation for ANN Models of Aircraft Motion Based on Multilayer Neural Networks 134
4.3 Application of ANN Models to Adaptive Control Problems for Nonlinear Dynamical Systems Operating Under Uncertainty Conditions 139
4.3.1. The Demand for Adaptive Systems 139
4.3.2. Model Reference Adaptive Control 140
4.3.3. Model Predictive Control 154
4.3.4. Adaptive Control of Angular Aircraft Motion Under Uncertainty Conditions 158
References 162

5. Semiempirical Neural Network Models of Controlled Dynamical Systems

5.1 Semiempirical ANN-Based Approach to Modeling of Dynamical Systems 165
5.2 Semiempirical ANN-Based Model Design Process 170
5.3 Semiempirical ANN-Based Model Derivatives Computation 177
5.4 Homotopy Continuation Training Method for Semiempirical ANN-Based Models 187
5.5 Optimal Design of Experiments for Semiempirical ANN-Based Models 192
References 196

6. Neural Network Semiempirical Modeling of Aircraft Motion

6.1 The Problem of Motion Modeling and Identification of Aircraft Aerodynamic Characteristics 199
6.2 Semiempirical Modeling of Longitudinal Short-Period Motion for a Maneuverable Aircraft 200
6.3 Semiempirical Modeling of Aircraft Three-Axis Rotational Motion 208
6.4 Semiempirical Modeling of Longitudinal Translational and Angular Motion for a Maneuverable Aircraft 216
References 225

A. Results of Computational Experiments With Adaptive Systems

Index 311

About the Authors

Yury V. Tiumentsev: PhD in Flight Dynamics and Control, DSc in System Analysis, Control, and Information Processing, Vice-President of the Russian Neural Network Society (RNNS). He is a Professor within the Department of Flight Dynamics and Control and the Department of Numerical Mathematics and Computer Programming, Moscow Aviation Institute (National Research University). His research interests are in artificial neural networks, neural network modeling and identification of dynamical systems, adaptive and intelligent systems, and mathematical modeling and computer simulation of nonlinear dynamical systems. He is author and coauthor of more than 130 publications and one monograph.

Mikhail V. Egorchev: PhD in Mathematical Modeling, Numerical Methods, and Software Engineering. At the time of writing the book he was a PhD student within the Department of Numerical Mathematics and Computer Programming of the MAI; currently he is a senior R&D Software Engineer at LLC RoboCV. His research activities are related to adaptive and optimal control methods for robotic applications. His current research interests are in artificial neural networks, neural network modeling and identification of dynamical systems, mathematical modeling and computer simulation of nonlinear dynamical systems, numerical optimization methods, and optimal control. He is author and coauthor of 20 articles.

Preface

One of the key elements of the development process for new engineering systems is the formation of mathematical and computer models that provide solutions to the problems of creating and using appropriate technical systems. As the complexity of the created systems grows, so do the requirements for their models, as well as for the funds raised for the development of these models.

At present, the possibilities of mathematical and computer modeling are lagging behind the needs of a number of engineering fields, such as aerospace technology, robotics, and control of complex production processes. Characteristic of technical systems from these areas is the high level of complexity of the objects and processes being modeled, their multidimensionality, nonlinearity, and nonstationarity, and the diversity and complexity of the functions implemented by the object being modeled. Solving the problems of modeling for objects of this kind is significantly complicated by the fact that the corresponding models have to be formed in the presence of multiple and heterogeneous uncertainties, such as incomplete and inaccurate knowledge of the characteristics and properties of the object being simulated, as well as the conditions in which the object will operate. In addition, the properties of the simulated object may undergo changes, including sharp and significant changes immediately during operation, for example, due to equipment failures and/or damage of its structure. In this case, an object model that was previously formed on the basis of its nominal ("intact") state becomes inadequate. If this model is used, for example, in an object control system, a critical situation arises.

In this regard, it is reasonable to search for new tools for modeling nonlinear controlled systems operating under significant and heterogeneous uncertainties. The presence of models of this kind opens up new opportunities for solving problems of control of complex dynamic systems.

The introductory chapter substantiates the need for a new approach to the formation of mathematical and computer models of controlled dynamical systems operating under conditions of multiple and heterogeneous uncertainties. We demonstrate that the control problems for dynamical systems under uncertainty conditions cause a need to give the property of adaptability to the dynamical system model. The traditional approach to mathematical and computer modeling of the dynamical systems does not satisfy this requirement. We can overcome this difficulty by applying techniques of neural network modeling. However, traditional ANN models, which belong to the black box class, do not allow for a complete solution to the task. This circumstance makes it necessary to expand the black box–type neural network models to the gray box class.

Chapter 1 deals with the modeling of controlled motion of nonlinear systems. It covers topics such as a dynamical system as an object of study, the formalization of the concept of the dynamical system, and the behavior and activities of such systems. The concept of adaptability is discussed, one of the most important concepts for the advanced dynamical systems being created, in particular for robotic aircraft of various classes. An approach is formulated to solve the problem of dynamical system modeling, a general scheme of the modeling process is given, and the main problems that need to be solved when forming the dynamical system model are identified.

Chapter 2 is about the ANN approach to the modeling and control of dynamical systems. The classes of ANN models for dynamical systems are considered in this chapter, including static (feedforward) networks and dynamic (feedback) networks. Possible ways to implement the three main elements of the process of forming ANN models are identified: (1) the generation of a potentially rich class of models that contains the ANN model being created as an element (class instance); (2) obtaining an informative data set required for the structural and parametric adjustment of the generated ANN model; (3) building learning algorithms that carry out structural adjustments and parametric tuning of the ANN model being formed. In addition, the issues of ensuring the adaptability of ANN models are considered, which are among the most important from the point of view of solving the problem of aircraft robotization.

In Chapter 3 we study the problem of modeling the controlled motion of dynamical systems, an aircraft in particular, using the traditional ANN modeling tools, in which the model is based only on experimental data on the behavior of the simulation object, i.e., it is purely empirical ("black box" model). We discuss two main strategies to the representation of dynamical systems. These are state space representations and input-output representations. We consider the problem of modeling and identification of dynamic systems, as well as the capabilities of feedforward networks and recurrent networks to solve this problem. Traditional (black box) networks are used to solve control problems using the simple problem of adjusting the dynamic properties of an aircraft as an example. The extension of this approach is the formation of an optimal ensemble of neurocontrollers for multimode dynamical system.

Chapter 4 gives examples of solving problems of motion modeling and control based on traditional-type ANN black box models. The ANN architecture of the NARX type is used, which allows to build both the aircraft motion model and control law for this motion. As an example, the problem of modeling a controlled longitudinal angular motion of an aircraft is considered. The corresponding motion models for aircraft of various classes are obtained for this problem, and their performance is evaluated. These models are then used to form adaptive control systems such as model reference adaptive control (MRAC) and model predictive control (MPC), for which a set of applied problems has been solved. The results obtained for these tasks allow us to estimate the potentialities of the ANN simulation for the considered range of problems. It is shown that in some cases the capabilities of this class of models are insufficient when solving problems of modeling controlled motion of an aircraft. As a consequence, an extension of this class of models is required, which is discussed in Chapters 5 and 6.

Chapter 5 proposes a variant of ANN modeling, which expands the possibilities of traditional dynamic ANN models by embedding into these models the available theoretical knowledge about the object of modeling. The resulting combined ANN models are called semiempirical ("gray box") models. The processes of formation of such models and the implementation of the main elements of these processes are considered. We illustrate their specificity using a demo example of a dynamical system. The same example, in combination with its sophisticated variants, is used for the initial experimental evaluation of the possibilities of semiempirical ANN modeling of controlled dynamic systems. Also we describe in this chapter the properties of semiempirical state-space continuous time ANN-based models. Then, we discuss the continuous time counterparts of Real-Time Recurrent Learning (RTRL) and backpropagation through time (BPTT) algorithms required for the computation of error function derivatives. We also provide a description of the homotopy continuation training method for semiempirical ANN-based models. Finally, we treat the topic of optimal design of experiments for semiempirical models of controlled dynamical systems.

Chapter 6 presents the results of computational experiments obtained in solving real-world applied problems related to the simulation of motion and the identification of the aerodynamic characteristics of a maneuverable aircraft. These results show how efficient the semiempirical approach is to modeling nonlinear controlled dynamical systems and to solving problems of identifying their characteristics. First, the simpler task of modeling the longitudinal short-period motion of a maneuverable aircraft is considered. Then the results of solving the problem of modeling the total angular motion of a maneuverable aircraft, as well as the problem of identifying aerodynamic characteristics for it (coefficients of lifting and lateral force, coefficients of angles of roll, yaw and pitch), obtained as nonlinear functions of several variables, are presented. Then, modeling of the longitudinal trajectory and angular motion of the aircraft is considered, which allows identifying the drag coefficient for it.

The Appendix presents the results of numerous simulation results for adaptive systems of the MRAC and MPC types, based on the use of ANN modeling, as applied to aircraft of various classes.

Yury Tiumentsev, Mikhail Egorchev
Moscow, Russian Federation
December 2018

Acknowledgment

We want to express our gratitude for the many useful discussions and support of the following Professors of the Moscow Aviation Institute: Vladimir Brusov, Vladimir Bobronnikov, Alexander Bortakovsky, Andrey Chernyshev, Alexander Efremov, Valery Grumondz, Mikhail Krasilshchikov, Nikolay Markin, Nikolay Morozov, Valery Ovcharenko, Andrey Panteleyev, Dmitry Reviznikov, and Valentin Zaitsev.

We would also like to express our gratitude to our colleagues in the Russian Neural Network Society for many years of fruitful cooperation. Among them, we wish to mention Galina Beskhlebnova, Alexander Dorogov, Vitaly Dunin-Barkovsky, Yury Kaganov, Boris Kryzhanovsky, Leonid Litinsky, Nikolay Makarenko, Olga Mishulina, Yury Nechayev, Vladimir Redko, Sergey Shumsky, Lev Stankevich, Dmitry Tarkhov, Sergey Terekhov, Alexander Vasiliev, and Vladimir Yakhno.

The results presented in this book were obtained with the significant participation of members of our research team, among whom Alexey Kondratiev, Dmitry Kozlov, Alexey Lukanov, and Alexander Yakovenko.

We express our sincere gratitude to the Elsevier Team that worked on our book for their valuable support and assistance: our Editorial Project Manager, Anna Valutkevich, our Senior Acquisitions Editor, Chris Katsaropoulos, our Editorial Project Manager, Leticia Melo Garcia de Lima, and our Copyrights Coordinator, Josy, Sheela Bernardine B., our Project Manager, Reference Content Production, Anitha Sivaraj, as well as the Elsevier TeX-Support Team.

Yury Tiumentsev, Mikhail Egorchev
December 2018

List of Acronyms

$\mathbb{DS}$	Deterministic system
$\mathbb{VS}$	Vague system
$\mathbb{CS}$	Controllable system
$\mathbb{AS}$	Adaptive system
$\mathbb{IS}$	Intelligent system
$\mathcal{SE}$	Stereotyped environment
$\mathcal{UE}$	Uncertain environment
$\mathcal{RE}$	Reacting environment
$\mathcal{AE}$	Adaptive environment
$\mathcal{IE}$	Intelligent environment
AC	Adjusting controller
ANN	Artificial neural network
AWACS	Airborne warning and control system
BPTT	Backpropagation through time
CMA-ES	Covariance Matrix Adaptive Evolution Strategy
DAE	Differential algebraic equation
DTDNN	Distributed Time Delay Neural Network
EKF	Extended Kalman filter
ENC	Ensemble of neural controllers
FB	Functional basis
FTDNN	Focused Time Delay Neural Network
GLONASS	Global Navigation Satellite System
GS	Gain scheduling
GPS	Global Positioning System
HRV	Hypersonic research vehicle
IVP	Initial value problem
KM	Kalman filter
LDDN	Layered Digital Dynamic Network
MAC	Mean aerodynamic chord
MLP	MultiLayer Perceptron
MDS	Multimode dynamical system
MIMO	Multiple Input Multiple Output
MISO	Multiple Input Single Output
MPC	Model predictive control
MRAC	Model reference adaptive control
MSE	Mean square error
NARMAX	Nonlinear AutoRegressive network with Moving Average and eXogeneous inputs
NARX	Nonlinear AutoRegressive network with eXogeneous inputs
NASA	National Aeronautical and Space Administration
NASP	National AeroSpace Plane
NC	Neural controller
NLP	Nonlinear programming
NM	Network model
ODE	Ordinary differential equation
PDE	Partial differential equation
PD, PI, PID	Proportional-differential, proportional-integral, proportional-integral-differential
RBF	Radial basis function
RLSM	Recursive least-squares method
RKNN	Runge–Kutta Neural Network
RM	Reference model
RMLP	Recurrent MultiLayer Perceptron
RMSE	Root-mean-square error
RTRL	Real-Time Recurrent Learning
SISO	Single Input Single Output
TDL	Time delay line, tapped delay line
TDNN	Time Delay Neural Network
UAV	Unmanned aerial vehicle

List of Notations

$\mathbb{U} = \mathbb{S}^* \cup \mathcal{E}^*$	universe of reasoning
$\mathbb{S}^*$	dynamical system
$\mathcal{E}^*$	environment
$\mathbb{K} = \langle \mathbb{S}, \mathcal{E} \rangle$	system-complex
$\mathbb{S}$	system-object, $\mathbb{S} \subset \mathbb{S}^*$
$\mathcal{E}$	system-environment, $\mathcal{E} \subset \mathcal{E}^*$
$\mathbf{x} = (x_1, \ldots, x_n) \in X$, $x_1 \in X_1 \subseteq \mathcal{R}, \ldots, x_n \in X_n \subseteq \mathcal{R}$, $X = X_1 \times \ldots \times X_n \subseteq \mathcal{R}^n$	state (phase) vector of system $\mathbb{S}$
$\mathcal{R}$	real number line
$\mathbf{y} = (y_1, \ldots, y_p) \in Y$	observation vector of system $\mathbb{S}$
$\mathbf{u} = (u_1, \ldots, u_m) \in U$	control vector of system $\mathbb{S}$
$\mathbf{w} = (w_1, \ldots, w_s) \in W$	vector of parameters for system $\mathbb{S}$
$\xi = (\xi_1, \ldots, \xi_q) \in \Xi$	vector of perturbations for system $\mathbb{S}$
$\zeta = (\zeta_1, \ldots, \zeta_q) \in Z$	vector of measurement noises for system $\mathbb{S}$
$t \in T \subseteq \mathcal{R}$	time
$\lambda(t_i)$	situation for time instant t_i
$\lambda_{ext}(t_i)$	external components of situation for time instant t_i
$\lambda_{int}(t_i)$	internal components of situation for time instant t_i
$\gamma(t_i)$	goal of the system for time instant t_i
D	aerodynamic drag
L	lift force
Y	side force
C_D	drag coefficient (wind axes)
C_L	lift coefficient (wind axes)
C_y	side force coefficient (wind axes)
D	axial force (body axes)
Z	normal force (body axes)
Y	lateral force (body axes)
C_x	axial force coefficient (body axes)
C_y	lateral force coefficient (body axes)
C_z	normal force coefficient (body axes)
L	rolling moment
M	pitching moment
N	yawing moment
C_l	rolling moment coefficient
C_m	pitching moment coefficient
C_n	yawing moment coefficient $C_{L_\alpha} = \dfrac{\partial C_L}{\partial \alpha}$; $C_{y_\beta} = \dfrac{\partial C_y}{\partial \beta}$; $C_{m_\alpha} = \dfrac{\partial C_m}{\partial \alpha}$; $C_{m_q} = \dfrac{\partial C_m}{\partial q}$
T	thrust of the engine
T_{ref}	specified thrust level
T_{cr}	current thrust level
T_{idle}	thrust of the engine in the idle mode
T_{mil}	thrust of the engine in the military mode
T_{max}	thrust of the engine in the maximum mode
P_c	command relative thrust value
P_a	actual relative thrust value
δ_{th}	relative position of the engine throttle
α	angle of attack
β	angle of sideslip
ϕ	roll angle
ψ	yaw angle
θ	pitch angle
p	angular velocity about x-axis
r	angular velocity about y-axis
q	angular velocity about z-axis
Δq_{turb}	increment in angular velocity q due to atmospheric turbulence
γ	flight path angle
δ_a	angle of aileron deflection
δ_e	angle of elevator deflection
δ_r	angle of rudder deflection
V	airspeed
M_0	free stream Mach number
H	altitude of flight
q_p	dynamic air pressure
ρ	air density
p	atmospheric pressure
g	acceleration due to gravity
W	wind speed
S	wing area
b	wing span
$\bar{c}$	mean aerodynamic chord (MAC)
m	aircraft mass
W	aircraft weight
I_x	moment of inertia in roll
I_y	moment of inertia in pitch
I_z	moment of inertia in yaw
I_{xz}	product of inertia about ox and oz axes
ω_{rm}	natural frequency of the reference model

Introduction

The world of controlled dynamical systems is diverse and multifaceted. Among the most important classes of such systems, traditionally difficult to study, are aircraft of various kinds. One of the challenging modern problems of aeronautical engineering is to create highly autonomous robotic unmanned aerial vehicles (UAVs) intended for the accomplishment of civil and military missions under a wide variety of conditions. These missions include patrolling, threat detection, and object protection; monitoring of power lines, pipelines, and forests; aerial photography, monitoring and survey of ice and fishing areas; performing various assembly operations and terrestrial and naval rescue operations; and providing assistance in natural and technogenic disaster recovery, various military operations, and many other scenarios.

Currently, these types of missions are carried out mostly by manned aircraft (airplanes and helicopters). The constantly growing number of applications of UAVs in this field is attractive due to the following reasons:

- UAVs achieve a significantly higher payload fraction compared to manned aircraft because they require neither flight crew nor life support systems and flight compartments;
- UAVs are capable of higher maneuverability compared to manned aircraft because of a higher limit g-load level;
- there is the possibility of creating small UAVs that are significantly less expensive to build and operate;
- UAVs are capable of accomplishing missions when human presence is undesirable or unacceptable (radioactive hazard, high-risk level, etc.).

Nowadays almost all UAVs are actually remotely piloted vehicles that require a human operator (or even an entire crew) located at the ground control station to manage UAV flight. However, UAVs will be truly effective only when they are capable to accomplish missions with a maximum degree of autonomy, i.e., with minimal human assistance, which is generally reduced to stating the mission goal, monitoring its accomplishment, and, sometimes, adjusting the mission during the flight. This is caused by potential vulnerability of wireless communication channels required for the remotely piloted vehicle. Moreover, the human operator ability to react to complex and rapidly changing situations has certain psychophysiological limitations (attentional capacity, reaction time). High autonomy refers not to the ability to follow a predefined flight schedule, but to a "smart" autonomous behavior that adapts to highly uncertain, dynamically changing situations. At present, a significant number of researchers from different countries are making an effort to solve the problem of smart autonomy; however there is no satisfactory solution yet for this problem.

Thus, the most important requirement for UAVs is that they must have a high level of independence in solving their tasks. To meet these requirements, a robotic UAV should be able:

- to achieve its goals in a highly dynamic environment with a significant number of heterogeneous uncertainties in it, taking into account possible counteraction;
- to adjust the goals, as well as to form new goals and sets of goals, based on the value and regulatory attitudes (motivation) laid down in the UAV behavior control system;

- to be able to assess the current situation on the basis of a multilateral perception of the external and internal environment, to be able to form a forecast of the development of the situation;
- to gain new knowledge, accumulate experience in solving various tasks, learn from this experience, and modify its behavior based on the knowledge gained and accumulated experience;
- to be able to learn how to solve problems not provided for by the original design of the system;
- to form teams that are able to solve some problem by interactions between their members.

In order for robotic UAVs to be able to accomplish difficult missions on the same efficiency level as a manned aircraft, a radical revision of the current approach to development and management of control algorithms for UAV behavior is needed. In robotics, the totality of all types of processes of functioning of the robot is usually called the behavior of the robot. Accordingly, bearing in mind the ever-increasing trends in the robotization of the UAV, it is accepted to talk about the task of the behavior control for UAVs as the implementation of all types of its functioning necessary to fulfill the abovementioned target tasks. The behavior control of the UAV includes the following elements:

- planning flight operations, managing its implementation, updating the plan when a situation changes;
- UAV motion control, including its trajectory motion (including guidance and navigation) and angular motion;
- management of the solution of target tasks (control of the action of observation and reconnaissance equipment, control of the actions for performing assembly operations, etc.);
- management of interaction with other aircraft, both unmanned and manned, when the task is accomplished by some team of aircraft, which includes the given UAV.

The control algorithms (formation of control actions, decision making for control) should use information about the mission goals and about the situation characterized by assessments of the current and predicted situation in which it performs the task of the UAV, as input data. This situation is made up of both external components (state of the environment, the state and actions of partners and opponents) and internal components (data on aircraft state, diagnostics data, and performance evaluations of the structure and aircraft systems). Means of obtaining this basic information should also be included in the complex of algorithms that implement the desired behavior of a robotic UAV.

The aforementioned requirements can only be fulfilled if the UAV's behavior control system possesses advanced mechanisms, which allow an adaptation to significantly changing situations with a high degree of uncertainty and also learning and knowledge acquisition based on current UAV activity for future use. Such mechanisms should allow the possibility to solve the following important tasks:

- obtaining situation awareness which involves current situation assessment and future situation prediction;
- synthesis and implementation of UAV behavior as an aggregation of purposeful reactions to a current and/or predicted situation.

The implementation of these mechanisms provides the ability to create adaptive and intelligent systems to control the behavior of UAVs. The use of such systems gives an opportunity to implement highly autonomous robotic UAVs, designed to effectively accomplish difficult missions under uncertainty conditions. Another important implication of adaptive and intelligent control of UAV behavior is the possibility to significantly increase survivability of an aircraft in case of severe airframe damage and onboard systems failures.

When implementing the above functions, both in the process of design and in the subsequent operation of various types of aircraft, a significant place is occupied by the analysis of the behavior of dynamical systems, the synthesis of control algorithms for them, and the identification of their unknown or inaccurately known characteristics. A crucial role in solving the problems of these three classes belongs to mathematical and computer models of dynamical systems.

The traditional classes of mathematical models for engineering systems are ordinary differential equations (ODEs) (for systems with lumped parameters) and partial differential equations (PDEs) (for systems with distributed parameters). As applied to controlled dynamical systems, ODEs are most widely used as a modeling tool. These models, in combination with appropriate numerical methods, are widely used in solving problems of synthesis and analysis of controlled motion of aircraft of various classes. Similar tools are also used to simulate the motion of dynamical systems of other types, including surface and underwater vehicles and ground moving vehicles.

Methods of forming and using models of the traditional type are by now sufficiently developed and successfully used to solve a wide range of tasks. However, in relation to modern and advanced engineering systems, a number of problems arise, the solutions of which cannot be provided by traditional methods. These problems are caused by the presence of various and numerous uncertainties in the properties of the corresponding system and in its operational conditions, which can be parried only if the system in question has the property of adaptability, i.e., if there are means of operational adjustment of the system and its model to the changing current situation. In addition, the requirements for the accuracy of models imposed on the basis of the specificity of the applied problem being solved in some cases exceed the capabilities of traditional methods.

As experience shows, the modeling tool that is most appropriate for this situation is the concept of an artificial neural network (ANN). Such an approach can be considered as an alternative to traditional methods of dynamical system modeling, which provides, i.a., the possibility of obtaining adaptive models. At the same time, traditional neural network dynamical system models, in particular, the models of the NARX and NARMAX classes, which are most often used for the simulation of controlled dynamical systems, are purely empirical ("black box"–type) models, i.e., based solely on experimental data on the behavior of an object. However, in tasks of the complexity level that is typical for aerospace technology, this kind of empirical models is very often not capable of achieving the required level of accuracy. In addition, due to the peculiarities of the structural organization of such models, they do not allow solving the problem of identifying the characteristics of the dynamical system (for example, the aerodynamic characteristics of an aircraft), which is a serious disadvantage of this class of models.

One of the most important reasons for the low efficiency of traditional-type ANN models in problems associated with complex engineering systems is that a purely empirical (black box) model is being formed, which should cover all the peculiarities of the dynamical system behavior. For this, it is necessary to build an ANN model of a sufficiently high dimension (that is, with a large number of adjustable parameters in it). At the same time, it is known from experience of ANN modeling that the larger the dimension of the ANN model, the greater the amount of training data required to configure it. As a result, with the amount of experimental data that can actually be obtained for complex engineering systems, it is not possible to train such models, providing a given level of accuracy.

To overcome this kind of difficulty, which is characteristic of traditional models, both in the form of differential equations and in the form of ANN models, it is proposed to use a combined approach. It is based on ANN modeling, due to the fact that only in this variant it is possible to get adaptive models. Theoretical knowledge about the object of modeling, existing in the form of ODEs (these are, for example, traditional models of aircraft motion), is introduced in a special way into the ANN model of the combined type (semiempirical ANN model). At the same time, a part of the ANN model is formed on the basis of the available theoretical knowledge and does not require further adjustment (training). Only those elements that contain uncertainties, such as the aerodynamic characteristics of the aircraft, are subject to adjustment and/or structural correction in the learning process of the generated ANN model.

The result of this approach is semiempirical ANN models, which allow us to solve problems inaccessible to traditional ANN methods. We can sharply reduce the dimensionality of the ANN model, which allows achieving the required accuracy using training sets that are insufficient in volume for traditional ANN models. Besides, this approach provides the ability to identify the characteristics of the dynamical system, described by nonlinear functions of many variables (for example, the dimensionless coefficients of aerodynamic forces and moments).

In subsequent chapters, we consider an implementation of this approach, as well as examples of its application for simulating the motion of an aircraft and identifying its aerodynamic characteristics.

Chapter 1 is devoted to a statement of the modeling problem for controlled motion of nonlinear dynamical systems. We consider the classes of problems, which arise from the processes of development and operation of dynamical systems (analysis, synthesis, and identification problems) and reveal the role of mathematical modeling and computer simulation in solving these problems. The next set of questions relates to the problem of the adaptability of dynamical systems. In this regard, we analyze the kinds of adaptation, the basic types of adaptive control schemes, and the role of models in the problem of adaptive control. The need for adaptability of the controlled object model is revealed, as well as the need for neural network implementation of adaptive modeling and control algorithms.

Chapter 2 presents the neural network approach to modeling and control of dynamical systems. The classes of ANN models for dynamical systems and their structural organization are considered in this chapter, including static (feedforward) networks and dynamic (recurrent) networks. The next significant problem that arises in the formation of ANN models of dynamical systems is related to the algorithms of their learning. In the second chapter, algorithms for learning dynamic ANN models are considered. The difficulties associated with such learning, as well as ways to overcome them, are analyzed. One of the fundamental requirements for the considered ANN models is giving them the property of adaptability. Methods for satisfying this requirement are considered, including the use of ANN models with interneurons and subnets of interneurons, as well as the incremental formation of ANN models. One of the critical problems when generating ANN models, especially dynamical system models, is an acquisition of training sets. In the second chapter, the specific features of processes needed to generate training sets for the ANN modeling of dynamical systems are analyzed. We consider direct and indirect approaches to the generation of these training sets. Algorithms for generating a set of test maneuvers and test excitation signals for the dynamical system required to obtain a representative set of training data are given.

In Chapter 3, we deal with the neural network black box approach to solving modeling problems associated with dynamical systems. We discuss state space representations and input-output representations for such systems. We attempt to show that using ANN technology we can solve the problem of appropriate representation of a nonlinear model of some dynamical system motion with high efficiency. Using such representation, we synthesize a neural controller that solves the problem of adjusting the dynamic properties of the controlled object (maneuverable aircraft). The next problem we solve in this chapter relates to designing control laws for multimode objects, in particular for airplanes. We consider here the concept of an ensemble of neural controllers (ENC) concerning the control problem for a multimode dynamical system (MDS) as well as the problem of optimal synthesis for the ENC.

Chapter 4 deals with black box neural network modeling of nonlinear dynamical systems for the example of aircraft controlled motion. First of all, we consider the design process for ANN empirical dynamical system models, which belong to the family of black box models. The basic types of such models are described, and approaches to taking into account disturbing actions on the dynamical system are analyzed. Then, we construct the ANN model of the aircraft motion based on a multilayer neural network. As a baseline model, a multilayered neural network with feedbacks and delay lines is considered, in particular, NARX- and NARMAX-type models. The training of such an ANN model in batch mode and in real-time mode is described. Then, the performance of the obtained ANN model of the aircraft motion is evaluated for an example problem of the longitudinal short-period aircraft motion modeling. The performance evaluation of the model is carried out using computational experiments. One of the most important applications of dynamical models is related to the problem of adaptive control for such systems. We consider the solution to the problem of adaptive fault-tolerant control for nonlinear dynamical systems operating under uncertainty conditions to demonstrate the potential capabilities of ANN models in this area. We apply both the model reference adaptive control (MRAC) and model predictive control (MPC) methods using empirical (black box)-type ANN models. Also, synthesis of neurocontrollers is carried out.

In Chapter 5 we discuss the hybrid, semiempirical ("gray box") neural network–based modeling approach. Semiempirical models rely on both the theoretical knowledge of the system and the experimental data on its behavior. As evidenced by the results of numerous computational experiments, such models possess high accuracy and computational speed. Also, the semiempirical modeling approach makes it possible to state and solve the identification problem for the characteristics of dynamical systems. That is a problem of great importance, and it is traditionally difficult to solve. These semiempirical ANN-based models possess the required adaptivity feature, just like the pure empirical ones. First, we describe the properties of semiempirical state space continuous time ANN-based models. Then, we outline the stages of the model design procedure and present an illustrative example. We discuss the continuous time counterparts of Real-Time Recurrent Learning (RTRL) and backpropagation through time (BPTT) algorithms required for the computation of error function derivatives. We also describe the homotopy continuation training method for semiempirical ANN-based models. Finally, we treat the topic of optimal design of experiments for semiempirical models of controlled dynamical systems.

Chapter 6 presents the simulation results that show how efficient the semiempirical approach is to the simulation of controlled dynamical systems and to the solution of problems of identifying their characteristics. First, the simpler task of modeling the longitudinal short-period motion of a maneuverable aircraft is considered. After that, the problem of modeling the total angular motion of a maneuverable aircraft is solved, as well as the ANN identification problem of aerodynamic characteristics for it (lift and lateral force coefficients, roll, yaw, and pitch moment coefficients) obtained as nonlinear functions of several variables. Then, another identification problem is solved, as a result of which we get the ANN representation for the aerodynamic drag coefficient of the maneuverable aircraft.

The Appendix presents the results of numerous computational experiments with adaptive systems of MRAC and MPC types with reference to aircraft of various classes. These results provide an adequate representation of topics such as (1) the effect of individual structural elements of these systems and the properties of these elements on the characteristics of the synthesized systems; (2) the capabilities of the systems under consideration from the point of view of ensuring their resistance to changes in the properties of the object due to failures and damages; (3) an assessment of the nature of the impact of atmospheric turbulence on the controlled system of the class under consideration; (4) an assessment of the possibilities of adaptation of the systems under consideration to uncertainties in the source data; and (5) assessment of the importance of adaptation mechanisms on the example of the problem of controlling the angular motion of the aircraft.

CHAPTER

1

The Modeling Problem for Controlled Motion of Nonlinear Dynamical Systems

1.1 THE DYNAMICAL SYSTEM AS AN OBJECT OF STUDY

Let us first formulate the concept of a dynamical system on an intuitive level, and then formalize it.

1.1.1 The General Concept of a Dynamical System

1.1.1.1 Concept of the System, Basic Postulates

The considerations presented below are based on some postulates that are assumptions based on the ideas of general systems theory [1–5].

Postulate 1. The World is a fragment of the Macrocosm, and this fragment is isolated in some way. This World is the *universe* $\mathbb{U}$ of our reasoning.

Postulate 2. The World (system-universe) consists of two systems: the *system* $\mathbb{S}^*$, which is the subject of our study, and the *environment* $\mathcal{E}^*$, which includes all other elements. In other words, we will assume that the world is arranged like this: there is a specific system $\mathbb{S}^*$, which is the subject of our study, and there is everything else not included in this system, that is, the environment $\mathcal{E}^*$.[1]

Postulate 3. Both the system $\mathbb{S}^*$ and the environment $\mathcal{E}^*$ are sets of interacting systems of lower (as compared to $\mathbb{S}^*$ and $\mathcal{E}^*$) hierarchical level which are, in turn, composed of systems of even lower hierarchical level, etc.

Postulate 4. The system $\mathbb{S}^*$ and the environment $\mathcal{E}^*$ *interact* with each other; in other words, $\mathbb{S}^* \rightleftarrows \mathcal{E}^*$. This interaction can be either *statical*, if the impact of the environment on the system and the response of the system to this effect are constant, or *dynamical*, if the impacts and/or reactions vary in time.[2] The classes of systems that correspond to these types of interaction are usually referred to as statical and dynam-

[1]Generally speaking, the separation of the system-universe $\mathbb{U}$ into the system $\mathbb{S}^*$ and the environment $\mathcal{E}^*$ is somewhat arbitrary. The boundary between $\mathbb{S}^*$ and $\mathcal{E}^*$ does not only depend on the problem being solved (that is, what we want to get from the system), but can be even *movable*, i.e., it can vary during the lifetime of the system $\mathbb{S}^*$. An example of this kind of situation arises in the case of object adaptation (see Section 1.2.1.3).

[2]The interaction of the system and the environment is *symmetric* in the sense that one can speak of the effect of the environment on the system with the response of the system to this effect, and the effect of the system on the environment with the response of the environment to this effect.

Neural Network Modeling and Identification of Dynamical Systems
https://doi.org/10.1016/B978-0-12-815254-6.00011-3

ical systems. The overwhelming majority of practically existing classes of systems are dynamical, and they will be the subject of our study.[3]

It should be noted that the universe $\mathbb{U}$ is a too general construction regarding the purposes of this book: both the system $\mathbb{S}^*$ and the environment $\mathcal{E}^*$ in the interpretation, as they are given above, are redundant. We need to reduce this redundancy to decrease the level of complexity of the problem being solved. This can be done based on the following considerations. In the universe $\mathbb{U}$, the system $\mathbb{S}^*$ is represented with all its inherent properties, while the environment $\mathcal{E}^*$, as noted above, is the part of the universe $\mathbb{U}$ that did not enter the system $\mathbb{S}^*$. For practical purposes, this composition of $\mathbb{S}^*$ and $\mathcal{E}^*$ is clearly redundant.

We introduce a partial representation for the system $\mathbb{S}^*$, which we denote as $\mathbb{S}$, and we denote the relation between the system and its partial representation as $\mathbb{S} \subset \mathbb{S}^*$.

When we talk about the "partial representation" $\mathbb{S}$ for $\mathbb{S}^*$, we mean that $\mathbb{S}$ takes into account only a part of the properties inherent to the system $\mathbb{S}^*$. In this case, only those properties of the system $\mathbb{S}^*$ are taken into account in $\mathbb{S}$ which are significant for the problem being solved. Different problems solved for the same system $\mathbb{S}^*$ require consideration of its various properties, i.e., each time only a part of the properties that $\mathbb{S}^*$ possesses is required. For example, if we solve the problem of analyzing the longitudinal motion of an aircraft, then it is not necessary to introduce into this model the variables associated with the aircraft lateral motion.

Similarly, we introduce a partial representation $\mathcal{E}$ for the environment $\mathcal{E}^*$, which interacts with the system $\mathbb{S}$. We will denote by $\mathcal{E} \subset \mathcal{E}^*$ the fact that $\mathcal{E}$ is a partial representation of the environment $\mathcal{E}^*$. This representation also depends significantly on the problem being solved. Because of this, the "radius of the environment" in some problems can be of the order of meters (for example, to take into account the effects of atmospheric disturbances on the aircraft), and in others it can be of the order of hundreds of kilometers (for example, tracking targets in systems such as AWACS).

Accordingly, instead of the universe

$$\mathbb{U} = \mathbb{S}^* \cup \mathcal{E}^*$$

we will consider the union of the "truncated" versions of the system $\mathbb{S} \subset \mathbb{S}^*$ and the environment $\mathcal{E} \subset \mathcal{E}^*$. We will call this combination the "System-Complex" $\mathbb{K}$:

$$\text{System-Complex } \mathbb{K} = \text{System-Object } \mathbb{S} + \text{System-Environment } \mathcal{E},$$

$$\mathbb{K} = \langle \mathbb{S}, \mathcal{E} \rangle, \quad \mathbb{S} \rightleftarrows \mathcal{E}.$$

1.1.1.2 *Concepts of System-Object and System-Environment*

The dynamical system $\mathbb{S}$ is a system whose state varies with time under the influence of some external and/or internal factors [1,6–19]. The source of external factors is the environment $\mathcal{E}$ in which the dynamical system operates, and the source of internal factors is the set of features that characterize the system as well as events occurring in the system (for example, failures and damage affecting the dynamic properties of the system).

In general, the dynamical system $\mathbb{S}$ is not isolated, but rather it operates in some environ-

[3]In fact, statical systems in nature, most likely, simply do not exist. Even those systems about which we are accustomed to think of as statical (more precisely, *not* dynamical), for example, buildings and structures of various kinds, in fact, are not. For example, television towers, skyscrapers, and other high-rise buildings are characterized by the presence of oscillations in their upper parts with a rather large amplitude (due to wind effects). So the "statical system" is just a simplification of a dynamical system in cases where the "dynamical" component can be neglected under the conditions of the problem being solved. Consequently, without loss of generality, we can only talk about dynamical systems.

ment $\mathcal{E}$ and interacts with this environment.[4] The response of the system to the influence of the environment can be either passive,[5] or active.[6]

Dynamical systems, actively interacting with the environment, are controllable dynamical systems. In this case, the nature of the interaction of the system with the environment depends on the properties of the system, as well as on the resources that it has. We can divide such resources into two types: internal and external. If we take as an example the task of controlling the movement of an aircraft, these resources will be as follows:

- internal resources are control laws, digital terrain map, tools of inertial navigation, etc.;
- external resources are systems of satellite navigation (GLONASS and GPS), tools for remote control and radio navigation, etc.

The less the system depends on external resources (information sources and control commands), the higher the degree of its autonomy. The study of highly autonomous control systems is becoming increasingly relevant at the present time due to the rapid development of unmanned vehicles of various classes, in particular unmanned aerial vehicles (UAVs) and unmanned cars.

[4]In most of the applied problems, only part of this interaction is taken into account, namely the influence of the environment $\mathcal{E}$ on the system $\mathbb{S}$. For example, it can be the influence of the gravitational field and/or atmosphere on the aircraft. However, in some cases it is also necessary to take into account the second half of this interaction, that is, the influence of $\mathbb{S}$ on $\mathcal{E}$. In particular, this kind of interaction generates a trail (wake turbulence) behind the aircraft. We must take into account such factors to solve problems related, for example, to formation flying of aircraft.

[5]Examples of passive interaction with the environment are such tasks as the flight of an artillery shell or an unguided rocket.

[6]In case of active interaction, the system generates and implements some reaction to the impact of the environment according to some "control law". For example, the elevator will be deflected as a reaction to a vertical gust of wind that has affected the aircraft.

We assume that the system $\mathbb{S}$ is completely defined if certain sets of variables (states, controls, disturbances, etc.) are given, as well as time and a rule that determines the next state of the system based on its current and/or previous states. We will represent the system $\mathbb{S}$ as an ordered triple

$$\begin{gathered}\mathbb{S}=\langle X^S, \Phi^S, W^S\rangle,\\ X^S=\{X_i^S\},\ i=1,\dots,N_S;\ \ \Phi^S=\{\Phi_j^S\},\ j=1,\dots,N_R;\\ W^S=\langle T^S,\ E^S\rangle,\ T^S\subseteq T.\end{gathered} \tag{1.1}$$

Here $X^S=\{X_i^S\}$, $i=1,\dots,N_S$, is the set of *structures* of the system $\mathbb{S}$ (the term "structure" is treated here in the general mathematical sense, as a collection of sets with the collection of relations defined on these sets); $\Phi^S=\{\Phi_j^S\}$, $j=1,\dots,N_R$, is the set of *rescripts* of the system $\mathbb{S}$ (the term "rescript" is used here as a generic name for transformations of all kinds, i.e., mappings, algorithms, inference procedures, etc.); $W^S=\langle T^S, E^S\rangle$ is the *clock* of the system, i.e., the set of instants of the system operating time ("system time"), T is the set of all possible time instants ("world time"), endowed with a structure of linear order (i.e., ordered by the relation $\leqslant$), and E^S is the activity mechanism ("clock generator") of the system $\mathbb{S}$.

The system-object $\mathbb{S}$ is not isolated; it interacts with the system-environment $\mathcal{E}$, represented by an ordered triple of the form

$$\begin{gathered}\mathcal{E}=\langle Q^{\mathcal{E}}, \Psi^{\mathcal{E}}, W^{\mathcal{E}}\rangle,\\ Q^{\mathcal{E}}=\{Q_i^{\mathcal{E}}\},\ i=1,\dots,M_{\mathcal{E}};\ \ \Psi^{\mathcal{E}}=\{\Psi_j^{\mathcal{E}}\},\ j=1,\dots,M_R;\\ W^{\mathcal{E}}=\langle T^{\mathcal{E}},\ E^{\mathcal{E}}\rangle,\ T^{\mathcal{E}}\subseteq T.\end{gathered} \tag{1.2}$$

Here $Q^{\mathcal{E}}=\{Q_i^{\mathcal{E}}\}$, $i=1,\dots,M_{\mathcal{E}}$, is a collection of *structures* of the environment $\mathcal{E}$; $\Psi^{\mathcal{E}}=\{\Psi_j^{\mathcal{E}}\}$, $j=1,\dots,M_R$, is the set of *rescripts* of the environment $\mathcal{E}$; $W^{\mathcal{E}}=\langle T^{\mathcal{E}}, E^{\mathcal{E}}\rangle$ is the clock of the environment $\mathcal{E}$, $T^{\mathcal{E}}\subseteq T$ is the set of instants of the environment operating time, T is the set of all possible time instants, and $E^{\mathcal{E}}$ is the activity mechanism of the environment $\mathcal{E}$.

1.1.1.3 The Uncertainties in the System-Complex

When solving problems related to system-complex $\mathbb{K}$, it is necessary to take into account various uncertainty factors:

- uncertainties, caused by uncontrolled disturbances acting on the system-object $\mathbb{S}$;
- incomplete and inaccurate knowledge of the properties and characteristics of the system-object $\mathbb{S}$ and the conditions in which it will operate (i.e., properties of the system-environment $\mathcal{E}$);
- uncertainties caused by a change in the properties of the system-object $\mathbb{S}$ due to equipment failures and structural damage.

We can distinguish the following typical classes of uncertainty factors:

- uncertainties of *parametric type* associated with variables describing the parameters of the system-object $\mathbb{S}$ (for example, the mass m and the moments of inertia I_x, I_y, I_z, I_{xz} for the case when the system-object $\mathbb{S}$ is an aircraft);
- uncertainties of *functional type* related to the characteristics of the system-object $\mathbb{S}$ (for example, the coefficients of the aerodynamic forces C_x, C_y, C_z, the moment coefficients C_l, C_m, C_n, and thrust of the power plant T for the case when the system-object $\mathbb{S}$ is an aircraft);
- uncertainties associated with the effects of system-environment $\mathcal{E}$ (for example, air density ρ; atmospheric pressure p; wind speed W; atmospheric turbulence).

Uncertainties of the parametric type are usually defined in the interval form as

$$\lambda \in [\lambda_{min}, \lambda_{max}],$$

where λ is one of the parameters m, I_x, I_y, I_z, I_{xz}, etc.; λ_{min} and λ_{max} are the minimum and maximum possible values for the parameter λ, respectively.

We can define uncertainties of a functional type in the form of parametric families of curves. For example, consider the following representation:

$$\varphi(x) = \varphi(x)_{nom} + \Delta\varphi(x),$$
$$\Delta\varphi(x) = w_0 + w_1 x + w_2 x^2,$$
$$w_0 \in [w_0^{min}, w_0^{max}], \quad w_1 \in [w_1^{min}, w_1^{max}],$$
$$w_2 \in [w_2^{min}, w_2^{max}],$$

where $\varphi(x)$ is one of the aircraft characteristics $C_x, C_y, C_z, C_l, C_m, C_n, T$; $\varphi(x)_{nom}$ is some nominal value of the characteristic $\varphi(x)$; $\Delta\varphi(x)$ is the deviation of the actual characteristic $\varphi(x)$ from its nominal value.

Uncertainties associated with environmental influences, i.e., air density ρ, atmospheric pressure p, wind speed W, atmospheric turbulence, etc., are usually defined according to generally accepted probabilistic models.

Let us explain the meaning of uncertainty caused by the change in the properties of the system-object $\mathbb{S}$ due to equipment failures and damage in the structure, by examples related to aircraft.

During the flight of aircraft, various kinds of abnormal (emergency) situations can occur, caused by failures of equipment and aircraft systems or damage to the airframe and power plant of the aircraft. Some of these failures and kinds of damage have a direct impact on the dynamical characteristics of aircraft as an object of modeling and control.

For these reasons, there arises a need for an adjustment of the aircraft control algorithms that would provide an opportunity to adapt to the changed dynamical properties of the aircraft. In this case it is extremely difficult, if possible at all, to anticipate all possible failures and damage as well as their combinations in advance. As for the damage to the airframe and power plant, it is impossible to foresee all feasible options in principle. Therefore, it seems appropriate for aircraft control to interpret possible sharp and unpredictable changes in the dynamical properties of aircraft due to failures and damage as another class of uncertainties. Compensation of the impact of these factors on the aircraft behavior is the responsibility of the adaptation mechanisms. These mechanisms must ensure fault-tolerant and damage-tolerant control of the aircraft, i.e., control that can adapt to a change in the dynamics of the controlled object generated by failures or damage. The result of such adaptation should be the restoration of the aircraft stability and controllability to a level acceptable from the flight safety point of view.

1.1.2 Classes of Dynamical Systems

In this section, we form a hierarchy of system-objects as a sequence of definitions of their classes, ordered by their level of capabilities.

The system-objects, considered below, differ from each other in their properties and, as a consequence, the level of potential capabilities.

We use the following features to differentiate systems of various classes:

- the presence/absence of uncertainties in the system that affect the properties of the system;
- the presence/absence of the possibility for the system to control its behavior as a way of actively responding to changes in the current and/or predicted situation;
- the presence/absence of the possibility for the system to adapt to changes in the properties of the object and/or environment;
- the presence/absence of goal-setting capabilities in the system.

1.1.2.1 Deterministic Systems

At the lowest level of the hierarchy of systems $\mathbb{S}$ lie the deterministic dynamical systems $\mathbb{DS}$,[7] i.e., those that respond to the same actions in the same way. The properties of such systems are either constant or vary according to some preserved relationship. Examples of such systems are an unguided missile with a detachable booster and an uncontrolled aircraft with a mass that varies due to fuel consumption.

We assume that the system $\mathbb{DS}$ is a triple

$$\begin{gathered}\mathbb{DS} = \langle X^{DS}, \Phi^{DS}, T^{DS} \rangle, \\ X^{DS} \subseteq X, T^{DS} \subseteq T, \\ \Phi^{DS} = \Phi^{DS}(\mathbf{x}, t), \mathbf{x} \in X^{DS}, \ t \in T^{DS}.\end{gathered} \tag{1.3}$$

Here X is the *phase space* (state space) of the system[8] (1.3), the elements ("points") $\mathbf{x} \in X$ of which are the possible *phase states* (phase vector) of the given system, $\mathbf{x} = (x_1, \ldots, x_n)^T$, $x_1 \in X_1, \ldots, x_n \in X_n$, $X = X_1 \times \ldots \times X_n$; $x_1, \ldots, x_n$ are the state variables (phase coordinates) of the system $\mathbb{DS}$; $X^{DS} \subseteq X$ is the range of admissible values of phase states of the system (1.3); T is the set of all possible time instants ("world time") endowed with a structure of linear order (that is, ordered by the relation $\leqslant$), $T^{DS} \subseteq T$ is the set of instants of the system operation time ("system time"); $\Phi^{DS} = \Phi^{DS}(\mathbf{x}, t)$, $\mathbf{x} \in X^{DS}$, $t \in T^{DS}$, is a rule that allows us to determine the state of the system (1.3) at each time instant $t \in T^{DS}$, given its states at all previous time instants.[9] Systems of the class $\mathbb{DS}$ are the subject of the study of the modern theory of dynamical systems (see, for example, [7,20]).

[7] DS is the abbreviation for the Deterministic System.

[8] The terminology adopted here goes back to the traditions of mechanics, the theory of dynamical systems, and also the theory of control and controlled systems. This seems quite logical, because of the fundamental role played by the concept of a dynamical system for the class of problems we consider.

[9] This means, generally speaking, that the rule Φ^S must have infinite memory in order to store all the states previously attained by the system $\mathbb{S}$. Most often, based on the specifics of the problem being solved, it can be argued that this requirement is excessive and it is sufficient to use the state prehistory of finite length; in a number of cases it can be assumed that all future states of the system $\mathbb{S}$ for $t > t_i$, $t, t_i \in T^S$, are determined only by its state $\mathbf{x}(t_i)$ at the given current time instant $t_i \in T^S$ (and, of course, by the rule Φ^S).

1.1.2.2 Systems With Uncertainties

Systems of class $\mathbb{DS}$ whose properties are deterministic represent a somewhat rare particular case of dynamical systems. Much more often, dynamical systems contain uncertainties of some kind from those listed in Section 1.1.1.3. In particular, these uncertainties may be due to incomplete and inaccurate knowledge of specific properties of the object. For example, for aircraft, we often have a situation where its aerodynamic characteristics are known not precisely and not wholly (in particular, there are no data on them for a part of the flight regimes).

The system $\mathbb{VS}$,[10] as well as the system $\mathbb{DS}$, remains uncontrollable and reacts to influences of the environment $\mathbb{VS} \rightleftarrows \mathcal{E}$, but it contains the uncertainties of the parametric and/or functional type associated with the system or the environment (see Section 1.1.1.3). Thus, we can define the system $\mathbb{VS}$ as follows:

$$\begin{gathered}\mathbb{VS} = \langle X^{VS}, T^{VS}, \Phi^{VS} \rangle, \\ X^{VS} \subseteq X, T^{VS} \subseteq T, \; \Phi^{VS} = \Phi^{VS}(\mathbf{x}, \boldsymbol{\xi}, t), \\ \mathbf{x} \in X^{VS}, \; \boldsymbol{\xi} \in \Xi, \; t \in T^{VS} \subseteq T.\end{gathered} \tag{1.4}$$

The notations for the system (1.4) are similar to those introduced above for the system (1.3). The difference between these systems lies in the fact that the rule Φ^{VS} includes the uncertainty factors $\boldsymbol{\xi} = (\xi_1, \ldots, \xi_q) \in \Xi$. As was shown in Section 1.1.1.3, usually systems of the class $\mathbb{VS}$ not only contain uncertainties due to the features of the system and the available information about them, but they also interact with the environment $\mathcal{E}$ that contains uncertainties itself (an example of such environment is the turbulent atmosphere in which an aircraft is flying). Therefore, in this case, the state of the dynamical system depends not only on the current state $\mathbf{x}(t_i)$ of the system and time, but also on the value of the external uncontrolled disturbances described by the part of the components of the vector $\boldsymbol{\xi}$, which takes values from some domain Ξ. The "uncontrollability" of an external perturbation means that for the system $\mathbb{VS}$ there is no complete a priori information about the characteristics of this disturbance. In this case, corresponding components of the vector $\boldsymbol{\xi}$ can take random or fuzzy values.

[10] VS is the abbreviation for the Vague System, i.e., for a system containing uncertainties of some kind.

1.1.2.3 Controllable Systems

As noted above, the dynamical system $\mathbb{S}$ interacts with the environment $\mathcal{E}$, i.e., perceives the impact of the environment and responds to it accordingly. The response of the system to environmental influences can be either passive or active.

Passive interaction is, for example, the movement of a stone or the flight of an artillery shell or an uncontrolled rocket under the influence of gravitational and aerodynamic forces. Exactly this kind of interaction is realized by the systems $\mathbb{DS}$ of the form (1.3) and $\mathbb{VS}$ of the form (1.4).

In the case of active interaction, the system, influenced by the environment, generates and implements the response of the system to this action; for example, it deflects the aircraft control surfaces to compensate for the disturbance. This means that if the dynamical system is able to actively interact with the environment, then it is a *controllable* system.

A controllable dynamical system $\mathbb{CS}$[11] actively responds to the effects of the environment and is capable of compensating the perturbations arising in the interaction $\mathbb{CS} \rightleftarrows \mathcal{E}$ within certain limits. We describe this system as follows:

$$\begin{gathered}\mathbb{CS} = \langle X^{CS}, T^{CS}, \Phi^{CS}, \rangle, \\ X^{CS} \subseteq X, \; T^{CS} \subseteq T, \; \Phi^{CS} = \Phi^{CS}(\mathbf{x}, \mathbf{u}, \boldsymbol{\xi}, t), \\ \mathbf{x} \in X^{CS}, \mathbf{u} \in U^{CS}, \; \boldsymbol{\xi} \in \Xi, \; t \in T^{CS} \subseteq T.\end{gathered} \tag{1.5}$$

As in the case of the $\mathbb{VS}$ system of the form (1.4), the notations for the system (1.5) are sim-

[11] CS is the abbreviation for the Controllable System.

ilar to those introduced above for the system (1.3). The difference between these systems again lies in the form of the rule Φ^{CS}. This rule for the system $\mathbb{CS}$ depends not only on the current state $\mathbf{x}(t_i)$ of the system, uncontrollable disturbances $\boldsymbol{\xi}$, and time t, but also on the so-called control variables $\mathbf{u} = (u_1, \ldots, u_m) \in U$, which take (instantaneous) values from some region U^{CS}.

So, the system of the class $\mathbb{CS}$ is a dynamical, controlled, purposeful system, regularly responding to the effects of the environment $\mathcal{E}$. In a more general case, a system of the class $\mathbb{CS}$ can be influenced by various kinds of uncertain factors $\boldsymbol{\xi} \in \Xi$. The system of this class is characterized by the fact that, unlike the systems of the next two hierarchical levels (adaptive and intelligent), the control objective is external to the system and is taken into account only at the stage of synthesis of the control law for this system.

1.1.2.4 *Adaptive Systems*

The most important difference between systems of class $\mathbb{CS}$ and systems of classes $\mathbb{VS}$ and $\mathbb{DS}$ is that systems $\mathbb{CS}$ can actively respond to influences of the environment $\mathcal{E}$. These possibilities, however, are limited by the fact that the rule Φ^{CS} (the control law of the system $\mathbb{CS}$) in (1.5) is assumed to be preserved. There is no way to modify this rule during the operation of the system (1.5). But if the region of uncertainty Ξ is "large enough," it can turn out that it will be impossible to form a control law such that it provides the required level of control quality for any values of $\boldsymbol{\xi} \in \Xi$. As a consequence, there is a need to change the form of the control law of the system (the Φ rule) during the process of the system operation. The systems $\mathbb{AS}$,[12] which have this kind of property, are called *adaptive* and in general they can be represented as follows:

$$\begin{gathered}
\mathbb{AS} = \langle X^{AS}, T^{AS}, \Phi^{AS}, \Psi^{AS}, \Gamma^{AS} \rangle, \\
X^{AS} \subseteq X,\ T^{AS} \subseteq T, \\
\Phi^{AS} = \Phi^{AS}(\mathbf{x}, \mathbf{u}, \boldsymbol{\xi}, t), \Psi^{AS} = \Psi^{AS}(\boldsymbol{\gamma}, t), \\
\Gamma^{AS} \subseteq \Gamma,\ \mathbf{x} \in X^{AS}, \mathbf{u} \in U^{AS}, \\
\boldsymbol{\xi} \in \Xi,\ \boldsymbol{\gamma} \in \Gamma, t \in T^{AS} \subseteq T.
\end{gathered} \tag{1.6}$$

In the definition of the adaptive system $\mathbb{AS}$, compared to a controllable system of the class $\mathbb{CS}$, two new elements have appeared. The first of these is the Ψ^{AS} rule ("adaptation rule") that describes how to modify the rule Φ^{AS}, which determines the current behavior of the system under consideration. The second new element is the set of goals Γ^{AS}, according to which the work of the rule Ψ^{AS} proceeds.

So, the system of the class $\mathbb{AS}$ is a dynamical, controlled, purposeful system with adjustable control law Φ^{AS} that regularly interacts with the environment and is also influenced by various kinds of uncertainty factors. The set of goals Γ^{AS} that organize behavior in systems of class $\mathbb{AS}$ remains unchanged during the system operation.

1.1.2.5 *Intelligent Systems*

In comparison with the systems of the class $\mathbb{CS}$, adaptive systems $\mathbb{AS}$ are able to modify their behavior directly in the process of the system functioning due to the influence that the rule Ψ^{AS} has on the control law Φ^{AS} which in turn determines the current behavior of the system $\mathbb{AS}$.

However, the characteristic feature of $\mathbb{AS}$ systems is that the rule Ψ^{AS} for modifying the behavior pattern is defined during the system design stage and remains fixed afterwards. The choice for the rule Ψ^{AS} was made according to the specific goals Γ^{AS}, the achievement of which was the purpose of the system $\mathbb{AS}$ development. Hence the behavior of the $\mathbb{AS}$ system will be adequate so long as the goals Γ^{AS} remain relevant in a changing situation.

[12] AS is the abbreviation for the Adaptive System.

In an environment with high levels of various uncertainties, a change in the situation may require us to use a higher level of adaptation, namely, the adaptation of goals (see Section 1.2.1.4). However, systems of the class $\mathbb{AS}$ lack the knowledge required to perform this operation, because they do not have a goal-setting mechanism, i.e., a mechanism that generates new goals when necessary. Also, $\mathbb{AS}$ systems have no tool of influence on the rule Ψ^{AS}, i.e., a mechanism that adjusts the rule of $\mathbb{AS}$ behavior modification.

So, the next step is to add a new property to the system $\mathbb{AS}$, namely, the goal-setting.[13] Systems $\mathbb{IS}$,[14] possessing this kind of property, we will call *intelligent*, and we will represent them as follows:

$$\begin{aligned} \mathbb{IS} = \langle X^{IS}, T, \Phi^{IS}, \Psi^{IS}, \Gamma^{IS}, \Omega^{IS}, \Sigma^{IS} \rangle, \\ X^{IS} \subseteq X,\ T^{IS} \subseteq T, \\ \Phi^{IS} = \Phi^{IS}(\mathbf{x}, \mathbf{u}, \boldsymbol{\xi}, t),\ \Psi^{AS} = \Psi^{AS}(\boldsymbol{\gamma}, t), \\ \Gamma^{AS} \subseteq \Gamma,\ \Omega^{IS} = \Omega^{IS}(\mathbf{x}, \mathbf{u}, \boldsymbol{\xi}, \boldsymbol{\gamma}, t), \\ \mathbf{x} \in X^{IS},\ \mathbf{u} \in U^{IS}, \boldsymbol{\xi} \in \Xi,\ \boldsymbol{\gamma} \in \Gamma, \\ \Sigma^{IS} \subseteq \Sigma, t \in T^{IS} \subseteq T. \end{aligned} \tag{1.7}$$

In comparison with the adaptive system $\mathbb{AS}$, the rule Ω^{IS} is added to the definition of the intelligent system $\mathbb{IS}$. This rule describes the method of generating $\boldsymbol{\gamma} \in \Gamma$ goals. Thus, the intelligent system $\mathbb{IS}$ has a way of changing the set of goals Γ, as well as another structure, Σ^{IS}, which specifies such elements as values, motives, etc., that guide the goal-setting process.

So, a system of class $\mathbb{IS}$ is a dynamical, controlled, purposeful system that has the tools (mechanisms) of goal-setting. Systems of this class and systems of classes $\mathbb{CS}$ and $\mathbb{AS}$ interact with the environment both in a regular way and under the influence of various kinds of uncertainty factors.

[13]Development of a goal-setting mechanism is one of the most important topics of modern robotics. However, no satisfactory solution to this problem has yet been discovered. The results available in this field are covered in [21].

[14]IS is the abbreviation for the Intelligent Systems.

1.1.3 Classes of Environments

Similarly to classes of systems $\mathbb{S}$, we can also introduce classes of the environment $\mathcal{E}$. In the following sections we define various environment classes in order of increasing complexity, assuming that they interact with a system of the general form $\mathbb{S}$.

1.1.3.1 *Regular Environments*

The *regular environment* $\mathcal{SE}$[15] lies at the lowest level of the environment hierarchy. It implements a regular (i.e., not containing any uncertainties) effect on the system $\mathbb{S}$. To some extent, the central gravitational field of some celestial body can be considered as an environment example of the class $\mathcal{SE}$.

1.1.3.2 *Environments With Uncertainties*

The *environment with uncertainties* $\mathcal{UE}$[16] is the next level of the environment hierarchy. It is characterized by the fact that the effect it has on the system $\mathbb{S}$ contains a number of uncertainty factors, i.e., factors that are unknown a priori. The system $\mathbb{S}$ cannot manage and measure these factors. An example of an environment of class $\mathcal{UE}$ is a turbulent atmosphere.

1.1.3.3 *Reacting Environments*

The environments of the classes $\mathcal{SE}$ and $\mathcal{UE}$ are *passive*; they act in a certain way on the system $\mathbb{S}$, however their own response to the actions of the system $\mathbb{S}$ is very limited and not purposeful. For example, the turbulent atmosphere

[15]SE is the abbreviation for the Stereotyped Environment, i.e., a regular environment that acts on the system $\mathbb{S}$ in some routine, standard manner that does not change during the lifetime T^S of the system $\mathbb{S}$.

[16]UE is the abbreviation for the Uncertain Environment, i.e., an environment containing some uncertainties (factors that the system $\mathbb{S}$ can neither control nor, in some cases, even measure).

affects the trajectory of the body and gets some response, the results of which appear as a disturbed wake behind the body.

Environments that belong to any of the subsequent classes (hierarchical levels) are *active*; they respond to the actions of the system $\mathbb{S}$ in one way or another. An important point to be emphasized is that the active interaction of the environment $\mathcal{E}$ with the system $\mathbb{S}$ might either assist the system to achieve its goals (an example is some air traffic control system as the environment that interacts with the transport plane during the flight), or impede these goals (an example is an air defense system that protects some object from airstrikes). Another possible scenario of the interaction of the environment with the system is that the environment "does not interfere" (i.e., does not help, but also does not counteract) with the activity of the system in it.

The first of the active environment classes is the *reacting* environment $\mathcal{RE}$.[17] Environments of the $\mathcal{RE}$ class undergo changes in response to the actions of the system $\mathbb{S}$. However, these changes are not purposeful. Examples of such environments are fragments of the natural environment (biosphere) within which a particular technical system operates.

1.1.3.4 *Adaptive Environments*

The next class of active environments is the class of *adaptive environments* $\mathcal{AE}$.[18] For environments of this class, an active and *purposeful* response to the actions of the system $\mathbb{S}$ is typical. However, the set of goals that guide the processes in the environment $\mathcal{AE}$, once chosen and specified, remains unchanged during the processes of interaction of the system $\mathbb{S}$ and the environment $\mathcal{AE}$. New goals cannot be generated in environments of the $\mathcal{AE}$ class.

An example of an adaptive environment is a highly automated air defense system, especially in its territorial version. This air defense system should be considered as an environment for aircraft that try to overcome it, e.g., aeroplanes, cruise missiles, etc.

1.1.3.5 *Intelligent Environments*

The third and the highest level of active environments is the class of *intelligent environments* $\mathcal{IE}$.[19] Environments of this class contain goal-setting mechanisms. They also have their interests ("freedom of the will," in a certain sense).

As an example, here, as in the case of adaptive environments, we can indicate a territorial air defense system, but of a higher hierarchical level. This system contains fighter aircraft, anti-aircraft missile systems, tools for detecting and tracking targets, etc., but it moreover includes human operators. This circumstance ensures the variability of the environment reactions in various situations, including nonstandard ones. Another example of a similar type is an air traffic control system, especially in a large airport area.

1.1.4 Interaction Between Systems and Environment

Above, the following classes of dynamical systems were introduced:

- deterministic system ($\mathbb{DS}$);
- system with uncertainties ($\mathbb{VS}$);
- controllable system ($\mathbb{CS}$);
- adaptive system ($\mathbb{AS}$);
- intelligent system ($\mathbb{IS}$).

[17] RE is the abbreviation for the Reacting Environment, i.e., the environment that responds in some way to the activity of the system $\mathbb{S}$.

[18] AE is the abbreviation for the Adaptive Environment, i.e., the environment that actively and *purposefully* reacts to the activity of the system $\mathbb{S}$ in a certain way.

[19] IE is the abbreviation for the Intelligent Environment, i.e., the environment that is capable not only of actively and purposefully reacting to the operation of the system $\mathbb{S}$ in a certain way, but also of performing the *generation of goals*.

According to the level of potential capabilities (in order of increasing capabilities), these classes of systems are arranged as follows:

$$\mathbb{DS} \subset \mathbb{VS} \subset \mathbb{CS} \subset \mathbb{AS} \subset \mathbb{IS}.$$

Similarly, the hierarchy of environmental classes can be structured as follows:

$$\mathcal{SE} \subset \mathcal{UE} \subset \mathcal{RE} \subset \mathcal{AE} \subset \mathcal{IE}.$$

As noted above, the system $\mathbb{S}$ should be considered in interaction with the environment $\mathcal{E}$. Symbolically, we will denote this assertion as follows:

$$\text{System-Complex } \mathbb{K} = \text{System-Object } \mathbb{S} + \text{System-Environment } \mathcal{E},$$

$$\mathbb{K} = \mathbb{S} \rightleftarrows \mathcal{E},$$

that is, at the most general level, we consider the system-complex $\mathbb{K}$, which consists of two interacting systems, namely, the system-object $\mathbb{S}$ and the system-environment $\mathcal{E}$, in which the system-object operates,[20] so we have

$$\mathbb{K} = \langle \mathbb{S}, \mathcal{E}, T, \Theta \rangle, \quad \mathbb{S} \leftrightarrows \mathcal{E}. \tag{1.8}$$

Here Θ is the law of the interaction between $\mathbb{S}$ and $\mathcal{E}$ in time T.

The specific form of the complex (1.8) is determined by the way its constituent parts $\mathbb{S}$, $\mathcal{E}$, T, Θ are defined. For example, the following variants are possible:

- the complex $\mathbb{K}_{\mathbb{DS}}^{\mathcal{SE}} = \langle \mathbb{DS}, \mathcal{SE} \rangle$, which includes an uncontrollable deterministic dynamical system $\mathbb{DS}$; this system regularly interacts with the deterministic environment $\mathcal{SE}$ (an example is an object moving in the gravitational field of a celestial body that does not have an atmosphere);
- the complex $\mathbb{K}_{\mathbb{DS}}^{\mathcal{UE}} = \langle \mathbb{DS}, \mathcal{UE} \rangle$, which includes an uncontrollable deterministic dynamical system $\mathbb{DS}$, interacting with the environment $\mathcal{UE}$ that contains uncertainty factors (an example is an uncontrolled missile moving in a turbulent atmosphere);
- the complex $\mathbb{K}_{\mathbb{CS}}^{\mathcal{SE}} = \langle \mathbb{CS}, \mathcal{SE} \rangle$, which includes a controllable deterministic dynamical system $\mathbb{CS}$ that regularly interacts with the deterministic environment $\mathcal{SE}$ (an example is an aircraft performing controlled movement in a quiet atmosphere);
- the complex $\mathbb{K}_{\mathbb{AS}}^{\mathcal{UE}} = \langle \mathbb{AS}, \mathcal{UE} \rangle$, which includes an adaptive dynamical system $\mathbb{AS}$, interacting with the environment $\mathcal{UE}$ that contains uncertainty factors (an example is an aircraft that operates in an environment with uncertainties[21] while being able to quickly adapt to them).

1.1.5 Formalization of the Dynamical System Concept

We now introduce the formalized concept of the system $\mathbb{S}$ in the form in which it will be used later. In the general case, in this description we have to define the following elements related to $\mathbb{S}$:

1) the set of variables (with the range of their admissible values) describing $\mathbb{S}$ and the conditions in which $\mathbb{S}$ operates;
2) the set of variables (with the range of their admissible values) describing the factors affecting the states of the system $\mathbb{S}$;
3) the time in which $\mathbb{S}$ is running;
4) law of the functioning of $\mathbb{S}$, that is, a set of rules,[22] according to which the collection of variables describing $\mathbb{S}$ changes with time.

[20]In the following text, for brevity, we will simply refer to System-Complex as the "complex", to System-Object as the "system" (dynamical system) or "object" ("plant"), and to System-Environment as just the "environment."

[21]The uncertainties that might arise in the problems of modeling the behavior of dynamical systems are diverse (see, for example, [22–27]).

[22]In the theory of dynamical systems [7,9], this set of rules is often also referred to as the evolution law of the system $\mathbb{S}$.

The set of variables describing the system $\mathbb{S}$ includes the following:

- the variables $x_1, \ldots, x_n$, describing the state of the system $\mathbb{S}$,
- the variables $y_1, \ldots, y_m$, describing the observation results for the state of system $\mathbb{S}$.

The variables $x_1, \ldots, x_n$, describing the state of the system $\mathbb{S}$, are combined into a collection (vector) x, called the state (state vector) of the given system, i.e.,

$$
\begin{gathered}
x = (x_1, \ldots, x_i, \ldots, x_n); \\
x_i \in X_i \subset \mathcal{R}; \;\; x \in R_X \subset X = X_1 \times \cdots \times X_n.
\end{gathered} \tag{1.9}
$$

Here X is the region of all possible states of the system $\mathbb{S}$ (the space of states[23]); R_X is the domain of all admissible states of the system $\mathbb{S}$; $\mathcal{R}$ is the set of real numbers.

The variables $y_1, \ldots, y_p$, describing the observation results for the state of the system $\mathbb{S}$, are combined into the observation vector, i.e.,

$$
\begin{gathered}
y = (y_1, \ldots, y_j, \ldots, y_p); \\
y_j \in Y_j \subset \mathcal{R}; \;\; y \in R_Y \subset Y = Y_1 \times \cdots \times Y_p.
\end{gathered} \tag{1.10}
$$

The list of variables that describe the factors affecting the state of the system $\mathbb{S}$ includes the following elements:

- the variables describing the effects on $\mathbb{S}$, both controlled and uncontrolled, including:
  - controls $u_1, \ldots, u_m$ represent controllable influences on $\mathbb{S}$ that may be varied in order to meet the control goals of the system;
  - disturbances $\xi_1, \ldots, \xi_q$ represent uncontrollable influences on $\mathbb{S}$ (they can have either known or unknown properties, and they can be either measurable, for example, as an outdoor air temperature in the problem of thermal regulation of the house, or immeasurable, for example, as an atmospheric turbulence);
  - measuring noise $\zeta_1, \ldots, \zeta_r$, describing the errors introduced by sensors;
- variables $w_1, \ldots, w_s$ (constant or varying), describing the properties of the system $\mathbb{S}$ (their values are directly or indirectly determined by the decisions made during the design of $\mathbb{S}$) and influencing the behavior of the system through the law of its evolution (including both constant and varying parameters of the law). Examples of constant (permanent) parameters of the system are aircraft wingspan, wing area, length, etc. Examples of varying parameters of the system are coefficients of aerodynamic forces and moments, which are nonlinear functions of several variables (state variables of the object and the environment, as well as control variables).

[23]The state space in the theory of dynamical systems is often also called the phase space, and the states of the system $\mathbb{S}$ are phase states.

The variables $u_1, \ldots, u_m$, describing the control actions on the state of the system $\mathbb{S}$, are combined into a collection u, called the control (control vector) of the given system, i.e.,

$$
\begin{gathered}
u = (u_1, \ldots, u_k, \ldots, u_m); \\
u_k \in U_k \subset \mathcal{R}; \;\; u \in R_U \subset U = U_1 \times \cdots \times U_m.
\end{gathered} \tag{1.11}
$$

Here U is the domain of *all possible values* of controls for the system $\mathbb{S}$; R_U is the domain of *all admissible values* of controls for the system $\mathbb{S}$.

Example 1 (Longitudinal aircraft motion during climb phase). A system of equations describing the motion of an aircraft during climb phase for unsteady ($\dot{V} \neq 0$) and nonrectilinear ($\dot{\gamma} \neq 0$) flight can be written in the form [28–34]

$$
\begin{aligned}
m\frac{dV}{dt} &= T - D - W \sin\gamma, \\
mV\frac{d\gamma}{dt} &= L - W \cos\gamma.
\end{aligned} \tag{1.12}
$$

Here D is the aerodynamic drag force; L is the lift; T is thrust of the power plant; γ is the flight path angle; V is airspeed; $W = mg$ is the weight of the aircraft.

In the climb problem, the state of the aircraft as a dynamical system is described by two variables: airspeed V and the flight path angle γ. Accordingly, the state vector $x \in X$ in this case has the form

$$x = (x_1, x_2) = (V, \gamma).$$

Example 2 (Curved aircraft flight in the horizontal plane). The system of equations of motion for an airplane that performs a turn with roll and sideslip can be written in the form [28–34]

$$\begin{aligned} m\frac{dV}{dt} &= T - D, \\ mV\frac{d\Psi}{dt} &= -Y\cos\phi - L\sin\phi + T\sin\beta\cos\phi, \\ 0 &= L\cos\phi - W. \end{aligned} \tag{1.13}$$

Here D is the aerodynamic drag force; L is the lift; Y is the lateral force; T is thrust of the power plant; Ψ is the yaw angle; V is the airspeed; β is the angle of sideslip; ϕ is the roll angle; $W = mg$ is the weight of the aircraft.

In the task of performing a turn with roll and sideslip, the state of the aircraft as a dynamical system is described by such variables as the airspeed V and the yaw angle Ψ. Accordingly, the state vector $x \in X$ in this case has the form

$$x = (x_1, x_2) = (V, \Psi).$$

Example 3 (Longitudinal angular motion of an aircraft). A system of equations describing the longitudinal short-period motion of an aircraft can be written in the form [28–34]

$$\begin{aligned} \dot{\alpha} &= q - \frac{\bar{q}S}{mV}C_L(\alpha, q, \delta_e) + \frac{g}{V}\cos\theta, \\ \dot{q} &= \frac{q_p S\bar{c}}{I_y}C_m(\alpha, q, \delta_e), \\ T^2\ddot{\delta}_e &= -2T\zeta\dot{\delta}_e - \delta_e + \delta_{e_{act}}, \end{aligned} \tag{1.14}$$

where α is the angle of attack, deg; θ is angle of pitch, deg; q is the angular velocity of the pitch, deg/sec; δ_e is the deflection angle of the controlled stabilizer, deg; C_L is the lift coefficient; C_m is pitch moment coefficient; m is mass of the aircraft, kg; V is the airspeed, m/sec; $q_p = \rho V^2/2$ is the dynamic air pressure, kg·m^{-1} sec^{-2}; ρ is air density, kg/m^3; g is the acceleration of gravity, m/sec^2; S is the wing area, m^2; $\bar{c}$ is mean aerodynamic chord of the wing, m; I_y is the moment of inertia of the aircraft relative to the lateral axis, kg·m^2; the dimensionless coefficients C_L and C_m are nonlinear functions of their arguments; T, ζ are the time constant and the relative damping coefficient of the actuator; $\delta_{e_{act}}$ is the command signal to the actuator of the all-turn controllable stabilizer (limited to ± 25 deg). In the model (1.14), the variables α, q, δ_e, and $\dot{\delta}_e$ are the states of the controlled object, the variable $\delta_{e_{act}}$ is the control. Here, $g(H)$ and $\rho(H)$ are the variables describing the state of the environment (gravitational field and atmosphere, respectively), where H is the altitude of the flight; m, S, $\bar{c}$, I_z, T, ζ are constant parameters of the simulated object, C_L and C_m are variable parameters of the simulated object.

We usually impose some limitations R_X on the suitable combinations of state values of the system $\mathbb{S}$, as well as constraints R_U on appropriate combinations of control values. As a general rule, we also have certain restrictions on the values of a combination of the state vector x and the control vector u, i.e.,

$$\begin{aligned} \langle x, u\rangle &\in R_{XU} \subset X \times U \\ &= X_1 \times \cdots \times X_n \times U_1 \times \cdots \times U_p. \end{aligned} \tag{1.15}$$

Taking into account the above definitions, the system $\mathbb{S}$ can be represented in the following general form:

$$\mathbb{S} = \langle \{U, \Xi, Z\}, \{F, G\}, \{X, Y\}, T\rangle, \tag{1.16}$$

where U are controllable influences on $\mathbb{S}$; Ξ and Z are uncontrollable influences on the states and outputs of the system $\mathbb{S}$, respectively; F and G are the rules governing the evolution of the state and output of the system $\mathbb{S}$ over time, respectively; X and Y are respectively the set of states and observable outputs of the system $\mathbb{S}$. Elements of these sets will be denoted as $u \in U$, $\xi \in \Xi$, $\zeta \in Z$, $x \in X$, $y \in Y$, $t \in T$.

In (1.16), T denotes the time interval on which the system $\mathbb{S}$ is considered. The time within this interval can be either continuous, that is, $T \subset \mathcal{R}$, or discrete. In the case of discrete time, the sequence of time instants is given by the following rule:

$$T = \{t_0, t_1, \dots, t_{i-1}, t_i, t_{i+1}, \dots, t_{N-1}, t_N\}, \quad t_N = t_f,$$
$$t_{i+1} = t_i + \Delta t, \quad i = 0, 1, \dots, N. \tag{1.17}$$

In the following text, unless otherwise specified, a discrete time (1.17) will be used. This approach seems quite natural for the range of problems under consideration. Systems $\mathbb{S}$ are usually described in continuous time by ordinary differential equations (ODEs) or differential-algebraic equations (DAEs). However, in order to numerically solve problems of the analysis, identification, and control synthesis for such systems, we need to approximate them with finite difference (discrete time) equations given by corresponding recurrent schemes (Runge–Kutta, Adams, etc.). Thus, in the process of solving these problems, there is a mandatory transition from continuous to discrete time.

Similarly, the onboard implementation of control systems for modern and advanced aircraft is also performed in the digital environment, so these systems will also operate in discrete time.

The notation $u(t)$ here and below denotes a vector variable u as a function of time $t \in T$. The designation $u(t_i)$ denotes the instantaneous value of this variable at the instant of discrete time $t_i \in [t_0, t_f]$. We will also use the abbreviated notation u_i instead of $u(t_i)$. The notation $u^{(j)}(t)$ denotes the jth component of the vector variable $u(t)$. Similarly, the $u^{(j)}(t_i)$ designation or, in the abbreviated form, $u_i^{(j)}$, denotes the instantaneous value of the jth component of the vector variable $u(t)$ at the instant of discrete time $t_i \in [t_0, t_f]$. The corresponding notation for the variables $x \in X$, $y \in Y$, $\xi \in \Xi$, $\zeta \in Z$ is introduced similarly.

As noted above, our object of study is a controllable dynamical system, operating under uncertainty conditions. We can divide these uncertainties into the following main types:

- uncertainties generated by uncontrolled disturbances acting on the object (for example, atmospheric turbulence, wind gusts);
- insufficient knowledge of the simulated object and the environment in which it operates (for example, insufficiently known or unknown aerodynamic characteristics of the aircraft);
- uncertainties caused by a change in the properties of the object due to equipment failures and structural damage (for example, combat and/or operational structure damage and aircraft equipment failures that change the object properties).

In order to describe the current (instantaneous) state of the complex $\mathbb{K}$, we introduce the concept of a *situation* that includes components describing the state of both the system $\mathbb{S}$ and the environment $\mathcal{E}$. The components describing the system $\mathbb{S}$ will be called internal, while components describing the environment $\mathcal{E}$ will be called external; hence

$$\begin{aligned}\text{Situation} &= \text{External-Situation} \\ &\quad + \text{Internal-Situation}\\ \lambda(t_i) &= \langle \lambda_{int}(t_i), \lambda_{ext}(t_i) \rangle, \\ \lambda(t_i) &\in \Lambda, \ \lambda_{int}(t_i) \in \Lambda_{int}, \ \lambda_{ext}(t_i) \in \Lambda_{ext}.\end{aligned}$$

Along with the concept of the situation, the idea of *situational awareness* plays an important role. While the concept of the situation describes objective reality (object + environment), the concept of situational awareness describes the degree of awareness of the system $\mathbb{S}$ about this reality, i.e., about the current situation. The situational awareness concept describes which data are obtained by observations and are available to the system $\mathbb{S}$ for generating control decisions. Awareness of some components of the situation is usually incomplete (we know their values inaccurately) or zero (their values are unknown). For one part of the components it is provided by direct observation (measurement), while for the other part, awareness is provided algorithmically, i.e., by estimation of their values based on the known values of the other components.

Thus, when we consider a system $\mathbb{S}$ with uncertainties, we actually assume an *incomplete situational awareness* for it, i.e., the fact that the values of some part of the internal or external components of the situation for $\mathbb{S}$ are unknown or known inexactly.

1.1.6 Behavior and Activity of Systems

The current state of the system $\mathbb{S}$ is described by the set (1.9) of variables $x_i \in X_i$ describing it in the problem being solved. This set is considered to be a tuple of length n, i.e.,

$$x = \langle x_1, x_2, \ldots, x_n \rangle, \quad x_i \in X_i, \quad i = 1, 2, \ldots, n,$$

or as a column vector $x = [x_1 x_2 \ldots x_n]^T$. Here,

$$x \in X, \; X \subseteq X_1 \times X_2 \times \ldots \times X_n.$$

The ranges X_i of continuous variables x_i are usually subsets of the set of real numbers $\mathcal{R}$.

Which variables x_i are included in the set (vector) x that describes the state of the system $\mathbb{S}$ depends on the nature of the given system and the problem being solved. In a context of different problems, the same system $\mathbb{S}$ may be described by different sets x that include different variables x_i. Examples 1 and 2 on page 18 may serve as an illustration of this assertion: the state of the same aircraft is described by different variables for the problem of longitudinal motion during climb phase and for the problem of curvilinear flight in the horizontal plane.

The point $x \in X \subseteq X$ in the state space is the state of the system $\mathbb{S}$ at some time instant $t \in T = [t_0, t_N]$. For the continuous time $t \in T$ case and a finite-dimensional state vector $x \in X \subseteq \mathcal{R}^n$, in order to specify states at all time instants we need to specify a vector function

$$\begin{aligned} x(t) &= (x_1(t), x_2(t), \ldots, x_n(t)) \\ &= [x_1(t) x_2(t) \ldots x_n(t)]^T. \end{aligned} \tag{1.18}$$

Taking into account the remarks made above regarding the transition from the continuous time $t \in [t_0, t_N]$ to the discrete time (1.17), instead of continuous phase trajectories (1.18) we will consider their discrete representation in the form of a set of sequences of the form

$$x(t_i) = \{(x_1(t_i), x_2(t_i), \ldots, x_n(t_i))\}, \quad i = 0, 1, \ldots, N. \tag{1.19}$$

The *behavior* of the system $\mathbb{S}$ is the sequence of its phase states $x(t_i) \in X$, tied to the corresponding time instants $t_i \in T$, i.e.,

$$\{\langle x(t_i), t_i \rangle\}, \quad t_i \in [t_0, t_f] \subseteq T, \quad i = 0, 1, \ldots, N. \tag{1.20}$$

The *activity* of the system $\mathbb{S}$ is a sequence of its purposeful actions, each of which is a response of the form

$$\langle situation, goal \rangle \Rightarrow action \Rightarrow result,$$

that is,

$$\{\langle \lambda(t_i), \gamma(t_i) \rangle\} \xrightarrow{\Phi^S} \lambda(t_{i+1}), \quad i = 0, 1, 2, \ldots, N, \tag{1.21}$$

or, equivalently,

$$\lambda(t_{i+1}) = \Phi^S(\langle \lambda(t_i), \gamma(t_i) \rangle), \quad i = 0, 1, 2, \ldots, N.$$

Here $\lambda(t_i) \in \Lambda$ is the current situation, $\gamma(t_i) \in \Gamma$ is the current goal, and Φ^S is the law of evolution of the system $\mathbb{S}$.

All kinds of systems $\mathbb{S}$ exhibit this behavior (1.20), whether they are controllable or not and irrespective of the uncertainty factors. In contrast, the activity (1.21) is only available for systems that include the formulation of control goals in one form or another. These are either permanent goals (adaptive systems $\mathbb{AS}$) or self-corrected goals (intelligent systems $\mathbb{IS}$).

Control of the behavior for the system $\mathbb{S}$ is a more general concept than control of its motion. A list of elements involved in the behavior control is presented in the introduction on page 2.

Robotics borrows the concept of the activity from the life sciences (biology, psychology, ethology). In recent years, research in the field of robotization of controllable dynamical systems of various kinds has been actively conducted. As a consequence, we need to expand the capabilities of dynamical system modeling and control tools from traditional motion control tasks to the tasks of behavior and activity control for such systems. These problems are especially relevant for highly autonomous robotic UAVs (so-called "smart UAVs") and unmanned cars.

The action executed by the system $\mathbb{S}$ does not necessarily depend only on the current (i.e., considered at the time instant $t_i \in T$) values of the situation $\lambda(t_i)$ and the goal $\gamma(t_i)$. In a more general case, it depends on the set of situations $\Lambda(t_i)$ and the set of goals $\Gamma(t_i)$ at a given time instant t_i, i.e.,

$$\langle \Lambda(t_i), \Gamma(t_i) \rangle \xrightarrow{\Phi^S} \lambda(t_{i+1}), \quad i = 0, 1, 2, \ldots, N;$$
$$\Lambda(t_i) \subset \Lambda^S, \quad \Gamma(t_i) \subset \Gamma^S. \tag{1.22}$$

Here the transition to the situation $\lambda(t_{i+1})$ depends not only on the current situation $\lambda(t_i)$ and the current goal $\gamma(t_i)$ at a given time instant t_i, but also on the past (prehistory) and future (forecast) states and goals, which are described by the corresponding sets: a set of situations $\Lambda(t_i)$ and a set of goals $\Gamma(t_i)$ for a given time instant t_i.

1.2 DYNAMICAL SYSTEMS AND THE PROBLEM OF ADAPTABILITY

One of the most important classes of dynamical systems is aircraft of various types. As already noted, it is necessary to ensure the control of the motion for modern and advanced aircraft under significant and diverse uncertainties in the values of their parameters and characteristics, flight regimes, and environmental influences. Moreover, during flight, a variety of abnormal situations may arise, in particular, equipment failures and structural damage, the consequences of which can be compensated in most cases by reconfiguration of the aircraft control system and reallocation of its control surfaces.

The presence of significant and diverse uncertainties is one of the most severe factors that complicates the solution of all three problems (analysis, synthesis, identification) for dynamical systems. The problem here is that the current situation for the system can change dramatically. Besides, the situation change can be unpredictable because of uncertainties. Under these conditions, the system must be able to adjust quickly to such changes in the situation, so it must be adaptive.

We need to clarify the meaning of the concept of *system adaptability*. By adaptive we mean such a system that can quickly adjust to a changing situation by varying some of its elements. We assume that such elements include the control laws implemented by the dynamical system, as well as the model of the controlled object. Most of the changes in these elements can affect both the parameter values and the structure of the control laws and/or models. In most cases, the changes in these elements relate to the values of their parameters. Sometimes, the structure of control laws and/or models is also subject to change. These topics are discussed in more detail in the next section.

1.2.1 Types of Adaptation

Following [35], we distinguish the types (hierarchical levels) of adaptation as:

- parametric adaptation;
- structural adaptation;
- adaptation of object;
- adaptation of control goals.

1.2.1.1 Parametric Adaptation

Parametric adaptation is performed by adjusting the value of the tuning parameters $\boldsymbol{\vartheta}(t_i) \in \Theta$ of the system $\mathbb{S}$, which are a subset of system parameters w; see page 17 (such parameters can be, for example, the controller gains).

In this case, we consider that the rule Φ^S depends not only on $\mathbf{x}$, $\mathbf{u}$, $\boldsymbol{\xi}$, t, as was indicated earlier, but also on $\boldsymbol{\vartheta}(t_i) \in \Theta$. This means that $\Phi^S = \Phi^S(\boldsymbol{\vartheta})$ is a parametric family of functions. If we specify some constant value of the vector $\boldsymbol{\vartheta}(t_i) \in \Theta$, we thereby select a function $\Phi^S = \Phi^S(\mathbf{x}, \mathbf{u}, \boldsymbol{\xi}, t)$ from this family. The rule $\Psi^S = \Psi^S(\boldsymbol{\lambda}, \boldsymbol{\gamma}, t)$ defines values $\boldsymbol{\vartheta}(t_i)$ transferred to $\Phi^S = \Phi^S(\boldsymbol{\vartheta}(t_i))$, which leads to a change in the nature of the response of the system $\mathbb{S}$ to the effects of the environment $\mathcal{E}$, i.e., to a change in behavior of this system.

Possible mechanisms for changing the values of the vector of the parameters $\boldsymbol{\vartheta}(t_i) \in \Theta$ for the $\mathbb{S}$ system are not discussed here; we will consider them in the following sections.

The values of the vector of the parameters $\boldsymbol{\vartheta}(t_i) \in \Theta$ for the system $\mathbb{S}$ can be piecewise constant, i.e., their values will remain constant not just for a single ordered pair $\langle \boldsymbol{\lambda}(t_i), \boldsymbol{\gamma}(t_i) \rangle \in \Lambda \times \Gamma$, but for the entire subdomain $\Lambda_i \times \Gamma_i \subset \Lambda \times \Gamma$ of such a domain. This approach is rather widely used in control systems. We refer to it as Gain Scheduling.

The adjustment might also be continuous, so that each pair $\langle \boldsymbol{\lambda}(t_i), \boldsymbol{\gamma}(t_i) \rangle \in \Lambda \times \Gamma$ corresponds to some value $\boldsymbol{\vartheta}(t_i) \in \Theta$.

This concept of parametric adaptation corresponds to the concept of *accommodation* in biology.

1.2.1.2 Structural Adaptation

However, sometimes the required behavior variability for the system $\mathbb{S}$ cannot be achieved by varying only the values of the system parameters $\boldsymbol{\vartheta}(t_i) \in \Theta$. The next hierarchical level of adaptive systems are systems capable of structural adaptation, i.e., systems capable to change their structure (a set of elements of the system $\mathbb{S}$, as well as the relations between these elements) according to the changing situation $\boldsymbol{\lambda}(t_i) \in \Lambda^S$ and the goal $\boldsymbol{\gamma}(t_i) \in \Gamma^S$.

In the simplest case the system **S** includes a set of structurally alternative variants of the rule $\Phi^S = \{\Phi_p\}^S$, $p = 1, \ldots, N^P$. At the time instant t_i, a rule with a certain index p is applied. The specific value of this index is determined by the switching rule $\Psi^S = \Psi^S(\boldsymbol{\lambda}, \boldsymbol{\gamma}, t)$. A more complicated, but more exciting example is the system **S** that undergoes evolutionary changes in the structure under the influence of the environment $\mathcal{E}$ (and, possibly, some other factors).

In biology, a mechanism of this kind is called adaptation (which means an irreversible evolutionary change in the genotype of the system). On the other hand, the accommodation discussed in Section 1.2.1.1 is a reversible adjustment of parameters.

1.2.1.3 Adaptation of Object

It is quite possible that no variations of the system $\mathbb{S}$ parameters $\boldsymbol{\vartheta}(t_i) \in \Theta$ or structure would allow us to achieve certain goals. This situation is quite natural, because the potential capabilities of any system are limited, and these limits are caused by the "design" of the system. If a case of this kind arises, then the next level of adaptation may be involved, namely an adaptation of the object.

In Section 1.1.1.1, we formulated the thesis that there is a particular system, which is the object of our study, and there is all the rest that we did not include in this system. We call this "all the rest" as the (external) environment. The adaptation of an object involves a revision of the boundary between the object and the environment.

The main idea of this level of adaptation is that the solution of the required target problem may be achieved not only by one system $\mathbb{S}$, as in the previous two cases, but by a team of such systems $\{\mathbb{S}_\mu\}$, $\mu = 1, \dots, N^\mu$, that interact with each other. Therefore, instead of a single rule $\Phi^S = \Phi^S(\mathbf{x}, \mathbf{u}, \boldsymbol{\xi}, t)$ we consider a set of interacting rules $\Phi^S = \{\Phi_\mu\}^S$, corresponding to different systems $\mathbb{S}_\mu$, $\mu = 1, \dots, N^\mu$.

Example 4. Let us assume that the problem to be solved involves an interception of aerial targets, including group targets. If the airspace area and the number of targets in the group are relatively small, then in some cases the problem can be solved by a single fighter-interceptor, possessing missile armament and a multichannel system for detecting and tracking targets. If these conditions are not met, the capabilities of one aircraft are not enough.

In order to solve this problem we need to use a group of systems aimed at a cooperative solution of a common problem. An example of such approach is the MiG-31 [36,37] interception complex, in which a group of four interacting aircraft of this type manages the airspace in front of them with a length of 800–900 km. In the variant MiG-31B, the interception complex has an additional opportunity to automatically transmit the data about the targets detected by the fighter to ground-based air defense systems in order to support the targeting of antiaircraft missiles. In this case an expansion of the interception complex has been performed by adjusting the boundary between this complex and the environment.

Example 5. A similar example of the adaptation of the object is the development of the first Soviet interception complex Su-9 [36]. This complex, adopted in 1960, was a response to the fact that a "sole" interceptor, even with missile weapons, cannot effectively intercept air targets at medium and long ranges. The adaptation reaction was to include, in addition to the Su-9 airplane with the onboard radar TsD-30 and the guided air-to-air missiles RS-2-US, the system of automated ground guidance "Vozdukh" ("Air").

1.2.1.4 *Adaptation of Control Goals*

If the adaptation of the system $\mathbb{S}$ does not allow us to solve the assigned problem, i.e., to ensure the achievement of the system goals, it is quite possible that these goals $\boldsymbol{\gamma} \in \Gamma^S$ are unreachable for $\mathbb{S}$. In this case, it remains possible to change control goals so that they become attainable. This operation is performed using the rule $\Omega^S = \Omega^S(\boldsymbol{\lambda}, \boldsymbol{\gamma}, t)$, based on the motivational set of elements $\Sigma^{IS} \subseteq \Sigma$. Adaptation of goals is, in essence, the adjustment of the needs of the subject of control.

Example 6. We can explain the essence of adaptation of control goals as follows. Suppose that a self-propelled vehicle delivered to some celestial body must examine a particular object. It may be found that the solution of the assigned task requires too much expenditure of resources, which jeopardizes the fulfillment of other tasks of the expedition. In this case, proceeding from general settings (for example, to get the maximum possible knowledge about the celestial body under study), the system $\mathbb{S}$ can replace one goal with another and find for study an object "similar" to one that it failed to investigate, or refuse altogether from this point of the program, switching to others.

1.2.2 General Characteristics of the Adaptive Control Problem

As is well known [38,39], the traditional control theory requires knowledge of the mathematical model of the object. We need to know parameters and characteristics of the object, as well as parameters and characteristics of the environment in which the object operates.

In practice, we can not always meet these requirements. Also, in the process of operation, the parameters and characteristics of the object and environment can vary considerably. In these cases, traditional methods often give unsatisfactory results.

For this reason, there is a need to build control systems that do not require a complete a priori knowledge of the control object and the conditions for its operation. Such a system must be able to adjust to the changing object properties and environmental conditions. These requirements are met by the adaptive systems [40–54], in which the current available information is used not only to generate a control action (as in conventional nonadaptive systems), but also for changing (adjusting) the control algorithm.

We usually distinguish two main classes of adaptive systems [41,44]:

- self-tuning systems, in which the structure of the control algorithm does not change during operation, but only its parameters change;
- self-organizing systems, in which the structure of the control algorithm changes during operation.

Incomplete knowledge of the parameters and characteristics of the control object and environment in which it operates is typical for adaptive systems. We treat this incomplete knowledge as additional uncertainty factors and include them in the corresponding class Ξ.

For example, we can define the uncertainty factors associated with the aircraft control problem through the following three sets of parameters:

$$\begin{aligned} W &= W_1 \times W_2 \times \ldots \times W_p, \\ V &= V_1 \times V_2 \times \ldots \times V_q, \\ \Xi &= \Xi_1 \times \Xi_2 \times \ldots \times \Xi_r, \end{aligned} \tag{1.23}$$

where W_i, V_j, Ξ_k are the ranges of the values of w_i (which define the possible values of the parameters of the aircraft), v_j (which define the possible values of the characteristics of the aircraft), and ξ_k (which define the possible values of the parameters of the atmosphere and atmospheric influences), respectively, i.e.,

$$\begin{aligned} &w_i \in W_i, \quad W_i = [w_i^{min}, w_i^{max}], \quad i = 1, 2, \ldots, p, \\ &v_j \in V_j, \quad V_j = [v_j^{min}, v_j^{max}], \quad j = 1, 2, \ldots, q, \\ &\xi \in \Omega_k, \quad \Xi_k = [\xi_k^{min}, \xi_k^{max}], \quad k = 1, 2, \ldots, r. \end{aligned} \tag{1.24}$$

A particular combination of the parameters w_i, v_j, ξ_k generates a tuple ω_s of length $p+q+r$, i.e.,

$$\begin{aligned} &\omega_s = \langle w_1, \ldots, w_p, v_1, \ldots, v_q, \xi_1, \ldots, \xi_r \rangle, \\ &s = 1, 2, \ldots, p \cdot q \cdot r. \end{aligned} \tag{1.25}$$

All possible combinations of values of the uncertainty factors characterizing control problems for some dynamical system[24] form a collection of Ω tuples ω_s as the Cartesian product of the sets W, V, Ξ, i.e.,

$$\Omega = W \times V \times \Xi, \quad \omega_s \in \Omega, \tag{1.26}$$

or as a subset of $\widetilde{\Omega} \subset \Omega$, if not all tuples ω_s are valid. For other types of dynamical systems, we define the uncertainty factors and their possible combinations in a similar way.

Having in mind this definition of the operating conditions for the considered dynamical system, we can formulate the problem of adaptive control as follows: the controller will be adaptive in the class Ξ, if after a finite time T_a, called the adaptation time, it will ensure the fulfillment of the control goal.

1.2.3 Basic Structural Variants of Adaptive Systems

As already noted, the control system is considered adaptive if the current information on

[24]For example, a flight control problem for aircraft.

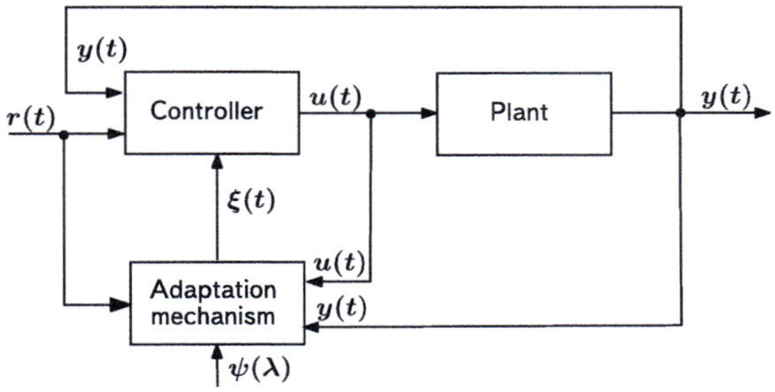

FIGURE 1.1 Scheme of a system with an adjustable control law implemented by the controller: $r(t)$ is the reference signal; $u(t)$ is the control; $y(t)$ is the output of the controlled object (plant); $\xi(t)$ is the adjusting effect for the controller; $\psi(\lambda)$, $\lambda \in \Lambda$ is additional information that should be taken into account when developing an adjusting action (for example, the speed and altitude of an aircraft in the problem of controlling its angular motion).

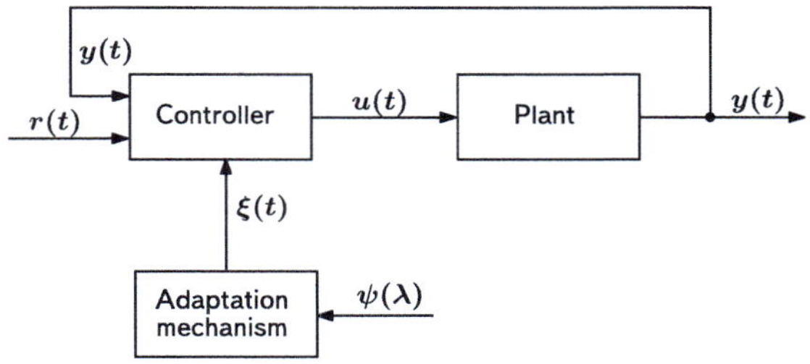

FIGURE 1.2 The scheme for adjusting the parameters of the control law implemented by the controller in accordance with the GS scheme: $r(t)$ is the reference signal; $u(t)$ is the control; $y(t)$ is the output of the controlled object; $\xi(t)$ is the adjusting effect for the controller; $\psi(\lambda)$, $\lambda \in \Lambda$ is additional information used to produce the adjusting action. From [56], used with permission from Moscow Aviation Institute.

the state of the system is used not only to generate a control action (as is the case of nonadaptive systems), but also to change (adjust) the control algorithm. In general, the structure of the adaptive system can be represented as shown in Fig. 1.1. We can see that the adjusting action $\xi(t)$ for the controller is generated using an adaptation mechanism. This mechanism uses as its inputs such signals as control $u(t)$, the output of the object $y(t)$, and some additional ("external") information $\psi(\lambda)$, $\lambda \in \Lambda$. We need to take these data into account in the development of an adjusting action. For example, in the problem of controlling the angular motion of an aircraft, the composition of these data can include the airspeed and altitude of flight.

Various types of this scheme are possible, differing from each other by the composition of the input data used to generate the adjusting action $\xi(t)$. One such option is that the adjustment is performed only based on the "external" data $\psi(\lambda)$, $\lambda \in \Lambda$; it is referred to as the Gain Scheduling (GS) approach. The principal difference of this approach (Fig. 1.2) from the full version of the adaptation scheme (Fig. 1.1) is that in the GS approach the values of adjusting effects ξ as a function of $\psi(\lambda)$ must be computed in advance (off-line). Then, this function remains unchanged during the control process. In the full version of the adaptation scheme, the adjustment algorithm operates on-line during the operation of the system, taking into account not only external data but other information concerning the state of the controlled object.

Despite the limited adaptation capabilities of the GS approach, it is used often in practice. For example, this approach was applied for control of test flights for the experimental hypersonic vehicle X-43A [55]. Gains in the longitudinal control law of this vehicle are scheduled with the angle of attack and Mach number as "external data" mentioned above.

Adaptive control schemes are usually divided into two main types: direct adaptive control and indirect adaptive control.

Direct adaptive control schemes are often based on the use of some reference model (RM) that specifies the required (desired) behavior of the system under consideration.[25] The struc-

[25]The purpose of the control is to make the dynamical system behavior as close as possible to the behavior defined by the reference model. The correction of the control goal mentioned in Section 1.1.1.1 can be carried out in this case by replacing one reference model with another.

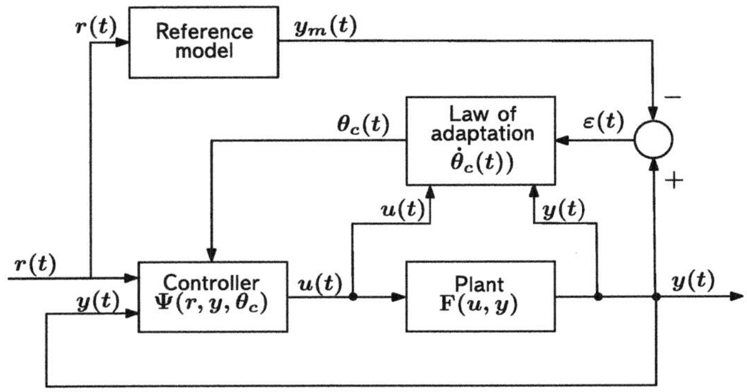

FIGURE 1.3 A direct adaptive control scheme: $r(t)$ is the reference signal; $u(t)$ is the control; $y(t)$ is the output of the controlled object; $y_m(t)$ is the output of the reference model; $\theta_c(t)$ are the adjustable parameters of the controller; $\varepsilon(t) = y(t) - y_m(t)$ is the difference between the outputs of the object and the reference model (From Gang Tao. Adaptive control design and analysis, Wiley-InterScience, 2003).

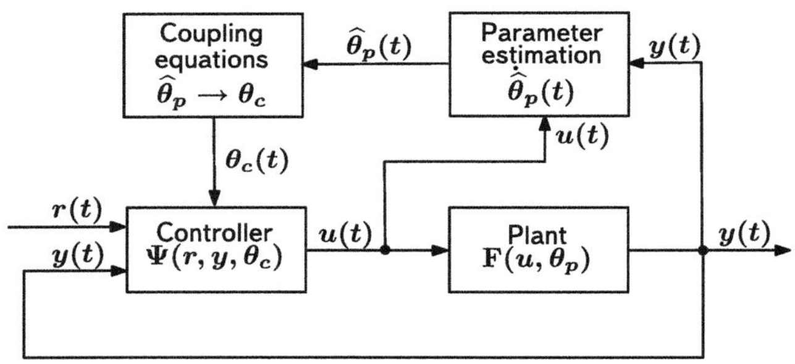

FIGURE 1.4 Indirect adaptive control scheme: $r(t)$ is the reference signal; $u(t)$ is the control; $y(t)$ is the output of the controlled object; $\theta_p(t)$ are the estimated parameters of the object; $\widehat{\theta}_p(t)$ is the estimate of object parameters; $\theta_c(t)$ are the adjustable parameters of the controller (From Gang Tao. Adaptive control design and analysis, Wiley-InterScience, 2003).

ture of such a system is shown in Fig. 1.3. In the direct adaptive control systems, the parameters of the controller $\theta_c(t)$ are adjusted by the algorithm implemented by the adaptation law, which computes the values of the derivative $\dot{\theta}_c(t)$ or the difference $\theta_c(t+1) - \theta_c(t)$. This computation is based directly on the tracking error value $\varepsilon(t) = y(t) - y_m(t)$.

In the indirect adaptive control systems, the parameters of the controller $\theta_c(t)$ are computed using the coupling equation that maps an estimate of object parameters $\widehat{\theta}_p(t)$ to values of parameters for the controller $\theta_c(t)$. Estimates $\widehat{\theta}_p(t)$ are produced operatively (on-line), in the process of object functioning, by computing the value of the derivative $\dot{\widehat{\theta}}_p(t)$ or the difference $\widehat{\theta}_p(t+1) - \widehat{\theta}_p(t)$. The structure of such a system is shown in Fig. 1.4.

In both direct and indirect adaptive control schemes, the basic idea is that the ideal values of the controller parameters (for direct adaptive control) or the object (for indirect adaptive control) are used as if they were parameters of a real controller or object, respectively. Because the ac-

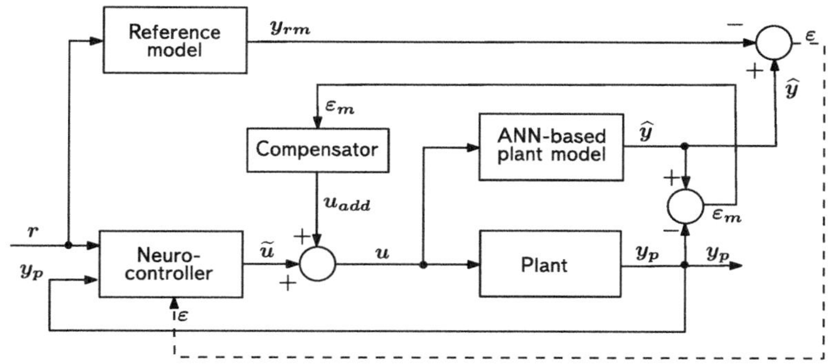

FIGURE 1.5 The general scheme of neural network–based model reference adaptive control (MRAC): $\widetilde{u}$ is control signal at the output of the neural controller; u_{add} is additional control from the compensator; u is the resultant control; y_p is output of the plant (controlled object); $\widehat{y}$ is the output of the neural network model of the plant; y_{rm} is output of the reference model; ε is the difference between the outputs of the plant and the reference model; ε_m is the difference between the outputs of the plant and the ANN model; r is the reference signal. From [56], used with permission from Moscow Aviation Institute.

tual values of these parameters inevitably differ from the ideal ones, an error arises that degrades the quality of control. One approach to compensating for this error is described below. It relies on interpretation of this error as some disturbing effect on the system and reduces this effect by introducing a compensating loop into the system.

1.2.4 The Role of Models in the Problem of Adaptive Control

We can illustrate the abovementioned critical role of models in solving problems related to dynamical systems, using the example of adaptive control.

Fig. 1.5 shows the general scheme of the neural network–based model reference adaptive control (MRAC). In this case, the role of the artificial neural network (ANN) plant model is to ensure that the computed error (the difference between outputs of the reference model and the ANN plant model) at the system output is converted to an error at the output of the neurocontroller, which is necessary to adjust the parameters of this controller.

The second example of the role of the dynamical system model in adaptive control is shown in Fig. 1.6, where the general scheme of neural network–based model predictive control (MPC) is presented.

Thus, models of the simulated object play a key role in solving problems related to adaptive systems. These models are required for a solution of some important subproblems, for example:

1. The subproblem of analyzing the behavior of a dynamical system as a part of the MPC scheme: the solution of this subproblem is necessary for prediction of the behavior of a dynamical system that is used to select an appropriate control action.
2. The subproblem of converting an error at the output of a dynamical system to an error at the output of a neurocontroller: when solving this subproblem, the object model plays the role of a "technological environment" for the specified transformation and the error transmission to the output of the neurocontroller.
3. Reconfiguration subproblem of the control system: here the model of the normative behavior of the object is used to detect an occurrence of an abnormal situation.

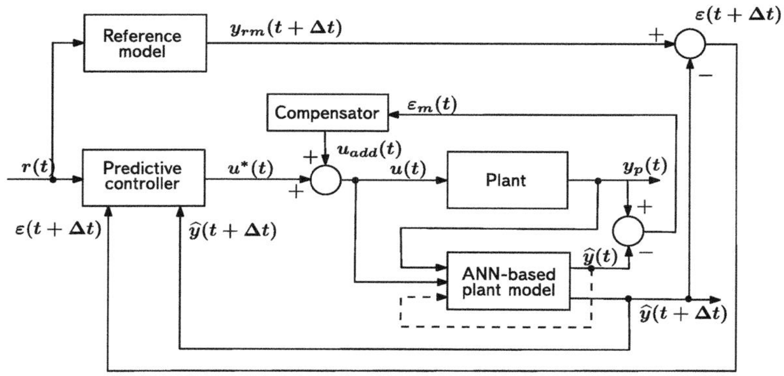

FIGURE 1.6 General scheme of neural network adaptive control with predictive model (MPC – model predictive control): u^* is control signal at the output of the optimization algorithm; u_{add} is additional control from the compensator; u is the resultant control; y_p is output of the plant (controlled object); $\hat{y}$ is the output of the neural network based model of the plant; y_{rm} is output of the reference model; e is the difference between the outputs of the plant and the reference model; r is the reference signal. From [56], used with permission from Moscow Aviation Institute.

1.3 A GENERAL APPROACH TO DYNAMICAL SYSTEM MODELING

1.3.1 A Scheme of the Modeling Process for Dynamical Systems

We need to answer the following three main questions regarding the system $\mathbb{S}$:

1) How does the system $\mathbb{S}$ respond to various kinds of effects applied to it?
2) How can we get the desired reactions (responses) from $\mathbb{S}$?
3) Given the empirical data for the reactions of the system $\mathbb{S}$ in response to certain effects, what might the structure of the system look like?

To obtain quantitative answers to these questions, we need to adopt some agreements related to the system $\mathbb{S}$ and the conditions in which it operates. In this case, it is required to clarify what the object of research is, and also what tasks will be performed for the given object.

As for the answer to the first of these two questions, the object of the study will be dynamical systems with lumped parameters, i.e., systems regarded as a rigid body or an interconnected set of rigid bodies (a set of rigid bodies with kinematic and/or dynamic links between them). This approach covers a vast range of applied problems from various scientific and engineering fields. Traditional mathematical models of such dynamical systems are systems of ODEs or DAEs.

In some cases the dynamical system should not be treated as a rigid body in order for the model to be adequate. In particular, rigid body assumption is violated if the structure elasticity plays an important role for the simulated object. In this case, we have to consider the dynamical system as a system with distributed parameters and describe this system by partial differential equations (PDEs). The approach to modeling of dynamical systems considered below can be extended to this case. This class of systems, however, is the subject of a separate study and we do not discuss it in this book.

To answer the question about classes of problems that we need to consider for a dynamical system as a simulated and/or controlled object, we introduce the definition of the system $\mathbb{S}$ as an ordered triple of the following form:

$$\mathbb{S} = \langle \mathrm{U}, \mathrm{P}, \mathrm{Y} \rangle, \tag{1.27}$$

where U is the input to the simulated/controlled object; P is the simulated/controlled object (plant); Y is the object's response to the input effects.

Inputs U include initial conditions, controls, and uncontrolled external disturbances for the object P. A simulated object P might be, in particular, an aircraft of some type. Outputs Y of the dynamical system $\mathbb{S}$ represent the observed reactions of the object P to the input actions U.

Based on these definitions, we can state three main classes of problems related to dynamical systems as follows:

1. $\langle \mathrm{U}, \mathrm{P}, \mathbf{Y} \rangle$ is the analysis problem for the behavior of the dynamical system (find **Y**, given U and P);
2. $\langle \mathbf{U}, \mathrm{P}, \mathrm{Y} \rangle$ is the synthesis problem for the dynamical system control (find **U**, given P and Y);
3. $\langle \mathrm{U}, \mathbf{P}, \mathrm{Y} \rangle$ is the identification problem for the dynamical system (find **P**, given U and Y).

Here problem 1 belongs to the class of direct problems, while problems 2 and 3 belong to the class of inverse problems of system dynamics. In this case, problem 3 is actually the problem of designing a model of a dynamical system, whereas problems 1 and 2 are solved using this model.

Another important question that we need to address for the considered dynamical systems is the problem of the uncertainties present in these systems as well as their environment.

As already noted in Section 1.1.1.2, uncertainties may be divided into the following types:

- uncertainties caused by uncontrolled disturbances acting on the object;
- insufficient knowledge of the simulated object and its environment;
- uncertainties caused by changes of the properties of the object due to failures of its equipment and structural damage.

Analysis of the problems associated with the development of complex systems operating under uncertainty conditions leads to the conclusion that we need to adopt an idea of adaptability. As shown above, models of simulated objects play a crucial role in the process of development for adaptive systems. These models are intended primarily for the use in onboard systems (for example, in aircraft control systems) in real or even advanced time, which imposes some requirements for them, namely increased accuracy and high computational speed and adaptability.

Traditional models (in the form of ODEs or DAEs) do not meet all of the requirements regarding accuracy and computational speed. They do not satisfy the requirement of adaptability at all. The ways of meeting these requirements can be as follows. Accuracy and computational speed of the model can be ensured by maximizing the use of knowledge and data about the simulated object. The adaptability of the model can be interpreted as the ability to rapidly restore its adequacy to the simulated object by means of structural and/or parametric adjustments.

The process of development for the dynamical system model is reduced to the solution of four main problems. Namely, we need to find:

1) a set of variables that describe the simulated object;
2) a class (family) of models, which includes the required (desired) model;
3) tools for selecting a particular model from a given class (the criterion of the adequacy and the search algorithm for this model);
4) a representative set of experimental data required to design and evaluate the model.

Problem 1, that is, the formation of a set of variables that describe the simulated object, is the subject of a separate study aimed at understanding how the meaningful interpretation

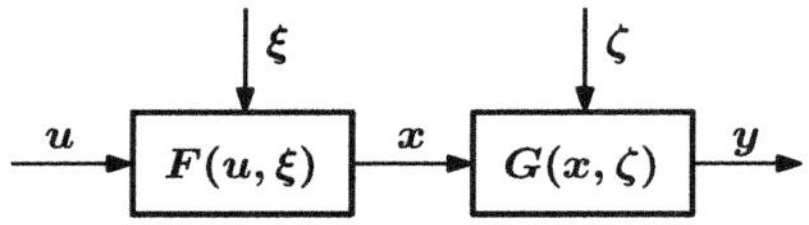

$$y = \Phi(u, \xi, \zeta) = G(F(u, \xi), \zeta)$$

FIGURE 1.7 General structure of the simulated dynamical system. From [56], used with permission from Moscow Aviation Institute.

of the problem is formalized in a mathematical model (see also remarks on Examples 1 and 2 on page 18). In the following we consider only problems 2, 3, and 4.

Let there be some dynamical system $\mathbb{S}$, which is an object of modeling (Fig. 1.7).

The system $\mathbb{S}$ perceives the controlled $u(t)$ and uncontrolled $\xi(t)$ impacts. Under these influences, $\mathbb{S}$ changes its state $x(t)$ according to the transformation rule (mapping) $F(u(t), \xi(t))$ that it implements. At the initial time $t = t_0$, the state of the system $\mathbb{S}$ takes the value $x(t_0) = x_0$.

The state $x(t)$ is perceived by the observer that transforms it into observation results $y(t)$ via mapping $G(x(t), \zeta(t))$. These observations $y(t)$ represent the output of the system $\mathbb{S}$. Sensors that perform measurements of the state introduce some error that we describe by additional uncontrolled influence $\zeta(t)$ ("measurement noise"). The composition of the maps $F(\cdot)$ and $G(\cdot)$ describes the relationship between the controlled input $u(t) \in U$ of the system $\mathbb{S}$ and its output $y(t) \in Y$, while taking into account the influence of uncontrolled disturbances $\xi(t)$ and $\zeta(t)$ on the system under consideration, i.e.,

$$y = \Phi(u(t), \xi(t), \zeta(t)) = G(F(u(t), \xi(t)), \zeta(t)).$$

Suppose that we have performed N_p observations for the system $\mathbb{S}$, i.e.,

$$\{y_i\} = \Phi(u_i, \xi, \zeta), \ i = 1, \dots, N_P, \tag{1.28}$$

and recorded the current value of the controlled input action $u_i = u(t_i)$ and the corresponding output $y_i = y(t_i)$ for each of them. The results $y(t_i)$, $t_i \in [t_0, t_f]$, of these observations together with the corresponding values of the controlled inputs u_i generate a set of N_P ordered pairs, i.e.,

$$\{\langle u_i, y_i \rangle\}, \ u_i \in U, \ y_i \in Y, \ i = 1, \dots, N_P. \tag{1.29}$$

Given the data (1.29), we need to find an approximation $\widehat{\Phi}(\cdot)$ for the mapping $\Phi(\cdot)$, implemented by the system $\mathbb{S}$ such that the following condition is satisfied:

$$\|\widehat{\Phi}(u(t), \xi(t), \zeta(t)) - \Phi(u(t), \xi(t), \zeta(t))\| \leqslant \varepsilon,$$
$$\forall u(t_i) \in U, \ \forall \xi(t_i) \in \Xi, \ \forall \zeta(t_i) \in Z, \ t \in [t_0, t_f]. \tag{1.30}$$

Thus, as follows from (1.30), it is necessary that the desired approximate mapping $\widehat{\Phi}(\cdot)$ possesses the required accuracy not only when reproducing the observations (1.29), but also for all valid values of $u_i \in U$. We refer to this property of the map $\widehat{\Phi}(\cdot)$ as *generalization*. The presence of entries $\forall \xi(t_i) \in \Xi$ and $\forall \zeta(t_i) \in Z$ in (1.30) means that $\widehat{\Phi}(\cdot)$ will have the required accuracy only provided that at any time $t \in [t_0, t_f]$ uncontrollable effects $\xi(t)$ on the system $\mathbb{S}$ and the measurement noise $\zeta(t)$ do not exceed the allowed limits.

The map $\Phi(\cdot)$ corresponds to the object under consideration (the dynamical system $\mathbb{S}$), and the map $\widehat{\Phi}(\cdot)$ will be referred to as the model of this object. We will also assume that we have data of the form (1.29) for the system $\mathbb{S}$, and possibly some knowledge about the "structure" of mapping $\Phi(\cdot)$, implemented by the system under consideration. In this case, the presence of data of the specified type is mandatory (at least, they are necessary for testing the model $\widehat{\Phi}(\cdot)$), while knowledge of the mapping $\Phi(\cdot)$ might be unavailable or unused for the development of the model $\widehat{\Phi}(\cdot)$.

It is required to clarify what is meant by the norm $\|\cdot\|$ in the expression (1.30), i.e., how to interpret the magnitude of the difference between the results given by the maps $\widehat{\Phi}(\cdot)$ and $\Phi(\cdot)$.

One possible definition of the residual (1.30) is the maximum deviation of $\widehat{\Phi}(\cdot)$ from $\Phi(\cdot)$, i.e.,

$$\|\widehat{\Phi}(u(t),\xi,\zeta)-\Phi(u,\xi,\zeta)\| = \max_{t_0\leqslant t\leqslant t_n} |\widehat{\Phi}(u(t),\xi,\zeta)-\Phi(u(t),\xi,\zeta)|. \quad (1.31)$$

The second, more common, approach is to estimate the values of the difference between $\widehat{\Phi}(\cdot)$ and $\Phi(\cdot)$ as the norm of the form

$$\|\widehat{\Phi}(u,\xi,\zeta)-\Phi(u,\xi,\zeta)\| = \int_{t_0}^{t_n} [\widehat{\Phi}(u(t),\xi,\zeta)-\Phi(u(t),\xi,\zeta)]^2 dt. \quad (1.32)$$

The number of experiments available for generation of the data set (1.29) is finite. Therefore, instead of (1.32), one of the possible finite-dimensional versions of the given expression should be used, for example, the standard deviation of the form

$$\|\widehat{\Phi}(u,\xi,\zeta)-\Phi(u,\xi,\zeta)\| = \frac{1}{N_P}\sum_{i=0}^{N_P}[\widehat{\Phi}(u_i,\xi,\zeta)-\Phi(u_i,\xi,\zeta)]^2 \quad (1.33)$$

or

$$\|\widehat{\Phi}(u,\xi,\zeta)-\Phi(u,\xi,\zeta)\| = \sqrt{\frac{1}{N_P}\sum_{i=0}^{N_P}[\widehat{\Phi}(u_i,\xi,\zeta)-\Phi(u_i,\xi,\zeta)]^2}. \quad (1.34)$$

Testing the mapping $\widehat{\Phi}(\cdot)$ in order to evaluate its generalization properties is performed using a set of ordered pairs similar to (1.29),

$$\{\langle \tilde{u}_j, \tilde{y}_j\rangle\},\ \tilde{u}\in U,\ \tilde{y}\in Y,\ i=1,\dots,N_T; \quad (1.35)$$

it is necessary that the condition $u_i \neq \tilde{u}_i$ is fulfilled, $\forall i \in \{1,\dots,N_P\}$, $\forall j \in \{1,\dots,N_T\}$, that is, all pairs in sets

$$\{\langle u_i, y_i\rangle\}_{i=1}^{N_P}, \quad \{\langle \tilde{u}_j, \tilde{y}_i\rangle\}_{j=1}^{N_T}$$

must be distinct.

The error on the test set (1.35) is computed in the same way as for the training set (1.29), i.e.,

$$\|\widehat{\Phi}(\tilde{u},\xi,\zeta)-\Phi(\tilde{u},\xi,\zeta)\| = \frac{1}{N_T}\sum_{j=0}^{N_T}[\widehat{\Phi}(\tilde{u}_j,\xi,\zeta)-\Phi(\tilde{u}_j,\xi,\zeta)]^2; \quad (1.36)$$

it can also be represented in the form

$$\|\widehat{\Phi}(\tilde{u},\xi,\zeta)-\Phi(\tilde{u},\xi,\zeta)\| = \sqrt{\frac{1}{N_T}\sum_{j=0}^{N_T}[\widehat{\Phi}(\tilde{u}_j,\xi,\zeta)-\Phi(\tilde{u}_j,\xi,\zeta)]^2}. \quad (1.37)$$

Now we can formulate the problem of the model synthesis for the dynamical system $\mathbb{S}$. It is required to construct a model $\widehat{\Phi}(\cdot)$, which will reproduce the mapping $\Phi(\cdot)$, implemented by the system $\mathbb{S}$, with the required level of accuracy. This means that the magnitude of the simulation error (1.36) or (1.37) on the test set (1.35) should not exceed the specified maximum allowed value ε in (1.30) for such model $\widehat{\Phi}(\cdot)$. The model synthesis procedure should be based on the data (1.29) used to adjust (learning) the model, and data (1.35) used to test the model. In addition, we may involve available knowledge about the simulated system $\mathbb{S}$.

It is assumed that in order to solve this problem, we need to select the optimal (in some sense) model $\widehat{\Phi}^*(\cdot)$ from some finite or infinite family (set) of options $\widehat{\Phi}_j(\cdot)$, $j=1,2,\dots$. In this regard, the following two questions arise:

- What is the given family of variants $\widehat{\Phi}^{(F)} = \{\widehat{\Phi}_j(\cdot)\}$, $j=1,2,\dots$?
- How can we choose $\widehat{\Phi}^*(\cdot)$ from the family $\widehat{\Phi}^{(F)}$, so that it satisfies the condition (1.30).

Abovementioned family of options for the model should meet the following two requirements, which, in general case, may contradict each other:

- a family of models $\{\widehat{\Phi}_j(\cdot)\}$, $j=1,2,\dots$, should be as rich as possible in order to "have a lot of options to choose from";

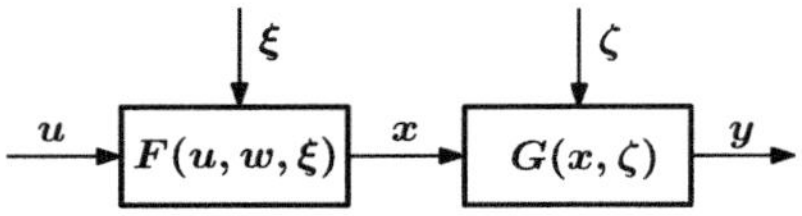

$$y = \Phi(u, w, \xi, \zeta) = G(F(u, w, \xi), \zeta)$$

FIGURE 1.8 Parametrization of the simulated dynamical system. From [56], used with permission from Moscow Aviation Institute.

- this family should be arranged in such a way as to simplify as much as possible the process of selection for the model $\widehat{\Phi}^*(\cdot)$.

As a basis for searching solutions that meet these requirements, the following sections implement an approach oriented on the effective structuring and parametrization of the desired model $\widehat{\Phi}(\cdot)$, implying an appropriate choice of its structure and the insertion of the required number of variable parameters (Fig. 1.8).

1.3.2 The Main Problems That Need to Be Solved During Design of a Model for a Dynamical System

During the design process for the dynamical system model, some problems need to be solved irrespective of the approach one takes. Namely, it is required to design:

- a set of variables that describe the simulated object;
- a class (family) of models that includes the required model;
- a representative (informative) set of experimental data for the development and testing of the model;
- tools for selection of a particular model from a given class (the criterion of adequacy for the model and the search algorithm).

Below is a brief description of these problems. In more detail, the methods of their solution within the framework of the developed approach are considered in Chapters 2, 3, and 5, and examples of applications of these methods can be found in Chapters 4 and 6.

1.3.2.1 *Design of a Set of Values Describing the Modeled Object*

The first thing that needs to be done during the design of the dynamical system model is to define a set of variables describing the given system. The related considerations were discussed above, on page 30. We assume that this problem has already been solved, i.e., we already have decided which variables should be taken into account in the simulation.

1.3.2.2 *Design of a Family of Models That Includes the Desired Model*

To solve the problem of dynamical system modeling, we first need to design a set (family) of model options $\widehat{\Phi}^{(F)} = \widehat{\Phi}_j(\cdot)$, $j = 1, 2, \ldots$, among which we then choose the best (in some sense) option $\widehat{\Phi}^*(\cdot)$. As already noted, solving this part of the dynamical system modeling problem requires us to answer the following two questions:

- What is the desired family of options $\widehat{\Phi}^{(F)} = \{\widehat{\Phi}_j(\cdot)\}$, $j = 1, 2, \ldots$?
- How can we select such an option $\widehat{\Phi}^*(\cdot)$ from the family $\widehat{\Phi}^{(F)}$ that satisfies the condition $\|\widehat{\Phi}(u(t), \xi, \zeta) - \Phi(u(t), \xi, \zeta)\| \leqslant \varepsilon, \quad t \in [t_0, t_f]$, $\forall u \in U, \xi \in \Xi, \zeta \in Z$?

The main ideas that provide answers to these questions are as follows:

- key properties of the model family $\widehat{\Phi}^{(F)}$ are its effective structuring and parametrization;
- key algorithms for choosing $\widehat{\Phi}^*(\cdot)$ from the set $\widehat{\Phi}^{(F)}$ are machine learning algorithms (including neural network learning algorithms).

1.3.2.3 *Design of a Representative Set of Experimental Data for the Development and Testing of the Model*

One of the most critical components of the process of dynamical system model design is the acquisition of a data set, which possesses the necessary completeness in regard to describing the behavior of the system under consideration. As will be shown later, the success of solving the modeling problem depends to a large extent on how informative the available training set is.

1.3.2.4 *Design of Tools for Selection of a Particular Model From a Given Class*

After the family of models for the dynamical system has been selected and a representative set of data describing its behavior has been obtained, it is necessary to define a tool that allows us to "extract" a specific model from this family that satisfies a particular set of requirements. As such an instrument in the framework of this approach, it is quite natural to use the tools of neural network learning.

REFERENCES

[1] Mesarovic MD, Takahara Y. General systems theory: Mathematical foundations. New York, NY: Academic Press; 1975.
[2] Lin Y. General systems theory: A mathematical approach. Systems science and engineering, vol. 12. New York, NY: Kluwer Academic Publishers; 2002.
[3] Skyttner L. General systems theory: Problems, perspectives, practice. 2nd ed. Singapore: World Scientific; 2005.
[4] van Gigch JP. Applied general systems theory. 2nd ed. New York, NY: Harper & Row, Publishers; 1978.
[5] Boyd DW. Systems analysis and modeling: A macro-to-micro approach with multidisciplinary applications. San Diego, CA: Academic Press; 2001.
[6] Kalman RE, Falb PL, Arbib MA. Topics in mathematical system theory. New York, NY: McGraw Hill Book Company; 1969.
[7] Katok A, Hasselblatt B. Introduction to the modern theory of dynamical systems. Encyclopedia of mathematics and its applications, vol. 54. Cambridge, Mass: Cambridge University Press; 1995.
[8] Hasselblatt B, Katok A. A first course in dynamics with a panorama of recent developments. Cambridge: Cambridge University Press; 2003.
[9] Ljung L, Glad T. Modeling of dynamic systems. Englewood Cliffs, NJ: Prentice-Hall; 1994.
[10] Holmgren RA. A first course in discrete dynamical systems. New York, NY: Springer; 1994.
[11] Pearson PK. Discrete-time dynamic models. New York–Oxford: Oxford University Press; 1999.
[12] Steeb WH, Hardy Y, Ruedi S. The nonlinear workbook. 3rd ed. Singapore: World Scientific; 2005.
[13] Khalil HK. Nonlinear systems. 3rd ed. Upper Saddle River, NJ: Prentice Hall; 2002.
[14] Hinrichsen D, Pritchard AJ. Mathematical systems theory I: Modeling, state space analysis, stability and robustness. Berlin, Heidelberg: Springer; 2005.
[15] Bennett BS. Simulation fundamentals. London, New York: Prentice Hall; 1995.
[16] Fishwick PA, editor. Handbook of dynamic system modeling. London, New York: Chapman & Hall/CRC; 2007.
[17] Kulakowski BT, Gardner JF, Shearer JL. Dynamic modeling and control of engineering systems. 3rd ed. Cambridge: Cambridge University Press; 2007.
[18] Marinca V, Herisanu N. Nonlinear dynamical systems in engineering: Some approximate approaches. Berlin, Heidelberg: Springer-Verlag; 2011.
[19] Ogata K. System dynamics. 4th ed. Upper Saddle River, New Jersey: Prentice Hall; 2004.
[20] Arnold VI. Mathematical methods of classical mechanics. 2nd ed. Graduate texts in mathematics, vol. 60. Berlin: Springer; 1989.
[21] Glazunov YT. Goal-setting modeling. Izhevsk: Regular and Chaotic Dynamics; 2012 (in Russian).
[22] Liu B. Theory and practice of uncertain programming. Studies in fuzziness and soft computing, vol. 102. Berlin: Springer; 2002.
[23] Martynyuk AA, Martynyuk-Chernenko YA. Uncertain dynamic systems: Stability and motion control. London: CRC Press; 2012.
[24] Ayyub BM, Klir GJ. Uncertainty modeling and analysis in engineering and the sciences. London, New York: Chapman & Hall/CRC; 2006.
[25] Klir GJ. Uncertainty and information: Foundations of generalized information theory. Hoboken, New Jersey: John Wiley & Sons, Inc.; 2006.
[26] Klir GJ, Yuan B. Fuzzy sets and fuzzy logic: Theory and applications. Upper Saddle River, New Jersey: Prentice Hall; 1995.
[27] Piegat A. Fuzzy modeling and control. Studies in fuzziness and soft computing, vol. 69. Berlin: Springer; 2001.
[28] Etkin B, Reid LD. Dynamics of flight: Stability and control. 3rd ed. New York, NY: John Wiley & Sons, Inc.; 2003.
[29] Boiffier JL. The dynamics of flight: The equations. Chichester, England: John Wiley & Sons; 1998.
[30] Roskam J. Airplane flight dynamics and automatic flight control. Part I. Lawrence, KS: DAR Corporation; 1995.
[31] Roskam J. Airplane flight dynamics and automatic flight control. Part II. Lawrence, KS: DAR Corporation; 1998.

[32] Cook MV. Flight dynamics principles. Amsterdam: Elsevier; 2007.
[33] Hull DG. Fundamentals of airplane flight mechanics. Berlin: Springer; 2007.
[34] Stevens BL, Lewis FL, Johnson E. Aircraft control and simulation: Dynamics, control design, and autonomous systems. 3rd ed. Hoboken, New Jersey: John Wiley & Sons, Inc.; 2016.
[35] Rastrigin LA. Adaptation of complex systems: Methods and applications. Riga: Zinatne; 1981 (in Russian).
[36] Ilyin VE. Fighter aircraft. Moscow: Victoria-AST; 1997 (in Russian).
[37] Gordon Y. Mikoyan MiG-31. Famous Russian aircraft. Hinckley, England: Midland Publishing; 2005.
[38] Nise NS. Control systems engineering. 6th ed. New York, NY: John Wiley & Sons; 2011.
[39] Coughanowr DR, LeBlank SE. Process systems analysis and control. 3rd ed. New York, NY: McGraw-Hill; 2009.
[40] Astolfi A. Nonlinear and adaptive control: Tools and algorithms for the user. London: Imperial College Press; 2006.
[41] Astolfi A, Karagiannis D, Ortega R. Nonlinear and adaptive control with applications. Berlin: Springer; 2008.
[42] Gros C. Complex and adaptive dynamical systems: A primer. Berlin: Springer; 2008.
[43] Ioannou P, Fidan B. Adaptive control tutorial. Philadelphia, PA: SIAM; 2006.
[44] Ioannou P, Sun J. Robust adaptive control. Englewood Cliffs, NJ: Prentice Hall; 1995.
[45] Ioannou P, Sun J. Optimal, predictive, and adaptive control. Englewood Cliffs, NJ: Prentice Hall; 1994.
[46] Sastry S, Bodson M. Adaptive control: Stability, convergence, and robustness. Englewood Cliffs, NJ: Prentice Hall; 1989.
[47] Spooner JT, Maggiore M, Ordóñez R, Passino KM. Stable adaptive control and estimation for nonlinear systems: Neural and fuzzy approximator techniques. New York, NY: John Wiley & Sons, Inc.; 2002.
[48] Tao G. Adaptive control design and analysis. New York, NY: John Wiley & Sons, Inc.; 2003.
[49] Widrow B, Walach E. Adaptive inverse control: A signal processing approach. Hoboken, New Jersey: John Wiley & Sons, Inc.; 2008.
[50] Farrell JA, Polycarpou MM. Adaptive approximation based control: Unifying neural, fuzzy and traditional adaptive approximation approaches. Hoboken, New Jersey: John Wiley & Sons, Inc.; 2006.
[51] French M, Szepesváry C, Rogers E. Performance of nonlinear approximate adaptive controllers. Chichester, England: John Wiley & Sons, Inc.; 2003.
[52] Hovakimyan N, Cao C. $\mathcal{L}_1$ adaptive control theory: Guaranteed robustness with fast adaptation. Philadelphia, PA: SIAM; 2010.
[53] Lavretsky E, Wise KA. Robust and adaptive control with aerospace applications. London: Springer-Verlag; 2013.
[54] Tyukin I. Adaptation in dynamical systems. Cambridge: Cambridge University Press; 2007.
[55] Davidson J, et al. Flight control laws for NASA's Hyper-X research vehicle. AIAA–99–4124, 11.
[56] Brusov VS, Tiumentsev YuV. Neural network modeling of aircraft motion. Moscow: MAI; 2016 (in Russian).

CHAPTER

2

Dynamic Neural Networks: Structures and Training Methods

2.1 ARTIFICIAL NEURAL NETWORK STRUCTURES

2.1.1 Generative Approach to Artificial Neural Network Design

2.1.1.1 The Structure of the Generative Approach

The generative approach is widely used in applied and computational mathematics. This approach, extended by the ideas of ANN modeling, is very promising as a flexible tool for the formation of dynamical system models.

The generative approach is interpreted further as follows. We can treat the class of models which contains the desired (generated) dynamical system model as a collection of tools producing dynamical system models that satisfy some specified requirements. There are two main requirements for this set of tools. Firstly, it must generate a potentially rich class of models (i.e., it must provide extensive choice possibilities) and, secondly, it should have as many as possible simple "arrangement," so that the implementation of this class of models is not an "unbearable" problem. These two requirements, generally speaking, are mutually exclusive. How and by what tools to ensure an acceptable balance between them is discussed later in this section.

To generate any model, we need to have at our disposal:

- a basis, i.e., a set of elements from which models are formed;
- the rules used to form models by appropriately combining the elements of the basis:
  - rules for the structuring of models;
  - rules for parametric adjustment of generated models.

One of the generative approach variants[1] is that the desired dependence $y(x)$ is represented as a linear combination of the basis functions $\varphi_i(x),\ i = 1, \ldots, n$, i.e.,

$$y(x) = \varphi_0(x) + \sum_{i=1}^{n} \lambda_i \varphi_i(x), \quad \lambda_i \in \mathcal{R}. \tag{2.1}$$

The set of functions $\{\varphi_i(x)\},\ i = 1, \ldots, n$, we will call the functional basis (FB). The expression of the form (2.1) is a decomposition (expansion) of the function $y(x)$ with respect to the functional basis $\{\varphi_i(x)\}_{i=1}^{n}$.

We will further consider the generation of the FB expansion by varying the adjustable parameters (the coefficients λ_i in the expansion (2.1)) as

[1]Examples of other variants are generative grammars from the theory of formal grammars and languages [1–3], a syntactic approach to the description of patterns in the theory of pattern recognition [4–7].

Neural Network Modeling and Identification of Dynamical Systems
https://doi.org/10.1016/B978-0-12-815254-6.00012-5

tools to produce solutions (each particular combination of λ_i provides some solution). The rule for combining FB elements in the case of (2.1) is a weighted summation of these items.

This technique is widely used in traditional mathematics. In the general form, the functional expansion can be represented as

$$y(x) = \varphi_0(x) + \sum_{i=1}^{n} \lambda_i \varphi_i(x), \quad \lambda_i \in \mathcal{R}. \tag{2.2}$$

Here the basis is a set of functions $\{\varphi_i(x)\}_{i=0}^{n}$, and the rule for combining the elements of a basis is a weighted summation. The required expansion is a linear combination of the functions $\varphi_i(x)$, $i = 1, \dots, n$, as elements of the FB.

Here we present some examples of functional expansions, often used in mathematical modeling.

Example 2.1. We have the Taylor series expansion, i.e.,

$$F(x) = a_0 + a_1(x - x_0) + a_2(x - x_0)^2 + \cdots + a_n(x - x_0)^n + \cdots. \tag{2.3}$$

The basis of this expansion is $\{(x - x_0)^i\}_{i=0}^{\infty}$, and the rule for combining FB elements is a weighted summation.

Example 2.2. We have the Fourier series expansion, i.e.,

$$F(x) = \sum_{i=0}^{\infty} (a_i \cos(ix) + b_i \sin(ix)). \tag{2.4}$$

The basis of this expansion is $\{\cos(ix), \sin(ix)\}_{i=0}^{\infty}$, and the rule for combining FB elements is a weighted summation.

Example 2.3. We have the Galerkin expansion, i.e.,

$$y(x) = u_0(x) + \sum_{i=1}^{n} c_i u_i(x). \tag{2.5}$$

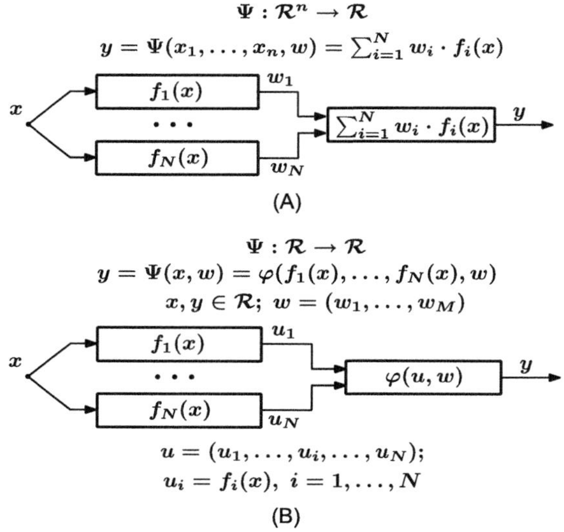

FIGURE 2.1 Functional dependence on one variable as (A) a linear and (B) a nonlinear combination of the FB elements $f_i(x)$, $i = 1, \dots, n$. From [109], used with permission from Moscow Aviation Institute.

The basis of this expansion is $\{u_i(x)\}_{i=0}^{n}$, and the rule for combining FB elements is a weighted summation.

In all these examples, the generated solutions are represented by linear combinations of basis elements, parametrized by the corresponding weights associated with each FB element.

2.1.1.2 Network Representation of Functional Expansions

We can give a network interpretation for functional expansions, which allows us to identify similarities and differences between their variants. Such description provides a further simple transition to ANN models and also allows to establish interrelations between traditional-type models and ANN models.

The structural representation of the functional dependence on one variable as a linear and nonlinear combination of the elements of the basis $f_i(x)$, $i = 1, \dots, n$, is shown in Fig. 2.1A and Fig. 2.1B, respectively.

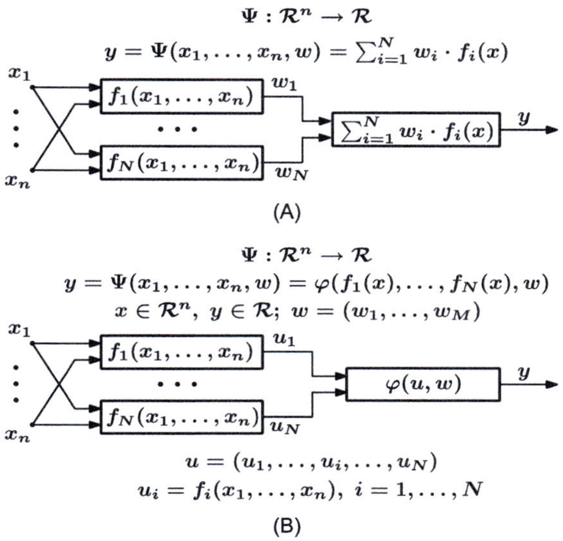

FIGURE 2.2 Scalar-valued functional dependence on several variables as (A) a linear and (B) a nonlinear combination of the elements of the basis $f_i(x_1, \ldots, x_n)$, $i = 1, \ldots, N$. From [109], used with permission from Moscow Aviation Institute.

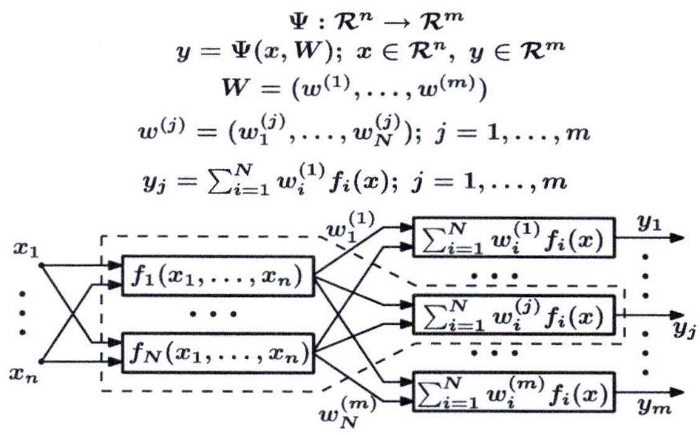

FIGURE 2.3 Vector-valued functional dependence on several variables as a linear combination of the elements of the basis $f_i(x_1, \ldots, x_n)$, $i = 1, \ldots, N$. From [109], used with permission from Moscow Aviation Institute.

Similarly, for a scalar-valued functional dependence on several variables as a linear and nonlinear combination of elements of the basis $f_i(x_1, \ldots, x_n)$, $i = 1, \ldots, N$, the structural representation is given, respectively, in Fig. 2.2A and Fig. 2.2B.

Vector-valued functional dependence on several variables as a linear combination of the elements of the basis $f_i(x_1, \ldots, x_n)$, $i = 1, \ldots, N$, in the network representation is shown in Fig. 2.3. The nonlinear combination is represented in a similar way, namely, we use nonlinear combining rules $\varphi_i(f_1(x), \ldots, f_m(x))$, $i = 1, \ldots, m$, $x = x_1, \ldots, x_m$, instead of the linear ones $\Sigma_{i=1}^N(\cdot)$.

We have written the traditional functional expansions mentioned above in general form as

$$y(x) = F(x_1, x_2, \ldots, x_n) = \sum_{i=0}^{m} \lambda_i \varphi_i(x_1, x_2, \ldots x_n). \tag{2.6}$$

Here, the function $F(x_1, x_2, \ldots, x_n)$ is a (linear) combination of the elements of the basis $\varphi_i(x_1, x_2, \ldots x_n)$.

An expansion of the form (2.6) has the following features:

- the resulting decomposition is one-level;
- functions $\varphi_i : \mathcal{R}^n \to \mathcal{R}$ as elements of the basis have limited flexibility (with variability of such types as displacement, compression/stretching) or are fixed.

Such limited flexibility of the traditional functional basis together with the one-level nature of the expansion sharply reduces the possibility of obtaining some "right" model.[2]

2.1.1.3 Multilevel Adjustable Functional Expansions

As noted in the previous section, the possibility of obtaining a "right" model is limited by a single-level structure and an inflexible basis for traditional expansions. For this reason, it is quite natural to build a model that overcomes these shortcomings. It must have the required level of flexibility (and the needed variability in the gen-

[2] At the intuitive level, a "right model" is a model with generalizing properties that are adequate to the application problem that we solve; see also Section 1.3 of Chapter 1.

$$\mathbf{x} = (x_1, \dots, x_N); \quad \mathbf{v} = (v_1, \dots, v_K); \quad \mathbf{u} = (u_1, \dots, u_M)$$

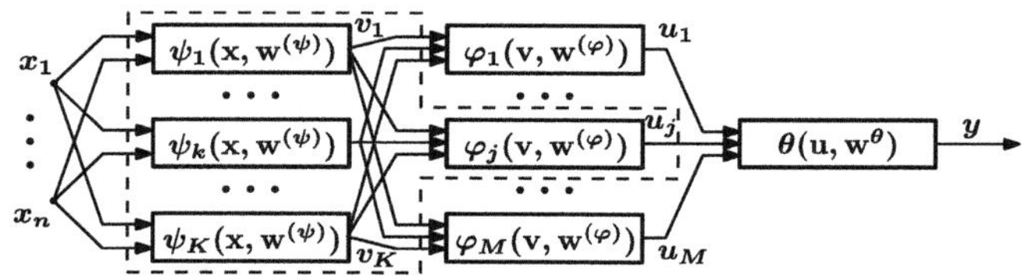

$$y = \theta(u_1, \dots, u_j, \dots, u_M, \mathbf{w}^{(\theta)})$$

$$u_j = \varphi_j(v_1, \dots, v_k, \dots, v_K, \mathbf{w}^{(\varphi_j)})$$

$$v_k = \psi_k(x_1, \dots, x_i, \dots, x_N, \mathbf{w}^{(\psi_k)})$$

j-th basis function:

$$u_j = \varphi_j(v_1, \dots, v_K, \mathbf{w}^{(\varphi_j)}) =$$
$$= \varphi_j(\psi_1(x_1, \dots, x_N, \mathbf{w}^{(\psi_1)}), \dots, \psi_K(x_1, \dots, x_N, \mathbf{w}^{(\psi_K)}), \mathbf{w}^{(\varphi_j)})$$

FIGURE 2.4 Multilevel adjustable functional expansion. From [109], used with permission from Moscow Aviation Institute.

erated variants of the required model), by forming it as a multilevel network structure and by appropriate parametrization of the elements of this structure.

Fig. 2.4 shows how we can construct a multilevel adjustable functional expansion. We see that in this case, the adjustment of the expansion is carried out not only by varying the coefficients of the linear combination, as in expansions of the type (2.6). Now the elements of the functional basis are also parametrized. Therefore, in the process of solving the problem, the basis is adjusted to obtain a dynamical system model which is acceptable in the sense of the criterion (1.30).

As we can see from Fig. 2.4, the transition from a single-level decomposition to a multilevel one consists in the fact that each element $\varphi_j(\mathbf{v}, \mathbf{w}^{\varphi})$, $j = 1, \dots, M$, is decomposed using some functional basis $\{\psi_k(\mathbf{x}, \mathbf{w}^{\psi})\}$, $j = 1, \dots, K$. Similarly, we can construct the expansion of the elements $\psi_k(\mathbf{x}, \mathbf{w}^{\psi})$ for another FB, and so on, the required number of times. This approach gives us the network structure with the required number of levels, as well as the required parametrization of the FB elements.

2.1.1.4 Functional and Neural Networks

Thus, we can interpret the model as an expansion on the functional basis (2.6), where each element $\varphi_i(x_1, x_2, \dots x_n)$ transforms n-dimensional input $x = (x_1, x_2, \dots x_n)$ in the scalar output y.

We can distinguish the following types of elements of the functional basis:

- the FB element as an integrated (one-stage) mapping $\varphi_i : \mathcal{R}^n \to \mathcal{R}$ that directly transforms some n-dimensional input $x = (x_1, x_2, \dots x_n)$ to the scalar output y;
- the FB element as a compositional (two-stage) mapping of the n-dimensional input $x = (x_1, x_2, \dots x_n)$ to the scalar output y.

In the two-stage (compositional) version, the mapping $\mathcal{R}^n \to \mathcal{R}$ is performed in the first stage, "compressing" the vector input $x = (x_1, x_2, \dots x_n)$ to the intermediate scalar output v, which in the second stage is additionally processed by the output mapping $\mathcal{R} \to \mathcal{R}$ to obtain the output y (Fig. 2.5).

Depending on which of these FB elements are used in the formation of network models (NMs), the following basic variants of these models are obtained:

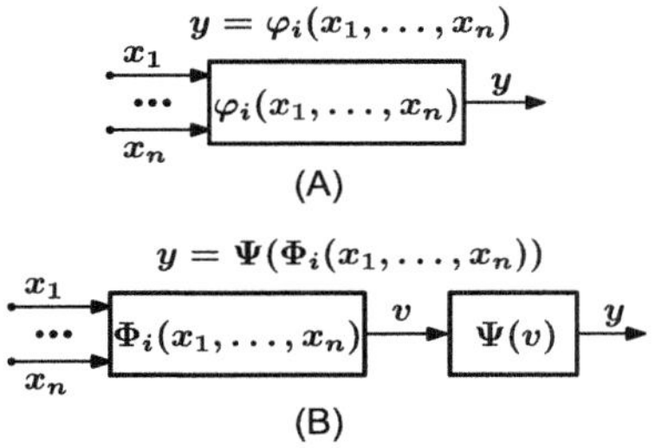

FIGURE 2.5 An element of an FB that transforms the n-dimensional $x = (x_1, x_2, \dots x_n)$ input into a scalar output y. (A) A one-stage mapping $\mathcal{R}^n \to \mathcal{R}$. (B) A two-stage (compositional) mapping $\mathcal{R}^n \to \mathcal{R} \to \mathcal{R}$. From [109], used with permission from Moscow Aviation Institute.

- The one-stage mapping $\mathcal{R}^n \to \mathcal{R}$ is an element of functional networks.
- The two-stage mapping $\mathcal{R}^n \to \mathcal{R} \to \mathcal{R}$ is an element of neural networks.

The element of compositional type, i.e., the two-stage mapping of the n-dimensional input to the scalar output, is a neuron; it is specific for functional expansion of the neural network type and is a "brand feature" of such expansions, in other words, ANN models of all kinds.

2.1.2 Layered Structure of Neural Network Models

2.1.2.1 *Principles of Layered Structural Organization for Neural Network Models*

We assume that the ANN models in the general case have a layered structure. This assumption means that we divide the entire set of neurons constituting the ANN model into disjoint subsets, which we will call *layers*. For these layers we introduce the notation $L^{(0)}, L^{(1)}, \dots, L^{(p)}, \dots, L^{(N_L)}$.

The layered organization of the ANN model determines the activation logic of its neurons. This logic will be different for different structural variants of the network. The following specificity takes place in the operation of the layered ANN model[3]: the neurons, of which the ANN model consists, operate layer by layer, i.e., until all the neurons of the pth layer work, the neurons of the $(p+1)$th layer do not come into operation. We will consider below the general variant that defines the rules for activating neurons in ANN models.

In the simplest variant of the structural organization of layered networks, all the layers $L^{(p)}$, numbered from 0 to N_L, are activated in the order of their numbers. This variant means that until all the neurons in the layer with the number p work, the neurons from the $(p+1)$th layer are waiting. In turn, the pth layer can start operating only if all the neurons of the $(p-1)$th layer have already worked.

Visually, we can represent such a structure as a "stack of layers," ordered by their numbers. In the simplest version, this "stack" looks like it is shown in Fig. 2.6A. Here the $L^{(0)}$ layer is the input layer, the elements of which are components of the ANN input vector.

Any layer $L^{(p)}$, $1 \leqslant p < N_L$, is connected with two adjacent layers: it gets its inputs from the previous layer $L^{(p-1)}$, and it transfers its outputs to the subsequent layer $L^{(p+1)}$. The exception is the layer $L^{(N_L)}$, the latter in the ANN (the output layer), which does not have a layer following it. The outputs of the layer $L^{(N_L)}$ are the outputs of the network as a whole. The layers $L^{(p)}$ with numbers $0 < p < N_L$ are called *hidden*.

Since the ANN shown in Fig. 2.6A is a feedforward network, all links between its layers go strictly sequentially from the layer $L^{(0)}$ to the layer $L^{(N_L)}$ without "hopping" (bypassing) through adjacent layers and backward (feedback) links. A more complicated ANN structure version with bypass connection is shown in Fig. 2.6B.

[3]For the case when the layers are in the order of their numbers and there are no feedbacks between the layers. In this case, the layers will operate sequentially and only once.

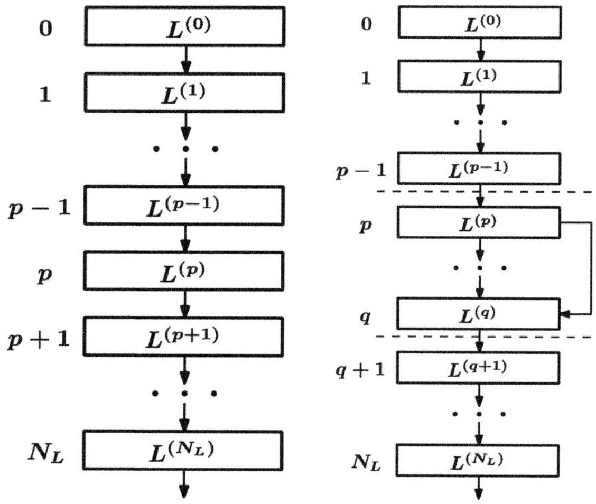

FIGURE 2.6 Variants of the structural organization for layered neural networks with sequential numbering of layers (feedforward networks). (A) Without bypass connections. (B) With bypass connections ($q > p + 1$). From [109], used with permission from Moscow Aviation Institute.

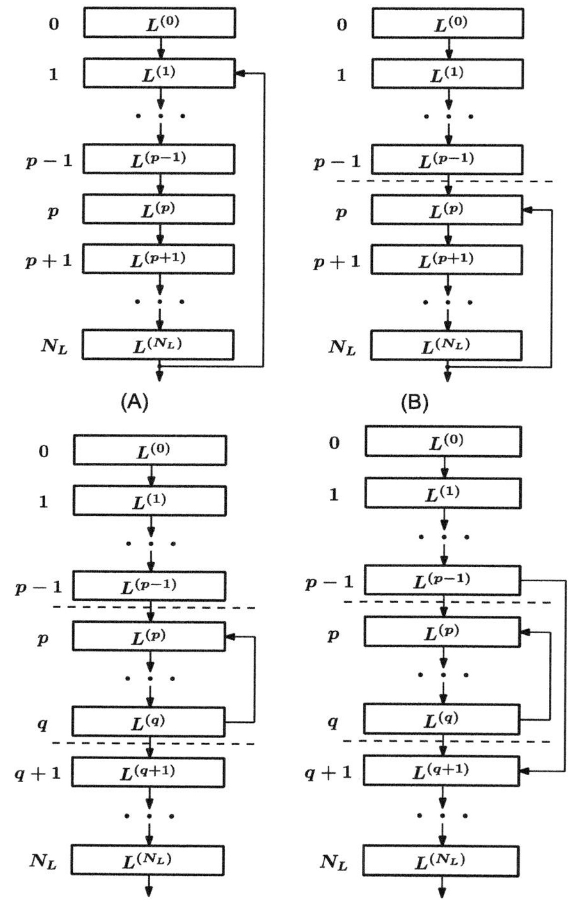

FIGURE 2.7 Variants of the structural organization for layered neural networks with sequential numbering of layers. (A) A network with a feedback from the output layer $L^{(N_L)}$ to the first processing layer $L^{(1)}$. (B) A network with feedback from the output layer $L^{(N_L)}$ to an arbitrary layer $L^{(p)}$, $1 < p < N_L$. (C) A network with feedback from the layer $L^{(q)}$, $1 < q < N_L$ to the layer $L^{(p)}$, $1 < p < N_L$. (D) An example of a network with feedback from the layer $L^{(q)}$, $1 < q < N_L$ to the layer $L^{(p)}$, $1 < p < N_L$ and bypass connection from the layer $L^{(p-1)}$ to the layer $L^{(q+1)}$. From [109], used with permission from Moscow Aviation Institute.

We also assume for networks of this type that any pair of neurons between which there is a connection refers to different layers. In other words, neurons within any of the processing layers $L^{(p)}$, $p = 1, \ldots, N_L$, have no connections with each other. Variants in which such relationships, called lateral ones, are available in neural networks require separate consideration.

We can complicate the structure of the connections of the layered network in comparison with the scheme shown in Fig. 2.6.

The first of the possible variants of such complication is the insertion of feedback into the ANN structure. This feedback transfers the received output of the network (i.e., the output of the layer $L^{(N_L)}$) "back" to the input of the ANN. More precisely, we move the network output to the input of its first processing layer $L^{(1)}$, as shown in Fig. 2.7A.

In Fig. 2.7B another way of introducing feedback into a layered network is shown, in which the feedback goes from the output layer $L^{(N_L)}$ to an arbitrary layer $L^{(p)}$, $1 < p < N_L$. This variant can also be treated as a composition (serial connection) of a feedforward network (layers $L^{(1)}, \ldots, L^{(p-1)}$) and a feedback network of the type shown in Fig. 2.7A ($L^{(p)}, \ldots, L^{(N_L)}$).

The most general way of introducing feedback into a "stack of layers"–type structure is shown in Fig. 2.7C. Here the feedback comes from some hidden layer $L^{(q)}$, $1 < q < N_L$, to the layer $L^{(p)}$, $1 \leqslant p < N_L$, $q > p$. Similar to the case shown in Fig. 2.7A, this variant can be treated as a serial connection of a feedforward neural network (layers $L^{(1)}, \ldots, L^{(p-1)}$), the network with feedback (layers $L^{(p)}, \ldots, L^{(q)}$), and another feedforward network (layers $L^{(q+1)}, \ldots, L^{(N_L)}$). The operation of such a network can, for example, be interpreted as follows. The recurrent subnet (the layers $L^{(p)}, \ldots, L^{(q)}$) is the main part of the ANN as a whole. One feedforward subnet (layers $L^{(1)}, \ldots, L^{(p-1)}$) preprocesses the data entering the main subnet, while the second subnet (layers $L^{(q+1)}, \ldots, L^{(N_L)}$) performs some postprocessing of the data produced by the main recurrent subnet.

Fig. 2.7D shows an example of a generalization of the structure shown in Fig. 2.7C, for the case in which, in addition to strictly consecutive connections between the layers of the network, there are also bypass connections.

In all the ANN variants shown in Fig. 2.6, the strict sequence of layers is preserved unchanged. The layers are activated one after the other in the order specified by forward and backward connections in the considered ANN. For a feedforward network, this means that any neuron from the layer $L^{(p)}$ receives its inputs only from neurons from the layer $L^{(p-1)}$ and passes its outputs to the layer $L^{(p+1)}$, i.e.,

$$L^{(p-1)} \rightarrow L^{(p)} \rightarrow L^{(p+1)}, \; p \in \{0, 1, \ldots, N_L\}. \tag{2.7}$$

At the same time (simultaneously) two or more layers cannot be executed ("fired"), even if there is such a technical capability (the network is executed on some parallel computing system) due to the sequential operation logic of the ANN layers noted above.

The use of feedback introduces cyclicity into the order of operation for the layers. We can implement this cyclicity for all layers, beginning with $L^{(1)}$ and up to $L^{(N_L)}$, or for some of them, for some range of numbers $p_1 \leqslant p \leqslant p_2$. The implementation depends on which layers of the ANN we cover by feedback. However, in any case, some strict sequence of operation of the layers is preserved. If one of the ANN layers started its work, then, until this work is completed, no other layer will be launched for processing.

The rejection of this kind of strict firing sequence for the ANN layers leads to the appearance of parallelism in the network at the level of its layers. In the most general case, we allow for any neuron from the layer $L^{(p)}$ and any neuron from the layer $L^{(q)}$ to establish a connection of any type. Namely, we allow forward, backward (for these cases $p \neq q$), or lateral (in this case $p = q$) connections. Here, for the time being, it is still considered that a layered organization like "stack of layers" is used.

Variants of the ANN structural organization shown in Fig. 2.7 use the same "stack of layers" scheme for ordering the layers of the network. Here, at each time interval, the neurons of only one layer work. The remaining layers either have already worked or are waiting for their turn. This approach applies to both feedforward networks and recurrent networks.

The following variant allows us to refuse the "stack of layers" scheme and to replace it with more complex structures. As an example illustrating structures of this kind, we show in Fig. 2.8 two variants of the structures of an ANN with parallelism in them at the layer level.[4]

Consider the schemes shown in Fig. 2.7 and Fig. 2.8. Obviously, to activate a neuron from some pth layer, it must first get the values of all its inputs it "waits for" until that moment. For paralleling the work of neurons, we must meet the same conditions. Namely, all neurons that have a complete set of inputs at a given mo-

[4]If we refuse the "stack of layers" scheme, some layers in the ANN can work in parallel, i.e., simultaneously with each other, if there is such a technical possibility.

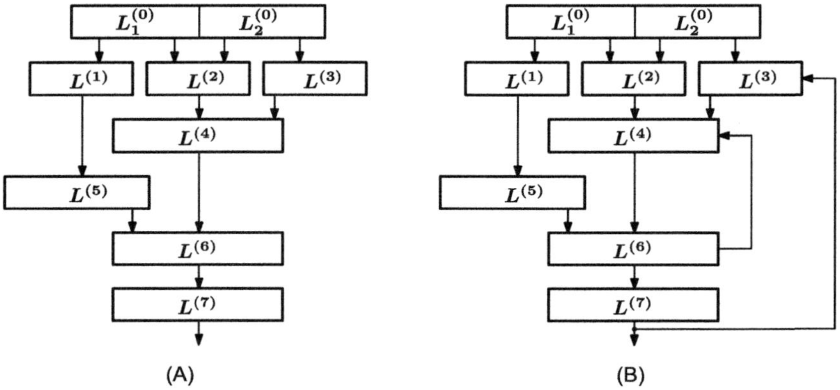

FIGURE 2.8 An example of a structural organization for a layered neural network with layer-level parallelism. (A) Feedforward ANN. (B) ANN with feedbacks.

ment of time can operate independently from each other in an arbitrary order or parallel if there is such a technical capability.

Suppose we have an ANN organized according to the "stack of layers" scheme. The logic of neuron activation (i.e., the sequence and conditions of neuron operation) in this ANN ensures the absence of conflicts between them. If we introduce a parallelism at the layer level in the ANN, we need to add some additional synchronization rules to provide such conflict-free network operation.

Namely, a neuron can work as soon as it is ready to operate, and it will be ready as soon as it receives the values for all its inputs. Once the neuron is ready for functioning, we should start it immediately, as soon as it becomes possible. This is significant because the outputs of this neuron are required to ensure the operational readiness for other neurons that follow.

For the particular ANN, it is possible to specify (to generate) a set of cause-and-effect relations (chains) that provide the ability to monitor the operational conditions for different neurons to prevent conflicts between them.

For layered feedforward networks with the structures shown in Fig. 2.7, the cause-and-effect chains will have a strictly linear structure, without branches and cycles. In structures with parallelism at the layer level for networks, as shown in Fig. 2.8, both forward "jumps" and feedbacks can be present. Such structures bring nonlinearity to the cause-and-effect chains; in particular, they provide tree structures and cycles.

The cause-and-effect chain should show which neurons transmit signals to some analyzed neuron. In other words, it is required to show which neural predecessors should work so that a given neuron receives a complete set of input values. As noted above, this is a necessary condition for the readiness to operate a given neuron. This condition is the causal part of the chain. Also, the chain indicates which neurons will get the output of this "current neuron." This indication will be the "effect" part of the cause-and-effect chain.

In all the considered variants of the ANN structural organization, only forward and backward links were contained, i.e., connections between pairs of neurons in which the neurons from this pair belong to different layers.

The third kind of connections that are possible between neurons in the ANN is lateral connections, in which the two neurons, between which the connection is established, belong to the same layer. One example of an ANN with lateral connections is the Recurrent MultiLayer Perceptron (RMLP) network [8–10].

2.1.2.2 *Examples of Layered Structural Organization for Neural Network Models*

Examples of structural organization options for static-type ANN models (i.e., without TDL elements and/or feedbacks) are shown in Fig. 2.9. Network ADALINE [11] is a single-layer (i.e., without hidden layers) linear ANN model. Its structure is shown in Fig. 2.9A. A more general variant of feedforward neural network (FFNN) is MLP (MultiLayer Perceptron) [10,11], which is a nonlinear network with one or more hidden layers (Fig. 2.9B).

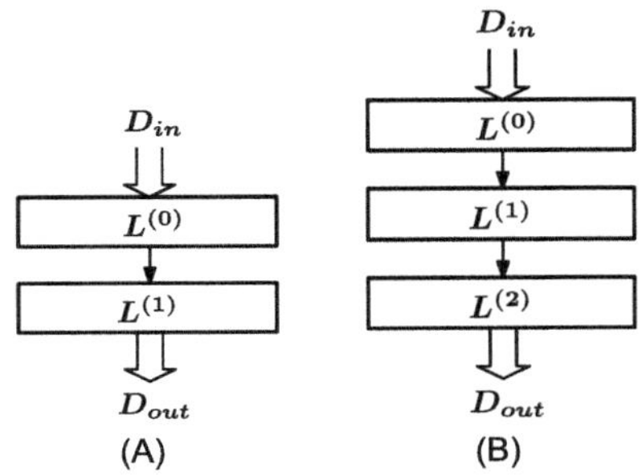

FIGURE 2.9 Examples of a structural organization for feedforward neural networks. (A) ADALINE (Adaptive Linear Network). (B) MLP (MultiLayer Perceptron). D_{in} are source (input) data; D_{out} are output data (results); $L^{(0)}$ is input layer; $L^{(1)}$ is output layer.

Dynamic networks can be divided into two classes [12–19]:

- feedforward networks, in which the input signals are fed through delay lines (TDL elements);
- recurrent networks in which feedbacks exist, and there may also be TDL elements at the inputs of the network.

Examples of the structural organization of the ANN models of the dynamic type of the first type (i.e., with TDL elements at the network inputs, but without feedbacks) are shown in Fig. 2.10.

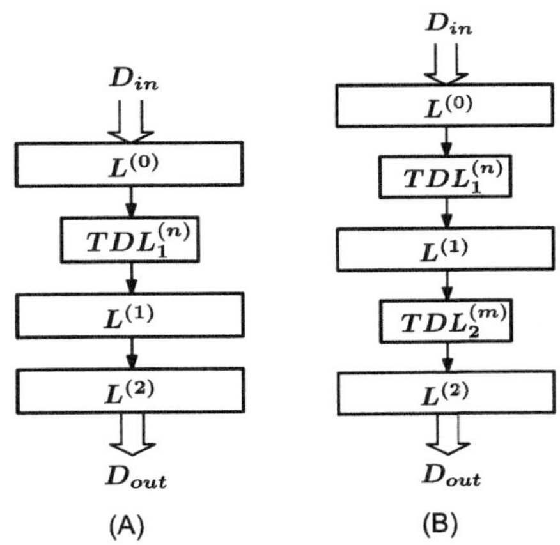

FIGURE 2.10 Examples of a structural organization for feedforward dynamic neural networks. (A) TDNN (Time Delay Neural Network). (B) DTDNN (Distributed Time Delay Neural Network). D_{in} are source (input) data; D_{out} are output data (results); $L^{(0)}$ is input layer; $L^{(1)}$ is hidden layer; $L^{(2)}$ is output layer; $TDL_1^{(n)}$ and $TDL_2^{(m)}$ are tapped delay lines (TDLs) of order n and m, respectively.

Typical variants of ANN models of this type are the Time Delay Neural Network (TDNN) [10,20–27], whose structure is shown in Fig. 2.10A (similarly, in the structural plan, the Focused Time Delay Neural Network [FTDNN] is organized) as well as the Distributed Time Delay Neural Network (DTDNN) network [28] (see Fig. 2.10B).

Examples of the structural organization of dynamic ANN models of the second kind, that is, of recurrent neural networks (RNN), are shown in Figs. 2.11–2.13.

Classical examples of recurrent networks, from which, to a large extent, this area of research began to develop, are the Jordan network [14,15] (Fig. 2.11A), the Elman network [10, 29–32] (Fig. 2.11B), the Hopfield network [10,11] (Fig. 2.12A), and the Hamming network [11,28] (Fig. 2.12B).

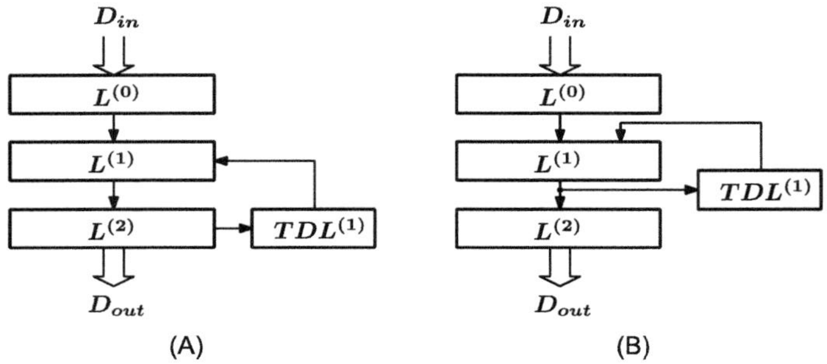

FIGURE 2.11 Examples of a structural organization for feedforward dynamic neural networks. (A) Jordan network. (B) Elman network. D_{in} are source (input) data; D_{out} are output data (results); $L^{(0)}$ is input layer; $L^{(1)}$ is hidden layer; $L^{(2)}$ is output layer; $TDL^{(1)}$ is tapped delay line (TDL) of order 1.

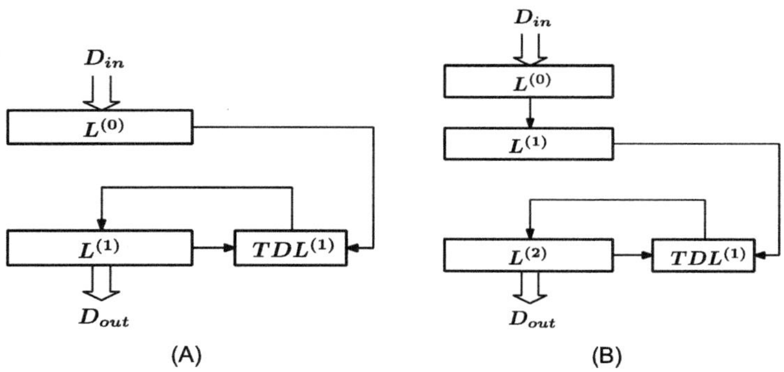

FIGURE 2.12 Examples of a structural organization for feedforward dynamic neural networks. (A) Hopfield network. (B) Hamming network. D_{in} are source (input) data; D_{out} are output data (results); $L^{(0)}$ is input layer; $L^{(1)}$ is hidden layer; $L^{(2)}$ is output layer; $TDL^{(1)}$ is tapped delay line (TDL) of order 1.

In Fig. 2.13A the ANN model Nonlinear AutoRegression with eXternal inputs (NARX) [33–41] is shown, which is widely used in modeling and control tasks for dynamical systems. The same structural organization has a variant of this network, expanded by the composition of the parameters considered. This is the ANN model Nonlinear AutoRegression with Moving Average and eXternal inputs (NARMAX) [42, 43].

In Fig. 2.13B we can see an example of an ANN model with the Layered Digital Dynamic Network (LDDN) structure [11,28]. Networks with a structure of this type can have practically any topology of forward and backward connections, that is, in a certain sense, this structural organization of the neural network is the most common.

The set of Figs. 2.14–2.17 allows us to specify the structural organization of the layers of the ANN model: the input layer (Fig. 2.14) and working (hidden and output) layers (Fig. 2.15). In Fig. 2.16 the structure of the TDL element is presented, and in Fig. 2.17 the structure of the neuron as the main element that is part of the working layers of the ANN model is shown.

One of the most popular static neural network architectures is a Layered Feedforward

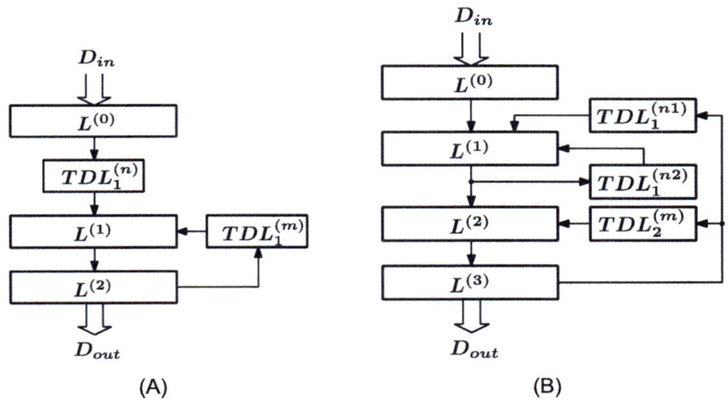

FIGURE 2.13 Examples of a structural organization for feedforward dynamic neural networks. (A) NARX (Nonlinear AutoRegression with eXternal inputs). (B) LDDN (Layered Digital Dynamic Network). D_{in} are source (input) data; D_{out} are output data (results); $L^{(0)}$ is input layer; $L^{(1)}$ is hidden layer; $L^{(2)}$ is output layer for NARX network and hidden layer for LDDN; $L^{(3)}$ is output layer for LDDN; $TDL_1^{(m)}$, $TDL_2^{(m)}$, $TDL_1^{(n1)}$, $TDL_1^{(n2)}$ are tapped delay lines of order $m, m, n1$, and $n2$ respectively.

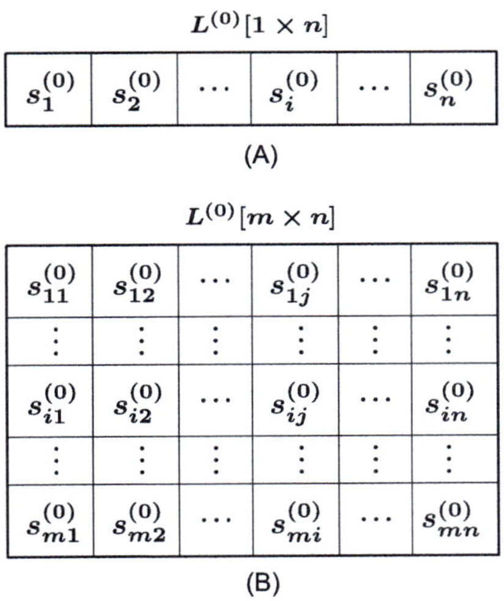

FIGURE 2.14 ANN input layer as a data structure. (A) One-dimensional array. (B) Two-dimensional array. $s_i^{(0)}$, $s_{ij}^{(0)}$ are numeric or character variables.

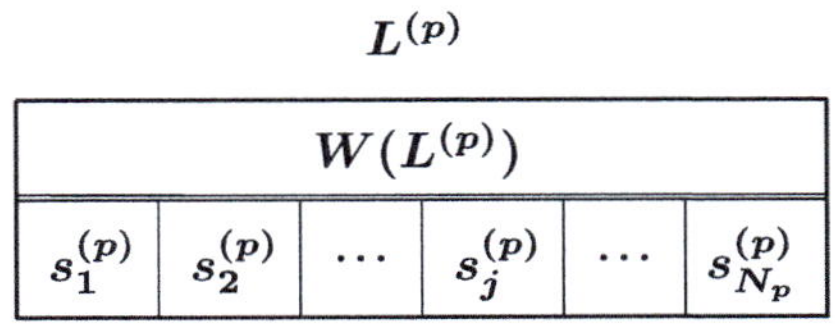

FIGURE 2.15 General structure of the operational (hidden and output) ANN layers: $s_i^{(0)}$ is the ith neuron of the pth ANN layer; $W(L^{(p)})$ is a matrix of synaptic weights for connection entering the neurons of the $L^{(p)}$ layer.

Neural Network (LFNN). We introduce the following notation: $L \in \mathbb{N}$ is the total number of layers; $S^l \in \mathbb{N}$ is the number of neurons within the lth layer; $S^0 \in \mathbb{N}$ is the number of network inputs; and $\mathbf{a}_i^0 \in \mathbb{R}$ is the value of the ith input. For each ith neuron of an lth layer we denote the following: $\mathbf{n}_i^l$ is the weighted sum of neuron inputs; $\boldsymbol{\varphi}_i^l : \mathbb{R} \to \mathbb{R}$ is the activation function; and $\mathbf{a}_i^l \in \mathbb{R}$ is the output of an activation function (the neuron state). Outputs $\mathbf{a}_i^L$ of activation functions of Lth layer neurons are called the network outputs. Also, $\mathbf{W} \in \mathbb{R}^{n_w}$ is the total vector of network parameters, which consists of biases $\mathbf{b}_i^l \in \mathbb{R}$ and connection weights $\mathbf{w}_{i,j}^l \in \mathbb{R}$. Thus, the layered feedforward neural network is a parametric function family, mapping the network inputs $\mathbf{a}^0$ and parameters $\mathbf{W}$ to the outputs $\mathbf{a}^L$ according

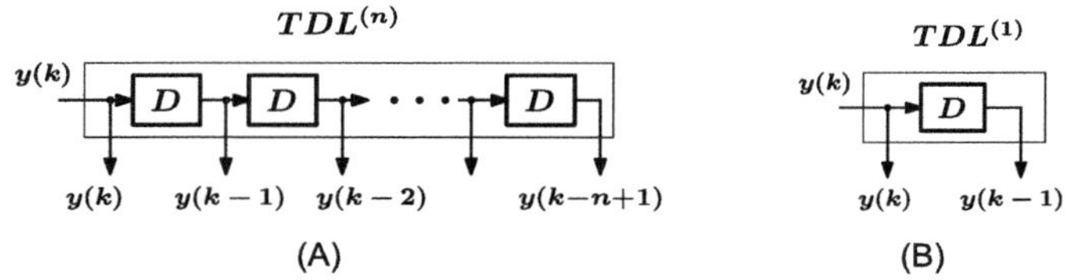

FIGURE 2.16 Tapped delay lines (TDLs) as ANN structural elements. (A) TDL of order n. (B) TDL of order 1. D is delay (memory) element.

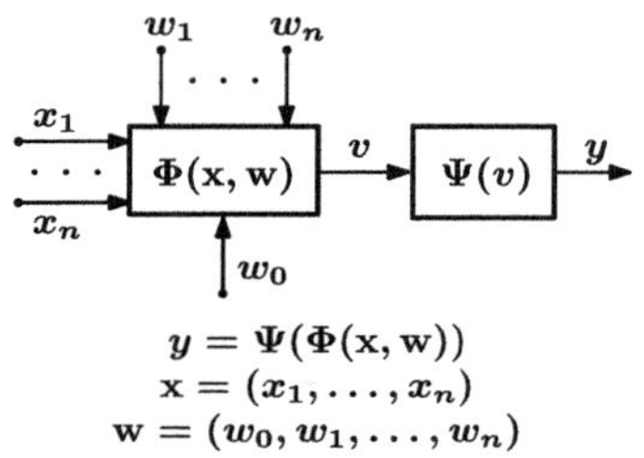

FIGURE 2.17 General structure of a neuron within operational (hidden and output) ANN layers. $\Phi(\mathbf{x}, \mathbf{w})$ is input mapping $\mathbb{R}^n \to \mathbb{R}^1$ (aggregating mapping) parametrized with synaptic weights $\mathbf{w}$; $\Psi(v)$ is output mapping $\mathbb{R}^1 \to \mathbb{R}^1$ (activation function); $\mathbf{x} = (x_1), \ldots, x_n$ are neuron inputs; $\mathbf{w} = (w_1), \ldots, w_n$ are synaptic weights; v is output of the aggregating mapping; y is output of the neuron.

to the following equations:

$$\left.\begin{aligned} \mathbf{n}_i^l &= \mathbf{b}_i^l + \sum_{j=1}^{S^{l-1}} \mathbf{w}_{i,j}^l \mathbf{a}_j^{l-1} \\ \mathbf{a}_i^l &= \varphi_i^l\left(\mathbf{n}_i^l\right) \end{aligned}\right\} l = 1, \ldots, L, \; i = 1, \ldots, S^l. \tag{2.8}$$

The Lth layer is called the output layer, while all the rest are called the hidden layers, since they are not directly connected to network outputs.

Common examples of activation functions for the hidden layer neurons are the hyperbolic tangent function,

$$\varphi_i^l(\mathbf{n}_i^l) = \mathrm{th}(\mathbf{n}_i^l) = \frac{e^{\mathbf{n}_i^l} - e^{-\mathbf{n}_i^l}}{e^{\mathbf{n}_i^l} + e^{-\mathbf{n}_i^l}}, \quad l = 1, \ldots, L-1, \; i = 1, \ldots, S^l, \tag{2.9}$$

and the logistic sigmoid function,

$$\varphi_i^l(\mathbf{n}_i^l) = \mathrm{logsig}(\mathbf{n}_i^l) = \frac{1}{1 + e^{-\mathbf{n}_i^l}}, \quad l = 1, \ldots, L-1, \; i = 1, \ldots, S^l. \tag{2.10}$$

The hyperbolic tangent is more suitable for function approximation problems, since it has a symmetric range $[-1, 1]$. On the other hand, the logistic sigmoid is frequently used for classification problems, due to its range $[0, 1]$. Identity functions are frequently used as activation functions for output layer neurons, i.e.,

$$\varphi_i^L(\mathbf{n}_i^L) = \mathbf{n}_i^L, \; i = 1, \ldots, S^L. \tag{2.11}$$

2.1.2.3 *Input–Output and State Space ANN-Based Models for Deterministic Nonlinear Controlled Discrete Time Dynamical Systems*

Nonlinear AutoRegressive network with eXogeneous inputs [44] (NARX). One popular class of models for deterministic nonlinear controlled discrete time dynamical systems is a class of input–output nonlinear autoregressive neural network based models, i.e.,

$$\hat{\mathbf{y}}(t_k) = \mathbf{F}(\hat{\mathbf{y}}(t_{k-1}), \ldots, \hat{\mathbf{y}}(t_{k-l_y}), \mathbf{u}(t_{k-1}), \ldots, \mathbf{u}(t_{k-l_u}), \mathbf{W}), \quad k \geqslant \max\left\{l_u, l_y\right\}, \tag{2.12}$$

where $\mathbf{F}(\cdot, \mathbf{W})$ is a static neural network, and l_u and l_y are the number of past controls and past outputs used for prediction. (See Fig. 2.18.)

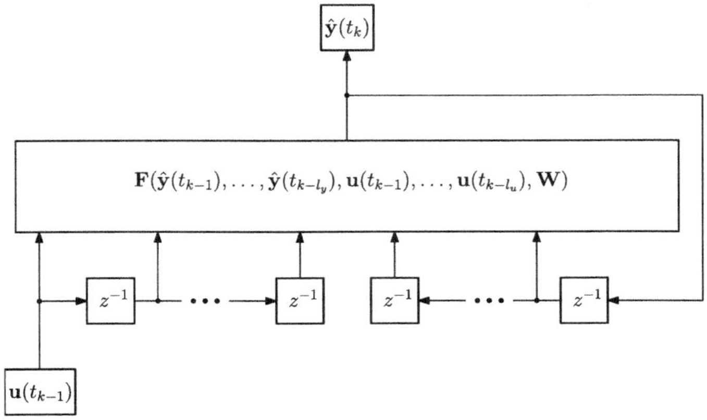

FIGURE 2.18 Nonlinear AutoRegressive network with eXogeneous inputs.

The input–output modeling approach has serious drawbacks: first, the minimum time window size required to achieve the desired accuracy is not known beforehand; second, in order to learn the long-term dependencies one might need an arbitrarily large time window; third, if a dynamical system is nonstationary, the optimal time window size might change over time.

Recurrent neural network. An alternative class of models for deterministic nonlinear controlled discrete time dynamical systems is a class of state-space neural network–based models, usually referred to as the recurrent neural networks, i.e.,

$$\begin{aligned} \mathbf{z}(t_{k+1}) &= \mathbf{F}(\mathbf{z}(t_k), \mathbf{u}(t_k), \mathbf{W}), \\ \hat{\mathbf{y}}(t_k) &= \mathbf{G}(\mathbf{z}(t_k), \mathbf{W}), \end{aligned} \tag{2.13}$$

where $\mathbf{z}(t_k) \in \mathbb{R}^{n_z}$ are the state variables (also called the context units), $\hat{\mathbf{y}}(t_k) \in \mathbb{R}^{n_y}$ are the predicted outputs, $\mathbf{W} \in \mathbb{R}^{n_w}$ is the model parameter vector, and $\mathbf{F}(\cdot, \mathbf{W})$ and $\mathbf{G}(\cdot, \mathbf{W})$ are static neural networks. (See Fig. 2.19.) One particular case of a state-space recurrent neural network (2.13) is the Elman network [30]. In general, the optimal number of state variables n_z is unknown. Usually, one simply selects n_z large enough to be able to represent the unknown dynamical system with the required accuracy.

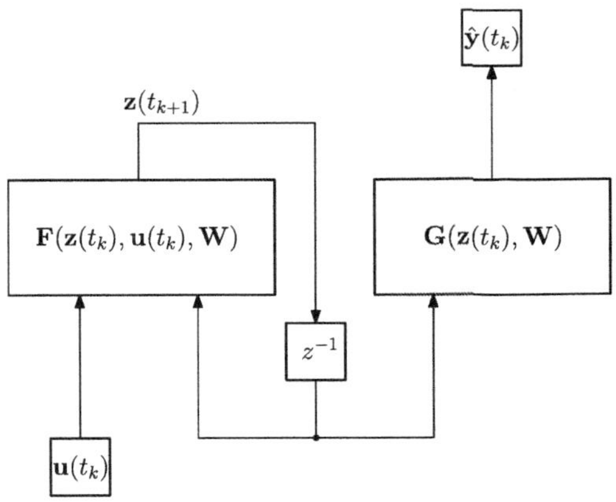

FIGURE 2.19 Recurrent neural network in state space.

2.1.3 Neurons as Elements From Which the ANN Is Formed

The set L_Σ of all elements (neurons) included in the ANN is divided into subsets (layers), i.e.,

$$L^{(0)}, L^{(1)}, \ldots, L^{(p)}, \ldots, L^{(N_L)}, \tag{2.14}$$

or, in a more concise notation,

$$\begin{aligned} L^{(p)}, &\quad p = 0, 1, \ldots, N_L, \\ L^{(p)}, L^{(q)}, L^{(r)}; &\quad p, q, r \in \{0, 1, \ldots, N_L\}, \end{aligned} \tag{2.15}$$

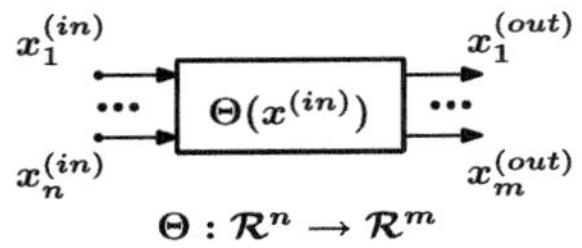

FIGURE 2.20 Neuron as a module converting n-dimensional input vector into m-dimensional output vector. From [109], used with permission from Moscow Aviation Institute.

where N_L is the number of layers into which the set of ANN elements is divided; p, q, r are the indices used to number the arbitrary ("current") ANN layer.

In the list (2.14) $L^{(0)}$ is the input (zero) layer, the purpose of which is to "distribute" the input data to the neuron elements, which perform the primary data processing. Layers $L^{(1)}, \ldots, L^{(N_L)}$ ensure the processing of the inputs of the ANN into the outputs.

Suppose that in the ANN there are N_L layers $L^{(p)}$, $p = 0, 1, \ldots, N_L$.

The layer $L^{(p)}$ has $N_L^{(p)}$ elements of the neuron elements $S_j^{(p)}$, i.e.,

$$L^{(p)} = \{S_j^{(p)}\}, \quad j = 1, \ldots, N_L^{(p)}. \tag{2.16}$$

The element $S_j^{(p)}$ has $N_j^{(p)}$ inputs $x_{i,j}^{(p,q)}$ and $M_j^{(p)}$ outputs $x_{j,k}^{(p,q)}$.

The connections of the element $S_j^{(p)}$ with other elements of the network can be represented as a set of tuples showing where the outputs of the element $S_j^{(p)}$ are transferred.

Thus, a single neuron as a module of the ANN (Fig. 2.20) is a mapping of the n-dimensional input vector $x^{(in)} = (x_1^{(in)}, \ldots, x_n^{(in)})$ into the m-dimensional output vector $x^{(out)} = (x_1^{(out)}, \ldots, x_m^{(out)})$, i.e., $x^{(out)} = \Theta(x^{(in)})$.

The mapping Θ is formed as a composition of the following primitive mappings (Fig. 2.21):

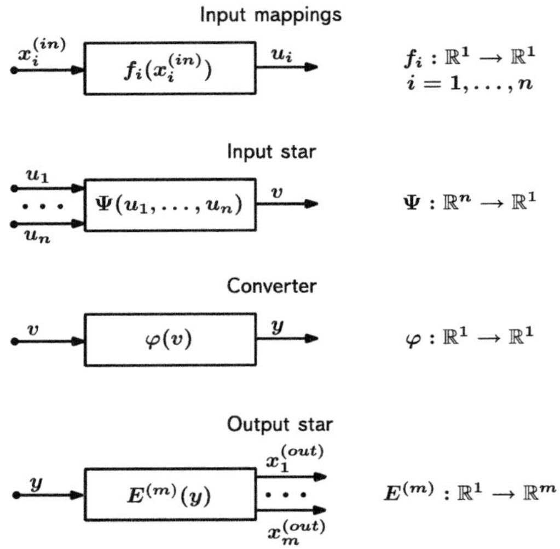

FIGURE 2.21 The primitive mappings of which the neuron consists. From [109], used with permission from Moscow Aviation Institute.

1) set of input mappings $f_i(x_i^{(in)})$:

$$f_i : \mathcal{R} \to \mathcal{R}; \quad u_i = f_i(x_i^{(in)}), \; i = 1, \ldots, n; \tag{2.17}$$

2) aggregating mapping ("input star") $\varphi(u_1, \ldots, u_n)$:

$$\varphi : \mathcal{R}^n \to \mathcal{R}; \quad v = \varphi(u_1, \ldots, u_n); \tag{2.18}$$

3) converter (activation function) $\Psi(v)$:

$$\psi : \mathcal{R} \to \mathcal{R}; \quad y = \psi(v); \tag{2.19}$$

4) output mapping ("output star") $E^{(m)}$:

$$E^{(m)} : \mathcal{R} \to \mathcal{R}^m; \quad E^{(m)}(y) = \{x_j^{(out)}\}, \; j = 1, \ldots, m, \; x_j^{(out)} = y, \; \forall j \in \{j = 1, \ldots, m\}. \tag{2.20}$$

The relations (2.20) are interpreted as follows: mapping $E^{(m)}(y)$ generates as a result an m-

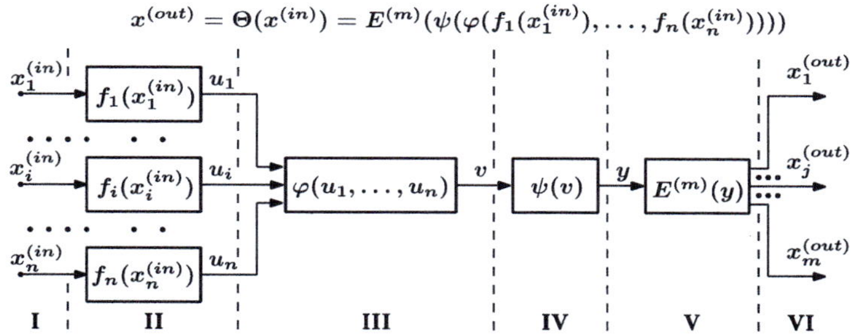

FIGURE 2.22 Structure of the neuron. I – input vector; II – input mappings; III – aggregating mapping; IV – converter; V – the output mapping; VI – output vector. From [109], used with permission from Moscow Aviation Institute.

$$x^{(out)} = \Theta(x^{(in)}) = E^{(m)}(\psi(\varphi(f_1(x_1^{(in)}), \ldots, f_n(x_n^{(in)}))))$$

I $\quad x^{(in)} = (x_1^{(in)}, \ldots, x_i^{(in)}, \ldots, x_n^{(in)})$

$\Downarrow$

II $\quad u_1 = f_1(x_1^{(in)}), \ldots, u_n = f_n(x_n^{(in)})$

$\Downarrow$

III $\quad v = \varphi(u_1, \ldots, u_n)$

$\Downarrow$

IV $\quad y = \psi(v)$

$\Downarrow$

V $\quad x^{(out)} = E^{(m)}(y)$

$\Downarrow$

VI $\quad x^{(out)} = (x_1^{(out)}, \ldots, x_j^{(out)}, \ldots, x_m^{(out)})$

FIGURE 2.23 The sequence of transformations (primitive mappings) realized by the neuron. I – input vector; II – input mappings; III – aggregating mapping; IV – converter (activation function); V – the output mapping; VI – output vector. From [109], used with permission from Moscow Aviation Institute.

element ordered set $\{x_j^{(out)}\}$, each element of which takes the value $x_j^{(out)} = y$.

The Θ map is formed as a composition of the mappings $\{f_i\}$, ψ, φ, and $E^{(m)}$ (Fig. 2.22), i.e.,

$$\begin{aligned} x^{(out)} &= \Theta(x^{(in)}) \\ &= E^{(m)}(\psi(\varphi(f_1^{(in)}(x_1^{(in)}), \ldots, f_n^{(in)}(x_n^{(in)})))). \end{aligned} \quad (2.21)$$

The interaction of primitive mappings forming a neuron is shown in Fig. 2.23.

2.1.4 Structural Organization of a Neuron

A separate neuron element $S_j^{(p)}$ of the neural network structure (i.e., the jth neuron from the pth layer) is an ordered pair of the form

$$S_j^{(p)} = \langle \Theta_j^{(p)}, R_j^{(p)} \rangle, \quad (2.22)$$

where $\Theta_j^{(p)}$ is the transformation of the input vector of dimension $N_j^{(p)}$ into the output vector of dimension $M_j^{(p)}$; $R_j^{(p)}$ is the connection of the output of the element $S_j^{(p)}$ with other neurons of the considered ANN (with neurons from other layers, they are direct and inverse relations; with neurons from the same layer, they are lateral connections).

The transformation $\Theta_j^{(p)}(x_{i,j}^{(r,p)})$ is the composition of the primitives from which the neuron consists, i.e.,

$$\Theta_j^{(p)}(x_{i,j}^{(r,p)}) = \Theta(\psi(\varphi(f_{i,j}^{(r,p)}(x_{i,j}^{(r,p)})))). \quad (2.23)$$

The connections $R_j^{(p)}$ of the neuron $S_j^{(p)}$ are the set of ordered pairs showing where the outputs

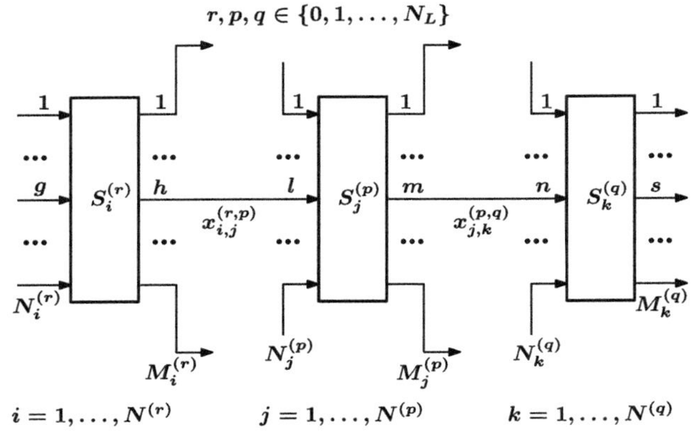

FIGURE 2.24 The numeration of the inputs/outputs of neurons and the notation of signals ($x_{i,j}^{(r,p)}$ and $x_{j,k}^{(p,q)}$), transmitted via interneuron links; it is the basic level of the description of the ANN. $S_i^{(r)}$, $S_j^{(p)}$, and $S_k^{(q)}$ are neurons of the ANN (ith in the rth layer, jth in the pth layer, and kth in the qth layer, respectively); $N_i^{(r)}$, $N_j^{(p)}$, $N_k^{(q)}$ are the number of inputs and $M_i^{(r)}$, $M_j^{(p)}$, $M_k^{(q)}$ are the number of outputs in the neurons $S_i^{(r)}$, $S_j^{(p)}$, and $S_k^{(q)}$, respectively; $x_{i,j}^{(r,p)}$ is the signal transferred from the output of the ith neuron from the rth layer to the input of the jth neuron from the pth layer; $x_{j,k}^{(p,q)}$ is the signal transferred from the output of the jth neuron from the pth layer to the input of the kth neuron from the qth layer; g, h, l, m, n, s are the numbers of the neuron inputs/outputs; N_L is the number of layers in the ANN; $N^{(r)}$, $N^{(p)}$, $N^{(q)}$ is the number of neurons in the layers with numbers r, p, q, respectively. From [109], used with permission from Moscow Aviation Institute.

of a given neuron go, i.e.,

$$R_j^{(p)} = \{\langle q, k\rangle\}, \quad q \in \{1, \ldots, N_L\},\ k \in \{1, \ldots, N_L^q\}. \tag{2.24}$$

Inputs/outputs of neurons are described as follows.

In the variant with the maximum detail of the description (extended level of the ANN description), which provides the possibility of representing any ANN structure, we use the notation of the form $x_{(i,l),(j,m)}^{(r,p)}$. It identifies the signal transmitted from the neuron $S_i^{(r)}$ (ith neuron from the rth layer) to $S_j^{(p)}$ (jth neuron from the pth layer), and the outputs of the ith neuron in the rth layer and the inputs of the jth neuron of the pth layer are renumbered; according to their numbering, l is the serial number of the output of the element $S_i^{(r)}$, and m is the entry serial number of the element $S_j^{(p)}$. Such a detailed representation is required in cases where the order of the input/output quantities is important, i.e., the set of these quantities is interpreted as a vector. For example, this kind of representation is used in the compressive mapping of the RBF neuron, which realizes the calculation of the distance between two vectors.

In the variant, when a complete specification of the neuron's connections is not required (this is the case when the result of "compression" (or "aggregation") $\varphi : \mathcal{R}^n \to \mathcal{R}$ does not depend on the order of the input components), we can use a simpler notation for the input/output signals of the neuron, which has the form $x_{i,j}^{(r,p)}$. In this case, it is simply indicated that the connection goes from the ith neuron of the rth layer to the jth neuron of the pth layer, without specifying the serial numbers of the input/output components.

The system of numbering of the neuron inputs/outputs in the ANN, as well as the interneuron connections, is illustrated in Fig. 2.24

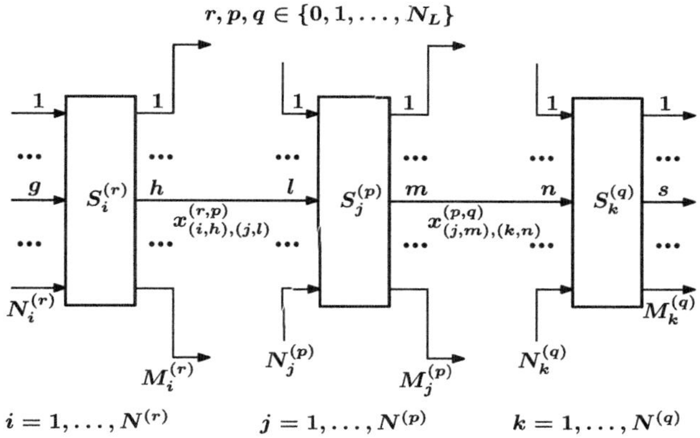

FIGURE 2.25 The numbering of the inputs/outputs of neurons and the designations of signals ($x^{(r,p)}_{(i,h),(j,l)}$ and $x^{(p,q)}_{(j,m),(k,n)}$), transmitted through interneuronal connections; it is the extended level of the description of the ANN. $S_i^{(r)}$, $S_j^{(p)}$, and $S_k^{(q)}$ are the neurons of the ANN (ith in the rth layer, jth in the pth layer, and kth in the qth layer, respectively); $N_i^{(r)}$, $N_j^{(p)}$, $N_k^{(q)}$ are the number of inputs and $M_i^{(r)}$, $M_j^{(p)}$, $M_k^{(q)}$ are the number of outputs in the neurons $S_i^{(r)}$, $S_j^{(p)}$, and $S_k^{(q)}$, respectively; $x^{(r,p)}_{(i,h),(j,l)}$ is the signal transferred from the hth output of the ith neuron from the rth layer on the lth input of the jth neuron from the pth layer; $x^{(r,p)}_{(i,h),(j,l)}$ is the signal transferred from the mth exit of the jth neuron from the pth layer to the nth input of the kth neuron from the qth layer; g, h, l, m, n, s are the numbers of the neuron inputs/outputs; N_L is the number of layers in the ANN; $N^{(r)}$, $N^{(p)}$, $N^{(q)}$ are the number of neurons in the layers with numbers r, p, q, respectively.

for the baseline level of the ANN description and in Fig. 2.25 for the advanced level.

2.2 ARTIFICIAL NEURAL NETWORK TRAINING METHODS

After an appropriate neural network structure has been selected, one needs to determine the values of its parameters in order to achieve the desired input–output behavior. The process of parameter modification is usually called *learning* or *training*, when referred to neural networks. Thus, the ANN learning algorithm is a sequence of actions which modifies the parameters so that the network would be able to solve some specific task.

There are several major approaches to the neural network learning problem:

- unsupervised learning;
- supervised learning;
- reinforcement learning.

The features of these approaches are as follows.

In the case of *unsupervised learning*, only the inputs are given, and there are no prescribed output values. Unsupervised learning aims at discovering inherent patterns in the data set. This approach is usually applied to clustering and dimensionality reduction problems.

In the case of a *supervised learning*, the desired network behavior is explicitly defined by a training data set. Each training example associates some input with a specific desired output. The goal of the learning is to find such values of the neural network parameters that the actual network outputs would be as close as possible to the desired ones. This approach is usually applied to classification, regression, and system identification problems.

If a training data set is not known beforehand, but rather presented sequentially one example at a time, and a neural network is expected to operate and learn simultaneously, then it is said to perform *incremental learning*. Additionally, if the environment is assumed to be nonstationary, i.e., the desired response to some input may vary over time, then the training data set becomes inconsistent and a neural network needs to perform *adaptation*. In this case, we face a stability-plasticity dilemma: if the network lacks plasticity, then it cannot rapidly adapt to changes; on the other hand, if it lacks stability, then it forgets the previously learned data.

Another variation of supervised learning is *active learning*, which assumes that the neural network itself is responsible for the data set acquisition. That is, the network selects a new input and queries an external system (for example, some sensor) for the desired outputs that correspond to this input. Hence, a neural network is expected to "explore" the environment by interacting with it and to "exploit" the obtained data by minimizing some objective. In this paradigm, finding a balance between exploration and exploitation becomes an important issue. *Reinforcement learning* takes the idea of active learning one step further by assuming that the external system cannot provide the network with examples of desired behavior – instead, it can only score the previous behavior of the network. This approach is usually applied to intelligent control and decision making problems.

In this book, we cover only the supervised learning approach and focus on the modeling and identification problem for dynamical systems. Section 2.3.1 treats the training methods for static neural networks with applications to function approximation problems. These methods constitute the basis for dynamic neural network training algorithms, discussed in Section 2.3.3. For a discussion of unsupervised methods, see [10]. Reinforcement learning methods are presented in the books [45–48].

We need to mention that the actual goal of the neural network supervised learning is *not* to achieve a perfect match of predictions with the training data, but to perform highly accurate predictions on the independent data during the network operation, i.e., the network should be able to *generalize*. In order to evaluate the generalization ability of a network, we split all the available experimental data into *training set* and *test set*. The model learns only on the training set, and then it is evaluated on an independent test set. Sometimes, yet another subset is reserved – the so-called *validation set*, which is used to select the model hyperparameters (such as the number of layers or neurons).

2.2.1 Overview of the Neural Network Training Framework

Suppose that the network parameters are represented by a finite-dimensional vector $\mathbf{W} \in \mathbb{R}^{n_w}$. The supervised learning approach implies a minimization of an error function (also called objective function, loss function, or cost function), which represents the deviation of actual network outputs from their desired values. We define a total error function $\bar{E} : \mathbb{R}^{n_w} \to \mathbb{R}$ to be a sum of individual errors for each of the training examples, i.e.,

$$\bar{E}(\mathbf{W}) = \sum_{i=1}^{P} E^{(p)}(\mathbf{W}). \tag{2.25}$$

The error function (2.25) is to be minimized with respect to neural network parameters $\mathbf{W}$. Thus, we have an unconstrained nonlinear optimization problem:

$$\underset{\mathbf{W}}{\text{minimize}} \quad \bar{E}(\mathbf{W}). \tag{2.26}$$

In order for the minimization problem to make sense, we require the error function to be bounded from below.

Minimization is carried out by means of various iterative numerical methods. The optimization methods can be divided into global and local ones, according to the type of minimum they seek for. Global optimization methods seek for an approximate global minimum, whereas local methods seek for a precise local minimum. Most of the global optimization methods have a stochastic nature (e.g., simulated annealing, evolutionary algorithms, particle swarm optimization) and the convergence is achieved almost surely and only in the limit. In this book we focus on the local deterministic gradient-based optimization methods, which guarantee a rapid convergence to a local solution under some reasonable assumptions. In order to apply these methods, we also require the error function to be sufficiently smooth (which is usually the case with neural networks provided all the activation functions are smooth). For more detailed information on local optimization methods, we refer to [49–52]. Metaheuristic global optimization methods are covered in [53,54].

Note that the local optimization methods require an initial guess $\mathbf{W}^{(0)}$ for parameter values. There are various approaches to the initialization of network parameters. For example, the parameters may be sampled from a Gaussian distribution, i.e.,

$$\mathbf{W}_i \sim \mathcal{N}(0,1),\ i = 1,\dots,n_w. \tag{2.27}$$

The following alternative initialization method for layered feedforward neural networks (2.8), called Xavier initialization, was suggested in [55]:

$$\begin{aligned} \mathbf{b}_i^l &= 0, \\ \mathbf{w}_{i,j}^l &\sim \mathcal{U}\left(-\sqrt{\frac{6}{S^{l-1}+S^l}}, \sqrt{\frac{6}{S^{l-1}+S^l}}\right). \end{aligned} \tag{2.28}$$

Optimization methods may also be classified by the order of error function derivatives used to guide the search process. Thus, zero-order methods use only the error function values; first-order methods rely on the first derivatives (gradient $\nabla \bar{E}$); second-order methods also utilize the second derivatives (Hessian $\nabla^2 \bar{E}$).

The basic descent method has the form

$$\mathbf{W}^{(k+1)} = \mathbf{W}^{(k)} + \alpha^{(k)}\mathbf{p}^{(k)}, \quad \bar{E}(\mathbf{W}^{(k+1)}) < \bar{E}(\mathbf{W}^{(k)}), \tag{2.29}$$

where $\mathbf{p}^{(k)}$ is a search direction and $\alpha^{(k)}$ represents a step length, also called the learning rate. Note that we require each step to decrease the error function. In order to guarantee the error function decrease for arbitrarily small step lengths, we need the search direction to be a descent direction, that is, to satisfy $\mathbf{p}^{(k)T}\nabla \bar{E}(\mathbf{W}^{(k)}) < 0$.

The simplest example of a first-order descent method is the *gradient descent (GD) method*, which utilizes the negative gradient search direction, i.e.,

$$\mathbf{p}^{(k)} = -\nabla \bar{E}(\mathbf{W}^{(k)}). \tag{2.30}$$

The step lengths may be assigned beforehand $\forall s$ $\alpha^{(k)} \equiv \alpha$, but if the step α is too large, the error function might actually increase, and then the iterations would diverge. For example, in the case of a convex quadratic error function of the form

$$\bar{E}(\mathbf{W}) = \frac{1}{2}\mathbf{W}^T\mathbf{A}\mathbf{W} + \mathbf{b}^T\mathbf{W} + c, \tag{2.31}$$

where $\mathbf{A}$ is a symmetric positive definite matrix with a maximum eigenvalue of $\lambda_{\max}$, the step length must satisfy

$$\alpha < \frac{2}{\lambda_{\max}},$$

in order to guarantee the convergence of gradient descent iterations. On the other hand, a small step α would result in a slow convergence. In order to circumvent this problem, we can perform a step length adaptation: we take a "trial" step, evaluate the error function and check whether it has decreased or not. If it has decreased, then we accept this trial step, and we increase the step length. Otherwise, we reject the trial step, and decrease the step length. An alternative approach is to perform a *line search* for an optimal step length which provides the maximum possible reduction of the error function along the search direction, i.e.,

$$\alpha^{(k)} = \underset{\alpha>0}{\operatorname{argmin}}\, \bar{E}\left(\mathbf{W}^{(k)} + \alpha \mathbf{p}^{(k)}\right). \tag{2.32}$$

The GD method combined with this exact line search is called the *steepest gradient descent*. Note that the global minimum of this univariate function is hard to find; in fact, even a search for an accurate estimate of a local minimum would require many iterations. Fortunately, we do not need to find an exact minimum along the specified direction – the convergence of an overall minimization procedure may be obtained if we guarantee a sufficient decrease of an error function at each iteration. If the search direction is a descent direction and if the step lengths satisfy the Wolfe conditions

$$\begin{aligned}\bar{E}\left(\mathbf{W}^{(k)} + \alpha^{(k)}\mathbf{p}^{(k)}\right) &\leqslant \bar{E}\left(\mathbf{W}^{(k)}\right) \\ &\quad + c_1 \alpha^{(k)} \nabla \bar{E}(\mathbf{W}^{(k)})^T \mathbf{p}^{(k)}, \\ \nabla \bar{E}\left(\mathbf{W}^{(k)} + \alpha^{(k)}\mathbf{p}^{(k)}\right)^T \mathbf{p}^{(k)} &\geqslant c_2 \nabla \bar{E}(\mathbf{W}^{(k)})^T \mathbf{p}^{(k)},\end{aligned} \tag{2.33}$$

for $0 < c_1 < c_2 < 1$, then the iterations converge to a stationary point $\lim_{s\to\infty} \|\nabla \bar{E}(\mathbf{W}^{(k)})\| = \mathbf{0}$ from an arbitrary initial guess (i.e., we have a global convergence to a stationary point). Note that there always exist intervals of step lengths which satisfy the Wolfe conditions. This justifies the use of inexact line search methods, which require fewer iterations to find an appropriate step length which provides a sufficient reduction of an error function. Unfortunately, the GD method has a linear convergence rate, which is very slow.

Another important first-order method is the *nonlinear conjugate gradient (CG) method*. In fact, it is a family of methods which utilize search directions of the following general form:

$$\begin{aligned}\mathbf{p}^{(0)} &= -\nabla \bar{E}(\mathbf{W}^{(0)}), \\ \mathbf{p}^{(k)} &= -\nabla \bar{E}(\mathbf{W}^{(k)}) + \beta^{(k)} \mathbf{p}^{(k-1)}.\end{aligned} \tag{2.34}$$

Depending on the choice of the scalar $\beta^{(k)}$, we obtain several variations of the method. The most popular expressions for $\beta^{(k)}$ are the following:

- the Fletcher–Reeves method:

$$\beta^{(k)} = \frac{\nabla \bar{E}(\mathbf{W}^{(k)})^T \nabla \bar{E}(\mathbf{W}^{(k)})}{\nabla \bar{E}(\mathbf{W}^{(k-1)})^T \nabla \bar{E}(\mathbf{W}^{(k-1)})}; \tag{2.35}$$

- the Polak–Ribière method:

$$\beta^{(k)} = \frac{\nabla \bar{E}(\mathbf{W}^{(k)})^T \left(\nabla \bar{E}(\mathbf{W}^{(k)}) - \nabla \bar{E}(\mathbf{W}^{(k-1)})\right)}{\nabla \bar{E}(\mathbf{W}^{(k-1)})^T \nabla \bar{E}(\mathbf{W}^{(k-1)})}; \tag{2.36}$$

- the Hestenes–Stiefel method:

$$\beta^{(k)} = \frac{\nabla \bar{E}(\mathbf{W}^{(k)})^T \left(\nabla \bar{E}(\mathbf{W}^{(k)}) - \nabla \bar{E}(\mathbf{W}^{(k-1)})\right)}{\left(\nabla \bar{E}(\mathbf{W}^{(k)}) - \nabla \bar{E}(\mathbf{W}^{(k-1)})\right)^T \mathbf{p}^{(k-1)}}. \tag{2.37}$$

Irrespective of the particular $\beta^{(k)}$ selected, the first search direction $\mathbf{p}^{(0)}$ is simply the negative gradient direction. If we assume that the error function is convex and quadratic (2.31), then the method generates a sequence of conjugate search directions (i.e., ${\mathbf{p}^{(i)}}^T \mathbf{A} \mathbf{p}^{(j)} = 0$ for $i \neq j$). If we also assume that the line searches are exact, then the method converges within n_w iterations. In the general case of a nonlinear error function, the convergence rate is linear; however, a twice differentiable error function with nonsingular Hessian is approximately quadratic in the neighborhood of the solution, which results in fast convergence. Note also that the search directions lose conjugacy, hence we need to perform so-called "restarts," i.e., to assign $\beta^{(k)} \leftarrow 0$. For example, we might reset $\beta^{(k)}$ if the consecutive directions are nonorthogonal $\frac{\left|\mathbf{p}^{(k)^T}\mathbf{p}^{(k-1)}\right|}{\|\mathbf{p}^{(k)}\|^2} > \varepsilon$. In the case of the Polak–Ribière method, we should also reset $\beta^{(k)}$ if it becomes negative.

The basic second-order method is Newton's method:

$$\mathbf{p}^{(k)} = -\left[\nabla^2 \bar{E}(\mathbf{W}^{(k)})\right]^{-1} \nabla \bar{E}(\mathbf{W}^{(k)}). \tag{2.38}$$

If the Hessian $\nabla^2 \bar{E}(\mathbf{W}^{(k)})$ is positive definite, the resulting search direction $\mathbf{p}^{(k)}$ is a descent direction. If the error function is convex and quadratic, Newton's method with a unit step length $\alpha^{(k)} = 1$ finds the solution in a single step. For a smooth nonlinear error function with positive definite Hessian at the solution, the convergence is quadratic, provided the initial guess lies sufficiently close to the solution. If a Hessian turns out to have negative or zero eigenvalues, we need to modify it in order to obtain a positive definite approximation $\mathbf{B}$ – for example, we might add a scaled identity matrix, so we have

$$\mathbf{B}^{(k)} = \nabla^2 \bar{E}(\mathbf{W}^{(k)}) + \mu^{(k)}\mathbf{I}. \tag{2.39}$$

The resulting damped method may be viewed as hybrid of the ordinary Newton method (for $\mu^{(k)} = 0$) and a gradient descent (for $\mu^{(k)} \to \infty$).

Note that the Hessian computation is very computationally expensive; hence there have been proposed various approximations. If we assume that each individual error is a quadratic form,

$$E^{(p)}(\mathbf{W}) = \frac{1}{2}\mathbf{e}^{(p)}(\mathbf{W})^T \mathbf{e}^{(p)}(\mathbf{W}), \tag{2.40}$$

then the gradient and Hessian may be expressed in terms of the error Jacobian as follows:

$$\begin{aligned}
\nabla E^{(p)}(\mathbf{W}) &= \frac{\partial \mathbf{e}^{(p)}(\mathbf{W})}{\partial \mathbf{W}}^T \mathbf{e}^{(p)}(\mathbf{W}), \\
\nabla^2 E^{(p)}(\mathbf{W}) &= \frac{\partial \mathbf{e}^{(p)}(\mathbf{W})}{\partial \mathbf{W}}^T \frac{\partial \mathbf{e}^{(p)}(\mathbf{W})}{\partial \mathbf{W}} \\
&\quad + \sum_{i=1}^{n_e} \frac{\partial^2 \mathbf{e}_i^{(p)}(\mathbf{W})}{\partial \mathbf{W}} \mathbf{e}_i^{(p)}(\mathbf{W}).
\end{aligned} \tag{2.41}$$

Then, the Gauss–Newton approximation to the Hessian is obtained by discarding the second-order terms, i.e.,

$$\nabla^2 E^{(p)}(\mathbf{W}) \approx \mathbf{B}^{(p)} = \frac{\partial \mathbf{e}^{(p)}(\mathbf{W})}{\partial \mathbf{W}}^T \frac{\partial \mathbf{e}^{(p)}(\mathbf{W})}{\partial \mathbf{W}}. \tag{2.42}$$

The resulting matrix $\mathbf{B}$ can turn out to be degenerate, so we might modify it by adding a scaled identity matrix as mentioned above in (2.39). Then we have

$$\mathbf{B}^{(p)} = \frac{\partial \mathbf{e}^{(p)}(\mathbf{W})}{\partial \mathbf{W}}^T \frac{\partial \mathbf{e}^{(p)}(\mathbf{W})}{\partial \mathbf{W}} + \mu^{(k)}\mathbf{I}. \tag{2.43}$$

This technique leads us to the Levenberg–Marquardt method.

A family of *quasi-Newton* methods estimate the inverse Hessian by accumulating the changes of gradients. These methods construct an inverse Hessian approximation $\mathbf{H} \approx \nabla^2 \bar{E}(\mathbf{W})^{-1}$ so as to satisfy the secant equation:

$$\begin{aligned}
\mathbf{H}^{(k+1)}\mathbf{y}^{(k)} &= \mathbf{s}^{(k)}, \\
\mathbf{s}^{(k)} &= \mathbf{W}^{(k+1)} - \mathbf{W}^{(k)}, \\
\mathbf{y}^{(k)} &= \nabla \bar{E}(\mathbf{W}^{(k+1)}) - \nabla \bar{E}(\mathbf{W}^{(k)}).
\end{aligned} \tag{2.44}$$

However, for $n_w > 1$ this system of equations is underdetermined and there exists an infinite number of solutions. Thus, additional constraints are imposed, giving rise to various quasi-Newton methods. Most of them require that the inverse Hessian approximation $\mathbf{H}^{(k+1)}$

be symmetric and positive definite, and also minimize the distance to the previous estimate $\mathbf{H}^{(k)}$ with respect to some norm: $\mathbf{H}^{(k+1)} = \underset{\mathbf{H}}{\operatorname{argmin}} \|\mathbf{H} - \mathbf{H}^{(k)}\|$. One of the most popular variations of quasi-Newton methods is the *Broyden–Fletcher–Goldfarb–Shanno (BFGS) algorithm*:

$$\mathbf{H}^{(k+1)} = \left(\mathbf{I} - \rho^{(k)} \mathbf{s}^{(k)} \mathbf{y}^{(k)T}\right) \mathbf{H}^{(k)} \left(\mathbf{I} - \rho^{(k)} \mathbf{y}^{(k)} \mathbf{s}^{(k)T}\right) + \rho^{(k)} \mathbf{s}^{(k)} \mathbf{s}^{(k)T}, \quad (2.45)$$

$$\rho^{(k)} = \frac{1}{\mathbf{y}^{(k)T} \mathbf{s}^{(k)}}.$$

The initial guess for inverse Hessian may be selected in different ways: for example, if $\mathbf{H}^{(0)} = \mathbf{I}$, then the first search direction corresponds to that of GD. Note that if $\mathbf{H}^{(0)}$ is positive definite and a step length is selected by a line search so as to satisfy the Wolfe conditions (2.33), then all the resulting $\mathbf{H}^{(k)}$ will also be positive definite. The BFGS algorithm has a superlinear rate of convergence. We should also mention that the inverse Hessian approximation contains n_w^2 elements, and when the number of parameters n_w is large, it might not fit in the memory. In order to circumvent this issue, a *limited memory BFGS* (L-BFGS) [56] version of the algorithm has been proposed, which stores only the $m \ll n_w$ most recent vector pairs $\left\{\left\langle \mathbf{s}^{(j)}, \mathbf{y}^{(j)} \right\rangle\right\}_{j=k-m}^{k-1}$ and uses them to compute the search direction without explicit evaluation of $\mathbf{H}^{(k)}$.

Another strategy, alternative to line search methods, is a family of trust region methods [57]. These methods repeatedly construct some local model $\bar{M}^{(k)}$ of the error function, which is assumed to be valid in a neighborhood of current point $\mathbf{W}^{(k)}$, and minimize it within this neighborhood. If we utilize a second-order Taylor series approximation of the error function as the model

$$\bar{E}(\mathbf{W}^{(k)} + \mathbf{p}) \approx \bar{M}^{(k)}(\mathbf{p}) = \bar{E}(\mathbf{W}^{(k)}) + \mathbf{p}^T \nabla \bar{E}(\mathbf{W}^{(k)}) + \frac{1}{2} \mathbf{p}^T \nabla^2 \bar{E}(\mathbf{W}^{(k)}) \mathbf{p} \quad (2.46)$$

and a ball of radius $\Delta^{(k)}$ as a trust region, we obtain the following constrained quadratic optimization subproblems:

$$\begin{aligned} \underset{\mathbf{p}}{\text{minimize}} \quad & \bar{M}^{(k)}(\mathbf{p}) \\ \text{subject to} \quad & \|\mathbf{p}\| \leqslant \Delta^{(k)}. \end{aligned} \quad (2.47)$$

The trust region radius is adapted based on the ratio of predicted and actual reduction of an error function, i.e.,

$$\rho^{(k)} = \frac{\bar{E}(\mathbf{W}^{(k)}) - \bar{E}(\mathbf{W}^{(k)} + \mathbf{p}^{(k)})}{\bar{M}(\mathbf{0}) - \bar{M}(\mathbf{p}^{(k)})}. \quad (2.48)$$

If this ratio is close to 1, the trust region radius is increased by some factor; if the ratio is close to 0, the radius is decreased. Also, if $\rho^{(k)}$ is negative or very small, the step is rejected. We could also use ellipsoidal trust regions of the form $\|\mathbf{D}\mathbf{p}\| \leqslant \Delta^{(k)}$, where $\mathbf{D}$ is a nonsingular diagonal matrix.

There exist various approaches to solving the constrained quadratic optimization subproblem (2.47). One of these techniques [58] relies on the *linear conjugate gradient* method in order to solve the system of linear equations

$$\nabla^2 \bar{E}(\mathbf{W}^{(k)}) \mathbf{p} = -\nabla \bar{E}(\mathbf{W}^{(k)}),$$

with respect to $\mathbf{p}$. This results in the following iterations:

$$
\begin{aligned}
\mathbf{p}^{(k,0)} &= \mathbf{0}, \\
\mathbf{r}^{(k,0)} &= \nabla \bar{E}(\mathbf{W}^{(k)}), \\
\mathbf{d}^{(k,0)} &= -\nabla \bar{E}(\mathbf{W}^{(k)}), \\
\alpha^{(k,s)} &= \frac{\mathbf{r}^{(k,s)^T} \mathbf{r}^{(k,s)}}{\mathbf{d}^{(k,s)^T} \nabla^2 \bar{E}(\mathbf{W}^{(k)}) \mathbf{d}^{(k,s)}}, \\
\mathbf{p}^{(k,s+1)} &= \mathbf{p}^{(k,s)} + \alpha^{(k,s)} \mathbf{d}^{(k,s)}, \\
\mathbf{r}^{(k,s+1)} &= \mathbf{r}^{(k,s)} + \alpha^{(k,s)} \nabla^2 \bar{E}(\mathbf{W}^{(k)}) \mathbf{d}^{(k,s)}, \\
\beta^{(k,s+1)} &= \frac{\mathbf{r}^{(k,s+1)^T} \mathbf{r}^{(k,s+1)}}{\mathbf{r}^{(k,s)^T} \mathbf{r}^{(k,s)}}, \\
\mathbf{d}^{(k,s+1)} &= \mathbf{r}^{(k,s+1)} + \beta^{(k,s+1)} \mathbf{d}^{(k,s)}.
\end{aligned}
$$

The iterations are terminated prematurely either if they cross the trust region boundary, $\|\mathbf{p}^{(k,s+1)}\| \geqslant \Delta^{(k)}$, or if a nonpositive curvature direction is discovered, $\mathbf{d}^{(k,s)^T} \nabla^2 \bar{E}(\mathbf{W}^{(k)}) \mathbf{d}^{(k,s)} \leqslant 0$. In these cases, a solution corresponds to the intersection of the current search direction with the trust region boundary. It is important to note that this method does not require one to compute the entire Hessian matrix; instead, we need only the Hessian vector products of the form $\nabla^2 \bar{E}(\mathbf{W}^{(k)}) \mathbf{d}^{(k,s)}$, which may be computed more efficiently by reverse-mode automatic differentiation methods described below. Such *Hessian-free* methods have been successfully applied to neural network training [59,60].

Another approach to solving (2.47) [61,62] replaces the subproblem with an equivalent problem of finding both the vector $\mathbf{p} \in \mathbb{R}^{n_w}$ and the scalar $\mu \geqslant 0$ such that

$$
\begin{aligned}
\left(\nabla^2 \bar{E}(\mathbf{W}^{(k)}) + \mu \mathbf{I}\right) \mathbf{p} &= -\nabla \bar{E}(\mathbf{W}^{(k)}), \\
\mu(\Delta - \|\mathbf{p}\|) &= 0,
\end{aligned}
\tag{2.49}
$$

where $\nabla^2 \bar{E}(\mathbf{W}^{(k)}) + \mu \mathbf{I}$ is positive semidefinite. There are two possibilities. If $\mu = 0$, then we have $\mathbf{p} = -\left[\nabla^2 \bar{E}(\mathbf{W}^{(k)})\right]^{-1} \nabla \bar{E}(\mathbf{W}^{(k)})$ and $\|\mathbf{p}\| \leqslant \Delta$. If $\mu > 0$, then we define $\mathbf{p}(\mu) = -\left[\nabla^2 \bar{E}(\mathbf{W}^{(k)}) + \mu \mathbf{I}\right]^{-1} \nabla \bar{E}(\mathbf{W}^{(k)})$ and solve a one-dimensional equation $\|\mathbf{p}(\mu)\| = \Delta$ with respect to μ.

Note that since the error function (2.25) is a summation of errors for each individual training example, its gradient as well as its Hessian may also be represented as summations of gradients and Hessians of these errors, i.e.,

$$
\nabla \bar{E}(\mathbf{W}) = \sum_{p=1}^{P} \nabla E^{(p)}(\mathbf{W}), \tag{2.50}
$$

$$
\nabla^2 \bar{E}(\mathbf{W}) = \sum_{p=1}^{P} \nabla^2 E^{(p)}(\mathbf{W}). \tag{2.51}
$$

In the case the neural network has a large number of parameters n_w and the data set contains a large number of training examples P, computation of the total error function value $\bar{E}$ as well as its derivatives can be time consuming. Thus, even for a simple GD method, each update of the weights takes a lot of time. Then, we might apply a *stochastic gradient descent (SGD) method*, which randomly shuffles training examples, iterates over them, and updates the parameters using the gradients of individual errors $E^{(p)}$:

$$
\begin{aligned}
\mathbf{W}^{(k,p)} &= \mathbf{W}^{(k,p-1)} - \alpha^{(k)} \nabla E^{(p)}(\mathbf{W}^{(k,p-1)}), \\
\mathbf{W}^{(k+1,0)} &= \mathbf{W}^{(k,P)}.
\end{aligned}
\tag{2.52}
$$

In contrast, the usual gradient descent is called the batch method. We need to mention that although the (k,p)th step decreases the error for the pth training example, it may increase the error for the other examples. On the one hand it allows the method to escape some local minima, but on the other hand it becomes difficult to converge to a final solution. In order to circumvent this issue, we might gradually decrease the step lengths $\alpha^{(k)}$. Also, in order to achieve a "smoother" convergence we could perform the weight updates based on random subsets of training examples, which is called a "minibatch" strategy. The stochastic or minibatch approach may also be applied to other optimization methods; see [63].

We should also mention that in the case of the batch or minibatch update strategy, the computation of total error function values, as well as its derivatives, can be efficiently parallelized. In order to do that, we need to divide the data set into multiple subsets, compute partial sums of the error function and its derivatives over the training examples of each subset in parallel, and then sum the results. This is not possible in the case of stochastic updates. In the case of an SGD method, we can parallelize the gradient computations by neurons of each layer.

Finally, we note that any iterative method requires a stopping criterion used to terminate the procedure. One simple option is a test based on first-order necessary conditions for a local minimum, i.e.,

$$\left\| \nabla E(\mathbf{W}^{(k)}) \right\| < \varepsilon_g. \tag{2.53}$$

We can also terminate iterations if it seems that no progress is made, i.e.,

$$\begin{aligned} E(\mathbf{W}^{(k)}) - E(\mathbf{W}^{(k+1)}) &< \varepsilon_E, \\ \left\| \mathbf{W}^{(k)} - \mathbf{W}^{(k+1)} \right\| &< \varepsilon_w. \end{aligned} \tag{2.54}$$

In order to prevent an infinite loop in the case of algorithm divergence, we might stop when a certain maximum number of iterations has been performed, i.e.,

$$k \geqslant \bar{k}. \tag{2.55}$$

2.2.2 Static Neural Network Training

In this subsection, we consider the function approximation problem. The problem is stated as follows. Suppose that we wish to approximate an unknown mapping $\mathbf{f}\colon \mathcal{X} \to \mathcal{Y}$, where $\mathcal{X} \subset \mathbb{R}^{n_x}$ and $\mathcal{Y} \subset \mathbb{R}^{n_y}$. Assume we are given an experimental data set of the form

$$\left\{ \left\langle \mathbf{x}^{(p)}, \tilde{\mathbf{y}}^{(p)} \right\rangle \right\}_{p=1}^{P}, \tag{2.56}$$

where $\mathbf{x}^{(p)} \in \mathcal{X}$ represent the input vectors and $\tilde{\mathbf{y}}^{(p)} \in \mathcal{Y}$ represent the observed output vectors. Note that in general the observed outputs $\tilde{\mathbf{y}}^{(p)}$ do not match the true outputs $\mathbf{y}^{(p)} = \mathbf{f}(\mathbf{x}^{(p)})$. We assume that the observations are corrupted by an additive Gaussian noise, i.e.,

$$\tilde{\mathbf{y}}^{(p)} = \mathbf{y}^{(p)} + \boldsymbol{\eta}^{(p)}, \tag{2.57}$$

where $\boldsymbol{\eta}^{(p)}$ represent the sample points of a zero-mean random vector $\boldsymbol{\eta} \sim \mathcal{N}(\mathbf{0}, \boldsymbol{\Sigma})$ with diagonal covariance matrix

$$\boldsymbol{\Sigma} = \begin{pmatrix} \sigma_1^2 & & 0 \\ & \ddots & \\ 0 & & \sigma_{n_y}^2 \end{pmatrix}.$$

The approximation is to be performed using a layered feedforward neural network of the form (2.8). Under the abovementioned assumptions on the observation noise, it is reasonable to utilize a least-squares error function. Thus, we have a total error function $\bar{E}$ of the form (2.25) with the individual errors

$$E^{(p)}(\mathbf{W}) = \frac{1}{2} \left(\tilde{\mathbf{y}}^{(p)} - \hat{\mathbf{y}}^{(p)} \right)^{\mathcal{T}} \boldsymbol{\Omega} \left(\tilde{\mathbf{y}}^{(p)} - \hat{\mathbf{y}}^{(p)} \right), \tag{2.58}$$

where $\hat{\mathbf{y}}^{(p)}$ represent the neural network outputs given the corresponding inputs $\mathbf{x}^{(p)}$ and weights $\mathbf{W}$. The diagonal matrix $\boldsymbol{\Omega}$ of fixed "error weights" has the form

$$\boldsymbol{\Omega} = \begin{pmatrix} \omega_1 & & 0 \\ & \ddots & \\ 0 & & \omega_{n_y} \end{pmatrix},$$

where ω_i are usually taken to be inversely proportional to noise variances.

We need to minimize the total approximation error $\bar{E}$ with respect to the neural network parameters $\mathbf{W}$. If activation functions of all the neurons are smooth, then the error function is also smooth. Hence, the minimization can be carried out using any of the optimization methods described in Section 2.2.1. However, in order to apply those methods, we need an efficient algorithm to compute the gradient and Hessian of the error function with respect to the parameters. As mentioned above, the total error gradient $\nabla \bar{E}$ and Hessian $\nabla^2 \bar{E}$ may be expressed in terms of the individual error gradients $\nabla E^{(p)}$ and Hessians $\nabla^2 E^{(p)}$. Thus, all that remains is to compute the derivatives of $E^{(p)}$. For notational convenience, in the remainder of this section we omit the training example index p.

There exist several approaches to computation of error function derivatives:

- numeric differentiation;
- symbolic differentiation;
- automatic (or algorithmic) differentiation.

The *numeric differentiation* approach relies on the derivative definition and approximates it via finite differences. This method is very simple to implement, but it suffers from truncation and roundoff errors. It is especially inaccurate for higher-order derivatives. Also, it requires many function evaluations: for example, in order to estimate the error function gradient with respect to n_w parameters using the simplest forward difference scheme we require error function values at $n_w + 1$ points.

Symbolic differentiation transforms a symbolic expression for the original function (usually represented in the form of a computational graph) into symbolic expressions for its derivatives by applying a chain rule. The resulting expressions may be evaluated at any point accurately to working precision. However, these expressions usually end up to have many identical subexpressions, which leads to duplicate computations (especially in the case we need the derivatives with respect to multiple parameters). In order to avoid this, we need to simplify the expressions for derivatives, which presents a nontrivial problem.

The *automatic differentiation* technique [64] computes function derivatives at a point by applying the chain rule to the corresponding numerical values instead of symbolic expressions. This method produces accurate derivative values, just like the symbolic differentiation, and also allows for a certain performance optimization. Note that automatic differentiation relies on the original computational graph for the function to be differentiated. Thus, if the original graph makes use of some common intermediate values, they will be efficiently reused by the differentiation procedure. Automatic differentiation is especially useful for neural network training, since it scales well to multiple parameters as well as higher-order derivatives. In this book, we adopt the automatic differentiation approach.

Automatic differentiation encompasses two different modes of computation: forward and reverse. *Forward mode* computes sensitivities of all variables with respect to input variables: it starts with the intermediate variables that explicitly depend on the input variables (most deeply nested subexpressions) and proceeds "forward" by applying the chain rule, until the output variables are processed. *Reverse mode* computes sensitivities of output variables with respect to all variables: it starts with the intermediate variables on which the output variables explicitly depend (outermost subexpressions) and proceeds "in reverse" by applying the chain rule, until the input variables are processed. Each mode has its own advantages and disadvantages. The forward mode allows to compute function values as well as its derivatives of multiple orders in a single pass. On the other hand, in order to compute the rth-order derivative using the reverse mode, one needs the derivatives of all the lower orders $s = 0, \dots, r-1$ beforehand. Computational complexity of first-order derivatives computation in the forward mode is proportional to the number of inputs, while in the reverse mode it is proportional to the number of outputs. In our case, there is only one output (the scalar error) and multiple inputs; therefore reverse mode is significantly faster than the forward mode. As shown in [65], under realistic assumptions the error function gradient can be computed in reverse mode at a cost of five function evaluations or less. Also note that in the ANN field the forward and reverse computation modes are usually referred to as forward propagation and backward propagation (or backpropagation).

In the rest of this subsection we present automatic differentiation algorithms for the computation of gradient, Jacobian, and Hessian of the squared error function (2.58) in the case of a layered feedforward neural network (2.8). All these algorithms rely on the fact that the derivatives of activation functions are known. For example, the derivatives of hyperbolic tangent activation functions (2.9) are

$$\left.\begin{aligned} \varphi_i^{l\prime}(\mathbf{n}_i^l) &= 1-\left(\varphi_i^l(\mathbf{n}_i^l)\right)^2 \\ \varphi_i^{l\prime\prime}(\mathbf{n}_i^l) &= -2\varphi_i^l(\mathbf{n}_i^l)\varphi_i^{l\prime}(\mathbf{n}_i^l) \end{aligned}\right\} \quad \begin{aligned} l&=1,\dots,L-1, \\ i&=1,\dots,S^l, \end{aligned} \tag{2.59}$$

while the derivatives of a logistic function (2.10) equal

$$\left.\begin{aligned} \varphi_i^{l\prime}(\mathbf{n}_i^l) &= \varphi_i^l(\mathbf{n}_i^l)\left(1-\varphi_i^l(\mathbf{n}_i^l)\right) \\ \varphi_i^{l\prime\prime}(\mathbf{n}_i^l) &= \varphi_i^{l\prime}(\mathbf{n}_i^l)\left(1-2\varphi_i^l(\mathbf{n}_i^l)\right) \end{aligned}\right\} \quad \begin{aligned} l&=1,\dots,L-1, \\ i&=1,\dots,S^l. \end{aligned} \tag{2.60}$$

Derivatives of the identity activation functions (2.11) are simply

$$\left.\begin{aligned} \varphi_i^{L\prime}(\mathbf{n}_i^L) &= 1 \\ \varphi_i^{L\prime\prime}(\mathbf{n}_i^L) &= 0 \end{aligned}\right\} \quad i=1,\dots,S^L. \tag{2.61}$$

Backpropagation algorithm for error function gradient. First, we perform a forward pass to compute the weighted sums $\mathbf{n}_i^l$ and activations $\mathbf{a}_i^l$ for all neurons $i=1,\dots,S^l$ of each layer $l=1,\dots,L$, according to equations (2.8).

We define the error function sensitivities with respect to weighted sums $\mathbf{n}_i^l$ to be as follows:

$$\boldsymbol{\delta}_i^l \triangleq \frac{\partial E}{\partial \mathbf{n}_i^l}. \tag{2.62}$$

Sensitivities for the output layer neurons are obtained directly, i.e.,

$$\boldsymbol{\delta}_i^L = -\omega_i\left(\tilde{\mathbf{y}}_i - \mathbf{a}_i^L\right)\varphi_i^{L\prime}(\mathbf{n}_i^L), \tag{2.63}$$

while sensitivities for the hidden layer neurons are computed during a backward pass:

$$\boldsymbol{\delta}_i^l = \varphi_i^{l\prime}(\mathbf{n}_i^l)\sum_{j=1}^{S^{l+1}}\boldsymbol{\delta}_j^{l+1}\mathbf{w}_{j,i}^{l+1}, \quad l=L-1,\dots,1. \tag{2.64}$$

Finally, the error function derivatives with respect to parameters are expressed in terms of sensitivities, i.e.,

$$\begin{aligned} \frac{\partial E}{\partial \mathbf{b}_i^l} &= \boldsymbol{\delta}_i^l, \\ \frac{\partial E}{\partial \mathbf{w}_{i,j}^l} &= \boldsymbol{\delta}_i^l\mathbf{a}_j^{l-1}. \end{aligned} \tag{2.65}$$

In a similar manner, we can compute the derivatives with respect to network inputs, i.e.,

$$\frac{\partial E}{\partial \mathbf{a}_i^0} = \sum_{j=1}^{S^1}\boldsymbol{\delta}_j^1\mathbf{w}_{j,i}^1. \tag{2.66}$$

Forward propagation for network outputs Jacobian. We define the pairwise sensitivities of weighted sums to be as follows:

$$\boldsymbol{\nu}_{i,j}^{l,m} \triangleq \frac{\partial \mathbf{n}_i^l}{\partial \mathbf{n}_j^m}. \tag{2.67}$$

Pairwise sensitivities for neurons of the same layer are obtained directly, i.e.,

$$\begin{aligned} \boldsymbol{\nu}_{i,i}^{l,l} &= 1, \\ \boldsymbol{\nu}_{i,j}^{l,l} &= 0, \quad i \neq j. \end{aligned} \tag{2.68}$$

Since the activations of neurons of some layer m do not affect the neurons of preceding layers $l < m$, the corresponding pairwise sensitivities are identically zero, i.e.,

$$\boldsymbol{\nu}_{i,j}^{l,m} = 0, \quad m > l. \tag{2.69}$$

The remaining pairwise sensitivities are computed during the forward pass, along with the weighted sums $\mathbf{n}_i^l$ and activations $\mathbf{a}_i^l$, i.e.,

$$\boldsymbol{\nu}_{i,j}^{l,m} = \sum_{k=1}^{S^{l-1}} \mathbf{w}_{i,k}^l \boldsymbol{\varphi}_k^{l-1\prime}(\mathbf{n}_k^{l-1}) \boldsymbol{\nu}_{k,j}^{l-1,m}, \quad l = 2, \ldots, L. \tag{2.70}$$

Finally, the derivatives of neural network outputs with respect to parameters are expressed in terms of pairwise sensitivities, i.e.,

$$\begin{aligned} \frac{\partial \mathbf{a}_i^L}{\partial \mathbf{b}_j^m} &= \boldsymbol{\varphi}_i^{L\prime}(\mathbf{n}_i^L) \boldsymbol{\nu}_{i,j}^{L,m}, \\ \frac{\partial \mathbf{a}_i^L}{\partial \mathbf{w}_{j,k}^m} &= \boldsymbol{\varphi}_i^{L\prime}(\mathbf{n}_i^L) \boldsymbol{\nu}_{i,j}^{L,m} \mathbf{a}_k^{m-1}. \end{aligned} \tag{2.71}$$

If we additionally define the sensitivities of weighted sums with respect to network inputs,

$$\boldsymbol{\nu}_{i,j}^{l,0} \triangleq \frac{\partial \mathbf{n}_i^l}{\partial \mathbf{a}_j^0}, \tag{2.72}$$

then we obtain the derivatives of network outputs with respect to network inputs. First, we compute the additional sensitivities during the forward pass, i.e.,

$$\begin{aligned} \boldsymbol{\nu}_{i,j}^{1,0} &= \mathbf{w}_{i,j}^1, \\ \boldsymbol{\nu}_{i,j}^{l,0} &= \sum_{k=1}^{S^{l-1}} \mathbf{w}_{i,k}^l \boldsymbol{\varphi}_k^{l-1\prime}(\mathbf{n}_k^{l-1}) \boldsymbol{\nu}_{k,j}^{l-1,0}, \quad l = 2, \ldots, L. \end{aligned} \tag{2.73}$$

Then, the derivatives of network outputs with respect to network inputs are expressed in terms of additional sensitivities, i.e.,

$$\frac{\partial \mathbf{a}_i^L}{\partial \mathbf{a}_j^0} = \boldsymbol{\varphi}_i^{L\prime}(\mathbf{n}_i^L) \boldsymbol{\nu}_{i,j}^{L,0}. \tag{2.74}$$

Backpropagation algorithm for error gradient and Hessian [66]. First, we perform a forward pass to compute the weighted sums $\mathbf{n}_i^l$ and activations $\mathbf{a}_i^l$ according to Eqs. (2.8), and also to compute the pairwise sensitivities $\boldsymbol{\nu}_{i,j}^{l,m}$ according to (2.68)–(2.70).

We define the error function second-order sensitivities with respect to weighted sums to be as follows:

$$\boldsymbol{\delta}_{i,j}^{l,m} \triangleq \frac{\partial^2 E}{\partial \mathbf{n}_i^l \partial \mathbf{n}_j^m}. \tag{2.75}$$

Next, during a backward pass we compute the error function sensitivities $\boldsymbol{\delta}_i^l$ as well as the second-order sensitivities $\boldsymbol{\delta}_{i,j}^{l,m}$. According to Schwarz's theorem on equality of mixed partials, due to continuity of second partial derivatives of an error function with respect to weighted sums, we have $\boldsymbol{\delta}_{i,j}^{l,m} = \boldsymbol{\delta}_{j,i}^{m,l}$. Hence, we need to compute the second-order sensitivities only for the case $m \leqslant l$.

Second-order sensitivities for the output layer neurons are obtained directly, i.e.,

$$\boldsymbol{\delta}_{i,j}^{L,m} = \omega_i \left[\left(\boldsymbol{\varphi}_i^{L\prime}(\mathbf{n}_i^L) \right)^2 - \left(\tilde{\mathbf{y}}_i - \mathbf{a}_i^L \right) \boldsymbol{\varphi}_i^{L\prime\prime}(\mathbf{n}_i^L) \right] \boldsymbol{\nu}_{i,j}^{L,m}, \tag{2.76}$$

while second-order sensitivities for the hidden layer neurons are computed during a backward pass, i.e.,

$$\begin{aligned} \boldsymbol{\delta}_{i,j}^{l,m} &= \boldsymbol{\varphi}_i^{l\prime}(\mathbf{n}_i^l) \sum_{k=1}^{S^{l+1}} \mathbf{w}_{k,i}^{l+1} \boldsymbol{\delta}_{k,j}^{l+1,m} \\ &\quad + \boldsymbol{\varphi}_i^{l\prime\prime}(\mathbf{n}_i^l) \boldsymbol{\nu}_{i,j}^{l,m} \sum_{k=1}^{S^{l+1}} \mathbf{w}_{k,i}^{l+1} \boldsymbol{\delta}_k^{l+1}, \\ l &= L-1, \ldots, 1. \end{aligned} \tag{2.77}$$

Due to continuity of second partial derivatives of an error function with respect to network parameters, the Hessian matrix is symmetric. Therefore, we need to compute only the lower-triangular part of the Hessian matrix. The error function second derivatives with respect to parameters are expressed in terms of second-order sensitivities. We have

$$\begin{aligned}
\frac{\partial^2 E}{\partial \mathbf{b}_i^l \partial \mathbf{b}_k^m} &= \boldsymbol{\delta}_{i,k}^{l,m}, \\
\frac{\partial^2 E}{\partial \mathbf{b}_i^l \partial \mathbf{w}_{k,r}^m} &= \boldsymbol{\delta}_{i,k}^{l,m} \mathbf{a}_r^{m-1}, \\
\frac{\partial^2 E}{\partial \mathbf{w}_{i,j}^l \partial \mathbf{b}_k^m} &= \boldsymbol{\delta}_{i,k}^{l,m} \mathbf{a}_j^{l-1} + \boldsymbol{\delta}_i^l \varphi_j^{l-1\prime}(\mathbf{n}_j^{l-1}) \boldsymbol{\nu}_{j,k}^{l-1,m}, \\
& \quad l > 1, \\
\frac{\partial^2 E}{\partial \mathbf{w}_{i,j}^1 \partial \mathbf{b}_k^1} &= \boldsymbol{\delta}_{i,k}^{1,1} \mathbf{a}_j^0, \\
\frac{\partial^2 E}{\partial \mathbf{w}_{i,j}^l \partial \mathbf{w}_{k,r}^m} &= \boldsymbol{\delta}_{i,k}^{l,m} \mathbf{a}_j^{l-1} \mathbf{a}_r^{m-1} \\
& \quad + \boldsymbol{\delta}_i^l \varphi_j^{l-1\prime}(\mathbf{n}_j^{l-1}) \boldsymbol{\nu}_{j,k}^{l-1,m} \mathbf{a}_r^{m-1}, \quad l > 1, \\
\frac{\partial^2 E}{\partial \mathbf{w}_{i,j}^1 \partial \mathbf{w}_{k,r}^1} &= \boldsymbol{\delta}_{i,k}^{1,1} \mathbf{a}_j^0 \mathbf{a}_r^0.
\end{aligned} \tag{2.78}$$

If we additionally define the second-order sensitivities of the error function with respect to network inputs,

$$\boldsymbol{\delta}_{i,j}^{l,0} \triangleq \frac{\partial^2 E}{\partial \mathbf{n}_i^l \partial \mathbf{a}_j^0}, \tag{2.79}$$

then we obtain error function second derivatives with respect to network inputs. First, we compute the additional second-order sensitivities during the backward pass, i.e.,

$$\begin{aligned}
\boldsymbol{\delta}_{i,j}^{L,0} &= \omega_i \left[\left(\varphi_i^{L\prime}(\mathbf{n}_i^L) \right)^2 - \left(\tilde{\mathbf{y}}_i - \mathbf{a}_i^L \right) \varphi_i^{L\prime\prime}(\mathbf{n}_i^L) \right] \boldsymbol{\nu}_{i,j}^{L,0}, \\
\boldsymbol{\delta}_{i,j}^{l,0} &= \varphi_i^{l\prime}(\mathbf{n}_i^l) \sum_{k=1}^{S^{l+1}} \mathbf{w}_{k,i}^{l+1} \boldsymbol{\delta}_{k,j}^{l+1,0} \\
& \quad + \varphi_i^{l\prime\prime}(\mathbf{n}_i^l) \boldsymbol{\nu}_{i,j}^{l,0} \sum_{k=1}^{S^{l+1}} \mathbf{w}_{k,i}^{l+1} \boldsymbol{\delta}_k^{l+1}, \quad l = L-1, \ldots, 1.
\end{aligned} \tag{2.80}$$

Then, the second derivatives of the error function with respect to network inputs are expressed in terms of additional second-order sensitivities, i.e.,

$$\begin{aligned}
\frac{\partial^2 E}{\partial \mathbf{a}_i^0 \partial \mathbf{a}_j^0} &= \sum_{k=1}^{S^1} \mathbf{w}_{k,i}^1 \boldsymbol{\delta}_{k,j}^{1,0}, \\
\frac{\partial^2 E}{\partial \mathbf{b}_i^l \partial \mathbf{a}_k^0} &= \boldsymbol{\delta}_{i,k}^{l,0}, \\
\frac{\partial^2 E}{\partial \mathbf{w}_{i,j}^l \partial \mathbf{a}_k^0} &= \boldsymbol{\delta}_{i,k}^{l,0} \mathbf{a}_j^{l-1} + \boldsymbol{\delta}_i^l \varphi_j^{l-1\prime}(\mathbf{n}_j^{l-1}) \boldsymbol{\nu}_{j,k}^{l-1,0}, \quad l > 1, \\
\frac{\partial^2 E}{\partial \mathbf{w}_{i,j}^1 \partial \mathbf{a}_j^0} &= \boldsymbol{\delta}_{i,j}^{1,0} \mathbf{a}_j^0 + \boldsymbol{\delta}_i^1, \\
\frac{\partial^2 E}{\partial \mathbf{w}_{i,j}^1 \partial \mathbf{a}_k^0} &= \boldsymbol{\delta}_{i,k}^{1,0} \mathbf{a}_j^0, \quad j \neq k.
\end{aligned} \tag{2.81}$$

2.2.3 Dynamic Neural Network Training

Traditional dynamic neural networks, such as the NARX and Elman networks, represent controlled discrete time dynamical systems. Thus, it is natural to utilize them as models for discrete time dynamical systems. However, they can also be used as models for the continuous time dynamical systems under the assumption of a uniform time step Δt. In this book we focus on the latter problem. That is, we wish to train the dynamic neural network so that it can perform accurate closed-loop multistep-ahead prediction of the dynamical system behavior. In this subsection, we discuss the most general state space form (2.13) of dynamic neural networks.

Assume we are given an experimental data set of the form

$$\left\{\left\{\left\langle \mathbf{u}^{(p)}(t_k), \tilde{\mathbf{y}}^{(p)}(t_k)\right\rangle\right\}_{k=0}^{K^{(p)}}\right\}_{p=1}^{P}, \tag{2.82}$$

where P is the total number of trajectories, $K^{(p)}$ is the number of time steps for the corresponding trajectory, $t_k = k\Delta t$ are the discrete time instants, $\mathbf{u}^{(p)}(t_k)$ are the control inputs, and $\tilde{\mathbf{y}}^{(p)}(t_k)$ are the observed outputs. We will also denote the total duration of the pth trajectory by $\bar{t}^{(p)} = K^{(p)}\Delta t$.

Note that in general the observed outputs $\tilde{\mathbf{y}}^{(p)}(t_k)$ do not match the true outputs $\mathbf{y}^{(p)}(t_k)$. We assume that the observations are corrupted by an additive white Gaussian noise, i.e.,

$$\tilde{\mathbf{y}}^{(p)}(t) = \mathbf{y}^{(p)}(t) + \boldsymbol{\eta}^{(p)}(t). \tag{2.83}$$

That is, $\boldsymbol{\eta}^{(p)}(t)$ represents a stationary Gaussian process with zero mean and a covariance function $K_\eta(t_1, t_2) = \delta(t_2 - t_1)\Sigma$, where

$$\boldsymbol{\Sigma} = \begin{pmatrix} \sigma_1^2 & & 0 \\ & \ddots & \\ 0 & & \sigma_{n_y}^2 \end{pmatrix}.$$

The individual errors $E^{(p)}$ for each trajectory have the following form:

$$E^{(p)}(\mathbf{W}) = \sum_{k=1}^{K^{(p)}} e(\tilde{\mathbf{y}}^{(p)}(t_k), \mathbf{z}^{(p)}(t_k), \mathbf{W}), \tag{2.84}$$

where $\mathbf{z}^{(p)}(t_k)$ are the model states and $e^{(p)}: \mathbb{R}^{n_y} \times \mathbb{R}^{n_z} \times \mathbb{R}^{n_w} \to \mathbb{R}$ represents the model prediction error at time instant t_k. Under the abovementioned assumptions on the observation noise, it is reasonable to utilize the instantaneous error function e of the following form:

$$e(\tilde{\mathbf{y}}, \mathbf{z}, \mathbf{W}) = \frac{1}{2}\left(\tilde{\mathbf{y}} - \mathbf{G}(\mathbf{z}, \mathbf{W})\right)^T \boldsymbol{\Omega} \left(\tilde{\mathbf{y}} - \mathbf{G}(\mathbf{z}, \mathbf{W})\right), \tag{2.85}$$

where $\boldsymbol{\Omega} = \mathrm{diag}(\omega_1, \ldots, \omega_{n_y})$ is the diagonal matrix of error weights, usually taken inversely proportional to the corresponding variances of measurement noise.

We need to minimize the total prediction error $\bar{E}$ with respect to the neural network parameters $\mathbf{W}$. Again, the minimization can be carried out using any of the optimization methods described in Section 2.2.1, provided we can compute the gradient and Hessian of the error function with respect to the parameters. Just like in the case of static neural networks, the total error gradient $\nabla\bar{E}$ and Hessian $\nabla^2\bar{E}$ may be expressed in terms of the individual error gradients $\nabla E^{(p)}$ and Hessians $\nabla^2 E^{(p)}$. Thus, we describe the algorithms for computation of the derivatives for $E^{(p)}$ and omit the trajectory index p.

Again, we have two different computation modes: forward-in-time and reverse-in-time, each with its own advantages and disadvantages. The forward-in-time approach theoretically allows one to work with infinite duration trajectories, i.e., to perform online adaptation as the new data arrive. In practice, however, each iteration is more computationally expensive as compared to the reverse-in-time approach. The reverse-in-time approach is only applicable when the whole training set is available beforehand, but it works significantly faster.

BackPropagation through time algorithm (BPTT) [67–69] for error function gradient. First, we perform a forward pass to compute the predicted states $\mathbf{z}(t_k)$ for all time steps t_k, $k = 1, \ldots, K$, according to equations (2.13). We also compute the error $E(\mathbf{W})$ according to (2.84) and (2.85).

We define the error function sensitivities with respect to model states at time step t_k to be as follows:

$$\boldsymbol{\lambda}(t_k) = \frac{\partial E(\mathbf{W})}{\partial \mathbf{z}(t_k)}. \tag{2.86}$$

Error function sensitivities are computed during a backward-in-time pass, i.e.,

$$\begin{aligned}
\boldsymbol{\lambda}(t_{K+1}) &= \mathbf{0}, \\
\boldsymbol{\lambda}(t_k) &= \frac{\partial e(\tilde{\mathbf{y}}(t_k), \mathbf{z}(t_k), \mathbf{W})}{\partial \mathbf{z}} \\
&\quad + \frac{\partial \mathbf{F}(\mathbf{z}(t_k), \mathbf{u}(t_k), \mathbf{W})}{\partial \mathbf{z}}^T \boldsymbol{\lambda}(t_{k+1}), \\
&\quad k = K, \ldots, 1.
\end{aligned} \tag{2.87}$$

Finally, the error function derivatives with respect to parameters are expressed in terms of sensitivities, i.e.,

$$\begin{aligned}
\frac{\partial E(\mathbf{W})}{\partial \mathbf{W}} = \sum_{k=1}^{K} & \frac{\partial e(\tilde{\mathbf{y}}(t_k), \mathbf{z}(t_k), \mathbf{W})}{\partial \mathbf{W}} \\
& + \frac{\partial \mathbf{F}(\mathbf{z}(t_{k-1}), \mathbf{u}(t_{k-1}), \mathbf{W})}{\partial \mathbf{W}}^T \boldsymbol{\lambda}(t_k).
\end{aligned} \tag{2.88}$$

First-order derivatives of the instantaneous error function (2.85) have the form

$$\begin{aligned}
\frac{\partial e(\tilde{\mathbf{y}}, \mathbf{z}, \mathbf{W})}{\partial \mathbf{W}} &= -\frac{\partial \mathbf{G}(\mathbf{z}, \mathbf{W})}{\partial \mathbf{W}}^T \boldsymbol{\Omega} \left(\tilde{\mathbf{y}} - \mathbf{G}(\mathbf{z}, \mathbf{W})\right), \\
\frac{\partial e(\tilde{\mathbf{y}}, \mathbf{z}, \mathbf{W})}{\partial \mathbf{z}} &= -\frac{\partial \mathbf{G}(\mathbf{z}, \mathbf{W})}{\partial \mathbf{z}}^T \boldsymbol{\Omega} \left(\tilde{\mathbf{y}} - \mathbf{G}(\mathbf{z}, \mathbf{W})\right).
\end{aligned} \tag{2.89}$$

Sine the mappings **F** and **G** are represented by layered feedforward neural networks, their derivatives can be computed as described in Section 2.2.2.

Real-Time Recurrent Learning algorithm (RTRL) [68–70] for network outputs Jacobian. The model state sensitivities with respect to network parameters are computed during the forward-in-time pass, along with the states themselves. We have

$$\begin{aligned}
\frac{\partial \mathbf{z}(t_0)}{\partial \mathbf{W}} &= \mathbf{0}, \\
\frac{\partial \mathbf{z}(t_k)}{\partial \mathbf{W}} &= \frac{\partial \mathbf{F}(\mathbf{z}(t_{k-1}), \mathbf{u}(t_{k-1}), \mathbf{W})}{\partial \mathbf{W}} \\
&\quad + \frac{\partial \mathbf{F}(\mathbf{z}(t_{k-1}), \mathbf{u}(t_{k-1}), \mathbf{W})}{\partial \mathbf{z}} \frac{\partial \mathbf{z}(t_{k-1})}{\partial \mathbf{W}}, \\
&\quad k = 1, \ldots, K.
\end{aligned} \tag{2.90}$$

The gradient of the individual trajectory error function (2.84) equals

$$\begin{aligned}
\frac{\partial E(\mathbf{W})}{\partial \mathbf{W}} = \sum_{k=1}^{K} & \frac{\partial e(\tilde{\mathbf{y}}(t_k), \mathbf{z}(t_k), \mathbf{W})}{\partial \mathbf{W}} \\
& + \frac{\partial \mathbf{z}(t_k)}{\partial \mathbf{W}}^T \frac{\partial e(\tilde{\mathbf{y}}(t_k), \mathbf{z}(t_k), \mathbf{W})}{\partial \mathbf{z}}.
\end{aligned} \tag{2.91}$$

A Gauss–Newton Hessian approximation may be obtained as follows:

$$\begin{aligned}
\frac{\partial^2 E(\mathbf{W})}{\partial \mathbf{W}^2} \approx \sum_{k=1}^{K} & \frac{\partial^2 e(\tilde{\mathbf{y}}(t_k), \mathbf{z}(t_k), \mathbf{W})}{\partial \mathbf{W}^2} \\
& + \frac{\partial^2 e(\tilde{\mathbf{y}}(t_k), \mathbf{z}(t_k), \mathbf{W})}{\partial \mathbf{W} \partial \mathbf{z}} \frac{\partial \mathbf{z}(t_k)}{\partial \mathbf{W}} \\
& + \frac{\partial \mathbf{z}(t_k)}{\partial \mathbf{W}}^T \frac{\partial^2 e(\tilde{\mathbf{y}}(t), \mathbf{z}(t_k), \mathbf{W})}{\partial \mathbf{z} \partial \mathbf{W}} \\
& + \frac{\partial \mathbf{z}(t_k)}{\partial \mathbf{W}}^T \frac{\partial^2 e(\tilde{\mathbf{y}}(t_k), \mathbf{z}(t_k), \mathbf{W})}{\partial \mathbf{z}^2} \frac{\partial \mathbf{z}(t_k)}{\partial \mathbf{W}}.
\end{aligned} \tag{2.92}$$

The corresponding approximations to second-order derivatives of the instantaneous error function have the form

$$\begin{aligned}
\frac{\partial^2 e(\tilde{\mathbf{y}}, \mathbf{z}, \mathbf{W})}{\partial \mathbf{W}^2} &\approx \frac{\partial \mathbf{G}(\mathbf{z}, \mathbf{W})}{\partial \mathbf{W}}^T \boldsymbol{\Omega} \frac{\partial \mathbf{G}(\mathbf{z}, \mathbf{W})}{\partial \mathbf{W}}, \\
\frac{\partial^2 e(\tilde{\mathbf{y}}, \mathbf{z}, \mathbf{W})}{\partial \mathbf{W} \partial \mathbf{z}} &\approx \frac{\partial \mathbf{G}(\mathbf{z}, \mathbf{W})}{\partial \mathbf{W}}^T \boldsymbol{\Omega} \frac{\partial \mathbf{G}(\mathbf{z}, \mathbf{W})}{\partial \mathbf{z}}, \\
\frac{\partial^2 e(\tilde{\mathbf{y}}, \mathbf{z}, \mathbf{W})}{\partial \mathbf{z}^2} &\approx \frac{\partial \mathbf{G}(\mathbf{z}, \mathbf{W})}{\partial \mathbf{z}}^T \boldsymbol{\Omega} \frac{\partial \mathbf{G}(\mathbf{z}, \mathbf{W})}{\partial \mathbf{z}}.
\end{aligned} \tag{2.93}$$

Backpropagation through time algorithm for error gradient and Hessian. Second-order sensitivities of the error function are computed during a backward-in-time pass as follows:

$$\begin{aligned}
\frac{\partial \boldsymbol{\lambda}(t_{K+1})}{\partial \mathbf{W}} &= \mathbf{0},\\
\frac{\partial \boldsymbol{\lambda}(t_k)}{\partial \mathbf{W}} &= \frac{\partial^2 e(\tilde{\mathbf{y}}(t_k), \mathbf{z}(t_k), \mathbf{W})}{\partial \mathbf{z} \partial \mathbf{W}} \\
&+ \frac{\partial^2 e(\tilde{\mathbf{y}}(t_k), \mathbf{z}(t_k), \mathbf{W})}{\partial \mathbf{z}^2} \frac{\partial \mathbf{z}(t_k)}{\partial \mathbf{W}} \\
&+ \sum_{i=1}^{n_z} \lambda_i(t_{k+1}) \left[\frac{\partial^2 \mathbf{F}_i(\mathbf{z}(t_k), \mathbf{u}(t_k), \mathbf{W})}{\partial \mathbf{z} \partial \mathbf{W}} \right. \\
&\left. + \frac{\partial^2 \mathbf{F}_i(\mathbf{z}(t_k), \mathbf{u}(t_k), \mathbf{W})}{\partial \mathbf{z}^2} \frac{\partial \mathbf{z}(t_k)}{\partial \mathbf{W}} \right] \\
&+ \frac{\partial \mathbf{F}(\mathbf{z}(t_k), \mathbf{u}(t_k), \mathbf{W})}{\partial \mathbf{z}}^T \frac{\partial \boldsymbol{\lambda}(t_{k+1})}{\partial \mathbf{W}}, \\
& k = K, \dots, 1.
\end{aligned} \tag{2.94}$$

The Hessian of the individual trajectory error function (2.84) equals

$$\begin{aligned}
\frac{\partial^2 E(\mathbf{W})}{\partial \mathbf{W}^2} &= \sum_{k=1}^{K} \frac{\partial^2 e(\tilde{\mathbf{y}}(t_k), \mathbf{z}(t_k), \mathbf{W})}{\partial \mathbf{W}^2} \\
&+ \frac{\partial^2 e(\tilde{\mathbf{y}}(t_k), \mathbf{z}(t_k), \mathbf{W})}{\partial \mathbf{W} \partial \mathbf{z}} \frac{\partial \mathbf{z}(t_k)}{\partial \mathbf{W}} \\
&+ \sum_{i=1}^{n_z} \lambda_i(t_k) \left[\frac{\partial^2 \mathbf{F}_i(\mathbf{z}(t_{k-1}), \mathbf{u}(t_{k-1}), \mathbf{W})}{\partial \mathbf{W}^2} \right. \\
&\left. + \frac{\partial^2 \mathbf{F}_i(\mathbf{z}(t_{k-1}), \mathbf{u}(t_{k-1}), \mathbf{W})}{\partial \mathbf{W} \partial \mathbf{z}} \frac{\partial \mathbf{z}(t_{k-1})}{\partial \mathbf{W}} \right] \\
&+ \frac{\partial \mathbf{F}(\mathbf{z}(t_{k-1}), \mathbf{u}(t_{k-1}), \mathbf{W})}{\partial \mathbf{W}}^T \frac{\partial \boldsymbol{\lambda}(t_k)}{\partial \mathbf{W}}.
\end{aligned} \tag{2.95}$$

Second-order derivatives of the instantaneous error function (2.85) have the form

$$\begin{aligned}
\frac{\partial^2 e(\tilde{\mathbf{y}}, \mathbf{z}, \mathbf{W})}{\partial \mathbf{W}^2} &= \frac{\partial \mathbf{G}(\mathbf{z}, \mathbf{W})}{\partial \mathbf{W}}^T \boldsymbol{\Omega} \frac{\partial \mathbf{G}(\mathbf{z}, \mathbf{W})}{\partial \mathbf{W}} \\
&- \sum_{i=1}^{n_y} \frac{\partial^2 \mathbf{G}_i(\mathbf{z}, \mathbf{W})}{\partial \mathbf{W}^2} \omega_i \left(\tilde{\mathbf{y}}_i - \mathbf{G}_i(\mathbf{z}, \mathbf{W}) \right), \\
\frac{\partial^2 e(\tilde{\mathbf{y}}, \mathbf{z}, \mathbf{W})}{\partial \mathbf{W} \partial \mathbf{z}} &= \frac{\partial \mathbf{G}(\mathbf{z}, \mathbf{W})}{\partial \mathbf{W}}^T \boldsymbol{\Omega} \frac{\partial \mathbf{G}(\mathbf{z}, \mathbf{W})}{\partial \mathbf{z}} \\
&- \sum_{i=1}^{n_y} \frac{\partial^2 \mathbf{G}_i(\mathbf{z}, \mathbf{W})}{\partial \mathbf{W} \partial \mathbf{z}} \omega_i \left(\tilde{\mathbf{y}}_i - \mathbf{G}_i(\mathbf{z}, \mathbf{W}) \right), \\
\frac{\partial^2 e(\tilde{\mathbf{y}}, \mathbf{z}, \mathbf{W})}{\partial \mathbf{z}^2} &= \frac{\partial \mathbf{G}(\mathbf{z}, \mathbf{W})}{\partial \mathbf{z}}^T \boldsymbol{\Omega} \frac{\partial \mathbf{G}(\mathbf{z}, \mathbf{W})}{\partial \mathbf{z}} \\
&- \sum_{i=1}^{n_y} \frac{\partial^2 \mathbf{G}_i(\mathbf{z}, \mathbf{W})}{\partial \mathbf{z}^2} \omega_i \left(\tilde{\mathbf{y}}_i - \mathbf{G}_i(\mathbf{z}, \mathbf{W}) \right).
\end{aligned} \tag{2.96}$$

In the rest of this subsection, we discuss various difficulties associated with the recurrent neural network training problem. First, notice that a recurrent neural network which performs a K-step-ahead prediction may be "unfolded" in time to produce an equivalent layered feedforward neural network, comprised of K copies of the same subnetwork, one per time step. Each of these identical subnetworks shares a common set of parameters.

Given a large prediction horizon, the resulting feedforward network becomes very deep. Thus, it is natural that all the difficulties associated with deep neural network training are also inherent to recurrent neural network training. In fact, these problems become even more severe. They include the following:

1. Vanishing and exploding gradients [71–74]. Note that the sensitivity of a recurrent neural network (2.13) state at time step t_k with respect to its state at time step t_l ($l \ll k$) has the following form:

$$\frac{\partial \mathbf{z}(t_k)}{\partial \mathbf{z}(t_l)} = \prod_{r=l}^{k-1} \frac{\partial \mathbf{F}(\mathbf{z}(t_r), \mathbf{u}(t_r), \mathbf{W})}{\partial \mathbf{z}}. \tag{2.97}$$

If the largest (in absolute value) eigenvalues of $\frac{\partial \mathbf{F}(\mathbf{z}(t_r),\mathbf{u}(t_r),\mathbf{W})}{\partial \mathbf{z}}$ are less than 1 for all time steps t_r, $r = l, \ldots, k-1$, then the norm of sensitivity $\frac{\partial \mathbf{z}(t_k)}{\partial \mathbf{z}(t_l)}$ will decay exponentially with $k - l$. Hence, the terms of the error gradient which correspond to recent time steps will dominate the sum. This is the reason why gradient-based optimization methods learn short-term dependencies much faster than the long-term ones. On the other hand, a gradient explosion (exponential growth of its norm) corresponds to a situation when the eigenvalues exceed 1 at all time steps. The gradient explosion effect might lead to divergence of the optimization method, unless care is taken.

In particular, if the mapping $\mathbf{F}$ is represented by a layered feedforward neural network (2.8), then the Jacobian $\frac{\partial \mathbf{F}(\mathbf{z}(t_r),\mathbf{u}(t_r),\mathbf{W})}{\partial \mathbf{z}}$ corresponds to derivatives of network outputs with respect to its inputs, i.e.,

$$\frac{\partial \mathbf{a}^L}{\partial \mathbf{a}^0} = \operatorname{diag}\left\{\boldsymbol{\varphi}^{L'}(\mathbf{n}^L)\right\}\boldsymbol{\omega}^L \cdots \operatorname{diag}\left\{\boldsymbol{\varphi}^{1'}(\mathbf{n}^1)\right\}\boldsymbol{\omega}^1. \tag{2.98}$$

Assume that the derivatives of all the activation functions $\boldsymbol{\varphi}^l$ are bounded by some constant η^l. Denote by $\lambda^l_{\max}$ the eigenvalue with the largest magnitude for the weight matrix $\boldsymbol{\omega}^l$ of the lth layer. If the inequality $\prod_{l=1}^{L} \lambda^l_{\max}\eta^l < 1$ holds, then the largest (in magnitude) eigenvalue of a Jacobian matrix $\frac{\partial \mathbf{a}^L}{\partial \mathbf{a}^0}$ is less than one. Derivatives of the hyperbolic tangent activation function, as well as the identity activation function, are bounded by 1.

One of the possibilities to speed up the training is to use the second-order optimization methods [59,74]. Another option would be to utilize the Long-Short Term Memory (LSTM) models [72,75–80] specially designed to overcome the vanishing gradient effect by using the special memory cells instead of context neurons. LSTM networks have been successfully applied in speech recognition, machine translation, and anomaly detection. However, little attention has been attracted to applications of LSTM for dynamical system modeling problems [81].

2. Bifurcations of a recurrent neural network dynamics [82–84]. Since the recurrent neural network is a dynamical system itself, its phase portrait might undergo qualitative changes during the training. If these changes affect the actual predicted trajectories, this might lead to significant changes of the error in response to small changes of parameters (i.e., the gradient norm becomes very large), provided the duration of these trajectories is large enough.

   In order to guarantee a complete absence of bifurcations during the network training, we would need a very good initial guess for its parameters, so that the model would already possess the desired asymptotic behavior. Since this assumption is very unrealistic, it seems more reasonable to modify the optimization methods in order to enforce their stability.
3. Spurious valleys in the error surface [85–87]. These valleys are called spurious due to the fact that they do not depend on the desired values of outputs $\tilde{\mathbf{y}}(t_k)$. The location of these valleys is determined only by initial conditions $\mathbf{z}(t_0)$ and the controls $\mathbf{u}(t_k)$. Reasons for occurrence of such valleys have been investigated in some special cases. For example, if the initial state $\mathbf{z}(t_0)$ of (2.13) is a global repeller within some area of a parameter space, then an infinitesimal control $\mathbf{u}(t_k)$ causes the model states $\mathbf{z}(t_k)$ to tend to infinity, which in turn leads to an unbounded error growth. Now assume that this area of parameter space contains a line along which the connection weights between the controls $\mathbf{u}(t_k)$ and the neurons of $\mathbf{F}$ are identically zero, that is, the recurrent neural network (2.13) does not depend on controls. Parameters along this

line result in a stationary behavior of model states $\mathbf{z}(t_k) \equiv \mathbf{z}(t_0)$, and the corresponding prediction error remains relatively low. Such line represents a spurious valley in the error surface.

It is worth mentioning that this problem can be alleviated by the use of a large number of trajectories for the training. Since these trajectories have different initial conditions and different controls, the corresponding spurious valleys are also located in different areas of the parameter space. Hence, these valleys are smoothed out on a surface of a total error function (2.25). In addition, we might apply the regularization methods so as to modify the error function, which results in valleys "tilted" in some direction.

2.3 DYNAMIC NEURAL NETWORK ADAPTATION METHODS

2.3.1 Extended Kalman Filter

Another class of learning algorithms for dynamic networks can be built based on the concept of an extended Kalman filter.

The standard Kalman filter algorithm is designed to work with linear systems. Namely, the following model of the dynamical system in the state space is considered:

$$\begin{aligned}\mathbf{z}(t_{k+1}) &= \mathbf{F}^-(t_{k+1})\mathbf{z}(t_k) + \boldsymbol{\zeta}(t_k),\\ \tilde{\mathbf{y}}(t_k) &= \mathbf{H}(t_k)\mathbf{z}(t_k) + \boldsymbol{\eta}(t_k).\end{aligned}$$

Here $\boldsymbol{\zeta}(t_k)$ and $\boldsymbol{\eta}(t_k)$ are Gaussian noises with zero mean and covariance matrices $\mathbf{Q}(t_k)$ and $\mathbf{R}(t_k)$, respectively.

The algorithm is initialized as follows. For $k = 0$, set

$$\begin{aligned}\hat{\mathbf{z}}(t_0) &= E[\mathbf{z}(t_0)],\\ \mathbf{P}(t_0) &= E[(\mathbf{z}(t_0) - E[\mathbf{z}(t_0)])(\mathbf{z}(t_0) - E[\mathbf{z}(t_0)])^T].\end{aligned}$$

Then, for $k = 1, 2, \ldots$, the following values are calculated:

- state estimation

$$\hat{\mathbf{z}}^-(t_k) = \mathbf{F}^-(t_k)\hat{\mathbf{z}}(t_{k-1});$$

- estimation of the error covariance

$$\mathbf{P}^-(t_k) = \mathbf{F}^-(t_k)\mathbf{P}(t_{k-1})\mathbf{F}^-(t_k)^T + \mathbf{Q}(t_{k-1});$$

- the gain matrix

$$\begin{aligned}\mathbf{G}(t_k) &= \mathbf{P}^-(t_k)\mathbf{H}(t_k)^T\\ &\times \left[\mathbf{H}(t_k)\mathbf{P}^-(t_k)\mathbf{H}(t_k)^T + \mathbf{R}(t_k)\right]^{-1};\end{aligned}$$

- correction of the state estimation

$$\hat{\mathbf{z}}(t_k) = \hat{\mathbf{z}}^-(t_k) + \mathbf{G}(t_k)\left(\tilde{\mathbf{y}}(t_k) - \mathbf{H}(t_k)\hat{\mathbf{z}}^-(t_k)\right);$$

- correction of the error covariance estimation

$$\mathbf{P}(t_k) = (\mathbf{I} - \mathbf{G}(t_k)\mathbf{H}(t_k))\,\mathbf{P}^-(t_k).$$

However, the dynamic ANN model is a nonlinear system, so the standard Kalman filter algorithm is not suitable for them. If we use the linearization of the original nonlinear system, then in this case we can obtain an extended Kalman filter (EKF) suitable for nonlinear systems.

To obtain the EKF algorithm, the model in the state space is written in the following form:

$$\begin{aligned}\mathbf{z}(t_{k+1}) &= \mathbf{f}(t_k, \mathbf{z}(t_k)) + \boldsymbol{\zeta}(t_k),\\ \tilde{\mathbf{y}}(t_k) &= \mathbf{h}(t_k, \mathbf{z}(t_k)) + \boldsymbol{\eta}(t_k).\end{aligned}$$

Here $\boldsymbol{\zeta}(t_k)$ and $\boldsymbol{\eta}(t_k)$ are Gaussian noises with zero mean and covariance matrices $\mathbf{Q}(t_k)$ and $\mathbf{R}(t_k)$, respectively.

In this case

$$\mathbf{F}^-(t_{k+1}) = \left.\frac{\partial \mathbf{f}(t_k, \mathbf{z})}{\partial \mathbf{z}}\right|_{\mathbf{z}=\mathbf{z}(t_k)},$$

$$\mathbf{H}(t_k) = \left.\frac{\partial \mathbf{h}(t_k, \mathbf{z})}{\partial \mathbf{z}}\right|_{\mathbf{z}=\mathbf{z}(t_k)}.$$

The EKF algorithm is initialized as follows. For $k = 0$, set

$$\hat{\mathbf{z}}(t_0) = E[\mathbf{z}(t_0)],$$
$$\mathbf{P}(t_0) = E[(\mathbf{z}(t_0) - E[\mathbf{z}(t_0)])(\mathbf{z}(t_0) - E[\mathbf{z}(t_0)])^T].$$

Then, for $k = 1, 2, \ldots$, the following values are calculated:

- state estimation
$$\hat{\mathbf{z}}^-(t_k) = \mathbf{f}(t_k, \hat{\mathbf{z}}(t_{k-1}));$$
- estimation of the error covariance
$$\mathbf{P}^-(t_k) = \mathbf{F}^-(t_k)\mathbf{P}(t_{k-1})\mathbf{F}^-(t_k)^T + \mathbf{Q}(t_{k-1});$$
- the gain matrix
$$\mathbf{G}(t_k) = \mathbf{P}^-(t_k)\mathbf{H}(t_k)^T \times \left[\mathbf{H}(t_k)\mathbf{P}^-(t_k)\mathbf{H}(t_k)^T + \mathbf{R}(t_k)\right]^{-1};$$
- correction of the state estimation
$$\hat{\mathbf{z}}(t_k) = \hat{\mathbf{z}}^-(t_k) + \mathbf{G}(t_k)\left(\tilde{\mathbf{y}}(t_k) - \mathbf{h}(t_k, \hat{\mathbf{z}}^-(t_k))\right);$$
- correction of the error covariance estimation
$$\mathbf{P}(t_k) = (\mathbf{I} - \mathbf{G}(t_k)\mathbf{H}(t_k))\,\mathbf{P}^-(t_k).$$

We will assume that for an ideal ANN model the observed process is stationary, that is, $\mathbf{w}(t_{k+1}) = \mathbf{w}(t_k)$, but its states (weight $\mathbf{w}(t_k)$) are "corrupted" by the noise $\zeta(t_k)$.

The Kalman filter (KF) in its standard version is applicable only for systems whose observations are linear in the estimated parameters, while the neural network observation equation is nonlinear, i.e.,

$$\mathbf{w}(t_{k+1}) = \mathbf{w}(t_k) + \zeta(t_k),$$
$$\hat{\mathbf{y}}(t_k) = \mathbf{f}(\mathbf{u}(t_k), \mathbf{w}(t_k)) + \eta(t_k),$$

where $\mathbf{u}(t_k)$ are control actions, ζ is the process noise, and η is the observation noise; these noises are Gaussian random sequences with zero mean and covariance matrices $\mathbf{Q}$ and $\mathbf{R}$.

In order to use the Kalman filter, it is required to linearize the observation equation. It is possible to use statistical linearization, i.e., linearization with respect to the mathematical expectation. This gives

$$\mathbf{w}(t_{k+1}) = \mathbf{w}(t_k) + \zeta(t_k),$$
$$\hat{\mathbf{y}}(t_k) = \mathbf{H}(t_k)\mathbf{w}(t_k) + \eta(t_k),$$

where the observation matrix has the form

$$\mathbf{H}(t_k) = \left.\frac{\partial \hat{\mathbf{y}}}{\partial \mathbf{w}^T}\right|_{\substack{\mathbf{w}=\mathbf{w}(t_k)\\ \mathbf{z}=\mathbf{z}(t_k)}} = -\frac{\partial \mathbf{e}(t_k)}{\partial \mathbf{w}(t_k)^T} = -\mathbf{J}(t_k).$$

Here $\mathbf{e}(t_k)$ is the observation error vector at the kth estimation step, i.e.,

$$\mathbf{e}(t_k) = \tilde{\mathbf{y}}(t_k) - \hat{\mathbf{y}}(t_k) = \tilde{\mathbf{y}}(t_k) - \mathbf{f}(\mathbf{z}(t_k), \mathbf{w}(t_k)).$$

The extended Kalman filter equations for estimating $\mathbf{w}(t_{k+1})$ in the next step have the form

$$\mathbf{S}(t_k) = \mathbf{H}(t_k)\mathbf{P}(t_k)\mathbf{H}(t_k)^T + \mathbf{R}(t_k),$$
$$\mathbf{K}(t_k) = \mathbf{P}(t_k)\mathbf{H}(t_k)^T\mathbf{S}(t_k)^{-1},$$
$$\mathbf{P}(t_{k+1}) = (\mathbf{P}(t_k) - \mathbf{K}(t_k)\mathbf{H}(t_k)\mathbf{P}(t_k))\,e^{\beta} + \mathbf{Q}(t_k),$$
$$\mathbf{w}(t_{k+1}) = \mathbf{w}(t_k) + \mathbf{K}(t_k)\mathbf{e}(t_k)$$

Here β is the forgetting factor, which affects the significance of the previous steps.

Here $\mathbf{K}(t_k)$ is the Kalman gain, $\mathbf{S}(t_k)$ is the covariance matrix for state prediction errors $\mathbf{e}(t_k)$, and $\mathbf{P}(t_k)$ is the covariance matrix for weight estimation errors $(\hat{\mathbf{w}}(t_k) - \mathbf{w}(t_k))$.

There are alternative variants of the EKF algorithm, which may prove to be more effective in solving the problems under consideration, in particular,

$$\mathbf{P}^-(t_k) = \mathbf{P}(t_k) + \mathbf{Q}(t_k),$$
$$\mathbf{S}(t_k) = \mathbf{H}(t_k)\mathbf{P}^-(t_k)\mathbf{H}(t_k)^T + \mathbf{R}(t_k),$$
$$\mathbf{K}(t_k) = \mathbf{P}^-(t_k)\mathbf{H}(t_k)^T\mathbf{S}(t_k)^{-1},$$
$$\mathbf{P}(t_{k+1}) = (\mathbf{I} - \mathbf{K}(t_k)\mathbf{H}(t_k))\,\mathbf{P}^-(t_k) \times (\mathbf{I} - \mathbf{K}(t_k)\mathbf{H}(t_k))^T + \mathbf{K}(t_k)\mathbf{K}(t_k)^T,$$
$$\mathbf{w}(t_{k+1}) = \mathbf{w}(t_k) + \mathbf{K}(t_k)\mathbf{e}(t_k).$$

The variant of the EKF of this type is more stable in computational terms and has robustness to rounding errors, which positively affects the computational stability of the learning process of the ANN model as a whole.

As can be seen from the relationships determining the EKF, the key point is again the calculation of the Jacobian $\mathbf{J}(t_k)$ of network errors by adjusted parameters.

When learning a neural network, it is impossible to use only the current measurement in the EKF due to the unacceptably low accuracy of the search (the effect of the noise ζ and η); it is necessary to form a vector estimate on the observation interval, and then the update of the matrix $\mathbf{P}(t_k)$ is more correct.

As a vector of observations, we can take a sequence of values on a certain sliding interval, i.e.,

$$\hat{\mathbf{y}}(t_k) = \left[\hat{\mathbf{y}}(t_{i-l}), \hat{\mathbf{y}}(t_{i-l+1}), \ldots, \hat{\mathbf{y}}(t_i)\right]^T,$$

where l is the length of the sliding interval, the index i refers to the time point (sampling step), and the index k indicates the valuation number.

The error of the ANN model will also be a vector value, i.e.,

$$\mathbf{e}(t_k) = \left[\mathbf{e}(t_{i-l}), \mathbf{e}(t_{i-l+1}), \ldots, \mathbf{e}(t_i)\right]^T.$$

2.3.2 ANN Models With Interneurons

From the point of view of ensuring the adaptability of ANN models, the idea of an intermediate neuron (interneuron) and the subnetwork of such neurons (intersubnet) is very fruitful.

2.3.2.1 *The Concept of an Interneuron and an ANN Model With Such Neurons*

An effective approach to the implementation of adaptive ANN models, based on the concepts of an interneuron and a pretuned network, was proposed by A.I. Samarin [88]. As noted in this paper, one of the main properties of ANN models, which makes them an attractive tool for solving various applied problems, is that the network can change, adapting to the problem being solved. This kind of adjustment can be carried out in the following directions:

- the neural network can be trained, i.e., it can change the values of their tuning parameters (this is, as a rule, the synaptic weights of the neural network connections);
- the neural network can change its structural organization by adding or removing neurons and rebuilding the interneural connections;
- the neural network can be dynamically tuned to the solution of the current task by replacing some of its constituent parts (subnets) with previously prepared fragments, or by changing the values of the network settings and its structural organization on the basis of the previously prepared relationships linking the task to the required changes in the ANN model.

The first of these options leads to the traditional learning of ANN models, the second to the class of growing networks, and the third to networks with pretuning.

The most important limitation related to the peculiarities of the first of these approaches (ANN training) to the adjustment of the ANN models is that the network, before it started to be taught, is potentially suitable for a wide class of problems, but after the completion of the learning process it can already decide only a specific task; in the case of another task, it is necessary to retrain the network, during which the skill of solving the previous task is lost.

The second approach (growing networks) allows to cope with this problem only partially. Namely, if new training examples appeared that do not fit into the ANN model obtained according to the first of the approaches, then this model is built up with new elements, with the addition of appropriate links, after which the network is trained additionally, not affecting the previously constructed part of it.

The third of the approaches (networks with pretuning) is the most powerful and, accordingly, the most complicated one. Following this approach, it is necessary either to organize the process of dynamic (i.e., directly during the operation of the ANN model) replacement of the components of the model with their alternative versions prepared in advance, corresponding to the changed task, or to organize the ANN model in the form of an integrated system in which there are special structural elements, called interneurons and intersubnets, whose function is to act on the operational elements of the network in such a way that their current characteristics meet the specifics of the particular task being solved at the given moment.

2.3.2.2 *An Intersubnet as an Adaptation Tool for ANN Models*

We can formulate the concept of NM, which generalizes the notion of the ANN model. Some NM is a set of interrelated elements (NM elements) organized as network associations built according to certain rules from a very small number of primitives. One possible example of an NM element can be a single artificial neuron.

If this approach is followed by the principle of minimalism, then the most promising way, as noted earlier, is the formation of a very limited set of basic NM elements. Then a variety of specific types of NM elements required to produce ANN models are formed as particular cases of basic elements.

Processing elements of the NM can have two versions: working elements and intermediate elements. The most important difference between them is that the work elements convert the input data to the desired output of the ANN model, that is, in the desired result. In other words, the set of interacting work items implements the algorithm for solving the required application task. In contrast, intermediate elements of the NM do not participate directly in the above algorithm; their role is to act on the work items, for example, by adjusting the values of the parameters of the work items, which in turn changes the character of the transformation realized by this element.

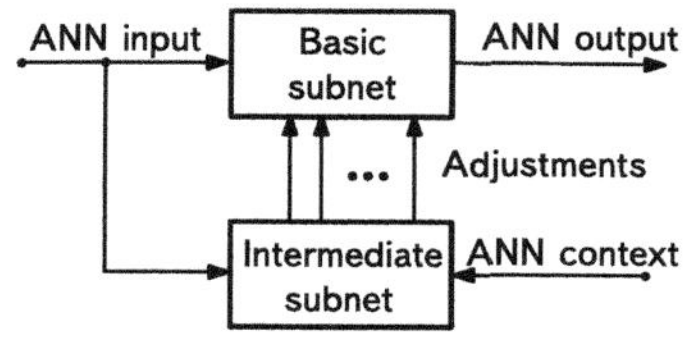

FIGURE 2.26 Intermediate elements in a network (compositional) model.

Thus, intermediate elements are introduced into the NM as a tool of contextual impact on the parameters and characteristics of the NM working elements. Using intermediate elements is the most effective way to make a network (composite) model adaptive. The functional role of the intermediate elements is illustrated in Fig. 2.26, in which the these elements are combined into an intermediate subnet (intersubnet). It is seen that the intermediate subnet receives the same inputs as the working subnet of the NM in question, which implements the basic algorithm for processing input data. In addition, an intersubnet can also receive some additional information, referred to herein as the NM context. According to the received initial data (inputs of the ANN model + context), the subnet introduces adjustments to the working subnet in such a way that this working subnetwork corresponds to the changed task. The training of the subnetwork is carried out in advance, at the stage of formation of the ANN model, so that the change of the task to be solved does not require additional training (and, especially, retraining) of the working subnet; only its reconfiguration is performed, which requires a short time.

2.3.2.3 *Presetting of ANN Models and Its Possible Variants*

We will distinguish two options for presetting: a strong presetting and a weak presetting.

Strong pretuning is oriented to the adaptation of the ANN model in a wide range of conditions. A characteristic architectural feature of the ANN model in this case is the presence of NM elements in the processing elements, along with the working elements, as well as insert elements affecting the parameters of the NM working elements. This approach allows implementing both parametric and structural adaptation of the ANN model.

Weak preadjustment does not use insert elements. With it, fragments of the ANN model are distinguished, which change as the conditions change and the fragments are adjusted according to a two-stage scheme. For example, let the problem of modeling the motion of an aircraft be solved. As the basis of the required model, a system of differential equations is used that describes the motion of an aircraft. This system, according to the scheme, which is presented in Section 5.2, is transformed into an ANN model. This is a general model, which should be refined in relation to a particular aircraft by specifying the specific values of its geometric, mass, inertial, and aerodynamic characteristics. The most difficult problem is the specification of the aerodynamic characteristics of the simulated aircraft due to incomplete and inaccurate knowledge of the corresponding quantities. In this situation, it is advisable to present these characteristics as a two-component structure: the first one is based on a priori knowledge (for example, on data obtained by experiments in a wind tunnel) and the second contains refining data obtained directly in flight. The presetting of the ANN model in this case is carried out due to the fact that during the transition from the simulation of one particular aircraft to another in the ANN model, a part of the description of the aerodynamic characteristics, based on a priori knowledge, is replaced. The clarifying part of this description is an instrument of adaptation of the ANN model, which is already implemented in the process of functioning of the modeled object.

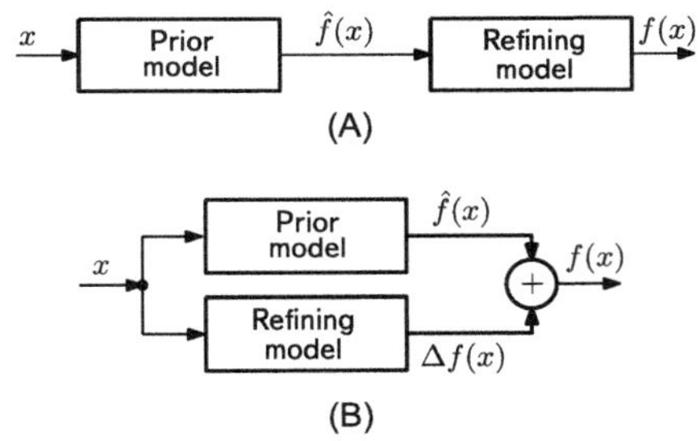

FIGURE 2.27 Structural options for presetting the ANN model. (A) A sequential variant. (B) A parallel version.

In both variants, both sequential and parallel, the a priori model is trained in off-line mode in advance using the available knowledge about the modeled object. The refinement model is adjusted already directly in the process of the object's operation on the basis of data received online.

In the sequential version (Fig. 2.27A), the output of the $\hat{f}(x)$ a priori model corresponding to this particular value of the input vector x is the input for the refinement model realizing the transformation $f(\hat{f}(x))$.

In the parallel version (Fig. 2.27B) the a priori and refinement models act independently of each other, calculating the estimate $\hat{f}(x)$ corresponding to this particular value of the input vector x and the initial knowledge of the modeled object, as well as the $\Delta f(x)$ correction for the same value of the input vector x, taking into account the data that became available for use in the process of object functioning. The required value of $f(x)$ is the sum of these components, i.e., $f(x) = \hat{f}(x) + \Delta f(x)$.

It should be emphasized that the neural network implementation of the a priori and refining models is, as a rule, different from the point of view of the attracted architectural solutions, although in a particular case it may be the same; for example, both models can be constructed in the form of multiperceptrons with sigmoid activation functions. This allows us to most effectively meet the requirements, which are different, generally speaking, for a priori and refining models. In particular, the main requirement for the a priori model is the ability to represent complex nonlinear dependencies with the required accuracy, while the time spent on learning such a model is uncritical, since this training is carried out in an autonomous (off-line) mode. At the same time, the refining model in its work must fit into the very rigid framework of the real (or even advanced) time scale. For this reason, in particular, in the vast majority of cases, the ANN architectures will be unacceptable, requiring full retraining, even with minor changes in the training data with which they work. In such a situation, an incremental approach to teaching and learning the ANN models is more appropriate, allowing not to retrain the entire network, but only to correct those elements that are directly related to the changed training data.

2.3.3 Incremental Formation of ANN Models

One of the tools for adapting ANN models is an incremental formation that exists in two variants: parametric and structural-parametric.

With the parametric version of the incremental formation, the structural organization of the ANN model is set immediately and fixed, after which it is incrementally adjusted (basic or additional learning) in several stages, for example, to extend the domain of operation modes of the dynamical system in which the model operates with the required accuracy.

For example, if we take a full spatial model of the aircraft motion, taking into account both its trajectory and angular motion, then in accordance with the incremental approach, first an off-line training of this model is carried out for a relatively small subdomain of the values of state and control variables, and then, in the online mode, an incremental learning process of the ANN model is performed, during which at each step a step-by-step extension of the subregion is performed. From here, the model is operational, in order to eventually expand the given subdomain to the full domain of the variables.

In the structural-parametric version of the incremental model formation procedure, at first, a "truncated" ANN model is constructed. This preliminary model has only a part of the state variables as its inputs, and it is trained on a dataset that covers only a subset of the domain of definition. This initial model is then gradually expanded by introducing new variables into it, followed by further training.

For example, the initial model is the model of the longitudinal angular motion of the aircraft, which is then expanded by adding trajectory longitudinal motion, after which lateral motion components are added to it, that is, the model is brought to the desired full model of the space motion in a few steps.

The structural-parametric variant of the incremental formation of ANN models allows us to start with a simple model, sequentially complicating it, for example, according to the scheme

$$
\begin{gathered}
\text{material point} \\
\Downarrow \\
\text{rigid body} \\
\Downarrow \\
\text{elastic body} \\
\Downarrow \\
\text{a set of coupled rigid and/or elastic bodies}
\end{gathered}
$$

This makes it possible to build up the model step-by-step in a structural sense.

2.4 TRAINING SET ACQUISITION PROBLEM FOR DYNAMIC NEURAL NETWORKS

2.4.1 Specifics of the Process of Forming Data Sets Required for Training Dynamic Neural Networks

Getting a training set that has the required level of informativeness is a critical step in solving the problem of forming the ANN model. If some features of dynamics (behavior) do not find reflection in the training set, they, accordingly, will not be reproduced by the model. In one of the fundamental guidelines for the identification of systems, this provision is formulated as the Basic Identification Rule: "If it is not in the data, it cannot be identified" (see [89], page 85).

The training data set required for the formation of the dynamical system ANN model should be informative (representative). For the time being we will assume that the training set is informative if the data contained in it are sufficient to produce an ANN model that, with the required level of accuracy, reproduces the behavior of the dynamical system over the entire range of possible values for the quantities and their derivatives that characterize this behavior. To ensure the fulfillment of this condition, when forming a training set, it is required to obtain data not only about changes in quantities, but also about the rate of their change, i.e., we can assume that the training set has the required informativeness if the ANN model obtained with its use reproduces the behavior of the system not only over the whole range of changes in the values of the quantities characterizing the behavior of the dynamical system but also their derivatives (and also all admissible combinations of both quantities and the values of their derivatives).

Such an intuitive understanding of the informativeness of the training set will be further refined.

2.4.2 Direct Approach to the Process of Forming Data Sets Required for Training Dynamic Neural Networks

2.4.2.1 *General Characteristics of the Direct Approach to the Forming of Training Data Sets*

We will clarify the concept of informative content of the training set, and we will also estimate its required volume to provide the necessary level of informativeness. First, we will perform these actions in the framework of a direct approach to solving the problem of the formation of a training set; in the next section, the concept will be extended to an indirect approach.

Consider a controllable dynamical system of the form

$$\dot{x} = F(x, u, t), \tag{2.99}$$

where $x = (x_1, x_2, \ldots, x_n)$ are the state variables, $u = (u_1, u_2, \ldots, u_m)$ are control variables, and $t \in T = [t_0, t_f]$ is time.

The variables $x_1, x_2, \ldots, x_n$ and $u_1, u_2, \ldots, u_m$, taken at a particular moment in time $t_k \in T$, characterize, respectively, the state of the dynamical system and the control actions on it at a given time. Each of these values takes values from the corresponding area, i.e.,

$$\begin{aligned} &x_1(t_k) \in X_1 \subset \mathcal{R}, \ldots, x_n(t_k) \in X_n \subset \mathcal{R};\\ &u_1(t_k) \in U_1 \subset \mathcal{R}, \ldots, u_n(t_k) \in U_m \subset \mathcal{R}. \end{aligned} \tag{2.100}$$

In addition, there are, as a rule, restrictions on the values of the combinations of these variables, i.e.,

$$\begin{aligned} x &= \langle x_1, \ldots, x_n \rangle \in R_X \subset X_1 \times \cdots X_n,\\ u &= \langle u_1, \ldots, u_m \rangle \in R_U \subset U_1 \times \cdots U_n, \end{aligned} \tag{2.101}$$

as well as on blends of these combinations,

$$\langle x, u \rangle \in R_{XU} \subset R_X \times R_U. \tag{2.102}$$

The example included in the training set should show the response of the DS to some combination of $\langle x, u\rangle$. By a reaction of this kind, we will understand the state $x(t_{k+1})$, to which the dynamical system (2.99) passes from the state $x(t_k)$ with the value $u(t_k)$ of the control action, written

$$\langle x(t_k), u(t_k)\rangle \xrightarrow{F(x,u,t)} x(t_{k+1}). \tag{2.103}$$

Accordingly, some example p from the training set P will include two parts, namely, the input (this is the pair $\langle x(t_k), u(t_k)\rangle$) and output (this is the reaction $x(t_{k+1})$) of the dynamical system.

2.4.2.2 *Informativity of the Training Set*

The training set should (ideally) show the dynamical system responses to any combinations of $\langle x, u\rangle$ satisfying the condition (2.102). Then, according to the Basic Identification Rule (see page 73), the training set will be informative, that is, allow to reproduce in the model all the specific behavior of the simulated DS.[5]

Let us clarify this situation. We introduce the notation

$$p_i = \{\langle x^{(i)}(t_k), u^{(i)}(t_k)\rangle, x^{(i)}(t_{k+1})\}, \tag{2.104}$$

where $p_i \in P$ is the ith example from the training set P. In this example

$$\begin{aligned} x^{(i)}(t_k) &= (x_1^{(i)}(t_k), \ldots, x_n^{(i)}(t_k)), \\ u^{(i)}(t_k) &= (u_1^{(i)}(t_k), \ldots, u_m^{(i)}(t_k)). \end{aligned} \tag{2.105}$$

The response $x^{(i)}(t_{k+1})$ of the considered dynamical system to the example p_i is

$$x^{(i)}(t_{k+1}) = (x_1^{(i)}(t_{k+1}), \ldots, x_n^{(i)}(t_{k+1})). \tag{2.106}$$

[5]It should be noted that the availability of an informative training set provides a *potential* opportunity to obtain a model that will be adequate to a simulated dynamical system. However, this potential opportunity must still be taken advantage of, which is a separate nontrivial problem, the successful solution of which depends on the chosen class of models and learning algorithms.

In a similar way, we introduce one more example of $p_j \in P$:

$$p_j = \{\langle x^{(j)}(t_k), u^{(j)}(t_k)\rangle, x^{(j)}(t_{k+1})\}. \tag{2.107}$$

The source data of the examples p_i and p_j will be considered as not coincident, i.e.,

$$x^{(i)}(t_k) \neq x^{(j)}(t_k), \quad u^{(i)}(t_k) \neq u^{(j)}(t_k).$$

In the general case, the dynamical system responses to the original data from these examples do not coincide, i.e.,

$$x^{(i)}(t_{k+1}) \neq x^{(j)}(t_{k+1}).$$

We introduce the concept of ε-proximity for a pair of examples p_i and p_j. Namely, we will consider examples of p_i and p_j ε-close if the following condition is satisfied:

$$\|x^{(i)}(t_{k+1}) - x^{(j)}(t_{k+1})\| \leqslant \varepsilon, \tag{2.108}$$

where $\varepsilon > 0$ is the predefined real number.

We select from the set of examples $P = \{p_i\}_{i=1}^{N_p}$ a subset consisting of such examples p_s for which the ε-proximity relation to the example p_s is satisfied, i.e.,

$$\|x^{(i)}(t_{k+1}) - x^{(j)}(t_{k+1})\| \leqslant \varepsilon, \ \forall s \in I_s \subset I. \tag{2.109}$$

Here I_s is the set of indices (numbers) of those examples for which ε-proximity is satisfied with respect to the example p_s, while $I_s \subset I = \{1, \ldots, N_p\}$.

We call an example p_i ε-representative[6] if for the whole collection of examples p_s, $\forall s \in I_s$, that is, for any example p_s, $s \in I_s$, the condition of ε-proximity is satisfied. Accordingly, we can now replace the collection of examples $\{p_s\}$, $s \in I_s$, by a single ε-representative p_i, and the error introduced by such a replacement will not exceed ε. Input parts of collections of examples

[6]This means that the example p_i is included in the set of examples $\{p_s\}$, $s \in I_s$.

$\{p_s\}$, $s \in I_s$, allocate the subdomain $R_{XU}^{(s)}$, $s \in I_s$, in the domain R_{XU} defined by the relation (2.102); in this case

$$\bigcup_{s=1}^{N_p} R_{XU}^{(s)} = R_{XU}. \tag{2.110}$$

Now we can state the task of forming a training set as a collection of ε-representatives that covers the domain R_{XU} (2.102) of all possible values of pairs $\langle x, u \rangle$.

The relation (2.110) is the ε-covering condition for the training set P of the domain R_{XU}. A set P carrying out an ε-covering of the domain R_{XU} will be called ε-informative or, for brevity, simply informative.

If the training set P has ε-informativity, this means that for any pair $\langle x, u \rangle \in R_{XU}$ there is at least one example $p_i \in P$ which is an ε-representative for a given pair.

With respect to the ε-covering (2.110) of the domain R_{XU}, the following two problems can be formulated:

1. Given the number of examples N_p in the training set P, find their distribution in the domain R_{XU} which minimizes the error ε.
2. A permissible error value ε is given; obtain a minimal collection of a number of N_p examples which ensures that ε is obtained.

2.4.2.3 *Example of Direct Formation of Training Set*

Suppose that the controlled object under consideration (plant) is a dynamical system described by a vector differential equation of the form [91,92]

$$\dot{x} = \varphi(x, u, t). \tag{2.111}$$

Here, $x = (x_1\ x_2\ \dots\ x_n) \in \mathcal{R}^n$ is the vector of state variables of the op-amp; $u = (u_1\ u_2\ \dots\ u_m) \in \mathcal{R}^m$ is a vector of control variables of the op-amp; $\mathcal{R}^n$, $\mathcal{R}^m$ are Euclidean spaces of dimension n and m, respectively; $t \in [t_0, t_f]$ is the time.

In Eq. (2.111), $\varphi(\cdot)$ is a nonlinear vector function of the vector arguments x, u and the scalar argument t. It is assumed to be given and belongs to some class of functions that admits the existence of a solution of Eq. (2.111) for given $x(t_0)$ and $u(t)$ in the considered part of the space of states for the plant.

The behavior of the plant, determined by its dynamic properties, can be influenced by setting a correction value for the control variable $\Delta u(x, u^*)$. The operation of forming the required value $\Delta u(x, u^*)$ for some time t_{i+1} from the values of the state vector x and the command control vector u^* at the time instant t_i

$$\Delta u(t_{i+1}) = \Psi(x(t_i), u^*(t_i)) \tag{2.112}$$

we will perform in the device, which we call the correcting controller (CC). We assume that the character of the transformation $\Psi(\cdot)$ in (2.112) is determined by the composition and values of the components of a certain parameter vector $w = (w_1\ w_2\ \dots\ w_{N_w})$. The set (2.111), (2.112) from the plant and CC is referred to as a controlled system.

The behavior of the system (2.111), (2.112) with the initial conditions $x_0 = x(t_0)$ under the control $u(t)$ is a multistep process if we assume that the values of this process $x(t_k)$ are observed at time instants t_k, i.e.,

$$\{x(t_k)\}, \ t_k = t_0 + k\Delta t, \quad k = 0, 1, \dots, N_t, \ \Delta t = \frac{t_f - t_0}{N_t}. \tag{2.113}$$

In the problem (2.111), (2.112), as a teaching example, generally speaking, we could use a pair

$$\langle (x_0^{(e)}, u^{(e)}(t)), \{x^{(e)}(t_k), \ k = 0, 1, \dots, N_t\} \rangle,$$

where $(x_0^{(e)}, u^{(e)}(t))$ is the initial state of the system (2.111) and the formed control law, respectively, and $\{x^{(e)}(t_k), \ k = 0, 1, \dots, N_t\}$ is the multistep process (2.113), which should be carried out given the initial state $x_0^{(e)}$ under the influence of some control $u^{(e)}(t)$ on the time interval $[t_0, t_f]$. Comparing the process $\{x^{(e)}(t_k)\}$ with the process $\{x(t_k)\}$, obtained for the same initial conditions $x_0^{(e)}$ and control $u^{(\ rusinde)}(t)$, in fact, that is, for some fixed value of the parameters w, it would be possible in some way to determine the distance between the required and actually implemented processes, and then try to minimize it by varying the values of the parameters w. This kind of "straightforward" approach, however, leads to a sharp increase in the amount of computation at the stage of training of the ANN and, in particular, at the stage of the formation of the corresponding training set.

There is, however, the possibility of drastically reducing these volumes of calculations if we take advantage of the fact that the state into which the system goes (2.111), (2.112) for the time $\Delta t = t_{i+1} - t_i$, depends only on its state $x(t_i)$ at time t_i, and also on the value $u(t_i)$ of the control action at the same instant of time. This circumstance gives grounds to replace the multistep process $\{x^{(e)}(t_k)\}$, $k = 0, 1, \ldots, N_t$, a set of N_t one-step processes, each of which consists in (2.111), (2.112) of one step in time of length Δt from some initial point $x(t_k)$.

In order to obtain a set of initial points $x_t(t_0)$, $u_t(t_0)$, which completely characterizes the behavior of the system (2.111), (2.112) on the whole range of admissible values $R_{XU} \subseteq X \times U$, $x \in X$, $u \in U$, we construct the corresponding grid.

Let the state variables x_i, $i = 1, \ldots, n$, in equation (2.111) take values from the ranges defined for each of them, i.e.,

$$x_i^{min} \leqslant x_i \leqslant x_i^{max}, \ i = 1, \ldots, n. \tag{2.114}$$

Similar inequalities hold for control variables u_j, $j = 1, \ldots, m$, in (2.111), i.e.,

$$u_j^{min} \leqslant u_j \leqslant u_j^{max}, \ j = 1, \ldots, m. \tag{2.115}$$

We define on these ranges a grid $\{\Delta^{(i)}, \Delta^{(j)}\}$ as follows:

$$\begin{aligned} \Delta^{(i)} : x_i^{(s_i)} &= x_i^{min} + s_i \Delta x_i \,, \\ i &= 1, \ldots, n; \ s_i = 0, 1, \ldots, N_i \,, \\ \Delta^{(j)} : u_j^{(p_j)} &= u_j^{min} + p_j \Delta u_j \,, \\ j &= 1, \ldots, m; \ p_i = 0, 1, \ldots, M_j . \end{aligned} \tag{2.116}$$

In expressions (2.116), we have

$$\Delta x_i = \frac{x_i^{max} - x_i^{min}}{N_i}, \ i = 1, \ldots, n,$$
$$\Delta u_j = \frac{u_j^{max} - u_j^{min}}{M_j}, \ j = 1, \ldots, m.$$

Here we denote the following: N_i is the number of segments divided by the range of values for the state variable x_i, $i = 1, \ldots, n$; M_j is the number of segments divided by the range of values for the control variable u_j, $j = 1, \ldots, m$.

The nodes of this grid are tuples of length $(n+m)$ of the form $\langle x_i^{(s_i)}, u_j^{(p_j)} \rangle$, where the components $x_i^{(s_i)}$, $i = 1, \ldots, n$, are taken from the corresponding $\Delta^{(i)}$, and the components $u_j^{(p_j)}$, $j = 1, \ldots, m$, from $\Delta^{(j)}$ in (2.116). If the domain R_{XU} is a subset of the Cartesian product $X \times U$, then this fact can be taken into account by excluding the "extra" tuples from the grid (2.116).

In [90] an example of the solution of the ANN modeling problem was considered, in which the training set was formed according to the method presented above. The source model of motion in this example is a system of equations of the following form:

$$\begin{aligned} m(\dot{V}_z - q V_x) &= Z \,, \\ I_y \dot{q} &= M \,, \end{aligned} \tag{2.117}$$

where Z is the aerodynamic normal force; M is the aerodynamic pitching moment; q is the pitch angular velocity; m is the aircraft mass; I_y is the pitch moment of inertia; V_x and V_z are the longitudinal and normal velocities, respectively.

Here, the force Z and moment M depend on the angle of attack α. However, in case of a rectilinear horizontal flight the angle of attack equals the pitch angle θ. The pitch angle, in turn, is related to velocity V_z and airspeed V by the following kinematic dependence:

$$V_z = V \sin\theta .$$

Thus, the system of equations (2.117) is closed.

The pitching moment M in (2.117) is a function of the all-moving stabilizer deflection angle, i.e., $M = M(\delta_e)$.

Thus, the system of equations (2.117) describes transient processes in angular velocity and pitch angle, which arise immediately after a violation of balancing corresponding to a steady horizontal flight.

So, in the particular case under consideration, the composition of the state and control variables is as follows:

$$\mathbf{x} = [V_z \; q]^T, \quad \mathbf{u} = [\delta_e]. \tag{2.118}$$

In terms of the problem (2.117), when the mathematical model of the controlled object of the inequality is approximated (2.114),

$$\begin{aligned} V_z^{\min} \leqslant V_z \leqslant V_z^{\max}, \\ q^{\min} \leqslant q \leqslant q^{\max}, \end{aligned} \tag{2.119}$$

the inequality (2.115) will be written as

$$\delta_e^{\min} \leqslant \delta_e \leqslant \delta_e^{\max}, \tag{2.120}$$

and the grid (2.116) is rewritten in the following form:

$$\begin{aligned} \Delta^{(V_z)} &: V_z^{(s_{V_z})} = V_z^{\min} + s_{V_z} \Delta V_z , \\ & s_{V_z} = 0, 1, \ldots, N_{V_z} , \\ \Delta^{(q)} &: q^{(s_q)} = q^{\min} + s_q \Delta q , \\ & s_q = 0, 1, \ldots, N_q , \\ \Delta^{(\delta_e)} &: \delta_e^{(p)} = \delta_e^{\min} + p_{\delta_e} \Delta\delta_e , \\ & p_{\delta_e} = 0, 1, \ldots, M_{\delta_e} . \end{aligned} \tag{2.121}$$

As noted above, each of the grid nodes (2.116) is used as the initial value $x_0 = x(t_0)$, $u_0 = u(t_0)$ for the system of equations (2.111); with these initial values, one step of integration is performed with the value Δt. These initial values $x(t_0)$, $u(t_0)$ constitute the input vector in the learning example, and the resulting value $x(t_0 + \Delta t)$ is the target vector, that is, vector-sample, showing the learning algorithm of the HC model, which should be the output value of the NS under given starting conditions $x(t_0)$, $u(t_0)$.

The formation of a learning set for solving the neural network approximation problem of the dynamical system (2.111) (in particular, in its particular version (2.117)) is a nontrivial task. As the computing experiment [90] has shown, the convergence of the learning process is very sensitive to the grid step Δx_i, Δu_j and the time step Δt.

We explain this situation by the example of the system (2.117), when

$$\Delta x_1 = \Delta V_z, \; \Delta x_2 = \Delta q, \; \Delta u_1 = \Delta\delta_e .$$

We represent, as shown in Fig. 2.28, the part of the grid $\{\Delta^{(V_z)}, \Delta^{(q)}\}$, whose nodes are used as initial values (the input part of the training example) to obtain the target part of the training example. In Fig. 2.28, the grid node is shown in a circle, and the cross is the state of the system (2.117), obtained by integrating its equations with a time step Δt with the initial conditions $(V_z^{(i)}, q^{(j)})$, for a fixed position of the stabilizer $\delta_e^{(k)}$.

In a series of computational experiments it was established that for $\Delta t = \text{const}$, the conditions of convergence of the learning process of the neural controller will be as follows:

$$\begin{aligned} V_z(t_0 + \Delta t) - V_z(t_0) < \Delta V_z , \\ q(t_0 + \Delta t) - q(t_0) < \Delta q , \end{aligned} \tag{2.122}$$

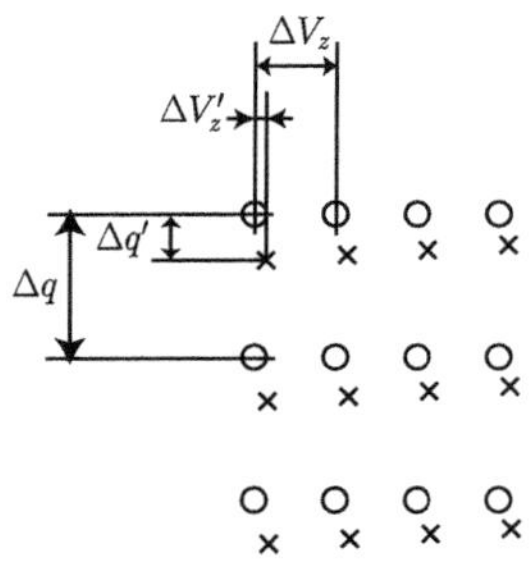

FIGURE 2.28 Fragment of the grid $\{\Delta^{(V_z)}, \Delta^{(q)}\}$ for $\delta_e =$ const. ○ – starting grid node; × – mesh target point; ΔV_z, Δq is the grid spacing for the state variables V_z and q, respectively; $\Delta V_z'$, $\Delta q'$ is the shift of the target point relative to the grid node that spawned it (From [90], used with permission from Moscow Aviation Institute).

where ΔV_z, Δq is the grid spacing (2.121) for the corresponding state variables for the given fixed value δ_e.

The grid $\{\Delta^{(V_z)}, \Delta^{(q)}\}$, constructed for some fixed point $\delta_e^{(p)}$ from $\Delta^{(\delta_e)}$, can be graphically depicted as shown in Fig. 2.29. Here, for each of the grid nodes (they are shown as circles), also the corresponding target points are represented (crosses). The set ("bundle") of such images, each for its value $\delta_e^{(p)} \in \Delta^{(\delta_e)}$, gives important information about the structure of the training set for the system (2.117), allowing, in some cases, to significantly reduce the volume of this set.

Now, after the grid is formed (2.116) (or (2.121), for the case of longitudinal short-period motion), you can build the corresponding training set, after which the problem of learning the network with the teacher can be solved. This task was done in [90]. The results obtained in this paper show that the direct method of forming training sets can be successfully used for problems of small dimension (determined by the dimensions of the state and control vectors, and also by the magnitude of the range of admissible values of the components of these vectors).

2.4.2.4 *Assessment of the Volume of the Training Set With a Direct Approach to Its Formation*

Let us estimate the volume of the training set, obtained with a direct approach to its formation. Let us first consider the simplest version of a direct one-step method of forming a training set, i.e., the one in which the reaction of DS (2.106) at the time instant t_{k+1} depends on the values of the state and control variables (2.105) only at time instant t_k.

Let us consider this question on a specific example related to the problem, which is solved in Section 6.2 (formation of the ANN model of longitudinal short-period motion of a maneuverable aircraft). The initial model of motion in the form of a system of ODEs is written as follows:

$$\begin{aligned} \dot{\alpha} &= q - \frac{\bar{q}S}{mV} C_L(\alpha, q, \delta_e) + \frac{g}{V}\cos\theta, \\ \dot{q} &= \frac{\bar{q}S\bar{c}}{I_y} C_m(\alpha, q, \delta_e), \\ T^2\ddot{\delta}_e &= -2T\zeta\dot{\delta}_e - \delta_e + \delta_{e_{act}}, \end{aligned} \tag{2.123}$$

where α is the angle of attack, deg; θ is the pitch angle, deg; q is the pitch angular velocity, deg/sec; δ_e is the all-moving stabilizer deflection angle , deg; C_L is the lift coefficient; C_m is pitching moment coefficient; m is mass of the aircraft, kg; V is the airspeed, m/sec; $\bar{q} = \rho V^2/2$ is the dynamic pressure, kg·m^{-1}sec^{-2}; ρ is air density, kg/m^3; g is the acceleration due to gravity, m/sec^2; S is the wing area, m^2; $\bar{c}$ is the mean aerodynamic chord of the wing, m; I_y is the moment of inertia of the aircraft relative to the lateral axis, kg·m^2; the dimensionless coefficients C_L and C_m are nonlinear functions of their arguments; T, ζ are the time constant and the relative damping coefficient of the actuator, and $\delta_{e_{act}}$ is the command signal to the actuator of the all-moving controllable stabilizer (limited to ± 25 deg). In the model (2.123), the variables α, q, δ_e, and $\dot{\delta}_e$ are the states of the controlled object, and the variable $\delta_{e_{act}}$ is the control.

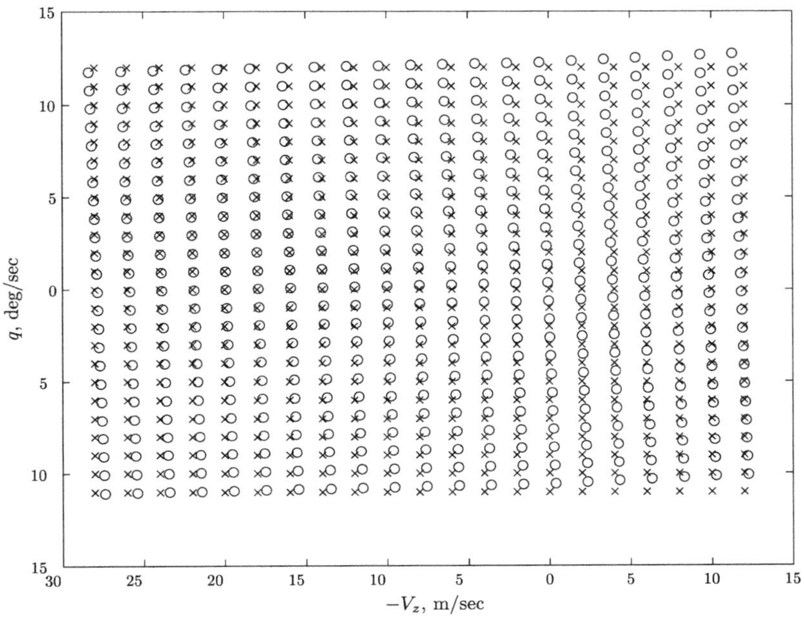

FIGURE 2.29 Graphic grid representation $\{\Delta^{(V_z)}, \Delta^{(q)}\}$ when $\delta_e = \text{const}$, combined with the target points; this grid sheet is built with $\delta_e = -8$ deg (From [90], used with permission from Moscow Aviation Institute).

Let us carry out a discretization of the considered dynamical system as it was described in the previous section. In order to reduce the dimension of the problem, we will only consider the variables α, q, and $\delta_{e_{act}}$, which directly characterize the behavior of the considered dynamical system, and treat the variables δ_e and $\dot{\delta}_e$ as "hidden" variables.

If the dependencies for δ_e and $\dot{\delta}_e$ are "hidden", then for the remaining variables α, q, and $\delta_{e_{act}}$ we set variables N_α, N_q, $M_{\delta_{e_{act}}}$, which are the number of counts for these variables. Assuming that all combinations of the values of these variables are admissible, the quantity $N_\Sigma = N_\alpha \cdot N_q \cdot M_{\delta_{e_{act}}}$, the number of examples in the problem book for different values of the number of samples N_α, N_q, $M_{\delta_{e_{act}}}$ (for simplicity, we assume that $N_\alpha = N_q = M_{\delta_{e_{act}}} = N$), is

$$
\begin{aligned}
N = 20 &: \ 20 \times 20 \times 20 = \ 8000, \\
N = 25 &: \ 25 \times 25 \times 25 = 15625, \\
N = 30 &: \ 30 \times 30 \times 30 = 27000.
\end{aligned}
\tag{2.124}
$$

If not only the variables α, q, and $\delta_{e_{act}}$, but also δ_e and $\dot{\delta}_e$ are required in the dynamical system model to be formed, then the estimates of the volume of training sets received take the form

$$
\begin{aligned}
N = 20 &: \ 20 \times 20 \times 20 \times 20 \times 20 = \ 3200000, \\
N = 25 &: \ 25 \times 25 \times 25 \times 25 \times 25 = \ 9765625, \\
N = 30 &: \ 30 \times 30 \times 30 \times 30 \times 30 = 25200000.
\end{aligned}
\tag{2.125}
$$

As we can see from these estimates, from the point of view of the volume of the training set, only the variants related to the dynamical system with state vectors and small-size controls and with a moderate number of samples with respect to these variables are acceptable (the first and second variants in (2.124)). Even a slight increase in the values of these parameters leads, as can be seen from (2.125)), to unacceptable values of the value of the training set.

In real-world applied problems, where the possibilities of ANN modeling are particularly in demand, the result is even more impressive.

In particular, in the full model of the angular motion of the aircraft (the corresponding ANN model for this case is considered in Section 6.3) of Chapter 6, we have 14 state variables and 3 control variables, hence the volume of the training set for it in the direct approach to its formation and at $N_w = N_q = M_{\delta_e} = 20$ will be $N_\Sigma = 20^{17} = 2 \cdot 10^{18}$, which, of course, is completely unacceptable.

Thus, the direct approach to the formation of training sets for modeling dynamical systems has a very small "niche," in which its application is possible – simple problems of low dimensionality. An alternative indirect approach is more well-suited for complex high dimensional problems. This approach is based on the application of a set of specially designed control signals to a dynamical system of interest. This approach is discussed in more detail in the next section. The indirect approach has its advantages and disadvantages. The indirect approach is the only viable option in situations where the training data acquisition is required to be performed in real or even in advanced time. However, in cases when there are no rigid time restrictions regarding the acquisition and processing of training data, the most appropriate approach is a mixed one, which is a combination of direct and indirect approaches.

2.4.3 Indirect Approach to the Acquisition of Training Data Sets for Dynamic Neural Networks

2.4.3.1 *General Characteristics of the Indirect Approach to the Acquisition of Training Data Sets*

As noted in the previous section, the indirect approach is based on the application of a set of specially designed control signals to the dynamical system, instead of direct sampling of the domain $R_{X,U}$ of feasible values of state and control variables.

With this approach, the actual motion of the dynamical system $(x(t), u(t))$ is composed of a program (test maneuver) of the motion $(x^*(t), u^*(t))$, generated by the control signal $u^*(t)$, as well as the motion $(\tilde{x}(t), \tilde{u}(t))$, generated by the additional perturbing action $\tilde{u}(t)$, i.e.,

$$x(t) = x^*(t) + \tilde{x}(t), \quad u(t) = u^*(t) + \tilde{u}(t). \quad (2.126)$$

Examples of test maneuvers include:

- a straight-line horizontal flight with a constant speed;
- a flight with a monotonically increasing angle of attack;
- a U-turn in the horizontal plane;
- an ascending/descending spiral.

Possible variants of the test perturbing actions $\tilde{u}(t)$ are considered below.

The type of test maneuver $(x^*(t), u^*(t))$ in (2.126) determines the obtained ranges for changing the values of the state and control variables; $\tilde{u}(t)$ is the variety of examples within these ranges.

What is the ideal form of a training set and how can it be obtained in practice using an indirect approach? We consider this issue in several stages, starting with the simplest version of the dynamical system and proceeding to more complex versions.

We first consider a simpler case of an uncontrolled dynamical system (Fig. 2.30).

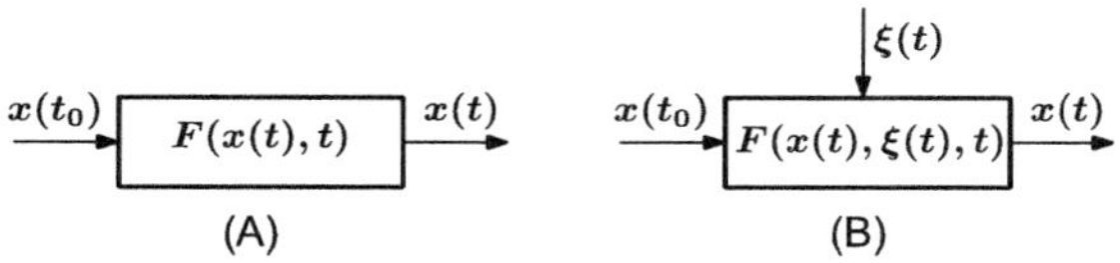

FIGURE 2.30 Uncontrolled dynamical system. (A) Without external influences. (B) With external influences.

Suppose that there is some dynamical system, that is, a system whose state varies with time. This dynamical system is uncontrollable, its behavior is affected only by the initial conditions and, possibly, by some external influences (the impact of the environment in which and in interaction with which the dynamical system realizes its behavior). An example of such a dynamical system can be an artillery shell, the flight path of which is affected by the initial conditions of shooting. The impact of the medium in this case is determined by the gravitational field in which the projectile moves, and also by the atmosphere.

The state of the dynamical system in question at a particular moment in time $t \in T = [t_0, t_f]$ is characterized by a set of values $x = (x_1, \ldots, x_n)$. The composition of this set of quantities, as noted above, is determined by the questions that are asked about the dynamical system in question.

The state of the dynamical system in question at a particular moment in time $t \in T = [t_0, t_f]$ is characterized by a set of values $x = (x_1, \ldots, x_n)$. The composition of this set of quantities, as noted above, is determined by the questions that are asked about the considered dynamical system.

At the initial instant of time $t \in T$, the state of the dynamical system takes on the value $x^0 = x(t_0) = (x_1^0, \ldots, x_n^0)$, where $x^0 = x(t_0) \in X$.

Since the variables $\{x_i\}_{i=1}^n$ describe exactly some dynamical system, they, according to the definition of the dynamical system, vary with time, that is, the dynamical system is characterized by a set of variables $\{x_i(t)\}_{i=1}^n$, $t \in T$. This set will be called the behavior (phase trajectory or trajectory in the state space) of the dynamical system.

The behavior of an (uncontrolled) dynamical system is determined, as already noted, by its initial state $\{x_i(t_0)\}_{i=1}^n$ and "the nature of the dynamical system," i.e., the way in which the variables x_i are related to each other in the evolution law (the law of functioning) of the dynamical system $F(x, t)$. This evolution law determines the state of the dynamical system at time $(t + \Delta t)$, if these states are known at previous time instants.

2.4.3.2 Formation of a Set of Test Maneuvers

It was noted above that the selected program motion (reference trajectory) as part of the test maneuver determines the range of values of the state variables in which the training data will be obtained. It is required to choose such a set of reference trajectories that covers the whole range of changes in the values of the state variables of the dynamical system. The required number of trajectories in such a collection is determined from the condition of ε-proximity of the phase trajectories of the dynamical system, i.e.

$$\|x_i(t) - x_j(t)\| \leqslant \varepsilon, \; x_i(t), x_j(t) \in X, \; t \in T. \tag{2.127}$$

We define a family of reference trajectories of the dynamical system,

$$\{x_i^*(t)\}_{i=1}^{N_R}, \quad x_i^*(t) \in X, \quad t \in T. \tag{2.128}$$

We assume that the reference trajectory $x_i^*(t)$, $i = 1, \ldots, N_R$, is an ε-representative of the family $X_i \subset X$ of the phase trajectories of the dynamical system in the domain $X_i \subset X$ if for each of the phase trajectories $x(t) \in X_i$ the following condition is satisfied:

$$\|x_i^*(t) - x(t)\| \leqslant \varepsilon, \ x_i^*(t) \in X_i, \ x(t) \in X_i, \ t \in T. \tag{2.129}$$

The family of reference trajectories $\{x_i^*(t)\}_{i=1}^{N_R}$ of the dynamical system must be such that

$$\bigcup_{i=1}^{N_R} X_i = X_1 \cup X_2 \cup \ldots \cup X_{N_R} = X, \tag{2.130}$$

where X is the family (collection) of all phase trajectories (trajectories in the state space) potentially realized by the dynamical system in question. This condition means that the family of reference trajectories $\{x_i^*(t)\}_{i=1}^{N_R}$ should together represent all potentially possible variants of the behavior of the dynamical system in question. This condition can be treated as a condition for completeness of the ε-covering by support trajectories of the domain of possible variants of the behavior of the dynamical system.

An optimal ε-covering problem for the domain X of possible variants of the dynamical system behavior can be stated, consisting in minimizing the number N_R of reference trajectories in the set $\{x_i^*(t)\}_{i=1}^{N_R}$, i.e.,

$$\{x_i^*(t)\}_{i=1}^{N_R^*} = \min_{N_R} \{x_i^*(t)\}_{i=1}^{N_R}, \tag{2.131}$$

that allows to minimize the volume of the training set while preserving its informativeness.

A desirable condition (but difficult to realize) is also the condition

$$\bigcap_{i=1}^{N_R} X_i = X_1 \cap X_2 \cap \ldots \cap X_{N_R} = \varnothing. \tag{2.132}$$

2.4.3.3 *Formation of Test Excitation Signal*

As already noted, the type of test maneuver in (2.126) determines the resulting ranges for changing the values of the state and control variables, while the kind of perturbation effect provides a variety of examples within these ranges. In the following sections, the questions of forming (with a given test maneuver) test excitatory influences in such a way as to obtain an informative set of training data for a dynamical system are considered.

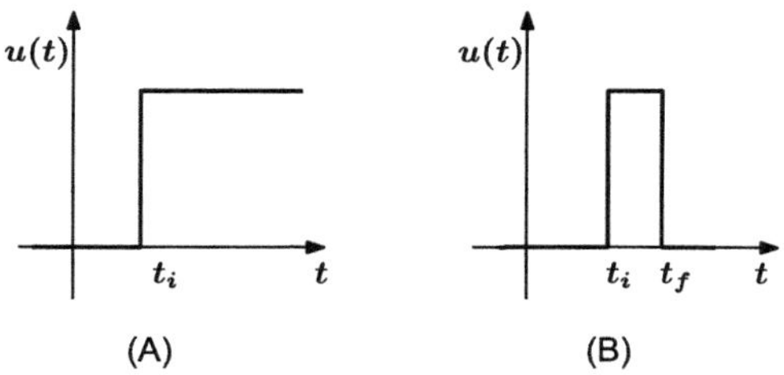

FIGURE 2.31 Typical test excitation signals used in the study of the dynamics of controllable systems. (A) Stepwise excitation. (B) Impulse excitation. From [109], used with permission from Moscow Aviation Institute.

TYPICAL TEST EXCITATION SIGNALS FOR THE IDENTIFICATION OF SYSTEMS

Elimination of uncertainties in the ANN model by refining (restoring) a number of elements included in it (for example, functions describing the aerodynamic characteristics of the aircraft) is a typical problem of identifying systems [44,93–99]. When solving identification problems for controllable dynamic systems, a number of typical test disturbances are used. Among them, the most common are the following impacts [89,100–103]:

- stepwise excitation;
- impulse excitation;
- doublet (signal type 1–1);
- triplet (signal type 2–1–1);
- quadruplet (signal type 3–2–1–1);
- random signal;
- polyharmonic signal.

Stepwise excitation (Fig. 2.31A) is a function $u(t)$ that changes at a certain moment in time t_i from $u = 0$ to $u = u^*$, i.e.,

$$u(t) = \begin{cases} 0, & t < t_i; \\ u^*, & t \geqslant t_i. \end{cases} \tag{2.133}$$

Let $u^* = 1$. Then (2.133) is the function of the unit jump $\sigma(t)$. With its use, you can define an-

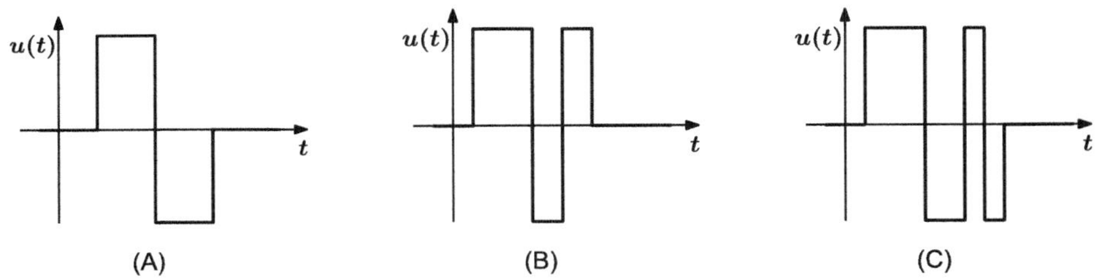

FIGURE 2.32 Typical test excitation signals used in the study of the dynamics of controllable systems. (A) Doublet (signal type 1–1). (B) Triplet (signal type 2–1–1). (C) Quadruplet (signal type 3–2–1–1). From [109], used with permission from Moscow Aviation Institute.

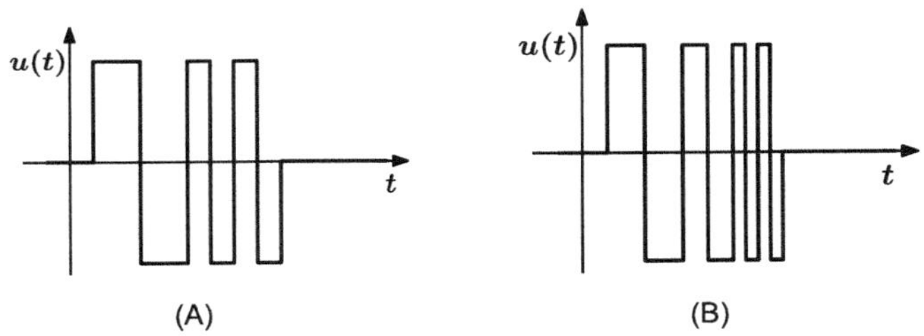

FIGURE 2.33 Modified versions of the test excitation signals used in the study of the dynamics of controllable systems. (A) Triplet (signal type 2–1–1). (B) Quadruplet (signal type 3–2–1–1). From [109], used with permission from Moscow Aviation Institute.

other kind of test action – a rectangular pulse (Fig. 2.31B):

$$u(t) = A(\sigma(t) - \sigma(t - T_r)), \tag{2.134}$$

where A is the pulse amplitude and $T_r = t_f - t_i$ is the pulse duration.

On the basis of the rectangular pulse signal (2.134), perturbing effects of oscillatory character are determined, consisting of a series of rectangular oscillations with a definite relationship between their periods. Among the most commonly used influences of this kind are the doublet (Fig. 2.32A), the triplet (Fig. 2.32B), and the quadruplet (Fig. 2.32C).

The doublet (also denoted as a signal of type 1–1) is one complete rectangular wave with a period $T = 2T_r$ equal to twice the duration of the rectangular pulse.

A triplet (signal of the type 2–1–1) is a combination of a rectangular pulse of duration $T = 2T_r$ and a complete rectangular oscillation with a period $T = 2T_r$.

A quadruplet (a signal of the type 3–2–1–1–1) is formed from a triplet by adding to its origin a rectangular pulse of width $T = T_r$. In addition, we can also use triplet and quadruplet variants in which each of the constituent parts of the signal is a full-period oscillation (see Fig. 2.33). We will designate them as signals of the type 2–1–1 and 3–2–1–1, respectively.

Another typical excitation signal is shown in Fig. 2.34A. Its values are kept constant for all time intervals $[t_i, t_{i+1})$, $i = 0, 1, \ldots, n-1$, and at time instances t_i it can be changed randomly. In more detail, a signal of this type will be considered below by the example of solving the problem of the ANN simulation of the longitudinal angular motion of an aircraft.

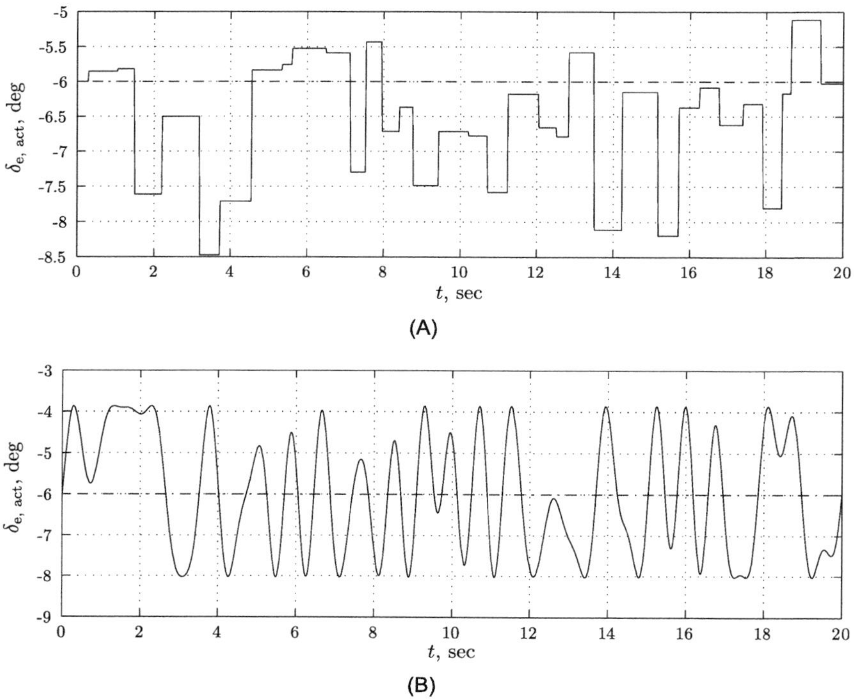

FIGURE 2.34 Test excitations as functions of time used in studying the dynamics of controlled systems. (A) A random signal. (B) A polyharmonic signal. Here, φ_{act} is the command signal for the elevator actuator (full-movable horizontal tail) of the aircraft from the example (2.123) on page 78. From [109], used with permission from Moscow Aviation Institute.

POLYHARMONIC EXCITATION SIGNALS FOR THE IDENTIFICATION OF SYSTEMS

To solve problems of identification of dynamic systems, including aircraft, frequency methods are successfully applied. The available results show [104–107] that for a given frequency range, it is possible to effectively estimate the parameters of dynamical system models in real time.

The determination of the composition of the experiments for modeling the dynamical system in the frequency domain is an important part of the identification problem solving process. The experiments should be carried out with the aid of excitation signals applied to the input of the dynamical system covering a certain predetermined frequency range.

In the case where dynamical system parameter estimation is performed in real time, it is desirable that the stimulating effects on the dynamical system are small. If this condition is met, then the response of the dynamical system (in particular, aircraft) to the effect of the exciting inputs will be comparable in intensity with the reaction, for example, to atmospheric turbulence. Then the test excitatory effects will be hardly distinguishable from the natural disturbances and will not cause unnecessary worries to the crew of the aircraft.

Modern aircraft, as one of the most important types of simulated dynamical system, have a significant number of controls (rudders, etc.). When obtaining the data required for frequency analysis and dynamical system identification, it is highly desirable to be able to apply a test excitation signal to all these organs at the same time to reduce the total time required for data collection.

Schröder's work [108] showed the promise of using a polyharmonic excitation signal for these purposes, which is a set of sinusoids shifted in phase relative to each other. Such a signal makes it possible to obtain an excitation signal with a rich frequency content and a small peak factor value (amplitude coefficient). Such a signal is referred to as a Schröder sweep.

The peak factor is the ratio of the maximum amplitude of the input signal to the energy of the input signal. Inputs with small peak factor values are effective in that they provide a good frequency content of the dynamical system response without large amplitudes of the output signal (reaction) of the dynamical system in the time domain.

The paper [107] proposes the development of an approach to the formation of a sweep signal by Schröder, which makes it possible to obtain such a signal for the case of several controls used simultaneously, with optimization of the peak factor values for them. This development is oriented to work in real time.

The excitation signals generated in [107] are mutually orthogonal in both the time and frequency domain; they are interpreted as perturbations that are additional to the values of the corresponding control inputs required for the realization of the given behavior of the dynamical system.

To generate test excitation signals, only a priori information is needed in the form of approximate estimates of the frequency band inherent in the dynamical system in question, as well as the relative effectiveness of the controls for correctly scaling the amplitudes of the input signals.

GENERATION OF A SET OF POLYHARMONIC EXCITATION SIGNALS

The mathematical model of the input perturbation signal $\mathbf{u}_j$ affecting the jth control is the harmonic polynomial

$$\begin{gathered}\mathbf{u}_j = \sum_{k \in I_k} A_k \sin\left(\frac{2\pi k t}{T} + \varphi_k\right), \\ I_k \subset K, \; K = \{1, 2, \ldots, M\},\end{gathered} \tag{2.135}$$

as a finite linear combination of the fundamental harmonic $A_1 \sin(\omega t + \varphi_1)$ and higher-order harmonics $A_2 \sin(2\omega t + \varphi_2)$, $A_3 \sin(3\omega t + \varphi_2)$, and so on.

The input effect for each of the m controls (for example, the steering surfaces of the aircraft) is formed as the sum of the harmonic signals (sinusoids), each of which has its own phase shift φ_k. The input signal $\mathbf{u}_j$ corresponding to the jth control body has the following form:

$$\begin{gathered}\mathbf{u}_j = \sum_{k \in I_k} A_k \cos\left(\frac{2\pi k \mathbf{t}}{T} + \varphi_k\right), \; j = 1, \ldots, m, \\ I_k \subset K, \; K = \{1, 2, \ldots, M\},\end{gathered} \tag{2.136}$$

where M is the total number of harmonically related frequencies; T is the time interval during which a test excitation signal acts on the dynamical system; A_k is the amplitude of the kth sinusoidal component. The expression (2.136) is written in discrete time for N samples

$$\mathbf{u}_j = \{u_j(0), u_j(1), \ldots, u_j(i), \ldots, u_j(N-1)\},$$

where $u_j(i) = u_j(t(i))$.

Each of the m inputs (disturbance effects) is formed from sinusoids with frequencies

$$\omega_k = \frac{2\pi k}{T}, \quad k \in I_k, \quad I_k \subset K, \quad K = \{1, 2, \ldots, M\},$$

where $\omega_M = 2\pi M/T$ is the upper boundary value of the frequency band of the exciting input signals (influences). The interval $[\omega_1, \omega_M]$ specifies the frequency range in which the dynamics of the aircraft under study is expected to lie.

If the phase angles φ_k in (2.136) are chosen randomly in the interval $(-\pi, \pi]$, then in general, the individual harmonic components (oscillations), being summed, can give at $t(i)$ a value of the amplitude of the sum signal $u_j(i)$ at which the conditions of proximity of the disturbed motion to the reference one are violated.

In (2.136), φ_k is the phase shift that must be selected for each of the harmonic components in such a way as to provide a small value of the peak factor[7] (amplitude factor) $\mathrm{PF}(\mathbf{u}_j)$, which is defined by the relation

$$\mathrm{PF}(\mathbf{u}_j) = \frac{(u_j^{max} - u_j^{min})}{2\sqrt{(\mathbf{u}_j^T \mathbf{u}_j)/N}} \tag{2.137}$$

or

$$\mathrm{PF}(\mathbf{u}_j) = \frac{(u_j^{max} - u_j^{min})}{2\,\mathrm{rms}(\mathbf{u}_j)} = \frac{||\mathbf{u}_j||_\infty}{||\mathbf{u}_j||_2}, \tag{2.138}$$

where the last equality is satisfied only in the case when $\mathbf{u}_j$ oscillates symmetrically with respect to zero. In the relations (2.137) and (2.138),

$$u_j^{min} = \min_i [u_j(i)], \quad u_j^{max} = \max_i [u_j(i)].$$

For an individual sinusoidal component in (2.135), if the value of the peak factor equals $\mathrm{PF} = \sqrt{2}$, then the value of the peak factor related to such a component $\mathrm{RPF}(\mathbf{u}_j)$ (relative peak factor,[8] relative amplitude factor) is defined as

$$\mathrm{RPF}(\mathbf{u}_j) = \frac{(u_j^{max} - u_j^{min})}{2\sqrt{2}\,\mathrm{rms}(\mathbf{u}_j)} = \frac{\mathrm{PF}(\mathbf{u}_j)}{\sqrt{2}}. \tag{2.139}$$

Minimizing the exponent (2.139) by selecting the appropriate phase shift values φ_k for all k allows to prevent the occurrence of the situation mentioned above, with the deviation of the disturbed motion from the reference to an invalid value.

[7]PF – Peak Factor.

[8]RPF – Relative Peak Factor.

GENERATION PROCEDURE FOR POLYHARMONIC EXCITATION SIGNALS

The procedure for forming a polyharmonic input for a given set of controls consists of the following steps.

1. Set the value of the time interval T, during which a disturbing effect will be applied to the input of the control object. The value of T determines the smallest value of the resolving power in frequency $\Delta f = 1/T$, as well as the minimum frequency limit $f_{min} \geqslant 2/T$.
2. Set the frequency range $[f_{min}, f_{max}]$, from which the frequencies of disturbing effects for the dynamical system under consideration will be selected. It corresponds to the frequency range of the expected reactions of this system to the applied effects. These effects cover the interval $[f_{min}, f_{max}]$ uniformly, with step Δf. The total number of used frequencies is

$$M = \left\lfloor \frac{f_{max} - f_{min}}{\Delta f} \right\rfloor + 1,$$

where $\lfloor \cdot \rfloor$ is the integer part of the real number.
3. Divide the set of indices $K = \{1, 2, \dots, M\}$ into approximately equal in number of elements subsets $I_j \subset K$, each of which determines the set of frequencies for the corresponding jth body management. This separation should be performed in such a way that the frequencies for different controls alternate. For example, for two controls, the set $K = \{1, 2, \dots, 12\}$ is divided according to this rule into subsets $I_1 = \{1, 3, \dots, 11\}$ and $I_2 = \{2, 4, \dots, 12\}$, and for three controls into subsets $I_1 = \{1, 4, 7, 10\}$, $I_2 = \{2, 5, 8, 11\}$, and $I_3 = \{3, 6, 9, 12\}$. This approach ensures the production of small peak factor values for individual input signals and also allows uniform coverage of the frequency range $[f_{min}, f_{max}]$ for each of these signals. If necessary, this kind of uniformity can be avoided, for example, if certain frequencies are to be empha-

sized, or, if necessary, some frequency components should be eliminated (in particular, for fear of causing undesired reaction of the control object). In the paper [106] it was established empirically that if the sets of I_j indices are formed in such a way that they contain numbers greater than 1, multiples of 2 or 3 (for example, $k = 2, 4, 6$ or $k = 5, 10, 15, 20$), then the phase shift for them can be optimized in such a way that the relative peak factor for the corresponding input action will be very close to 1, and in some cases it will be even less than 1. For the distribution of indices over subsets I_j, the following conditions must be satisfied:

$$\bigcup_j I_j = K, \quad K = \{1, 2, \ldots, M\}, \quad \bigcap_j I_j = \varnothing.$$

Each index $k \in K$ must be used exactly once. Compliance with this condition ensures mutual orthogonality of the input actions both in the time domain and in the frequency domain.

4. Generate, according to (2.136), the input action $\mathbf{u}_j$ for each of the controls used, and then calculate the initial phase angle values φ_k according to the Schröder method, assuming the uniformity of the power spectrum.
5. Find the phase angle values φ_k for each of the input actions $\mathbf{u}_j$ which minimize the relative peak factor for them.
6. For each of the input actions $\mathbf{u}_j$, perform a one-dimensional search procedure to find a constant time offset value such that the corresponding input signal starts at a zero value of its amplitude. This operation is equivalent to shifting the graph of the input signal along the time axis so that the point of intersection of this graph with the abscissa axis (i.e., with the time axis) coincides with the origin. The phase shift corresponding to such a displacement is added to the values of φ_k of all sinusoidal components (harmonics) of the considered input actions $\mathbf{u}_j$. It should be noted that to obtain a constant time shift of all components $\mathbf{u}_j$ their phase shifts will be different in magnitude, since each of the components has its own frequency different from the frequency of the other components. Since all components of the signal $\mathbf{u}_j$ are harmonics of the same fundamental frequency for the period of oscillations T, if the phase angles φ_k of all components are changed so that the initial value of the input signal was zero, then its value at the final moment of time will also be zero. In this case, the energy spectrum, orthogonality, and relative peak factor of the input signals remain unchanged.
7. Go back to step 5 and repeat the appropriate actions until either the relative peak factor reaches the prescribed value, or the limit number of iterations of the process is reached. For example, the target value of the relative peak factor can be set as 1.01, the maximum number of iterations 50.

There are a number of methods that allow to optimize the frequency spectrum of input (test) signals when solving the problem of estimating the parameters of a dynamic system. However, all these methods require a significant amount of computation, as well as a certain level of knowledge about the dynamical being investigated, usually tied to a certain nominal state of the system. With respect to the situation considered in this chapter, such methods are useless because the task is to identify the dynamics of the system in real time for various modes of its functioning that vary widely. In addition, the solution of the task of reconfiguring the control system in the event of failures and damage of the dynamical system requires the solution of the problem of identification with significant and unpredictable changes in the dynamics of the system. Under such conditions, the laborious calculation of the input effect optimized for the frequency spectrum does not make sense, and in some cases it is impossible, since it does not fit into real time. Instead, the frequency spectrum of all generated input influences is selected in such a way that it is uniform in a given frequency range in order to exert a sufficient excitatory effect on the dynamical system.

Step 6 of the process described above provides an input perturbation signal added to the main control action selected, for example, for balancing an airplane or for performing a predetermined maneuver.

REFERENCES

[1] Ollongren A. Definition of programming languages by interpreting automata. London, New York, San Francisco: Academic Press; 1974.

[2] Brookshear JG. Theory of computation: Formal languages, automata, and complexity. Redwood City, California: The Benjamin/Cummings Publishing Co.; 1989.

[3] Chiswell I. A course in formal languages, automata and groups. London: Springer-Verlag; 2009.

[4] Fu KS. Syntactic pattern recognition. London, New York: Academic Press; 1974.

[5] Fu KS. Syntactic pattern recognition and applications. Englewood Cliffs, New Jersey: Prentice Hall, Inc.; 1982.

[6] Fu KS, editor. Syntactic methods in pattern recognition, applications. Berlin, Heidelberg, New York: Springer-Verlag; 1977.

[7] Gonzalez RC, Thomason MG. Syntactic pattern recognition: An introduction. London: Addison-Wesley Publishing Company Inc.; 1978.

[8] Tutschku K. Recurrent multilayer perceptrons for identification and control: The road to applications. University of Würzburg, Institute of Computer Science, Research Report Series, Report No. 118; June 1995.

[9] Heister F, Müller R. An approach for the identification of nonlinear, dynamic processes with Kalman-filter-trained recurrent neural structures. University of Würzburg, Institute of Computer Science, Research Report Series, Report No. 193; April 1999.

[10] Haykin S. Neural networks: A comprehensive foundation. 2nd ed. Upper Saddle River, NJ, USA: Prentice Hall; 1998.

[11] Hagan MT, Demuth HB, Beale MH, De Jesús O. Neural network design. 2nd ed. PSW Publishing Co.; 2014.

[12] Graves A. Supervised sequence labelling with recurrent neural networks. Berlin, Heidelberg: Springer; 2012.

[13] Hammer B. Learning with recurrent neural networks. Berlin, Heidelberg: Springer; 2000.

[14] Kolen JF, Kremer SC. A field guide to dynamical recurrent networks. New York: IEEE Press; 2001.

[15] Mandic DP, Chambers JA. Recurrent neural networks for prediction: Learning algorithms, architectures and stability. New York, NY: John Wiley & Sons, Inc.; 2001.

[16] Medsker LR, Jain LC. Recurrent neural networks: Design and applications. New York, NY: CRC Press; 2001.

[17] Michel A, Liu D. Qualitative analysis and synthesis of recurrent neural networks. London, New York: CRC Press; 2002.

[18] Yi Z, Tan KK. Convergence analysis of recurrent neural networks. Berlin: Springer; 2004.

[19] Gupta MM, Jin L, Homma N. Static and dynamic neural networks: From fundamentals to advanced theory. Hoboken, New Jersey: John Wiley & Sons; 2003.

[20] Lin DT, Dayhoff JE, Ligomenides PA. Trajectory production with the adaptive time-delay neural network. Neural Netw 1995;8(3):447–61.

[21] Guh RS, Shiue YR. Fast and accurate recognition of control chart patterns using a time delay neural network. J Chin Inst Ind Eng 2010;27(1):61–79.

[22] Yazdizadeh A, Khorasani K, Patel RV. Identification of a two-link flexible manipulator using adaptive time delay neural networks. IEEE Trans Syst Man Cybern, Part B, Cybern 2010;30(1):165–72.

[23] Juang JG, Chang HH, Chang WB. Intelligent automatic landing system using time delay neural network controller. Appl Artif Intell 2003;17(7):563–81.

[24] Sun Y, Babovic V, Chan ES. Multi-step-ahead model error prediction using time-delay neural networks combined with chaos theory. J Hydrol 2010;395:109–16.

[25] Zhang J, Wang Z, Ding D, Liu X. H_∞ state estimation for discrete-time delayed neural networks with randomly occurring quantizations and missing measurements. Neurocomputing 2015;148:388–96.

[26] Yazdizadeh A, Khorasani K. Adaptive time delay neural network structures for nonlinear system identification. Neurocomputing 2002;77:207–40.

[27] Ren XM, Rad AB. Identification of nonlinear systems with unknown time delay based on time-delay neural networks. IEEE Trans Neural Netw 2007;18(5):1536–41.

[28] Beale MH, Hagan MT, Demuth HB. Neural network toolbox: User's guide. Natick, MA: The MathWorks, Inc.; 2017.

[29] Čerňanský M, Beňušková L. Simple recurrent network trained by RTRL and extended Kalman filter algorithms. Neural Netw World 2003;13(3):223–34.

[30] Elman JL. Finding structure in time. Cogn Sci 1990;14(2):179–211.

[31] Elman JL. Distributed representations, simple recurrent networks, and grammatical structure. Mach Learn 1991;7:195–225.

[32] Elman JL. Learning and development in neural networks: the importance of starting small. Cognition 1993;48(1):71–99.

[33] Chen S, Wang SS, Harris C. NARX-based nonlinear system identification using orthogonal least squares basis hunting. IEEE Trans Control Syst Technol 2008;16(1):78–84.
[34] Sahoo HK, Dash PK, Rath NP. NARX model based nonlinear dynamic system identification using low complexity neural networks and robust H_∞ filter. Appl Soft Comput 2013;13(7):3324–34.
[35] Hidayat MIP, Berata W. Neural networks with radial basis function and NARX structure for material lifetime assessment application. Adv Mater Res 2011;277:143–50.
[36] Wong CX, Worden K. Generalised NARX shunting neural network modelling of friction. Mech Syst Signal Process 2007;21:553–72.
[37] Potenza R, Dunne JF, Vulli S, Richardson D, King P. Multicylinder engine pressure reconstruction using NARX neural networks and crank kinematics. Int J Eng Res 2017;8:499–518.
[38] Patel A, Dunne JF. NARX neural network modelling of hydraulic suspension dampers for steady-state and variable temperature operation. Veh Syst Dyn: Int J Veh Mech Mobility 2003;40(5):285–328.
[39] Gaya MS, Wahab NA, Sam YM, Samsudin SI, Jamaludin IW. Comparison of NARX neural network and classical modelling approaches. Appl Mech Mater 2014;554:360–5.
[40] Siegelmann HT, Horne BG, Giles CL. Computational capabilities of recurrent NARX neural networks. IEEE Trans Syst Man Cybern, Part B, Cybern 1997;27(2):208–15.
[41] Kao CY, Loh CH. NARX neural networks for nonlinear analysis of structures in frequency domain. J Chin Inst Eng 2008;31(5):791–804.
[42] Billings SA. Nonlinear system identification: NARMAX methods in the time, frequency and spatio-temporal domains. New York, NY: John Wiley & Sons; 2013.
[43] Pearson PK. Discrete-time dynamic models. New York–Oxford: Oxford University Press; 1999.
[44] Nelles O. Nonlinear system identification: From classical approaches to neural networks and fuzzy models. Berlin: Springer; 2001.
[45] Sutton RS, Barto AG. Reinforcement learning: An introduction. Cambridge, Massachusetts: The MIT Press; 1998.
[46] Busoniu L, Babuška R, De Schutter B, Ernst D. Reinforcement learning and dynamic programming using function approximators. London: CRC Press; 2010.
[47] Kamalapurkar R, Walters P, Rosenfeld J, Dixon W. Reinforcement learning for optimal feedback control: A Lyapunov-based approach. Berlin: Springer; 2018.
[48] Lewis FL, Liu D. Reinforcement learning and approximate dynamic programming for feedback control. Hoboken, New Jersey: John Wiley & Sons; 2013.
[49] Gill PE, Murray W, Wright MH. Practical optimization. London, New York: Academic Press; 1981.
[50] Nocedal J, Wright S. Numerical optimization. 2nd ed. Springer; 2006.
[51] Fletcher R. Practical methods of optimization. 2nd ed. New York, NY, USA: Wiley-Interscience. ISBN 0-471-91547-5, 1987.
[52] Dennis J, Schnabel R. Numerical methods for unconstrained optimization and nonlinear equations. Society for Industrial and Applied Mathematics; 1996.
[53] Gendreau M, Potvin J. Handbook of metaheuristics. International series in operations research & management science. US: Springer. ISBN 9781441916655, 2010.
[54] Du K, Swamy M. Search and optimization by metaheuristics: Techniques and algorithms inspired by nature. Springer International Publishing. ISBN 9783319411927, 2016.
[55] Glorot X, Bengio Y. Understanding the difficulty of training deep feedforward neural networks. In: Teh YW, Titterington M, editors. Proceedings of the Thirteenth International Conference on Artificial Intelligence and Statistics. Proceedings of machine learning research, vol. 9. Chia Laguna Resort, Sardinia, Italy: PMLR; 2010. p. 249–56. http://proceedings.mlr.press/v9/glorot10a.html.
[56] Nocedal J. Updating quasi-Newton matrices with limited storage. Math Comput 1980;35:773–82.
[57] Conn AR, Gould NIM, Toint PL. Trust-region methods. Philadelphia, PA, USA: Society for Industrial and Applied Mathematics. ISBN 0-89871-460-5, 2000.
[58] Steihaug T. The conjugate gradient method and trust regions in large scale optimization. SIAM J Numer Anal 1983;20(3):626–37.
[59] Martens J, Sutskever I. Learning recurrent neural networks with Hessian-free optimization. In: Proceedings of the 28th International Conference on International Conference on Machine Learning. USA: Omnipress. ISBN 978-1-4503-0619-5, 2011. p. 1033–40. http://dl.acm.org/citation.cfm?id=3104482.3104612.
[60] Martens J, Sutskever I. Training deep and recurrent networks with Hessian-free optimization. In: Neural networks: Tricks of the trade. Springer; 2012. p. 479–535.
[61] Moré JJ. The Levenberg–Marquardt algorithm: Implementation and theory. In: Watson G, editor. Numerical analysis. Lecture notes in mathematics, vol. 630. Springer Berlin Heidelberg. ISBN 978-3-540-08538-6, 1978. p. 105–16.
[62] Moré JJ, Sorensen DC. Computing a trust region step. SIAM J Sci Stat Comput 1983;4(3):553–72. https://doi.org/10.1137/0904038.
[63] Bottou L, Curtis F, Nocedal J. Optimization methods for large-scale machine learning. SIAM Rev 2018;60(2):223–311. https://doi.org/10.1137/16M1080173.

[64] Griewank A, Walther A. Evaluating derivatives: Principles and techniques of algorithmic differentiation. 2nd ed. Philadelphia, PA, USA: Society for Industrial and Applied Mathematics. ISBN 0898716594, 2008.

[65] Griewank A. On automatic differentiation. In: Mathematical programming: Recent developments and applications. Kluwer Academic Publishers; 1989. p. 83–108.

[66] Bishop C. Exact calculation of the Hessian matrix for the multilayer perceptron. Neural Comput 1992;4(4):494–501. https://doi.org/10.1162/neco.1992.4.4.494.

[67] Werbos PJ. Backpropagation through time: What it does and how to do it. Proc IEEE 1990;78(10):1550–60.

[68] Chauvin Y, Rumelhart DE, editors. Backpropagation: Theory, architectures, and applications. Hillsdale, NJ, USA: L. Erlbaum Associates Inc.. ISBN 0-8058-1259-8, 1995.

[69] Jesus OD, Hagan MT. Backpropagation algorithms for a broad class of dynamic networks. IEEE Trans Neural Netw 2007;18(1):14–27.

[70] Williams RJ, Zipser D. A learning algorithm for continually running fully recurrent neural networks. Neural Comput 1989;1(2):270–80.

[71] Bengio Y, Simard P, Frasconi P. Learning long-term dependencies with gradient descent is difficult. Trans Neural Netw 1994;5(2):157–66. https://doi.org/10.1109/72.279181.

[72] Hochreiter S, Bengio Y, Frasconi P, Schmidhuber J. Gradient flow in recurrent nets: The difficulty of learning long-term dependencies. In: Kolen J, Kremer S, editors. A field guide to dynamical recurrent networks. IEEE Press; 2001. p. 15.

[73] Kremer SC. A field guide to dynamical recurrent networks. 1st ed. Wiley-IEEE Press. ISBN 0780353692, 2001.

[74] Pascanu R, Mikolov T, Bengio Y. On the difficulty of training recurrent neural networks. In: Proceedings of the 30th International Conference on International Conference on Machine Learning, vol. 28. JMLR.org; 2013. pp. III–1310–III–1318.

[75] Hochreiter S, Schmidhuber J. Long short-term memory. Neural Comput 1997;9:1735–80.

[76] Gers FA, Schmidhuber J, Cummins F. Learning to forget: Continual prediction with LSTM. Neural Comput 1999;12:2451–71.

[77] Gers FA, Schmidhuber J. Recurrent nets that time and count. In: Proceedings of the IEEE-INNS-ENNS International Joint Conference on Neural Networks. IJCNN 2000. Neural computing: new challenges and perspectives for the New Millennium, vol. 3; 2000. p. 189–94.

[78] Gers FA, Schraudolph NN, Schmidhuber J. Learning precise timing with LSTM recurrent networks. J Mach Learn Res 2003;3:115–43. https://doi.org/10.1162/153244303768966139.

[79] Graves A, Schmidhuber J. Framewise phoneme classification with bidirectional LSTM networks. In: Proceedings. 2005 IEEE International Joint Conference on Neural Networks, 2005, vol. 4; 2005. p. 2047–52.

[80] Greff K, Srivastava RK, Koutník J, Steunebrink BR, Schmidhuber J. LSTM: A search space odyssey. CoRR 2015;abs/1503.04069. http://arxiv.org/abs/1503.04069.

[81] Wang Y. A new concept using LSTM neural networks for dynamic system identification. In: 2017 American Control Conference (ACC), vol. 2017; 2017. p. 5324–9.

[82] Doya K. Bifurcations in the learning of recurrent neural networks. In: Proceedings of 1992 IEEE International Symposium on Circuits and Systems, vol. 6; 1992. p. 2777–80.

[83] Pasemann F. Dynamics of a single model neuron. Int J Bifurc Chaos Appl Sci Eng 1993;03(02):271–8. http://www.worldscientific.com/doi/abs/10.1142/S0218127493000210.

[84] Haschke R, Steil JJ. Input space bifurcation manifolds of recurrent neural networks. Neurocomputing 2005;64(Supplement C):25–38. https://doi.org/10.1016/j.neucom.2004.11.030.

[85] Jesus OD, Horn JM, Hagan MT. Analysis of recurrent network training and suggestions for improvements. In: Neural Networks, 2001. Proceedings. IJCNN '01. International Joint Conference on, vol. 4; 2001. p. 2632–7.

[86] Horn J, Jesus OD, Hagan MT. Spurious valleys in the error surface of recurrent networks: Analysis and avoidance. IEEE Trans Neural Netw 2009;20(4):686–700.

[87] Phan MC, Hagan MT. Error surface of recurrent neural networks. IEEE Trans Neural Netw Learn Syst 2013;24(11):1709–21. https://doi.org/10.1109/TNNLS.2013.2258470.

[88] Samarin AI. Neural networks with pre-tuning. In: VII All-Russian Conference on Neuroinformatics. Lectures on neuroinformatics. Moscow: MEPhI; 2005. p. 10–20 (in Russian).

[89] Jategaonkar RV. Flight vehicle system identification: A time domain methodology. Reston, VA: AIAA; 2006.

[90] Morozov NI, Tiumentsev YV, Yakovenko AV. An adjustment of dynamic properties of a controllable object using artificial neural networks. Aerosp MAI J 2002;(1):73–94 (in Russian).

[91] Krasovsky AA. Automatic flight control systems and their analytical design. Moscow: Nauka; 1973 (in Russian).

[92] Krasovsky AA, editor. Handbook of automatic control theory. Moscow: Nauka; 1987 (in Russian).

[93] Graupe D. System identification: A frequency domain approach. New York, NY: R.E. Krieger Publishing Co.; 1976.

[94] Ljung L. System identification: Theory for the user. 2nd ed. Upper Saddle River, NJ: Prentice Hall; 1999.
[95] Sage AP, Melsa JL. System identification. New York and London: Academic Press; 1971.
[96] Tsypkin YZ. Information theory of identification. Moscow: Nauka; 1995 (in Russian).
[97] Isermann R, Münchhoh M. Identification of dynamic systems: An introduction with applications. Berlin: Springer; 2011.
[98] Juang JN, Phan MQ. Identification and control of mechanical systems. Cambridge, MA: Cambridge University Press; 1994.
[99] Pintelon R, Schoukens J. System identification: A frequency domain approach. New York, NY: IEEE Press; 2001.
[100] Berestov LM, Poplavsky BK, Miroshnichenko LY. Frequency domain aircraft identification. Moscow: Mashinostroyeniye; 1985 (in Russian).
[101] Vasilchenko KK, Kochetkov YA, Leonov VA, Poplavsky BK. Structural identification of mathematical model of aircraft motion. Moscow: Mashinostroyeniye; 1993 (in Russian).
[102] Klein V, Morelli EA. Aircraft system identification: Theory and practice. Reston, VA: AIAA; 2006.
[103] Tischler M, Remple RK. Aircraft and rotorcraft system identification: Engineering methods with flight-test examples. Reston, VA: AIAA; 2006.
[104] Morelli EA, In-flight system identification. AIAA–98–4261, 10.
[105] Morelli EA, Klein V. Real-time parameter estimation in the frequency domain. J Guid Control Dyn 2000;23(5):812–8.
[106] Morelli EA, Multiple input design for real-time parameter estimation in the frequency domain, in: 13th IFAC Conf. on System Identification, Aug. 27–29, 2003, Rotterdam, The Netherlands. Paper REG-360, 7.
[107] Smith MS, Moes TR, Morelli EA, Flight investigation of prescribed simultaneous independent surface excitations for real-time parameter identification. AIAA–2003–5702, 23.
[108] Schroeder MR. Synthesis of low-peak-factor signals and binary sequences with low autocorrelation. IEEE Trans Inf Theory 1970;16(1):85–9.
[109] Brusov VS, Tiumentsev YuV. Neural network modeling of aircraft motion. Moscow: MAI; 2016 (in Russian).

CHAPTER

3

Neural Network Black Box Approach to the Modeling and Control of Dynamical Systems

3.1 TYPICAL PROBLEMS ASSOCIATED WITH DEVELOPMENT AND MAINTENANCE OF DYNAMICAL SYSTEMS

As noted earlier, the object of our research is a controlled dynamical system, operating under conditions of various uncertainties.

Among the main types of uncertainties that need to be considered when solving problems related to controlled dynamical systems are the following:

- uncertainties generated by uncontrolled disturbances acting on the object;
- incomplete and inaccurate knowledge of the properties and characteristics of the simulated object and the conditions in which it operates;
- uncertainties caused by a change in the properties of the object due to failures of its equipment and structural damage.

As shown in Chapter 1, the behavior of the system is largely determined by the current and/or predicted situation, including external and internal components, i.e.,

$$\text{Situation} = \text{External-Situation (Environment)} + \text{Internal-Situation (Object)}.$$

The main problem is that, due to the presence of uncertainties, the current situation for the dynamic system under consideration may change significantly and unpredictably. This circumstance we have to take into account both in modeling the system and in controlling its behavior.

In Chapter 1, the dynamical system $\mathbb{S}$ was defined as an ordered triple of the following form:

$$\mathbb{S} = \langle \mathrm{U}, \mathrm{P}, \mathrm{Y} \rangle, \tag{3.1}$$

where U is the input to the simulated/controlled object, P is the simulated/controlled object (plant), and Y is a response of the object to the input signal. In this definition:

- the input actions U are the initial conditions, controls, and uncontrolled external influences on the object P;
- the simulated/controlled object P is an aircraft or another type of controllable dynamical system;
- the outputs Y of the dynamical system $\mathbb{S}$ are the observed reactions of the object P to the input actions.

https://doi.org/10.1016/B978-0-12-815254-6.00013-7

Bearing in mind the definition (3.1), we can distinguish the following main classes of problems related to dynamical systems:

1. $\langle U, P, \mathbf{Y} \rangle$ is behavior analysis for a dynamical system (for U and P find Y);
2. $\langle \mathbf{U}, P, Y \rangle$ is control synthesis for a dynamical system (for P and Y find U);
3. $\langle U, \mathbf{P}, Y \rangle$ is identification for a dynamical system (for U and Y find P).

Problems 2 and 3 belong to the class of inverse problems. Problem 3 is related to the process of creating a model of some dynamical system, while problems 1 and 2 associate with using previously developed models.

3.2 NEURAL NETWORK BLACK BOX APPROACH TO SOLVING PROBLEMS ASSOCIATED WITH DYNAMICAL SYSTEMS

Traditionally, differential equations (for continuous time systems) or difference equations (for discrete time systems) are used as models of dynamical systems. As noted above, in some cases such models do not meet certain requirements, in particular, the requirement of adaptability, which is necessary in case the model is supposed to be applied in onboard control systems. An alternative approach is to use ANN models that are well suited for application of various adaptation algorithms.

In this section, we consider ANN models of the traditional empirical type, i.e., models of the black box type [1–11] for dynamical systems. In Chapter 5, we will extend these models to semiempirical (gray box) ones by embedding the available theoretical knowledge about the simulated system into the model.

3.2.1 Main Types of Models

There are two main approaches to the representation (description) of dynamical systems [12–14]:

- a representation of the dynamical system in the state space (state space representation);
- a representation of the dynamical system in terms of input–output relationships (input–output representation).

To simplify the description of approaches to the modeling of dynamical systems, we will assume that the system under consideration has a single output. The obtained results are generalized to dynamical systems with vector-valued output without any difficulties.

For the case of discrete time (most important for ANN modeling), we say about the model that it is a representation of a dynamical system in the state space if this model has the following form:

$$\begin{aligned} x(k) &= f(x(k-1), u(k-1), \xi_1(k-1)), \\ y(k) &= g(x(k), \xi_2(k)), \end{aligned} \tag{3.2}$$

where the vector $x(k)$ is the state vector (also called phase vector) of the dynamical system whose components are variables describing the state of the object at time instant t_k; the vector $u(k)$ contains the input control variables of the dynamical system as its components; vectors $\xi_1(k)$ and $\xi_2(k)$ describe disturbances that affect the dynamical system; the scalar variable $y(k)$ is the output of the dynamical system; $f(\cdot)$ and $g(\cdot)$ are a nonlinear vector-valued function and a scalar-valued function, respectively. The dimension of the state vector (that is, the number of state variables in this vector) is usually called the order of the model. State variables can be either available for observation and measurement of their values, or unobservable. As a special case, dynamical system output may be equal to one of its state variables. The disturbances $\xi_1(k)$ and $\xi_2(k)$ can affect the values of the dynamical system outputs and/or its states. In contrast to the input control actions, these disturbances are unobservable.

The design procedure for a state space dynamical system model involves finding approximate representations for the functions $f(\cdot)$ and $g(\cdot)$, using the available data on the system. In the case a model of the black box type is designed, that is, we use no a priori knowledge of the nature and features of the simulated system, such data are represented by sequences of values of the input and output variables of the system.

A dynamical system model is said to have an input–output representation (a representation of the system in terms of its inputs and outputs), if it has the following form:

$$y(k) = h(y(k-1), \ldots, y(k-n), u(k-1), \ldots, u(k-m), \xi(k-1), \ldots, \xi(k-p)), \tag{3.3}$$

where $h(\cdot)$ is a nonlinear function, n is the order of the model, m and p are positive integer constants, $u(k)$ is a vector of input control signals of the dynamical system, and $\xi(k)$ is the disturbance vector. The input–output representation can be considered a special case of the state space representation when all the components of the state vector are observable and treated as output signals of the dynamical system.

In the case the simulated system is linear and time invariant, the state space representation and the input–output representation are equivalent [12,13]. Therefore we can choose which of them is more convenient and efficient from the point of view of the problem being solved. In contrast, if the simulated system is nonlinear, the state space representation is more general and at the same time more reasonable in comparison with the input–output representation. However, the implementation of the model in the state space is usually somewhat more difficult than the input–output model because it requires to obtain an approximate representation for two mappings, $f(\cdot)$ and $g(\cdot)$ in (3.2) instead of a single mapping $h(\cdot)$ in (3.3).

The choice of the suitable model representation (state space or input–output model) is not the only design choice required to take when modeling a nonlinear dynamical system. The choice of method for taking disturbances into account also plays an important role. There are three possible options:

- disturbances affect the states of the dynamical system;
- disturbances affect the outputs of the dynamical system;
- disturbances affect both the states and the outputs of the dynamical system.

As shown in [14], the nature of the disturbance effect on the dynamical system significantly influences the optimal structure of the model being formed, the type of the required algorithm for its learning, and the operation mode of the generated model. In the next section, we consider these issues in more detail.

3.2.2 Approaches to Consideration of Disturbances Acting on a Dynamical System

As noted, the way in which we take into account the influence of disturbances in the model significantly influences both the structure of the model and its training algorithm.

3.2.2.1 Input–Output Representation of the Dynamical System

Let us first consider the case in which disturbance affects the state of the dynamical system. We assume that the required representation of the dynamical system has the following form:

$$y_p(k) = \psi(y_p(k-1), \ldots, y_p(k-n), u(k-1), \ldots, u(k-m)) + \xi(k), \tag{3.4}$$

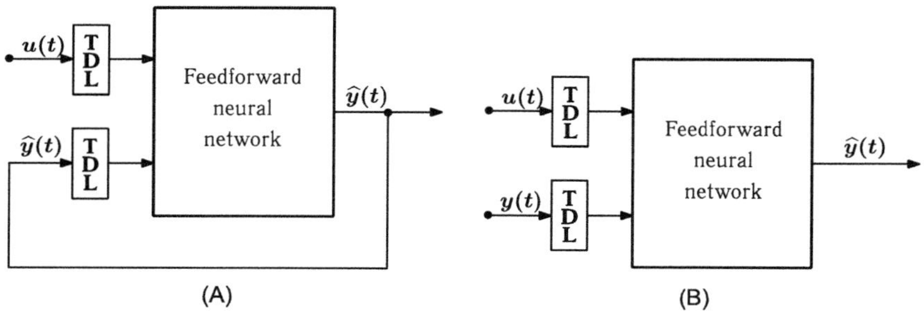

FIGURE 3.1 General structure of the NARX-model. (A) Model with parallel architecture. (B) Model with series-parallel architecture.

where $y_p(k)$ is the observed (measured) output of the process described by the dynamical system [12–14].

We assume that the output of the dynamical system is affected by additive noise, and the point of summation of the output signal and noise precedes the point from which the feedback signal comes. In this case, the output of the system at time step k will be affected by the noise signal both at time step k and also at n previous time steps.

In the case the function ψ is represented by a feedforward neural network, the representation (3.4) corresponds to a NARX-type model (Nonlinear Auto-Regressive network with eXogeneous inputs) [15–23], i.e., nonlinear autoregression with external inputs, in its series-parallel version (see Fig. 3.1B).

As noted above, we consider the case when the additive noise affecting the output of the dynamical system influences outputs not only directly at current time step k, but also via the outputs at n preceding time steps. The requirement to take into account previous outputs is imposed because ideally, the simulation error at step k should be equal to the noise value at the same time. Accordingly, when designing a model of a dynamical system, it is necessary to take into account system outputs at past time instants to compensate for the noise effects that have occurred. The corresponding ideal model can have the form of a feedforward neural network that implements the following mapping:

$$g(k) = \varphi_{NN}(y_p(k-1), \dots, y_p(k-n), u(k-1), \dots, u(k-m), w), \tag{3.5}$$

where w is a vector of parameters and $\varphi_{NN}(\cdot)$ is a function implemented by a feedforward network.

Suppose that the values of parameters w for the network are computed by training it in such a way that $\varphi_{NN}(\cdot) = \varphi(\cdot)$, i.e., this network accurately reproduces the outputs of the simulated dynamical system. In this case, for all time instants k, the relation

$$y_p(k) - g(k) = \xi(k), \quad \forall k \in \{0, N\},$$

will be satisfied, i.e., the simulation error is equal to the noise affecting the output of the dynamical system. This model can be called ideal in the sense that it accurately reflects the deterministic components of the dynamical system process and does not reproduce the noise that distorts the output signal of the system. The inputs of this model are the values of the control variables, as well as the measured outputs of the process implemented by the dynamical system. In this case, the ideal model, which is a one-step-ahead predictor, is trained as a feedforward neural network, and not as a recurrent network. Thus, in this case in order to obtain an optimal model, it is advisable to use supervised learning methods available for static ANN models.

Since the inputs of the predictor network include, in addition to the control values, the measured (observed) values of the outputs for the process implemented by the dynamical system, the output of the model of the considered type can be calculated only one time step ahead (accordingly, predictors of this type are usually called one-step-ahead predictors). If the generated model should reflect the behavior of the dynamical system on a time horizon exceeding one time step, we will have to feed back the outputs of the predictor at the previous time instants to its inputs at the current time step. In this case, the predictor will no longer have the properties of the ideal model due to the accumulation of the prediction error.

The second type of noise impact on a system that requires consideration corresponds to the case when noise affects the output of the dynamical system. In this case, the corresponding description of the process implemented by the dynamical system has the following form:

$$\begin{aligned} x_p(k) &= \varphi(x_p(k-1), \dots, x_p(k-n), \\ &\qquad u(k-1), \dots, u(k-m)), \\ y_p(k) &= x_p(k) + \xi(k). \end{aligned} \tag{3.6}$$

This structural organization of the model implies that additive noise is added directly to the output signal of the dynamical system (this is a parallel version of the NARX-type model architecture; see Fig. 3.1A). Thus, noise signal at some time step k affects only the dynamical system output at the same time instant k.

Since the output of the model at time step k depends on the noise only at the same instant of time, the optimal model does not require the values of the outputs of the dynamical system at the preceding instants; it is sufficient to use their estimates generated by the model itself. Therefore, an "ideal model" for this case is represented by a recurrent neural network that implements a mapping of the following form:

$$\begin{aligned} g(k) = \varphi_{NN}(&g(k-1), \dots, g(k-n), \\ &u(k-1), \dots, u(k-m), w), \end{aligned} \tag{3.7}$$

where, as in (3.5), w is a vector of parameters and $\varphi_{NN}(\cdot)$ is a function implemented by a feedforward network.

Again, let us suppose that the values of parameters w of the network are computed by training it in such a way that $\varphi_{NN}(\cdot) = \varphi(\cdot)$. We also assume that for the first n time points, the prediction error is equal in magnitude to the noise affecting the dynamical system. In this case, for all time instants k, $k = 0, \dots, n-1$, the relation

$$y_p(k) - g(k) = \xi(k), \quad \forall k \in \{0, n-1\},$$

will be satisfied. Therefore, the simulation error will be numerically equal to the noise affecting the output of the dynamical system, i.e., this model might be considered to be optimal in the sense that it accurately reflects the deterministic components of the process of the dynamical system operation and does not reproduce the noise that distorts the output signal of the system.

If the initial modeling conditions are not satisfied (exact output values at initial time steps are unavailable), but the condition $\varphi_{NN}(\cdot) = \varphi(\cdot)$ is satisfied and the model is stable with respect to the initial conditions, then the simulation error will decrease as the time step k increases.

As we can see from the above relations, the ideal model under the additive output noise assumption is a closed-loop recurrent network, as opposed to the case of state noise, when the ideal model is represented by a static feedforward network.

Accordingly, in order to train a parallel-type model, in general, it is required to apply methods designed for dynamic networks, which, of course, are more difficult in comparison with the learning methods for static networks. However, for the models of the type in question, learning methods can be proposed that take advantage of the specifics of these models to reduce the computational complexity of the conventional methods for dynamic networks learning. Possible ways of constructing such methods are discussed in Chapters 2 and 5.

Due to the nature of the impact of noise on the operation of parallel models, they can be used not only as one-step-ahead predictors, as is the case for serial-parallel models, but also as full-fledged dynamical system models that allow us to analyze the behavior of these systems over a time interval of the required duration, and not just one step ahead.

The last type of the noise influence on the simulated system is the case when the noise simultaneously effects both the outputs and states of the dynamical system. This corresponds to a model of the form

$$\begin{aligned} x_p(k) = \varphi(&x_p(k-1), \dots, x_p(k-n), \\ &u(k-1), \dots, u(k-m), \\ &\xi(k-1), \dots, \xi(k-p)), \\ y_p(k) = x_p&(k) + \xi(k). \end{aligned} \tag{3.8}$$

Such models belong to the class NARMAX (Nonlinear Auto-Regressive networks with Moving Average and eXogeneous inputs) [1, 24], i.e., represent nonlinear autoregression with moving average and external (control) inputs. In the case under consideration, the model being developed takes into account both the previous values of the measured outputs of the dynamical system and the previous values of the outputs estimated by the model itself. Because such a model is a combination of the two models considered earlier, it can be used only as a one-step-ahead predictor, similar to a model with noise affecting the states.

3.2.2.2 *State Space Representation of the Dynamical System*

In the previous section, we have discussed several ways to take into account the disturbances and demonstrated how this design choice influences the input–output model structure and its training procedure. Now we consider the state space representation of a dynamical system, which in the case of nonlinear system modeling, as noted above, is more general than the input–output representation [12–14].

Let us first consider the case when noise affects the output of the dynamical system. We assume that the required representation of the dynamical system has the following form:

$$\begin{aligned} x(k) &= \varphi(x(k-1), u(k-1)), \\ y(k) &= \psi(x(k)) + \xi(k). \end{aligned} \tag{3.9}$$

Since in this case the noise is present only in the observation equation, it does not affect the dynamics of the simulated object. Based on the arguments similar to those given above for the case of input–output representation of a dynamical system, the ideal model for the case under consideration is represented by a recurrent network defined by the following relations:

$$\begin{aligned} x(k) &= \varphi_{NN}(x(k-1), u(k-1)), \\ y(k) &= \psi_{NN}(x(k)), \end{aligned} \tag{3.10}$$

where $\varphi_{NN}(\cdot)$ is the exact representation of the function $\varphi(\cdot)$ and $\psi_{NN}(\cdot)$ is the exact representation of the function $\psi(\cdot)$.

Let us consider the noise assumption of the second type, namely, the case when noise affects the state of the dynamical system. In this case, the corresponding description of the process implemented by the dynamical system has the form

$$\begin{aligned} x(k) &= \varphi(x(k-1), u(k-1), \xi(k-1)), \\ y(k) &= \psi(x(k)). \end{aligned} \tag{3.11}$$

Based on the same considerations as for the input–output representation of the dynamical system, we can conclude that in this case the inputs of the ideal model, in addition to the controls u, must also include the state variables of the dynamical system. There are two possible situations:

- state variables are observable, hence they can be interpreted as outputs of the system, and the problem is reduced to the one previously considered for the input–output representation case; the ideal model will be a feedforward neural network, which can be used as a one-step-ahead predictor;
- state variables are not observable, and therefore an ideal model cannot be constructed; In this case, we should use the input–output representation (with some loss of generality of the model), or build some recurrent model, although it will not be optimal in this situation.

The last type of the noise influence on the simulated system is the case when the noise simultaneously affects both the outputs and the states of the dynamical system. This assumption leads to the following model:

$$\begin{aligned} x(k) &= \varphi(x(k-1), u(k-1), \xi_1(k-1)), \\ y(k) &= \psi(x(k), \xi_2(k)). \end{aligned} \tag{3.12}$$

Similar to the previous case, two situations are possible:

- if the state variables are observable, they can be interpreted as outputs of the dynamical system, and the problem is reduced to the one previously considered for the input–output representation case;
- if the state variables are not observable, the ideal model should include both states and the observed output of the system.

3.3 ANN-BASED MODELING AND IDENTIFICATION OF DYNAMICAL SYSTEMS

3.3.1 Feedforward Neural Networks for Modeling of Dynamical Systems

The most natural approach to implementing models of dynamical systems is the use of recurrent neural networks. Such networks themselves are dynamical systems, which determines the validity of this approach. However, dynamic networks are very difficult to learn. For this reason, it is advisable, in situations where possible, to use feedforward networks that are simpler in terms of their learning processes.

Feedforward networks can be used both in tasks of modeling dynamical systems and for control of such systems in two situations.

In the first one, we solve the problem of modeling some uncontrolled dynamical system, which implements the trajectory depending only on the initial conditions (and possibly disturbances acting on the system). For a single variant of the initial conditions, the solution of the problem will be a trajectory described by some function which is nonlinear in the general case. As is well known [14,25–29], the feedforward networks have universal approximating properties, i.e., the task of describing the behavior of a dynamical system, in this case, is reduced to the formation of appropriate architecture and the training of a feedforward neural network.

In real-world problems, the case of a single variant of the initial conditions is not typical. Usually, there is a range of relevant initial conditions for the dynamical system under consideration. In this case, we can enter the parametrization of the trajectories implemented by the system, where the parameters are the initial conditions. The simplest variant is to cover the range of reasonable values of the initial conditions with a finite set of their "typical values" and construct a bundle of trajectories corresponding to these initial conditions. In this case, we form this bundle in such a way that the distance between trajectories does not exceed a specific predetermined threshold value. Then, with the appearance of initial conditions that do not coincide with any of the available sets, we take from this set the value closest to the one presented. This approach is conceptually close to the one used in Chapter 5 to form a set of reference trajectories in the task of obtaining training data for

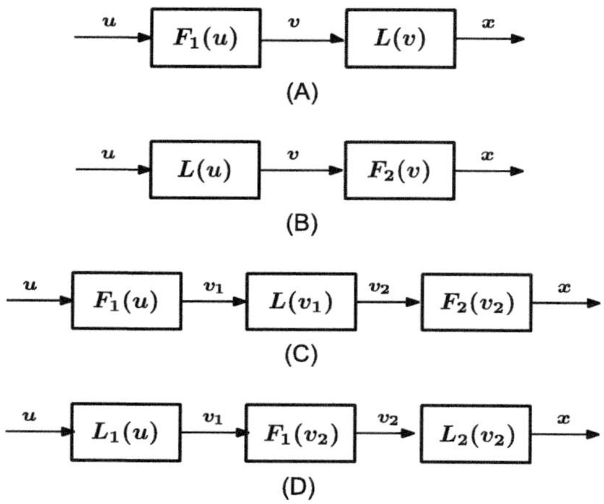

FIGURE 3.2 Block-oriented models of controllable dynamic systems. (A) Wiener model (N–L). (B) Hammerstein model (L–N). (C) Hammerstein–Wiener model (N–L–N). (D) Wiener–Hammerstein model (L–N–L). Here $F(\cdot)$, $F_1(\cdot)$, and $F_2(\cdot)$ are static nonlinearities (nonlinear functions); $L(\cdot)$, $F_1(\cdot)$, and $F_2(\cdot)$ are linear dynamical systems (differential equations or linear RNN).

dynamic ANN models. This approach, as well as some others, based on the use of feedforward neural networks, finds some application in solving problems of modeling and identification of dynamical systems [30–38].

The second situation is related to the block-oriented approach to system modeling. With this approach, the dynamical system is represented as a set of interrelated and interacting blocks. Some of these blocks will represent the realization of some functions that are nonlinear in the general case. These nonlinear functions can be realized in various ways, including in the form of a feedforward neural network. The value of the neural network approach in this case is that a particular kind of these ANN functions can be "recovered" on the basis of experimental data on the simulated system by using appropriate learning algorithms.

A typical example of such a block-oriented approach are the nonlinear controlled systems of Wiener, Hammerstein, Wiener–Hammerstein, and Hammerstein–Wiener type (Fig. 3.2) [39–50]. These systems are sets of blocks of the type "static nonlinearity" (realized as a nonlinear function) and "linear dynamical system" (realized as a system of linear differential equations or as a linear recurrent network). The Wiener model (Fig. 3.2A) contains a combination of one nonlinear block (N) of the first type and the next linear block (L) of the second type (structure of the form N–L), the Hammerstein model (Fig. 3.2B) is characterized by a structure of the type L–N. Combined variants of these structures consist of three blocks: two of the first type and one of the second type in the Hammerstein–Wiener model (structure of the type N–L–N, Fig. 3.2C) and two of the second type and one of the first type in the Wiener–Hammerstein model (a structure of the form L–N–L, Fig. 3.2D). This block-oriented approach is suitable for systems of classes SISO, MISO, and MIMO. Some works [39–50] show ANN implementations of models of these kinds.

Using the ANN approach to implement models of the abovementioned types in comparison with traditional approaches provides the following advantages:

- static nonlinearity (nonlinear function) can be of almost any complexity; in particular, it can be multidimensional (function of several variables);
- the $F(\cdot)$ transformations required to implement the block-based approach, which is often used to solve various applied problems, are formed by training on experimental data characterizing the behavior of the dynamical system under consideration, i.e., there is no need for a laborious process of forming such relationships before the beginning of the modeling process;
- a characteristic feature of ANN models is their "inherent adaptability," realized through the learning processes of networks, which provides, under certain conditions, the possibility of on-line adjustment of the model

immediately in the process of the dynamical system operating.

3.3.2 Recurrent Neural Networks for Modeling of Dynamical Systems

As already noted in Chapter 2, ANNs can be divided into two classes: static ANNs and dynamic ANNs.

Layered feedforward networks are static networks. Their characteristic feature is that their outputs depend only on their inputs, i.e., to calculate the outputs of such ANN, only the current values of the variables used as input are required.

In contrast, the output of dynamic ANNs depends not only on the current values of the inputs. In dynamic ANNs, when calculating their outputs, the current and/or previous values of the inputs, states, and outputs of the network are taken into account. Different architectures of dynamic ANNs use different combinations of these values (that is inputs, states, and outputs, their current and/or previous values). We give corresponding examples in Chapter 2. Dynamic networks of this kind appear because a memory (for example, as a TDL element) is introduced into their structure in some way, which allows us to save the values of the inputs, states, and outputs of the network for future use. The presence of memory in dynamic networks enables us to work with time sequences of values, which is fundamentally essential for ANN simulation of dynamical systems. Thus, it becomes possible to use some variable value (control variable as a function of time or state of the system) or a set of such values as the input of the ANN model. The response of the system and its corresponding model will also represent some set of variables, in other words, the trajectories of the system in its state space.

As for the possible options for dynamic networks used to model the behavior of controlled systems, there are two main directions:

1. Dynamic networks (models) derived from conventional feedforward networks by adding TDL elements to their inputs make it possible to take into account the dynamics of the change in the input (control) signal. This is because not only the current value of the input signal but also several values (prehistory) for several previous instants are fed to the input of the ANN model. Models of this type include networks such as TDNN (FTDNN) and DTDNN, discussed in Chapter 2. An example of using a TDNN-type model to solve a particular application problem is discussed below, in Section 2.4 of this chapter. The solution of some other application problems using networks of this type is also considered in [51–58].
2. Dynamic networks with feedbacks (recurrent networks) are a much more powerful tool for modeling controlled dynamical systems. This capability is provided by the fact that it becomes possible to take into account not only the prehistory of control signals (inputs) but also the prehistory for the outputs and internal states (outputs of hidden layers). Recurrent networks of types NARX [15–23] and NARMAX [1,24] are most often used to solve the problems of modeling, identification, and control of nonlinear dynamical systems. We discuss examples of using the NARX network to solve problems of simulation of aircraft motion in Chapter 4. A much more general version of the structural organization of ANN models for nonlinear dynamical systems is networks with LDDN architecture [29]. This architecture includes, as individual cases, almost any other neuroarchitecture (both feedforward and recurrent), including NARX and NARMAX. The LDDN architecture, as well as the learning algorithms for networks with such an architecture, allow, among other things, to build not only traditional-style ANN models (black box type) but also hybrid ANN models (gray box type). We discuss models of this type in

Chapter 5, and we present examples of their use in Chapter 6.

3.4 ANN-BASED CONTROL OF DYNAMICAL SYSTEMS

The development of complex controlled systems raises some problems that we cannot solve through traditional control theory alone. These tasks associate mainly with such uncertainty in the conditions of the system operation, which requires the implementation of decision-making procedures that are characteristic of a person with the ability to reason heuristically, to learn, to accumulate experience. The need for training arises when the complexity of the problem being solved or the level of uncertainty under its conditions does not allow for obtaining the required solutions in advance. Training in such cases makes it possible to accumulate information during the operation of the system and use it to develop solutions that meet the current situation dynamically. We call systems that implement such functions *intelligent control systems.*

In recent years, intelligent control, which studies methods and tools for constructing and implementing intelligent control systems, is an actively developing field of interdisciplinary research based on the ideas, techniques, and tools of traditional control theory, artificial intelligence, fuzzy logic, ANNs, genetic algorithms, and other search and optimization algorithms.

Complex aerospace systems, in particular, aircraft, entirely belong to these complex controlled systems. In the theory of automatic control and its aerospace applications, significant progress has been made in the last few decades. In particular, considerable progress has been made in the field of computing facilities, which allows carrying out a significant amount of calculations on board of the aircraft.

However, despite all these successes, the synthesis of control laws adequate to modern and advanced aircraft technology remains a challenging task so far. First of all, this is because a modern multipurpose highly maneuverable aircraft should operate under a wide range of flight conditions, masses, and inertial characteristics, with a significant nonlinearity of the aerodynamic characteristics and the dynamic behavior of aircraft.

For this reason, it seems relevant to try to expand the range of methods and tools traditionally used to solve aircraft control problems, basing these methods on the approaches offered by intelligent control. One such approach is based on the use of ANNs.

3.4.1 Adjustment of Dynamic Properties of a Controlled Object Using Artificial Neural Networks

In this section, an attempt is made to show that using ANN technology we can solve the problem of appropriate representation (approximation) of a nonlinear model of aircraft motion with high efficiency. Then, using such approximation, we can synthesize a neural controller that solves the problem of adjusting the dynamic properties of some controlled object (aircraft).

First, we state the problem of adjusting the dynamic properties of a controlled object (plant), based on an indirect assessment of these properties using the reference model. It is proposed to solve this problem by varying the values of the parameters of the controller, producing adjustive actions on the plant.

Then a structural diagram of varying the parameters of the adjusting (correcting) controller is proposed using a reference model of the behavior of the plant. We show that we have to replace the traditional model of aircraft motion in the form of a system of nonlinear ODEs with another model that would have substantially less computational complexity. We need this replacement to ensure the efficient operation of the proposed control scheme. As such an alternative model, it is suggested to use an ANN.

In the following sections, we describe the principal features of the ANN model for aircraft motion, and we propose the technology of its formation. The next step considers one of the possible applications of ANN models of dynamical systems. This is the synthesis of the control neural network (neurocontroller) to adjust the dynamic properties of the plant. We form the reference model of the aircraft motion, to the behavior of which the neurocontroller should try to lead the response of the original plant. We build an example of a neurocontroller, which produces signals for adjusting the behavior of the aircraft in a longitudinal short-period motion. This example is primarily based on the results of the ANN simulation of the object.

3.4.1.1 The Problem of Adjusting the Dynamic Properties of a Controlled Object

Let the considered controlled object (plant) be a dynamical system described by a vector differential equation of the form [59–61]

$$\dot{\mathbf{x}} = \varphi(\mathbf{x}, \mathbf{u}, t). \tag{3.13}$$

In Eq. (3.13), $\mathbf{x} = [x_1\ x_2\ \dots\ x_n]^T \in \mathcal{R}^n$ is the vector of state variables of the plant; $\mathbf{u} = [u_1\ u_2\ \dots\ u_m]^T \in \mathcal{R}^m$ is the vector of control variables of the plant; $\mathcal{R}^n$, $\mathcal{R}^m$ are Euclidean spaces of dimension n and m, respectively; $t \in [t_0, t_f]$ is the time.

In Eq. (3.13), $\varphi(\cdot)$ is a nonlinear vector function of the vector arguments **x**, **u** and the scalar argument t. It is assumed to be given and belongs to some class of functions that admits the existence of a solution of Eq. (3.13) for given $\mathbf{x}(t_0)$ and $\mathbf{u}(t)$ in the considered part of the plant state space.

The controlled object (3.13) is characterized by a set of its inherent dynamic properties [59, 61]. These properties are usually determined by the plant response to some typical test action. For example, when the plant is an airplane this action can be some stepwise deflection of its elevator by a prescribed angle. Dynamic properties are characterized by the stability of the motion of the plant and the quality of its transient processes.

The stability of the plant motion in the variable x_i, $i = 1, \dots, n$, is determined by its ability to return over time to some undisturbed value of this variable $x_i^{(0)}(t)$ after the disturbance disappears [61].

The nature of the plant transient processes that arise as a response to a stepwise action is estimated using the appropriate performance indices (quality indicators), which usually include the following [59,61]: transient time, maximum deviation in the transient process, overshoot, frequency of free oscillations, time of the first steady-state operation, and number of oscillations during the transient process.

Instead of these indices we use the indirect approach based on some reference model to evaluate the dynamic properties of the plant. It can be obtained using the abovementioned quality indicators for the transient processes of the plant and, possibly, some additional considerations, for example, pilots' assessments of the aircraft handling qualities.

Using the reference model, we can estimate the dynamic properties of the plant as follows:

$$I = \sum_{i=1}^{n} \int_0^{\infty} [x_i(t) - x_i^{(ref)}(t)]^2 dt \tag{3.14}$$

or

$$I = \sum_{i=1}^{n} \lambda_i \int_0^{\infty} [x_i(t) - x_i^{(ref)}(t)]^2 dt, \tag{3.15}$$

where λ_i are the weighting coefficients that establish the relative importance of the change for different state variables.

We could use the linear reference model

$$\dot{\mathbf{x}}^{(ref)} = A\mathbf{x}^{(ref)} + B\mathbf{u} \tag{3.16}$$

with matrices A and B matched appropriately (see, for example, [62]), as well as the original

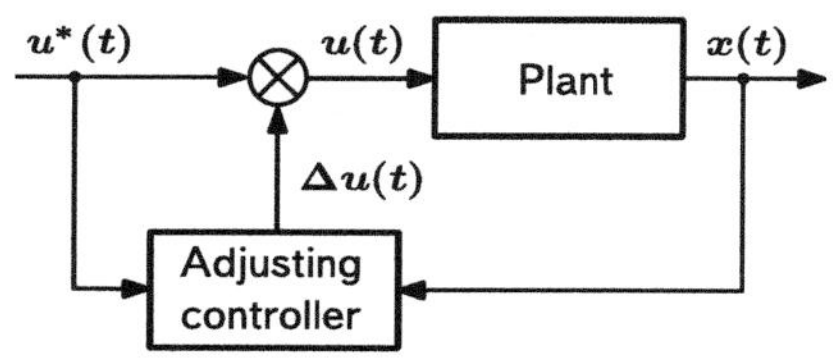

FIGURE 3.3 Structural diagram of adjusting the dynamic properties of the controlled object (plant). $\mathbf{x}$ is the vector of state variables of the plant; $\mathbf{u}^*$, $\Delta\mathbf{u}$ are the command and adjusting components of the plant control vector, respectively; $\mathbf{u} = \mathbf{u}^* + \Delta\mathbf{u}$ is the vector of control variables of the plant (From [99], used with permission from Moscow Aviation Institute).

nonlinear model (3.13), where the vector function $\varphi(\cdot)$, remaining nonlinear, is corrected by in order to get the required level of transient quality, i.e.,

$$\dot{\mathbf{x}}^{(ref)} = \varphi^{(ref)}(\mathbf{x}^{(ref)}, \mathbf{u}, t). \tag{3.17}$$

We will further use the indirect approach to evaluate the dynamic properties of the plant based on the nonlinear reference model (3.17).

Suppose there is a plant with the behavior described by (3.13), as well as the model of the desired behavior of the plant, given by (3.17).

The behavior of a plant, determined by its dynamic properties, can be affected by setting a correction value for the control variable $\Delta\mathbf{u}(\mathbf{x}, \mathbf{u}^*)$. The operation of forming the required value $\Delta\mathbf{u}(\mathbf{x}, \mathbf{u}^*)$ for some time t_{i+1} from the values of the state vector $\mathbf{x}$ and the command control vector $\mathbf{u}^*$ at time t_i,

$$\Delta\mathbf{u}(t_{i+1}) = \Psi(\mathbf{x}(t_i), \mathbf{u}^*(t_i)), \tag{3.18}$$

we will execute in the device, which we call *adjusting controller*. We assume that the character of the transformation $\Psi(\cdot)$ in (3.18) is determined by the composition and values of the components of some vector $\mathbf{w} = [w_1\, w_2 \ldots w_{N_w}]^T$. The combination (3.13), (3.18) from the plant and adjusting controller is referred to as the *controlled system* (Fig. 3.3).

The problem is to select the Ψ transformation implemented by the controller so that this controlled system would show the behavior closest to the behavior of the reference model. This task we call *the task of adjusting the dynamic properties* of the plant.

The task of adjusting the dynamic properties of some plant can be treated as a task of minimizing some *error function* $E(\mathbf{w})$, i.e.,

$$E(\mathbf{w}^*) = \min_{\mathbf{w}} E(\mathbf{w}),$$

where $\mathbf{w}^*$ is the value of the vector $\mathbf{w}$ that delivers the minimum of the function $E(\mathbf{w})$, which can be defined, for example, as

$$E(\mathbf{w}) = \int_{t_0}^{t_f} [x^{(ref)}(t) - x(\mathbf{w}, t)]^2 dt, \tag{3.19}$$

or as

$$E(\mathbf{w}) = \max_{t \in [t_0, t_f]} |x^{(ref)}(t) - x(\mathbf{w}, t)|. \tag{3.20}$$

A problem of this kind we can solve in two ways, differing in the approach to varying the parameters $\mathbf{w}$ in the adjusting controller.

Following the first approach, the selection of $\mathbf{w}$ is carried out autonomously, after which the obtained values of $\mathbf{w}$ are loaded into the adjusting controller and remain unchanged throughout the process of functioning of the controlled system.

Following the second approach, the selection of the coefficients $\mathbf{w}$ is carried out in the on-line mode, i.e., directly in the operation of the controlled system under consideration.

To specify the analysis, consider the longitudinal motion of the aircraft, i.e., its motion in a vertical plane without roll and sideslip.

The mathematical model of longitudinal motion, obtained by the projection of vector equations on the axis of the body-fixed coordinate system (see, for example, [63–66]), has the form

$$\begin{aligned} &m(\dot{V}_x - V_z q) = X, \\ &m(\dot{V}_z + V_x q) = Z, \\ &I_y \dot{q} = M_y, \\ &\dot{\theta} = q, \\ &\dot{H} = V \sin\theta, \end{aligned} \tag{3.21}$$

where X, Z are the projections of all forces acting on the aircraft on the Ox-axis and the Oz-axis, respectively; M_y is the projection of all the moments acting on the aircraft onto the Oy-axis; q is angular velocity of pitch; m is the mass of the aircraft; I_y is the moment of inertia of the aircraft relative to the Oy-axis; V is the airspeed; V_x, V_z are the projections of the airspeed on the Ox-axis and the Oz-axis, respectively; H is the altitude of flight.

The system of equations (3.21) can be simplified, based on the choice of the trajectory of motion and some physical features inherent in the aircraft. Let us first consider the steady horizontal flight of an airplane that occurs at a given altitude H with a given airspeed V. As is well known [63–66], in this case, from the solution of the system of equations

$$\begin{aligned} X(\alpha, V, H, T, \delta_e) &= 0, \\ Z(\alpha, V, H, T, \delta_e) &= 0, \\ M_y(\alpha, V, H, T, \delta_e) &= 0, \end{aligned}$$

we can find the angle of attack α_0, the thrust of the engine T_0, and the angle of deflection of the elevator (all-moving stabilizer) $\delta_e^{(0)}$, necessary for this flight. Suppose that at the time t_0, the deflection angle of the stabilizer (or the value of the corresponding command signal) has changed by the value $\Delta\delta_e$. The change in the position of the stabilizer disturbs the balance of the moments acting on the aircraft, as a result of which its angular position in space will change before this affects the change in the value of the aircraft velocity vector. This means that the study of transient processes with respect to the angular velocity of the pitch q and the pitch angle θ can be carried out using the assumption $V = \text{const}$. In this case, equations for $\dot{V}_x$ and $\dot{V}_z$ become equivalent to the equation $\dot{\theta} = q$, from which it follows that we can use the system of two equations, i.e., the equation for q and any of the above equivalent equations.

Here we choose the system of equations

$$\begin{aligned} m(\dot{V}_z + V_x q) &= Z, \\ I_y \dot{q} &= M_y. \end{aligned} \tag{3.22}$$

The system of equations (3.22) is closed, since the angle of attack α entering into the expressions for Z and M will be equal in the case under consideration to the pitch angle θ, which is related to V_z by the following kinematic dependence:

$$V_y = -V \sin\theta.$$

Thus, the system of equations (3.22) describes the transient processes concerning the angular velocity and the pitch angle, which occur immediately after breaking the balance corresponding to the steady horizontal flight.

Let us reduce the system of equations (3.22) to the Cauchy normal form, i.e.,

$$\begin{aligned} \frac{dV_z}{dt} &= \frac{Z}{m} - V_x q, \\ \frac{dq}{dt} &= \frac{M_y}{I_y}. \end{aligned} \tag{3.23}$$

In (3.23), the value of the pitch moment M_y is a function of the control variable. This variable is the deflection angle of the elevator (or all-turn stabilizer), that is, $M_y = M_y(\delta_e)$.

So, in the particular case under consideration, the composition of the state and control variables is as follows:

$$\mathbf{x} = [V_z \;\; q]^T, \quad \mathbf{u} = [\delta_e]. \tag{3.24}$$

As noted above, the analysis uses an indirect approach to estimating the dynamic properties of the plant based on the nonlinear reference model (3.17), which for the case of the system (3.23) takes the form

$$\frac{dV_z^{(ref)}}{dt} = \frac{Z}{m} - V_x q^{(ref)}, \qquad \frac{dq^{(ref)}}{dt} = \frac{M_y^{(ref)}}{I_y}. \tag{3.25}$$

The condition mentioned above for M_y in (3.23) also holds for (3.25).

The reference model (3.25) differs from the original model (3.23) by the expression for the moment of the pitch $M_y^{(ref)}$, which, in comparison with M_y in (3.23), adds additional damping so that the behavior of the control object becomes aperiodic.

With respect to the problem (3.23), (3.25), to simplify the discussion, we assume that the values of the parameters characterizing the plant (3.23) and its reference (unperturbed) motion (these are parameters I_y, m, V, H, etc.) remain unchanged.

With the same purpose, we assume that the values of the adjustment coefficients $\mathbf{w}$ are selected autonomously, are frozen, and do not change during the operating of the controlled system.

3.4.1.2 *Approximation of the Initial Mathematical Model of a Controlled Object Using an Artificial Neural Network*

In the adopted scheme for adjusting the dynamic properties of the plant (Fig. 3.3), the controlled system under consideration consists of a controlled object (plant) and an adjusting controller supplying corrective commands to the input of the plant. As noted above, we use the indirect approach based on the reference model to evaluate the dynamic properties of the plant.

Following the indirect approach, we can represent the structure of the adjusting (selection of values) of the parameters $\mathbf{w}$ in the adjusting controller as shown in Fig. 3.4.

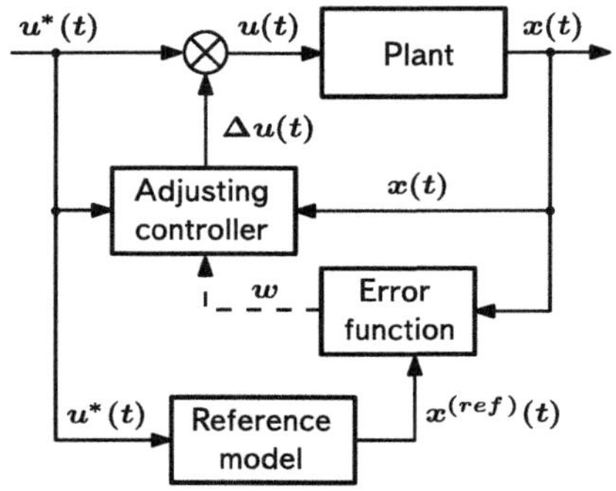

FIGURE 3.4 Tuning parameters of the adjusting controller. $\mathbf{x}$ is the vector of state variables of the plant; $\mathbf{x}^{(ref)}$ is the vector of reference model state variables; $\mathbf{u}^*$, $\Delta\mathbf{u}$ are the command and correction component of the plant control vector, respectively; $\mathbf{u} = \mathbf{u}^* + \Delta\mathbf{u}$ is the vector of control variables of the plant; $\mathbf{w}$ is a set of selectable parameters of the adjusting controller (From [99], used with permission from Moscow Aviation Institute).

The process of functioning of the system shown here begins at the time t_i from the same state for both the plant and the reference model, i.e., $\mathbf{x}(t_i) = \mathbf{x}^{(ref)}(t_i)$. Then, the same command signal $\mathbf{u}^*(t_i)$ is sent to the input of the plant and the reference model, for example, to implement the long-period component of the required motion. The quality of transient processes in the short-period motion caused by the resulting perturbation must correspond to the given $\mathbf{x}^{(ref)}(t_i) = \mathbf{x}(t_i)$ and $\mathbf{u}^*(t_i)$ of the reference model, which passes to the state $\mathbf{x}^{(ref)}(t_{i+1})$ after a period of time $\Delta t = t_{i+1} - t_i$. The state of the plant will become equal to $\mathbf{x}(t_{i+1})$ by the same time. Now we can find the mismatch between the outputs of the plant and the reference model $||\mathbf{x}(t_{i+1}) - \mathbf{x}^{(ref)}(t_{i+1})||$ and on this basis construct the error function $E(\mathbf{w})$. This operation is performed based on the following considerations (see also (3.19) and (3.20)).

The reference model in our control scheme is immutable and its output at the time t_{i+1} depends only on the reference model state at time t_i, that is, on $\mathbf{x}^{(ref)}(t_i)$, and also on the value of the command signal $\mathbf{u}^*(t_i)$ in the same moment of time.

In contrast to the reference model, the control action on the plant consists of the command signal $\mathbf{u}^*(t_i)$ and an additional signal

$$\Delta \mathbf{u}(t_i) = \Psi(\mathbf{x}(t_{i-1}), \mathbf{u}^*(t_i)),$$

in which the character of the function $\Psi(\cdot)$ depends, as above, on the composition and values of the parameters $\mathbf{w}$ in it.

So, the error function $E(\cdot)$ depends on the parameter vector $\mathbf{w}$, and by varying its components, we can choose the direction of their change so that $E(\mathbf{w})$ decreases.

As we can see from Fig. 3.4, the error function $E(\mathbf{w})$ is defined at the outputs of the plant. It has already been noted above that the goal of solving the problem of adjusting the dynamic properties of a plant is to minimize the function $E(\mathbf{w})$ with respect to the parameters $\mathbf{w}$, i.e.,

$$E(\mathbf{w}^*) = \min_{\mathbf{w}} E(\mathbf{w}). \tag{3.26}$$

Generally speaking, we could treat the problem (3.26) as a traditional optimization problem, namely, as a nonlinear programming problem (NLP), which has been well studied theoretically, for the solution of which there are a significant number of algorithms and software packages. With this approach, however, there is a circumstance that substantially limits its practical applicability.

Namely, the computational complexity of such algorithms (based, for example, on gradient search) is of the order of $\mathcal{O}(N_w^2)$ [67,68], i.e., it grows in proportion to the square of the number of variables in the problem being solved. Because of this, the solution of NLP problems with a large number of variables occurs, as a rule, with severe difficulties. Such a situation for traditional NLP problems can arise even when N_w is of the order of ten, especially in cases when even a single calculation of the objective function $E(\mathbf{w})$ is associated with significant computational costs.

At the same time, to track the complex nonlinear dynamics of the plant, a considerable number of "degrees of freedom" in the model used may be required. This number grows as the number of configurable parameters of the neurocontroller increases. Computational experiments show that even for relatively simple problems the necessary number of variables can be of the order of several tens.

To cope with this situation, we need a mathematical model for the adjusting controller, which has less computational complexity in solving the problem (3.26) than the traditional NLP problem mentioned above. One of the possible variants of such mathematical models is the ANN. The adjusting controller, implemented as the ANN, will hereafter be named *neurocontroller*.

More details on the main features of the structure and use of the ANN will be discussed below. For now we only note that using this approach to represent the mathematical model of the adjusting controller allows us to reduce the computational complexity of the problem (3.26) to about $\mathcal{O}(N_w)$ [14,28,29], i.e., it grows in proportion to the first power of the number of variables N_w. There is also the opportunity to reduce this complexity [27].

In the adopted scheme, as already noted, the minimized error function $E(\mathbf{w})$ is defined not at the outputs of the adjusting controller (realized, for example, in the ANN form), but at the outputs of the plant. But, as will be shown below, to organize the process of selecting the parameters of the ANN, it is necessary to know the error $\widetilde{E}(\mathbf{w})$ directly at the output of the adjusting controller.

Hence we need to solve the following problem. Let there be an output of the plant model, which differs from the desired ("reference") one. We must be able to answer the following question: how should the inputs of the plant model be changed so that its outputs change in the direction of reducing the error $E(\mathbf{w})$?

The inputs of the model which are adjusted in such a way become target outputs for the neurocontroller. The parameters $\mathbf{w}$ in the ANN vary to minimize the deviation of the current ANN outputs from these target ones, i.e., minimize the error $\widetilde{E}(\mathbf{w})$. Thus, there arises the need to solve the inverse problem of dynamics for some plant. If the plant model is a traditional nonlinear system of ODEs, then the solution of this problem is complicated to obtain. An alternative option is to use as a plant model some ANN, for which, as a rule, the solution of the inverse problem does not cause serious difficulties.

Thus, the neural network approach to the solution of the problem in question requires the use of two ANNs: one as the neurocontroller and the other as the plant model.

So, the first thing we need to be able to do to solve the problem of adjusting the dynamic properties of the plant in the way suggested above is to approximate the source system of differential equations (3.13) (or, concerning the particular problem in question, the system (3.23)). We can consider this problem as an ordinary task of identifying the mathematical model of the plant [59,69] for the case when the values of the outputs (state variables) of the plant are not obtained as a result of measurements but with the help of a numerical solution of the corresponding system of differential equations.

The approach consisting in the use of ANNs to approximate a mathematical model of a plant (a mathematical model of aircraft motion, in particular) is becoming increasingly widespread [4, 31,34,38,70–72].

The structure of such models, the acquisition of data for their training, as well as the learning algorithms, were considered in Chapter 2 for both feedforward and recurrent networks.

For the case of a plant of the form (3.23), i.e., for the aircraft performing the longitudinal short-period motion, a neural network approximating the motion model (3.23), after some computational experiments, has the form shown in Fig. 3.5.

The ANN inputs in Fig. 3.5 are two state variables: the vertical velocity V_z and the angular velocity of the pitch q in the body-fixed coordinate system at time t_i, and the control variable is

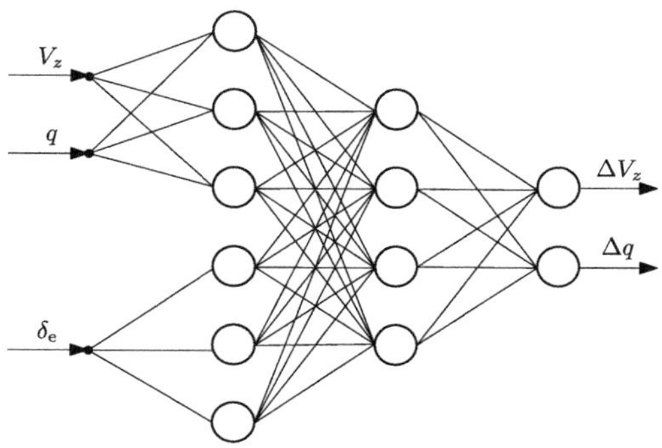

FIGURE 3.5 The neural network model of the short-period longitudinal motion of the aircraft. V_z, q are the values of the aircraft state variables at time t_i; δ_e is the value of the deviation angle of the stabilizer at time t_i; ΔV_z, Δq are the increments of the values of the aircraft state variables at time $t_i + \Delta t$ (From [99], used with permission from Moscow Aviation Institute).

the deflection angle of stabilizer δ_e for the time moment t_i. Values of the state variables V_z and q go to one group of neurons, and the value of the control variable δ_e goes to another group of neurons of the first hidden layer, which is the preprocessing layer of the input signals. The results of this preprocessing are applied to all four neurons of the second hidden layer. At the output of the ANN, the values of ΔV_z and Δq are increments of the values of the aircraft state variables at the time moment $t_i + \Delta t$. The neurons of the ANN hidden layers in Fig. 3.5 have activation functions of the Gaussian type, the output layer neurons are linear activation functions.

The model of the short-period aircraft motion (3.23) contains the deflection angle of the all-turn stabilizer δ_e as the control variable. In the model (3.23), the character of the process of forming the value δ_e is not taken into account. However, such a process, determined by the dynamic properties of a controlled stabilizer (elevator) actuator, can have a significant effect on the dynamic properties of the controlled system being created.

The dynamics of the stabilizer actuator in this problem is described by the following differen-

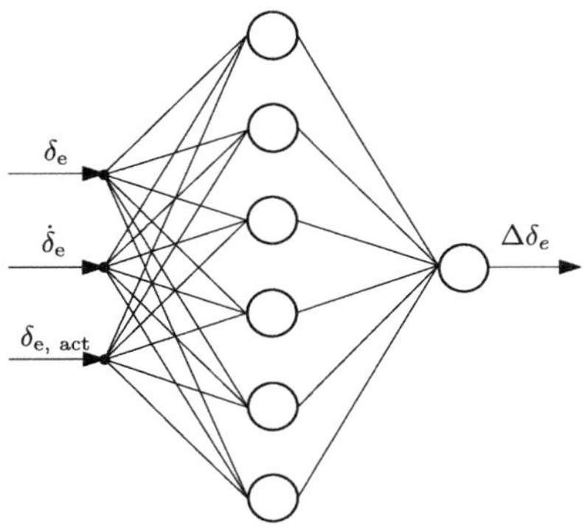

FIGURE 3.6 The neural network model of the elevator actuator. δ_e, $\dot{\delta}_e$, $\delta_{e,act}$ are the deflection angle of the stabilizer, the deflection speed of the stabilizer, and the command angle of the stabilizer deflection, respectively, for the time point t_i; $\Delta\delta_e$ is the increment of the value of the deflection angle of the stabilizer at the time moment $t_i + \Delta t$ (From [99], used with permission from Moscow Aviation Institute).

tial equations:

$$\begin{aligned}\dot{\delta}_e &= x,\\ \dot{x} &= \frac{1}{T_1^2}(\delta_{e,act} - 2\xi T_1 x - \delta_e).\end{aligned} \tag{3.27}$$

In (3.27), $\delta_{e,act}$ is the command value of the deflection angle of the stabilizer; T_1 is actuator time constant; ξ is the damping coefficient.

Using the same considerations as for the motion model (3.23) (see page 107), it is also necessary to construct a neural network approximation for the actuator model (3.27). Fig. 3.6 presents the structure of the neural network stabilizer actuator model, obtained during a series of computational experiments. In this ANN, the input layer contains three neurons, the only hidden layer includes six neurons with a Gaussian activation function, and in the output layer there is one neuron with a linear activation function.

Computational experiments in developing the neural network approximation technology for mathematical models of the form (3.23) were carried out regarding a maneuverable Su-17 aircraft [73]. (See Figs. 3.7 and 3.8.)

The first operation that needed to be done to perform these experiments was the generation of a training set. It is a pair of input–output matrices, the first of which specifies the set of all possible values of the aircraft variables, and the second is the change of the corresponding variables in a given time interval, assumed to be 0.01 sec.

The values of the parameters considered as constants in the model (3.23) were chosen as follows (the linear and angular velocities are given in the body-fixed coordinate system):

- $H = 5000$ m is altitude of flight;
- $T_a = 0.75$ is the relative thrust of the engine;
- $V_x = 235$ m/sec is the projection of the flight velocity V onto the Ox-axis of the body-fixed coordinate system.

The ranges of change of variables were accepted as follows (here the initial value, the step, and the final value of each of the variables are indicated):

- $q = -12 : 1 : 14$ deg/sec;
- $V_z = -28 : 2 : 12$ m/sec;
- $\delta_e = -26 : 1 : 22$ deg.

Thus, in the case under consideration, the training set is an input matrix of dimension 3×41013 values and its corresponding output is 2×41013. In this case, the input of the network is q, V_z, δ_e, and the output is the change of Δq and ΔV_z through the time interval $\Delta t = 0.01$ sec.

A comparison of modeling results with such a network and calculation results for the model (3.23) is shown in Fig. 3.9 (here only the model (3.23) is taken into account, not the dynamics of the actuator of the all-turn stabilizer) and in Fig. 3.10 (including the model (3.27), i.e., with dynamics of the stabilizer actuator).

The angle of attack, the changes of which in the transient process are shown in Figs. 3.9 and

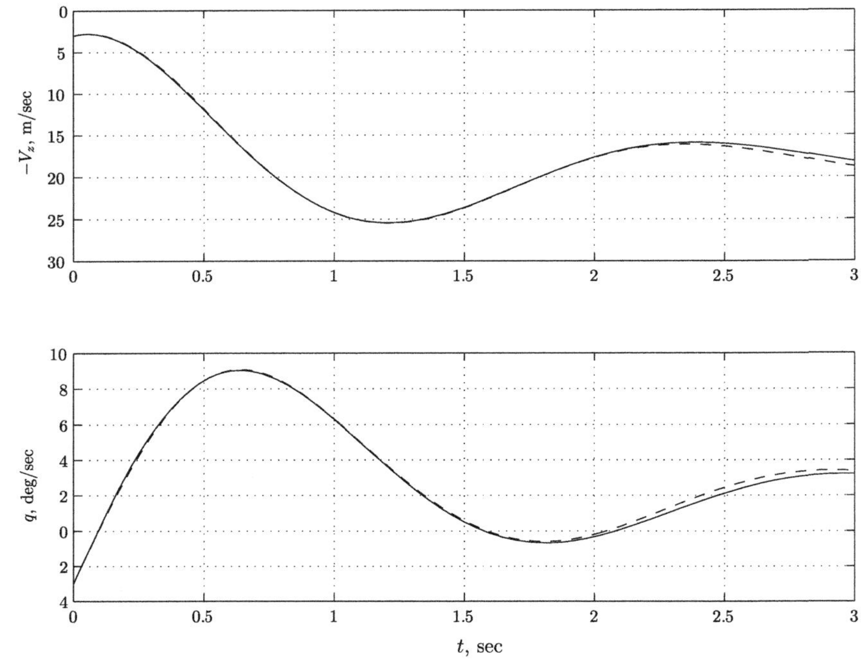

FIGURE 3.7 Comparison of the operation of the network (14 neurons, sigmoid activation function, shortened training set) and mathematical model (3.23). The solid line is model output (3.23); the dotted line is the output of the neural network model; the target mean square error is 1×10^{-7}; V_y is the component of the velocity vector along the Oy-axis; q is the angular velocity of the pitch; t is the time; the value of the deflection angle of the stabilizer δ_e is taken equal to -8 grad (From [99], used with permission from Moscow Aviation Institute).

3.10, was calculated according to the relation

$$\alpha = -\arctan(V_y / V_x).$$

Fig. 3.11 shows what will be the effect of the incorrect formation of the training set for the same ANN (see Chapter 2).

3.4.1.3 *Synthesis of a Neurocontroller That Provides the Required Adjustment of the Dynamic Properties of the Controlled Object*

The problem of neural network approximation of models of dynamical systems has a wide range of applications, including the formation of compact and fast mathematical models suitable for use on board aircraft and simulators in real time.

Besides, one more important application of such models is the construction of a neurocontroller intended to correct the dynamic properties of controlled objects.

Below we present the results of a computational experiment showing the possibilities of solving one type of such problems. In this experiment, in addition to the neural network model of the controlled object (see Fig. 3.5), the reference model for the motion of the aircraft (3.25) was used as well as the neurocontroller, shown in Fig. 3.12.

The neural controller is a control neural network, the input of which is given by the parameters q, V_z, and δ_e (angle of deflection of the all-turning horizontal tail), and the output is $\Delta\delta_{e,k}$ so that the behavior of the neural network model

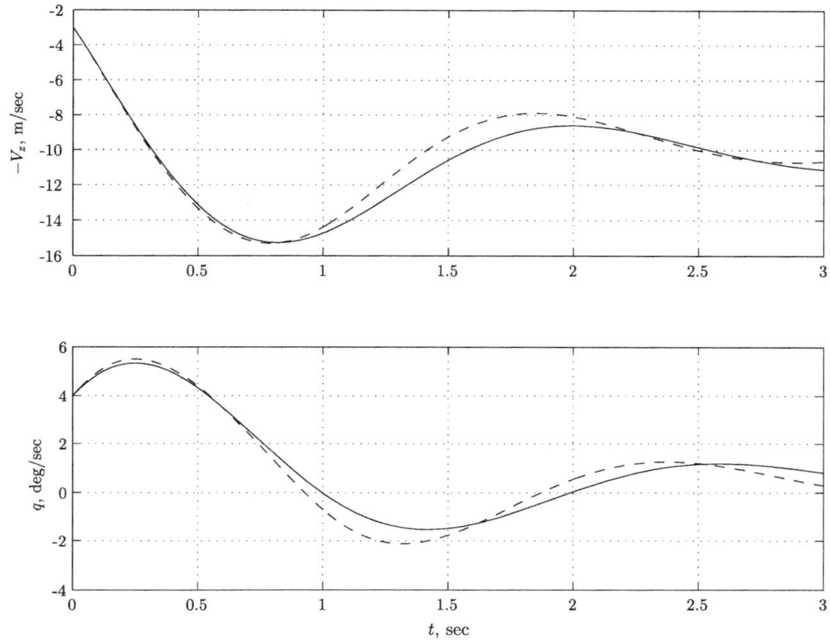

FIGURE 3.8 Comparison of the operation of the network (28 neurons, sigmoid activation function, full training set) and mathematical model (3.23). The solid line is model output (3.23); the dotted line is the output of the neural network model; the target mean square error is 1×10^{-8}; V_z is the component of the velocity vector along the Oy-axis; q is the angular velocity of the pitch; t is the time; the value of the deflection angle of the stabilizer δ_e is taken equal to -8 grad (From [99], used with permission from Moscow Aviation Institute).

is as close as possible to the behavior of the reference model.

To create a reference model, minor changes were made to the initial model of the Su-17 airplane motion by introducing an additional damping coefficient into it, which was selected in such a way that the nature of the transient processes had a pronounced aperiodic appearance.

The results of testing the reference model (3.25) in comparison with the original model (3.23) are shown in Fig. 3.13.

The generation of a training set for the task of synthesis of the neurocontroller occurred on the same principle as for the task of identifying a mathematical model.

When training the neurocontroller network, it was forbidden to change the weights $\mathbf{W}$ and the biases $\mathbf{b}$ of the neural network motion model which is the part of the combined network (the ANN plant model + neurocontroller).

It was allowed to vary only parameters for the network part that corresponded to the neurocontroller. Connections of neurons in the network were organized in such a way that the output of the neurocontroller $\Delta\delta_{e,k}$ was fed to the input of the neural network model δ_e as additions to the initial (command) position of the all-turning horizontal stabilizer, and input signals came simultaneously to the input of the neurocontroller and to the input of the neural network model.

Fig. 3.14 shows the result of testing the neurocontroller combined with the neural network model.

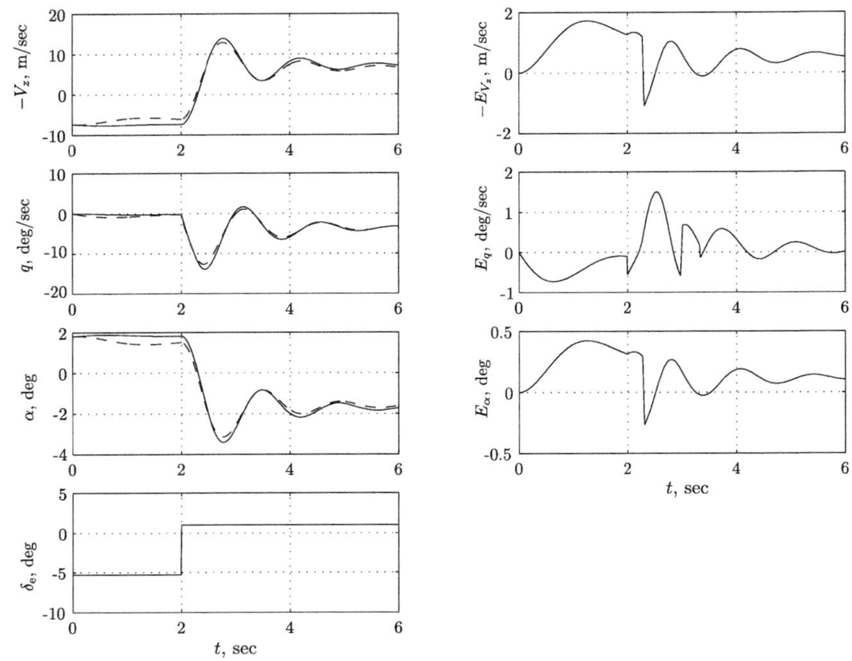

FIGURE 3.9 Comparison of the network operation with the preprocessing layer (without the stabilizer actuator model) and the mathematical model (3.23). The solid line is model (3.23) output; the dotted line is the output of the neural network model; V_z is the component of the velocity vector along the Oz-axis; q is the angular velocity of the pitch; α is the angle of attack; δ_e is the deflection angle of the stabilizer; t is the time; EV_z, Eq, and $E\alpha$ are the differences $|V_z - V_z^{(ref)}|$, $|q - q^{(ref)}|$, and $|\alpha - \alpha^{(ref)}|$, respectively (From [99], used with permission from Moscow Aviation Institute).

From the material presented in the previous sections, we can see that neural networks successfully cope with the problem of approximation of models of dynamic systems, as well as tasks of adjusting the dynamic properties of the controlled object toward a given reference model.

It should be emphasized that in the case under consideration, the ANN solves this task without even involving such a tool as adaptation, consisting in the operational adjustment of the synaptic neurocontroller weights directly during the flight of the aircraft. This kind of adaptation constitutes an important reserve for improving the quality of regulation, as well as the adaptability of the controlled system to the changing operating conditions [74–82].

3.4.2 Synthesis of an Optimal Ensemble of Neural Controllers for a Multimode Aircraft

Designing control laws for control systems for multimode objects, in particular for airplanes, remains a challenging task, despite significant advances in both control theory and in increasing the power of onboard computers that implement these control laws. This situation is due to a wide range of conditions in which the aircraft is used (airspeed and altitude, flight mass, etc.), for example, the presence of a large number of flight modes with artificially corrected aircraft dynamics, based on the requirement of the best solution of various tasks.

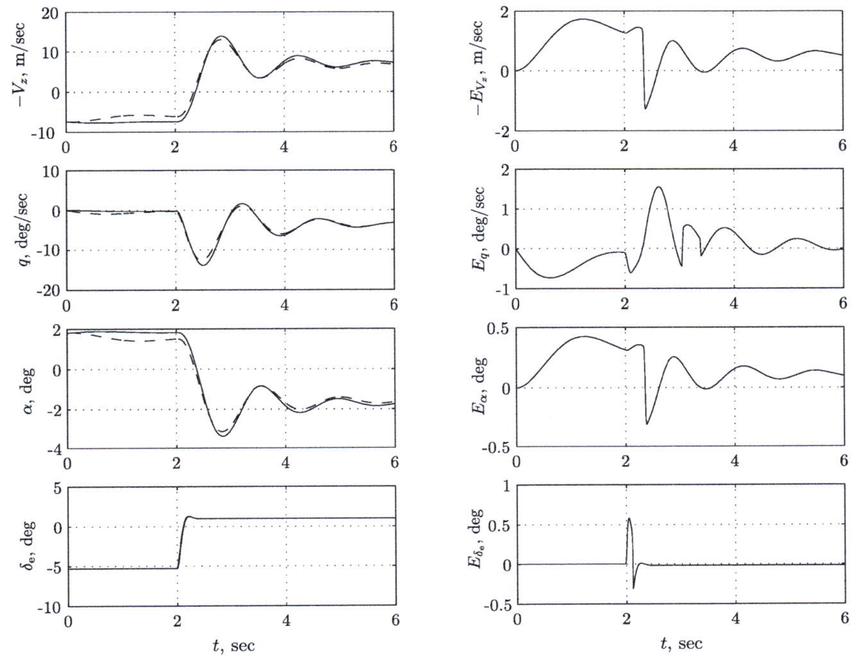

FIGURE 3.10 Comparison of the network operation with the preprocessing layer (with the stabilizer actuator model) and the mathematical model (3.23). The solid line is model (3.23) output; the dotted line is the output of the neural network model; V_z is the component of the velocity vector along the Oy-axis; q is the angular velocity of the pitch; α is the angle of attack; δ_e is the deflection angle of the stabilizer; t is the time; EV_z, Eq, and $E\alpha$ are the differences $|V_z - V_z^{(ref)}|$, $|q - q^{(ref)}|$, and $|\alpha - \alpha^{(ref)}|$, respectively (From [99], used with permission from Moscow Aviation Institute).

3.4.2.1 *Basic Approaches to the Use of Artificial Neural Networks in Control Problems*

There are three main approaches to the use of ANNs in control systems [83–86].

In the first approach ("conservative") the structural organization of the control system remains unchanged, i.e., as it was obtained when designing such a system using one of the traditional approaches. In this case, the ANN plays the role of a module for correcting specific parameters of the control system (for example, its gains) depending on the operating conditions of the system. In the second approach ("radical"), the whole control system or a functionally completed part is realized as an ANN system. The third approach ("compromise") is a combination of a conservative and radical approach, or more precisely, some compromise between them.

In the general case, the most effective (in terms of the applied problem) will, of course, be a radical approach. However, a less powerful conservative approach not only has the right to exist, but, moreover, is more preferable at present for a variety of reasons, both objective and subjective. Namely, following the conservative approach, it is possible to quickly achieve practically meaningful results, since this approach is not about creating a new control system from scratch, but about upgrading existing systems. Further, it is very difficult to imagine now the situation when on board (for example, an airplane) a control system based entirely on

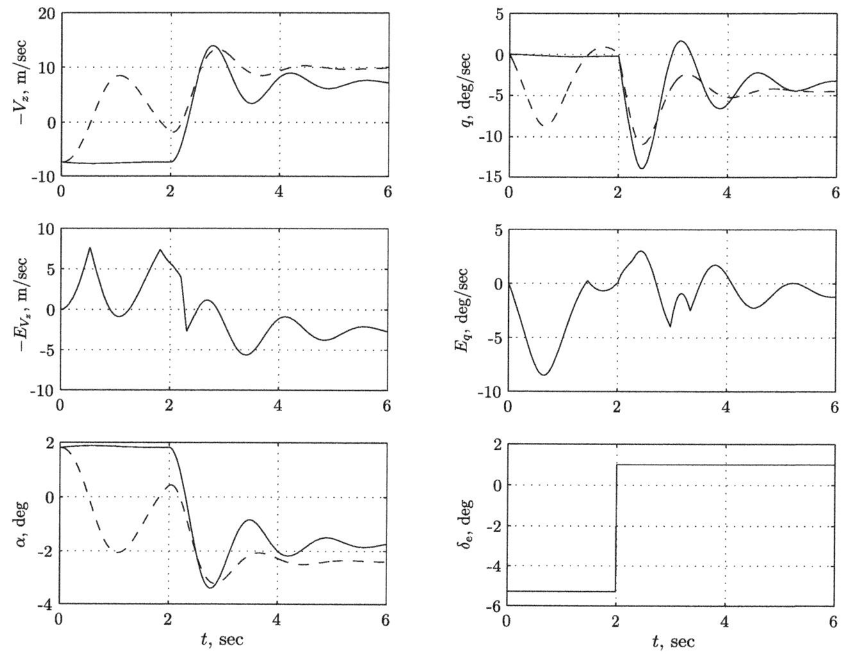

FIGURE 3.11 The effect of the incorrect formation of the training set on the example of the comparison of the network with the preprocessing layer (with regard to the stabilizer actuator model) and the mathematical model (3.23), (3.27). The solid line is output of the model (3.23); the dotted line is the output of the neural network model; V_z is the component of the velocity vector along the Oy-axis; q is the angular velocity of the pitch; α is the angle of attack; δ_e is the deflection angle of the stabilizer; t is the time; EV_z, Eq are the differences $|V_z - V_z^{(ref)}|$ and $|q - q^{(ref)}|$, respectively (From [99], used with permission from Moscow Aviation Institute).

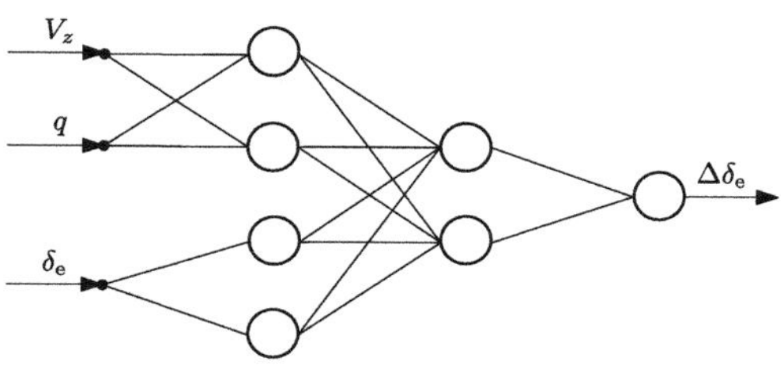

FIGURE 3.12 The neurocontroller in the control problem of the short-period longitudinal motion of the aircraft. V_z is the component of the velocity vector along the Oz-axis; q is the angular velocity of the pitch; δ_e is the deflection angle of the stabilizer; $\Delta\delta_{e,\,cc}$ is the angle of the adjusting deflection of the stabilizer (From [99], used with permission from Moscow Aviation Institute).

the ANN will be allowed. First, we must overcome a certain "novelty barrier" (albeit psychological, but quite realistic), to prove the right of the ANN to be present in the critical on-board systems, increasing (or, at least, not reducing) the effectiveness and safety of operation of the control facility.

In this regard, in the following sections, primary attention will be paid to the conservative approach to the use of the ANN as part of the control system. Then it will be shown how the formulated provisions are realized under radical and compromise approaches.

3.4.2.2 *Synthesis of Neurocontrollers and Ensembles of Neurocontrollers for Multimode Dynamical Systems*

Consider the concept of an ENC concerning the control problem for an MDS. To do this, we

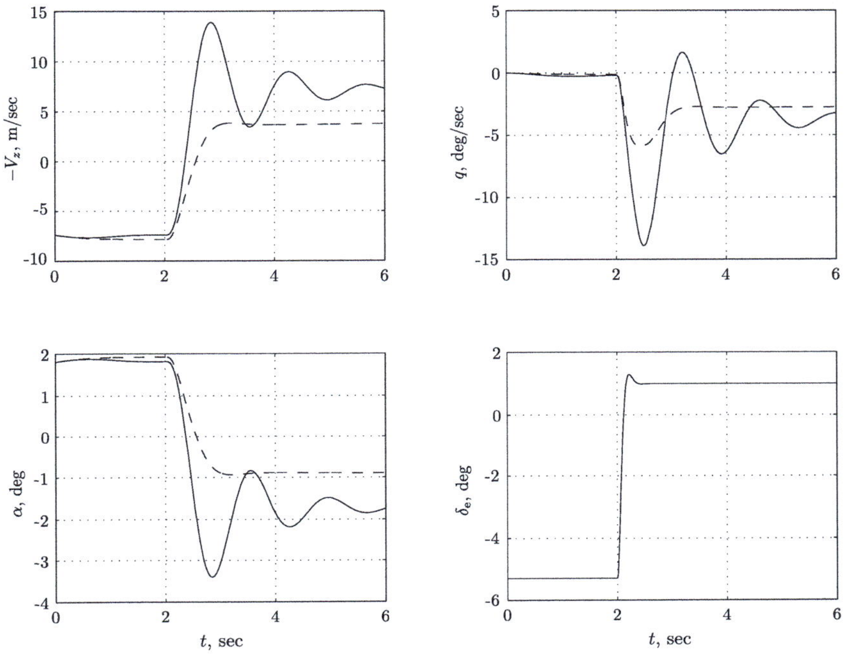

FIGURE 3.13 The character of the behavior of the reference aircraft motion model (3.25) in comparison with the model (3.23). The solid line is the model (3.23) output; the dashed line is the model (3.25) output; V_z is the component of the velocity vector along the Oz-axis; q is the angular velocity of the pitch; α is the angle of attack; δ_e is the deflection angle of the stabilizer (From [99], used with permission from Moscow Aviation Institute).

first create a model of such a system, then consider the construction of a neurocontroller for a single-mode dynamical system. On this basis, an ensemble of neurocontrollers is then formed to control the MDS.

MODEL OF CONTROLLED MULTIMODE DYNAMICAL SYSTEM

Consider a controlled dynamical system described by a vector differential equation,

$$\dot{\widetilde{\mathbf{x}}} = \Psi(\widetilde{\mathbf{x}}, \widetilde{\mathbf{u}}, \boldsymbol{\theta}, \boldsymbol{\lambda}, t), \tag{3.28}$$

where $\widetilde{\mathbf{x}} = (\widetilde{x}_1, \ldots, \widetilde{x}_n)^T \in \mathbf{X} \subset \mathcal{R}^n$ is the state vector of the dynamical system; $\widetilde{\mathbf{u}} = (\widetilde{u}_1, \ldots, \widetilde{u}_m)^T \in \mathbf{U} \subset \mathcal{R}^m$ is the control vector of the dynamical system; $\boldsymbol{\theta} = (\theta_1, \ldots, \theta_l)^T \in \boldsymbol{\Theta} \subset \mathcal{R}^l$ is the constant parameters vector of the dynamical system; $\boldsymbol{\lambda} = (\lambda_1, \ldots, \lambda_s)^T \in \boldsymbol{\Lambda} \subset \mathcal{R}^s$ is the vector of "external" parameters of the problem, the choice of which is not available for the designer of the system; $t \in [t_0, t_f]$ is a time.

Let

$$\begin{aligned} \widetilde{\mathbf{x}}^0 &= \widetilde{\mathbf{x}}^0[t, \boldsymbol{\theta}, \boldsymbol{\lambda}, t, (i, f)], \\ \widetilde{\mathbf{u}}^0 &= \widetilde{\mathbf{u}}^0[t, \boldsymbol{\theta}, \boldsymbol{\lambda}, t, (i, f)] \end{aligned} \tag{3.29}$$

be some reference motion of the system (3.28). In (3.29), following the work [87], (i, f) denotes the boundary conditions that the motion of the dynamical system (3.28) should satisfy.

We assume that the disturbed motion of the system (3.28) relative to the reference (program) motion (3.29) is described by vectors

$$\widetilde{\mathbf{x}} = \widetilde{\mathbf{x}}^0 + \mathbf{x}, \quad \widetilde{\mathbf{u}} = \widetilde{\mathbf{u}}^0 + \mathbf{u}. \tag{3.30}$$

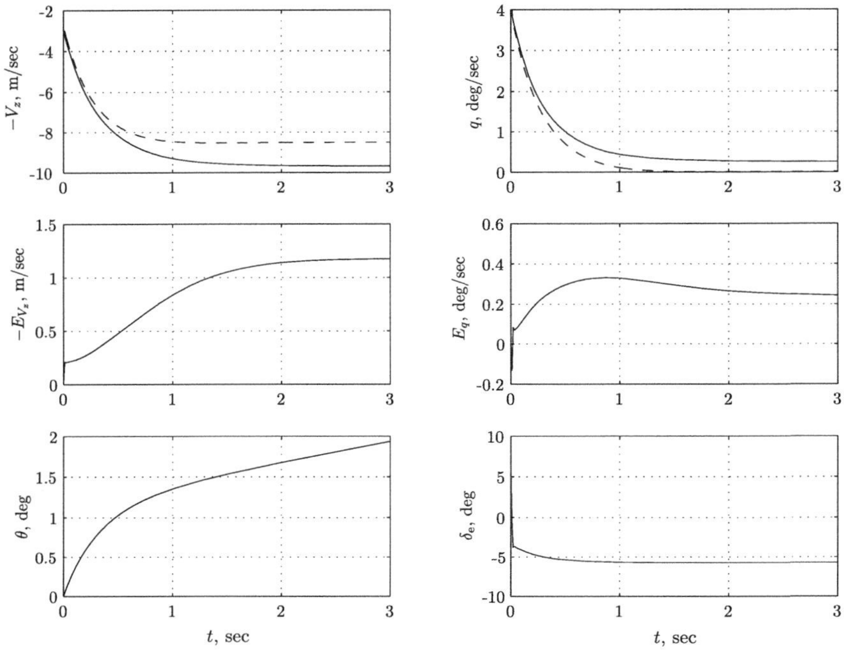

FIGURE 3.14 The results of testing the neurocontroller combined with the neural network model of the controlled object. The solid line is the output of the model (3.25); the dotted line is the output of the neural network model; V_z is the component of the velocity vector along the Oz-axis; q is the angular velocity of the pitch; θ is angle of pitch; δ_e is the deflection angle of the stabilizer; t is the time; EV_y, Eq are the differences $|V_z - V_z^{(ref)}|$ and $|q - q^{(ref)}|$, respectively (From [99], used with permission from Moscow Aviation Institute).

Assuming the norms of the vectors $||\mathbf{x}||$ and $||\mathbf{u}||$ to be small, we can get the linearized equations of the disturbed motion of the object (3.28),

$$\dot{\mathbf{x}} = A\mathbf{x} + B\mathbf{u}, \tag{3.31}$$

in which the elements of matrices A and B are functions of the parameters of program motion (3.29), elements $\lambda \in \Lambda \subset \mathcal{R}^s$, and, possibly, the time t, i.e.,

$$||a_{ij}|| = ||a_{i,j}(\widetilde{\mathbf{x}}^0, \widetilde{\mathbf{u}}^0, \boldsymbol{\lambda}, t)||,$$
$$||b_{ik}|| = ||b_{i,k}(\widetilde{\mathbf{x}}^0, \widetilde{\mathbf{u}}^0, \boldsymbol{\lambda}, t)||,$$
$$i, j \in \{1, \ldots, n\},$$
$$k \in \{1, \ldots, m\},$$

In the problem under consideration, it is the vector $\boldsymbol{\lambda} \in \boldsymbol{\Lambda}$ that is the source of uncertainty in the choice of the operation mode of the dynamical system, i.e., source of its multimode behavior. We need to clarify the nature of the uncertainty introduced by the vector $\boldsymbol{\lambda} \in \boldsymbol{\Lambda}$. Later, in the synthesis of the neurocontroller, this vector is assumed to be completely observable. However, during the synthesis process, we have no a priori data on the values of $\boldsymbol{\lambda}$ for each of the instants of time $t_0 \leqslant t_i \leqslant t_f$. These values become known only at the moment t_i, for which the corresponding control $\mathbf{u}(t_i)$ must be generated.

We assume that the system under consideration consists of a controlled object (plant), a command device (controller) producing control signals, and an actuator system generating control actions for a given control signal.

We describe the actuator system by the following equations:

$$\begin{aligned}\dot{z}_1 &= z_2,\\ \dot{z}_2 &= k_\vartheta \varphi(\sigma) - T_1\dot{z}_1 - T_2 z_1.\end{aligned} \tag{3.32}$$

Here T_1, T_2 are time constants; k_ϑ is the gain; $\varphi(\sigma)$ is the desired control law, i.e., some operation algorithm for the controller. The function $\varphi(\sigma)$ can take, for example, the following form:

$$\varphi(\sigma) = \sigma, \tag{3.33}$$

$$\varphi(\sigma) = \sigma + k_{n+1}\sigma^3, \tag{3.34}$$

$$\varphi(\sigma) = \sigma + k_{n+1}\sigma^3 + k_{n+2}\sigma^5, \tag{3.35}$$

$$\sigma = \sum_{j=1}^{n} k_j x_j.$$

The influence on the control quality for the disturbed motion is carried out through vectors $\boldsymbol{\theta} \in \boldsymbol{\Theta}$ for parameters of the plant, the command device, and the actuator system, as well as through the coefficients $\mathbf{k} = (k_1, \dots, k_n)^T \subset \mathcal{R}^n$ included in the control law.

The task in this case is to reduce the disturbed motion $(\widetilde{\mathbf{x}}, \widetilde{\mathbf{u}})$ to the reference one $(\widetilde{\mathbf{x}}^0, \widetilde{\mathbf{u}}^0)$ taking into account the uncertainty in the parameters $\boldsymbol{\lambda}$. We have to solve this problem in the best, in a certain sense, way using a control $\mathbf{u}$ added to the reference signal $\widetilde{\mathbf{u}}^0$. We only know about the $\boldsymbol{\lambda}$ parameters that they belong to some domain $\boldsymbol{\Lambda} \subset \mathcal{R}^s$. At best, we know the frequency $\rho(\boldsymbol{\lambda})$ with which one or another element of $\boldsymbol{\lambda} \subset \boldsymbol{\Lambda}$ will appear.

We term, following [88], the domain $\boldsymbol{\Lambda}$ as *external set* of the dynamical system (3.28). The system that should operate on a subset of the Cartesian product $\mathbf{X} \times \mathbf{U} \times \boldsymbol{\Lambda}$ is a *multimode dynamical system* (MDS). We can influence the control efficiency of such MDS by varying the values of $\mathbf{k} \in \mathbf{K}$ parameters of the command device. If the external set $\boldsymbol{\Lambda}$ of the considered MDS is "large enough," then, as a rule, there is a situation when we have no such $\mathbf{k} \in \mathbf{K}$ that would be equally suitable (in the sense of ensuring the required control efficiency) for all $\boldsymbol{\lambda} \subset \boldsymbol{\Lambda}$. The approach we could apply in this situation is to use different $\mathbf{k}$ for different $\boldsymbol{\lambda}$ [89–91]. In this case, the relationship $\mathbf{k} = \mathbf{k}(\boldsymbol{\lambda})$, $\forall \boldsymbol{\lambda} \subset \boldsymbol{\Lambda}$, is realized by the control system module, which is called the correction device or simply the corrector. We will call the combination of the command device and the corrector the controller.

NEUROCONTROLLER FOR A SINGLE-MODE DYNAMICAL SYSTEM AND ITS EFFICIENCY

The formation of the dependence $\mathbf{k} = \mathbf{k}(\boldsymbol{\lambda})$, $\forall \boldsymbol{\lambda} \subset \boldsymbol{\Lambda}$, implemented by the corrector, is a very time-consuming task. The traditional approach to the realization of dependence $\mathbf{k} = \mathbf{k}(\boldsymbol{\lambda})$ consists in its approximation or in interpolation according to the table of its values. For large dimensions of the vector $\boldsymbol{\lambda}$ and the large external set $\boldsymbol{\Lambda}$, the dependence $\mathbf{k}(\boldsymbol{\lambda})$ will be, as a rule, very complicated. It obstructs significantly an implementation of this dependence on board of the aircraft. To overcome this situation, we usually try to minimize the dimension of the $\boldsymbol{\lambda}$ vector. In this case, we usually take into account no more than two or three parameters, and in some cases we use only one parameter, for example, the dynamic air pressure in the task of controlling an aircraft motion. This approach, however, reduces the control efficiency since it does not take into account a number of factors affecting this efficiency.

At the same time, we know from the theory of ANNs (see, for example, [25–27]) that a feedforward neural network with one or two hidden layers can model (approximate) any continuous nonlinear functional relationship between N inputs and M outputs. Similar results were obtained for RBF networks, as well as for other types of ANNs (see, for example, [92]). Based on these results, it was proposed in [89] to use ANNs (MLP-type networks with two hidden layers) to synthesize the required continuous nonlinear mapping of the tuning parameters $\boldsymbol{\lambda}$ of the controller to the values of the control law coefficients, i.e., to form the dependence $\mathbf{k}(\boldsymbol{\lambda})$.

The neural network implementation of the dependence $\mathbf{k} = \mathbf{k}(\boldsymbol{\lambda})$ is significantly less critical to the complexity of this dependence, as well as to the dimensions of the vectors $\boldsymbol{\lambda}$ and $\mathbf{k}$. As a consequence, there is no need to minimize the number of controller tuning parameters. We have an opportunity to expand significantly the list of such parameters, including, for example, not only the dynamic air pressure (as mentioned above, in some cases it is the only tuning parameter), but also the Mach number, the angles of attack and sideslip, aircraft mass, and other variables influencing the controller coefficients on some flight regimes. In the same simple way, by introducing additional parameters, it is possible to take into account the change in the motion model (change in the type of aircraft dynamics) mentioned above.

Moreover, even a significant expansion of the list of controller tuning parameters does not lead to a significant complication of the synthesis processes for the control law and its use in the controller.

The variant of the correcting module based on the use of the ANN will be called the *neurocorrector*, and the aggregation of the controller and the neurocorrector we call *neurocontroller*.

We assume that the neurocontroller is an ordered five of the following form:

$$\Omega = (\boldsymbol{\Lambda}, \mathbf{K}, \mathbf{W}, \mathbf{V}, J), \tag{3.36}$$

where $\boldsymbol{\Lambda} \subset \mathcal{R}^s$ is the external set of the dynamical system, which is the domain of change in the values of input vectors of the neurocorrector; $\mathbf{K} \subset \mathcal{R}^n$ is the range of the values of the required controller coefficients, that is, the output vectors of the neurocorrector; $\mathbf{W} = \{W_i\}$, $i = 1, \dots, p+1$, is the set of matrices of the synaptic weights of the neurocorrector (here p is the number of hidden layers in the neurocorrector); $\mathbf{v} = (v_1, \dots, v_q) \in \mathbf{V} \subset \mathcal{R}^q$ is a set of additional variable parameters of the neurocorrector, for example, tuning parameters in the activation functions; J is the error functional, defined as the residual on the required and realized motion, which determines the nature of the neurocorrector training.

To assess the quality of the ANN control, it is necessary to have an appropriate performance index. This index (the optimality criterion of the neurocontroller) should obviously take into account not only the presence of variable parameters in the neurocontroller from the regions $\mathbf{W}$ and $\mathbf{V}$, but also the fact that the dynamical system with the given neurocontroller is multimode, that is, should be taken into account the presence of an external set $\boldsymbol{\Lambda}$.

In accordance with the approach proposed in [88], the formation of the optimality criterion of the neurocontroller on the domain $\boldsymbol{\Lambda}$ will be carried out on the basis of the efficiency evaluation of the neurocontroller "at the point," i.e., for a fixed value $\boldsymbol{\lambda}^* \in \boldsymbol{\Lambda}$, or, in other words, for a dynamic system in a single-mode version.

To do this, we construct a functional $J = J(\mathbf{x}, \mathbf{u}, \boldsymbol{\theta}, \boldsymbol{\lambda})$ or, taking into account that the vector $\mathbf{u} \in \mathbf{U}$ is uniquely determined by the vector $\mathbf{k}$ of the controller coefficients, $J(\mathbf{x}, \mathbf{k}, \boldsymbol{\theta}, \boldsymbol{\lambda})$. We assume that the control goal "at the point" is the maximum correspondence of the motion realized by the considered dynamical system to the motion determined by a certain reference model (the model of some "ideal" behavior of the dynamical system). This model can take into account both the desired nature of change in the state variables of the dynamical system and the various requirements for the nature of its operation (for example, the requirements for handling qualities of the aircraft).

Since we are discussing the control "at the point," the reference model can be local, defining the required character of the dynamical system operation for single value $\boldsymbol{\lambda} \in \boldsymbol{\Lambda}$. We will call these $\boldsymbol{\lambda}$ values operation modes. They represent characteristic points of the $\boldsymbol{\Lambda}$ region which are selected in one way or another.

As the reference we will use a linear model of the form

$$\dot{\mathbf{x}}_e = A_e \mathbf{x}_e + B_e \mathbf{u}_e, \tag{3.37}$$

where $(\mathbf{x}_e, \mathbf{u}_e)$ is the required "ideal" motion of the dynamical system.

Using the solutions of the systems (3.31) and (3.37), we can define a functional J estimating the deviation degree of the realized motion $(\mathbf{x}, \mathbf{u})$ from the required one $(\mathbf{x}_e, \mathbf{u}_e)$. As shown in [88, 93], all possible variants of such estimates can be reduced to one of two possible cases:

- guarantee criterion

$$J(\mathbf{x}, \mathbf{k}, \boldsymbol{\theta}, \boldsymbol{\lambda}) = \max_{t \in [t_0, t_f]} (|\mathbf{x}(t) - \mathbf{x}_e(t)|), \quad (3.38)$$

- integral criterion

$$J(\mathbf{x}, \mathbf{k}, \boldsymbol{\theta}, \boldsymbol{\lambda}) = \int_{t_0}^{t_f} \Xi((\mathbf{x}(t) - \mathbf{x}_e(t))^2) dt. \quad (3.39)$$

As we can see from (3.38), with the guaranteeing approach the largest of the deviations is $\Delta x_i = \mu_i(t)|x_i(t) - x_{ei}(t)|$, $i = 1, \ldots, n$, on the time interval $[t_0, t_f]$, as the measure of proximity of the real motion $(\mathbf{x}, \mathbf{u})$ to the required $(\mathbf{x}_e, \mathbf{u}_e)$. For the integral case, this measure is the integral of the square of the difference between $\mathbf{x}$ and $\mathbf{x}_e$. Here, the coefficients μ_i and the rule Ξ determine the relative importance (significance) of the deviations of Δx_i with respect to the corresponding state variables of the dynamical system for different instants of time $t \in [t_0, t_f]$.

In the case when, in accordance with the specifics of the applied task, it is necessary to take into account not only the deviations of the state variables of the controlled object, but also the required "costs" of control, the indicators (3.38) and (3.39) will take the following form:

- guarantee criterion

$$\begin{aligned} &J(\mathbf{x}, \mathbf{k}, \boldsymbol{\theta}, \boldsymbol{\lambda}) \\ &\quad = \max_{t \in [t_0, t_f]} (|\mathbf{x}(t) - \mathbf{x}_e(t)|, \boldsymbol{\mu}(t)|\mathbf{u}(t) - \mathbf{u}_e(t)|), \end{aligned} \quad (3.40)$$

- integral criterion

$$\begin{aligned} &J(\mathbf{x}, \mathbf{k}, \boldsymbol{\theta}, \boldsymbol{\lambda}) \\ &\quad = \int_{t_0}^{t_f} \Xi((\mathbf{x}(t) - \mathbf{x}_e(t))^2, (\mathbf{u}(t) - \mathbf{u}_e(t))^2) dt. \end{aligned} \quad (3.41)$$

It should be emphasized that it is the functional (3.38), (3.40) or (3.39), (3.41) that "directs" learning the ANN used in the neurocontroller (3.36), since it is its value that is minimized in the learning process of the neurocontroller.

A discussion of the approach used to get the $\mu_i(t)$, $\boldsymbol{\mu}(t)$ dependencies and the Ξ rule relates to the decision-making area for the vector-type efficiency criterion. This approach is based on the results obtained in [88,94] and it is beyond the scope of this book.

THE ENSEMBLE OF NEURAL CONTROLLERS AS A TOOL FOR TAKING INTO ACCOUNT THE MULTIMODE NATURE OF DYNAMICAL SYSTEMS

The dependence $\mathbf{k}(\boldsymbol{\lambda})$, including its ANN version, may be too complicated to implement on board of aircraft due to the limited computing resources that can be allocated to such implementation. If $\boldsymbol{\lambda}$ "does not change too much" when changing $\mathbf{k}$, we could try to find some "typical" value $\boldsymbol{\lambda}$, determine the corresponding $\mathbf{k}^*$ for it, and then replace $\mathbf{k}(\boldsymbol{\lambda})$ with this value $\mathbf{k}^*$. However, when $\boldsymbol{\lambda}$ significantly differs from its typical value, the quality of regulation of the controller obtained in this way may not meet the design requirements.

In order to overcome this difficulty, we can use a piecewise (piecewise-constant, piecewise-linear, piecewise-polynomial, etc.) variant of the approximation for $\mathbf{k}(\boldsymbol{\lambda})$. We will clarify considerations concerning the assessment of the quality of regulation in Section 3.4.2.3.

As a tool to implement this kind of approximation we introduce the *ensemble of neural controllers* (ENC)

$$\boldsymbol{\Omega} = (\Omega_0, \Omega_1, \ldots, \Omega_N), \quad (3.42)$$

where each of the neurocontrollers Ω_i, $i \in \{1, \dots, N\}$, is used on its own domain $\mathbf{D}_i \subset \mathbf{\Lambda}$, which we call the *area of specialization* of the neurocontroller Ω_i:

$$\mathbf{D}_i \subset \mathbf{\Lambda}, \ \bigcup_{i=1}^{N} \mathbf{D}_i = \mathbf{\Lambda}, \ \mathbf{D}_i \bigcap_{i \neq j} \mathbf{D}_j = \varnothing, \\ \forall i, j \in \{1, \dots, N\}.$$

We determine the rule of transition from one neurocontroller to another using the *distribution function* $E(\boldsymbol{\lambda})$, whose argument is the external vector $\boldsymbol{\lambda} \in \mathbf{\Lambda}$, and its integer values are the numbers of the specialization areas and, respectively, neurocontrollers operating them, i.e.,

$$\begin{aligned} E(\boldsymbol{\lambda}) &= i, \ i \in \{1, \dots, N\}, \\ \mathbf{D}_i &= \{\boldsymbol{\lambda} \in \mathbf{\Lambda} \mid E(\boldsymbol{\lambda}) = i\}. \end{aligned} \tag{3.43}$$

The distribution function $E(\boldsymbol{\lambda})$ is realized according to (3.43) by the Ω_0 element of the $\mathbf{\Omega}$ neurocontroller ensemble.

It should be emphasized that the ENC (3.42) is a set of *mutually agreed* neurocontrollers. All these neurocontrollers have the same current value of the external vector $\boldsymbol{\lambda} \in \mathbf{\Lambda}$.

In (3.42) there are two types of neurocontrollers. Neurocontrollers of the first type form a set $\{\Omega_1, \dots, \Omega_N\}$, members of which implement the corresponding control laws. The neurocontroller of the second type (Ω_0) is a kind of "conductor" of the ENC $\Omega_1, \dots, \Omega_N$. This neurocontroller for each current $\boldsymbol{\lambda} \in \mathbf{\Lambda}$ according to (3.43) produces the number i, $1 \leqslant i \leqslant N$, that is, indicates which of the neurocontrollers Ω_i, $i \in \{1, \dots, N\}$, have to control at a given $\boldsymbol{\lambda} \in \mathbf{\Lambda}$.

Thus, the ENC $\mathbf{\Omega}$ is a *mutually agreed set* of neurocontrollers, in which all neurocontrollers get the current value of the external vector $\boldsymbol{\lambda} \in \mathbf{\Lambda}$ as an input. Further, the neurocontroller Ω_0 by the current $\boldsymbol{\lambda}$ in accordance with (3.43) generates the number i, $1 \leqslant i \leqslant N$, indicating one of the neurocontrollers Ω_i, which should control for a given $\boldsymbol{\lambda} \in \mathbf{\Lambda}$.

3.4.2.3 *Optimization of an Ensemble of Neural Controllers for a Multimode Dynamical System*

For the ENC, it is very important to optimize it. The solution of this problem should ensure the minimization of the number of neurocontrollers in the ensemble for a given external set, that is, for a given region of MDS operation modes. If, for some reason, besides the external set of MDS, the number of neural controllers in the ENC is also specified, then its optimization allows choosing the values of the neurocontroller parameters so as to minimize the error generated by the ENC. The key problem here, as in any optimization problem, is the formation of an optimality criterion for the system under consideration.

FORMATION OF AN OPTIMALITY CRITERION FOR AN ENSEMBLE OF NEURAL CONTROLLERS FOR A MULTIMODE DYNAMICAL SYSTEM

One of the most important points in solving ENC optimization problems is the formation of the optimality criterion $F(\mathbf{\Lambda}, \mathbf{\Omega}, J, E(\boldsymbol{\lambda}))$, taking into account all the above mentioned features of the MDS and ENC. Based on the results obtained in [88], it is easy to show that such a criterion can be constructed knowing the way to calculate the efficiency of the considered system at the current $\boldsymbol{\lambda}$ point of the external set $\mathbf{\Lambda}$ for a fixed set $\{\Omega_i\}$, $i = 1, \dots, N$, of neural controllers in the ENC[1] as well as for the fixed distribution function $E(\boldsymbol{\lambda})$. In addition, we need to know the functional, which takes the form (3.38), (3.40) or (3.39), (3.41). A function describing the efficiency of the ENC under these assumptions,

$$f = f(\boldsymbol{\lambda}, \mathbf{\Omega}, J, E(\boldsymbol{\lambda})), \ \forall \boldsymbol{\lambda} \in \mathbf{\Lambda}, \tag{3.44}$$

we call *criterial function* of the ENC. Since (3.44) depends actually only on $\boldsymbol{\lambda} \in \mathbf{\Lambda}$, and all other arguments can be treated as parameters frozen in

[1] That is, for the given quantity N of neurocontrollers Ω_i in the ENC $\mathbf{\Omega}$ and the values of the parameters $\mathbf{W}$ and $\mathbf{V}$.

one way or another, to shorten the expressions in this section, we write, instead of (3.44),

$$f = f(\boldsymbol{\lambda}), \tag{3.45}$$

and instead of $F(\boldsymbol{\Lambda}, \boldsymbol{\Omega}, J, E(\boldsymbol{\lambda}))$, $F = F(\boldsymbol{\Lambda})$.

Thus, the problem of finding the value of the optimality criterion for the ENC is divided into two subtasks: first, we need to be able to calculate $f(\boldsymbol{\lambda})$, $\forall \boldsymbol{\lambda} \in \boldsymbol{\Lambda}$, taking into account the above assumptions; second, it is necessary to determine (generate) the Φ rule that would allow us to find $F(\boldsymbol{\Lambda})$, that is, the ENC optimality criterion on the whole external set $\boldsymbol{\Lambda}$, by $f(\boldsymbol{\lambda})$, i.e.,

$$F(\boldsymbol{\Lambda}) = \Phi[f(\boldsymbol{\lambda})]. \tag{3.46}$$

Let us first consider the question of constructing a criterial function $f(\boldsymbol{\lambda})$. We will assume that the ENC is defined if for each of the neurocontrollers $\{\Omega_i\}$, $i = 1, \ldots, N$, we know the set of synaptic weight matrices $\mathbf{W}^{(i)} = \{W_j^{(i)}\}$, $i = 1, \ldots, N$, $j = 1, \ldots, p^{(i)} + 1$, where $p^{(i)}$ is the number of hidden layers in the ANN used in the neurocontroller Ω_i, as well as the value of the vector $\mathbf{v}^{(i)} \in \mathbf{V}$ of additional adjusting parameters of this neurocontroller. Suppose that we also have an external set $\boldsymbol{\Lambda} \subset \mathcal{R}^s$ for the MDS. If we also freeze the "point" $\widetilde{\boldsymbol{\lambda}} \in \boldsymbol{\Lambda}$, then we get the usual problem of synthesizing the control law for a single-mode system. Solving this problem, we find $\widetilde{\mathbf{k}}^* \in \mathbf{K}$, that is, the optimal value of the regulator parameters vector (3.33)–(3.35) under the condition $\widetilde{\boldsymbol{\lambda}} \in \boldsymbol{\Lambda}$. In this case, the functional J takes the value

$$J^*(\widetilde{\boldsymbol{\lambda}}) = J(\widetilde{\boldsymbol{\lambda}}, \widetilde{\mathbf{k}}^*) = \min_{\mathbf{k} \in \mathbf{K}} J(\widetilde{\boldsymbol{\lambda}}, \widetilde{\mathbf{k}}).$$

If we apply the neurocontroller with the parameter vector $\widetilde{\mathbf{k}}^* \in \mathbf{K}$, which is optimal for the point $\widetilde{\boldsymbol{\lambda}} \in \boldsymbol{\Lambda}$, in the point $\boldsymbol{\lambda} \in \boldsymbol{\Lambda}$, then the functional J^* will be

$$J^* = J(\boldsymbol{\lambda}, \widetilde{\mathbf{k}}^*).$$

If $\widetilde{\mathbf{k}}^* \in \mathbf{K}$ is the point of absolute minimum of the functional J for $\widetilde{\boldsymbol{\lambda}} \in \boldsymbol{\Lambda}$, then the following inequality will be satisfied:

$$J(\boldsymbol{\lambda}, \widetilde{\mathbf{k}}^*) \geqslant J(\widetilde{\boldsymbol{\lambda}}, \widetilde{\mathbf{k}}^*), \ \forall \boldsymbol{\lambda} \in \boldsymbol{\Lambda}. \tag{3.47}$$

Based on the condition (3.47), we write the expression for the criterial function $f(\boldsymbol{\lambda}, \mathbf{k})$ in the form

$$f(\boldsymbol{\lambda}, \mathbf{k}) = J(\boldsymbol{\lambda}, \mathbf{k}) - J(\boldsymbol{\lambda}, \mathbf{k}^*), \tag{3.48}$$

where $J(\boldsymbol{\lambda}, \mathbf{k})$ is the value of the functional J for arbitrary admissible $\boldsymbol{\lambda} \in \boldsymbol{\Lambda}$ and $\mathbf{k} \in \mathbf{K}$, and $J(\boldsymbol{\lambda}, \mathbf{k}^*)$ is the minimal value of the functional J for the parameter vector of the neurocontroller, optimal for the given $\boldsymbol{\lambda} \in \boldsymbol{\Lambda}$.

Finding the value $J(\boldsymbol{\lambda}, \mathbf{k})$ does not cause any difficulties and can be performed by calculating it together with (3.31) by one of the numerical methods for solving the Cauchy problem for a system of ODEs of the first order. The calculation of the value $J(\boldsymbol{\lambda}, \mathbf{k}^*)$, that is, minimum of the functional $J(\boldsymbol{\lambda}, \mathbf{k})$ under the parameter vector $\mathbf{k}^* \in \mathbf{K}$, which is optimal for a given current $\boldsymbol{\lambda} \in \boldsymbol{\Lambda}$, is significantly more difficult, because in this case we need to solve the problem of synthesis of the optimal control law for a single-mode dynamical system. This problem in the case under consideration relates to the training of the ANN used in the correcting module of the corresponding neurocontroller.

We now consider the problem of determining the rule (construction function) Φ, which allows us to find the value of the optimality criterion $F(\boldsymbol{\Lambda})$ for the given ENC $\boldsymbol{\Omega}$, if we know $f(\boldsymbol{\lambda}, \mathbf{k})$, $\forall \boldsymbol{\lambda} \in \boldsymbol{\Lambda}$, $\forall \mathbf{k} \in \mathbf{K}$.

The construction function is assumed to be symmetric, i.e.,

$$\Phi(\eta, \nu) = \Phi(\nu, \eta),$$

and associative, i.e.,

$$\Phi(\eta, \Phi(\nu, \xi)) = \Phi(\nu, \Phi(\eta, \xi)) = \Phi(\xi, \Phi(\eta, \nu)).$$

These properties are attached to the construction function to ensure the independence of the criterion $F(\mathbf{\Lambda})$ from the order of combining the elements on the set $\mathbf{\Lambda}$.

The function $\Phi(\eta, \nu)$ gives a way to calculate the efficiency $F(\mathbf{\Lambda}_{\alpha\beta})$ of the ENC $\mathbf{\Omega}$ for the MDS with the external set

$$\mathbf{\Lambda}_{\alpha\beta} = \mathbf{\Lambda}_\alpha \bigcup \mathbf{\Lambda}_\beta,$$

by known efficiencies $f(\mathbf{\Lambda}_\alpha)$ and $f(\mathbf{\Lambda}_\beta)$, where $\mathbf{\Lambda}_\alpha \subset \mathbf{\Lambda}$, $\mathbf{\Lambda}_\beta \subset \mathbf{\Lambda}$, i.e.,

$$F(\mathbf{\Lambda}_\alpha \bigcup \mathbf{\Lambda}_\beta) = \Phi(f(\mathbf{\Lambda}_\alpha), f(\mathbf{\Lambda}_\beta)).$$

In the same way we can define on an exterior set

$$\mathbf{\Lambda}_{\alpha\beta\gamma} = (\mathbf{\Lambda}_\alpha \bigcup \mathbf{\Lambda}_\beta) \bigcup \mathbf{\Lambda}_\gamma, \ \mathbf{\Lambda}_\gamma \subset \mathbf{\Lambda},$$

the efficiency $F(\mathbf{\Lambda}_{\alpha\beta\gamma})$ for the system on it, i.e.,

$$\begin{aligned} F(\mathbf{\Lambda}_{\alpha\beta\gamma}) &= F((\mathbf{\Lambda}_\alpha \bigcup \mathbf{\Lambda}_\beta) \bigcup \mathbf{\Lambda}_\gamma) \\ &= \Phi(f(\mathbf{\Lambda}_\gamma), \Phi(\mathbf{\Lambda}_\alpha \bigcup \mathbf{\Lambda}_\beta)). \end{aligned}$$

Repeating this operation further, we get

$$\begin{aligned} &F(\mathbf{\Lambda}_1 \bigcup (\mathbf{\Lambda}_2 \bigcup (\mathbf{\Lambda}_3 \dots (\mathbf{\Lambda}_{N-1} \bigcup \mathbf{\Lambda}_N) \dots) \dots) \\ &\quad = \Phi(f(\mathbf{\Lambda}_1), \Phi(f(\mathbf{\Lambda}_2), \Phi(f(\mathbf{\Lambda}_3), \dots))). \end{aligned}$$

Bearing in mind that

$$\mathbf{\Lambda}_1 \bigcup (\mathbf{\Lambda}_2 \bigcup (\mathbf{\Lambda}_3 \dots (\mathbf{\Lambda}_{N-1} \bigcup \mathbf{\Lambda}_N) \dots)) = \mathbf{\Lambda},$$

we get

$$F(\mathbf{\Lambda}) = \Phi(f(\mathbf{\Lambda}_1), \Phi(f(\mathbf{\Lambda}_2), \Phi(f(\mathbf{\Lambda}_3), \dots))). \tag{3.49}$$

The method of constructing the rule Φ in (3.49), which allows us to know $f(\boldsymbol{\lambda}, \mathbf{k})$, $\forall \boldsymbol{\lambda} \in \mathbf{\Lambda}$, $\forall \mathbf{k} \in \mathbf{K}$, and find the value of the optimality criterion $F(\mathbf{\Lambda})$ of a multimode system of general form on the entire external set $\mathbf{\Lambda}$, is described in [88]. It is shown here that all possible types of estimating the effectiveness of a system on an external set $\mathbf{\Lambda}$ are reduced to two classes: guaranteed estimates, when the optimality criterion $F(\mathbf{\Lambda})$ takes the form

$$F(\mathbf{\Lambda}) = \max_{\substack{\boldsymbol{\lambda} \in \mathbf{\Lambda}, \\ \mathbf{k}=\text{const}}} [\rho(\boldsymbol{\lambda}) f(\boldsymbol{\lambda}, \mathbf{k})], \tag{3.50}$$

and integral, for which $F(\mathbf{\Lambda})$ is an expression of the form

$$F(\mathbf{\Lambda}) = \sum_{\substack{\boldsymbol{\lambda} \in \mathbf{\Lambda}, \\ \mathbf{k}=\text{const}}} [\rho(\boldsymbol{\lambda}) f(\boldsymbol{\lambda}, \mathbf{k})], \tag{3.51}$$

where $\rho(\boldsymbol{\lambda})$ is the degree of relative significance for the element $\boldsymbol{\lambda} \in \mathbf{\Lambda}$.

Taking into account the above, the criterial function $f(\boldsymbol{\lambda}, \mathbf{k})$ can be treated as a *degree of nonoptimality* of ENC for the MDS that operates on the $\boldsymbol{\lambda} \in \mathbf{\Lambda}$ mode. Then we can say that for a criterion of the form (3.50), the problem of minimizing the maximum degree of nonoptimality of ENC $\mathbf{\Omega}$ for the MDS with an external set $\mathbf{\Lambda}$ will be solved; we have

$$F_G^* = F_G(\mathbf{\Omega}^*) = \min_{\mathbf{\Omega}} \max_{\substack{\boldsymbol{\lambda} \in \mathbf{\Lambda}, \\ \mathbf{k}=\text{const}}} [\rho(\boldsymbol{\lambda}) f(\boldsymbol{\lambda}, \mathbf{k})]. \tag{3.52}$$

With the integral criterion, the ENC optimization problem reduces to minimizing the integral degree of nonoptimality of the ENC $\mathbf{\Omega}$ for an MDS acting on the external set $\mathbf{\Lambda}$, i.e.,

$$F_I^* = F_I(\mathbf{\Omega}^*) = \min_{\mathbf{\Omega}} \int\limits_{\substack{\boldsymbol{\lambda} \in \mathbf{\Lambda}, \\ \mathbf{k}=\text{const}}} [\rho(\boldsymbol{\lambda}) f(\boldsymbol{\lambda}, \mathbf{k})] d\boldsymbol{\lambda}. \tag{3.53}$$

Applied to the integral criterion (3.53), the Φ rule from (3.49) actually defines the weight function $\rho(\boldsymbol{\lambda})$, which specifies the relative importance of the elements $\boldsymbol{\lambda}$ of the external set $\mathbf{\Lambda}$. Accordingly, the integral criterion (3.51), (3.53) can be varied when it is formed within wide limits, corresponding to the specifics of the applied task.

Given the above formulations, the concept of mutual consistency of the neurocontrollers Ω_i, $i \in \{1, \ldots, N\}$, that are part of the ENC $\boldsymbol{\Omega}$, mentioned at the end of Section 3.4.2.2, can now be clarified.

Namely, the mutual consistency of the neurocontrollers in the ENC (3.42) is as follows:

- parameters of each of the neurocontrollers Ω_i, $i \in \{1, \ldots, N\}$, are selected taking into account all other neurocontrollers Ω_j, $j \neq i$, $i, j \in \{1, \ldots, N\}$, based on the requirements imposed by the optimality criterion of the MDS (3.50) or (3.51) for the ENC $\boldsymbol{\Omega}$ as a whole;
- it is guaranteed that each of the modes (tasks to be solved) $\boldsymbol{\lambda} \in \boldsymbol{\Lambda}$ will be worked out by the neurocontroller, the most effective of the available within the ENC (3.42), that is, such neurocontroller Ω_i, $i \in \{1, \ldots, N\}$, for which the value of the criterion function (degree of nonoptimality for neurocontroller Ω_i) $f_i(\boldsymbol{\lambda}, \mathbf{k})$, defined by expression (3.48), is the least for the given $\boldsymbol{\lambda} \in \boldsymbol{\Lambda}$ and $\mathbf{k} \in \mathbf{K}$.

OPTIMIZATION TASKS FOR AN ENSEMBLE OF NEURAL CONTROLLERS WITH A CONSERVATIVE APPROACH

With regard to optimization of the ENC, the following main tasks can be formulated:

1. The problem of optimal distribution for the ENC,

$$F(\boldsymbol{\Lambda}, \boldsymbol{\Omega}, J, E^*(\boldsymbol{\lambda})) = \min_{\substack{E(\boldsymbol{\lambda}),\\ N=\text{const},\\ k=\text{const}}} F(\boldsymbol{\Lambda}, \boldsymbol{\Omega}, J, E(\boldsymbol{\lambda})). \tag{3.54}$$

2. The problem of the optimal choice of the parameters for neurocontrollers included in the ENC,

$$F(\boldsymbol{\Lambda}, \boldsymbol{\Omega}^*, J, E^*(\boldsymbol{\lambda})) = \min_{\substack{E(\boldsymbol{\lambda},\boldsymbol{\Omega}),\\ N=\text{const}}} F(\boldsymbol{\Lambda}, \boldsymbol{\Omega}, J, E(\boldsymbol{\lambda})). \tag{3.55}$$

3. General optimization problem for the ENC,

$$F(\boldsymbol{\Lambda}, \boldsymbol{\Omega}^*, J, E^*(\boldsymbol{\lambda})) = \min_{\substack{E(\boldsymbol{\lambda},\boldsymbol{\Omega}),\\ N=\text{var}}} F(\boldsymbol{\Lambda}, \boldsymbol{\Omega}, J, E(\boldsymbol{\lambda})). \tag{3.56}$$

In the problem of the optimal distribution (3.54) there is a region $\boldsymbol{\Lambda}$ of the $\boldsymbol{\lambda}$ operation modes of the dynamical system (the external set of the system) and N given by some neurocontrollers $\Omega_i, i = 1, \ldots, N$. It is required to assign to each neurocontroller Ω_i the domain of its specialization $\mathbf{D}_i \subset \boldsymbol{\Lambda}$,

$$\mathbf{D}_i = \mathbf{D}(\Omega_i) = \{\boldsymbol{\lambda} \in \boldsymbol{\Lambda} \mid E(\boldsymbol{\lambda}) = i\},\ i \in \{1, \ldots, N\},$$

$$\bigcup_{i=1}^{N} \mathbf{D}_i = \boldsymbol{\Lambda},\ \mathbf{D}_j \bigcap_{j \neq k} \mathbf{D}_k = \varnothing,\ \forall j, k \in \{1, \ldots, N\},$$

where the use of this neurocontroller Ω_i is preferable to the use comparing all other neurocontrollers Ω_j, $j \neq i$, $i, j \in \{1, \ldots, N\}$. The division of the $\boldsymbol{\Lambda}$ domain into the $\mathbf{D}_i \subset \boldsymbol{\Lambda}$ specialization domains is given by the distribution function $E(\boldsymbol{\lambda})$ defined on the set $\boldsymbol{\Lambda}$ and takes integer values $1, 2, \ldots, N$. The function $E(\boldsymbol{\lambda})$ assigns to each $\boldsymbol{\lambda} \in \boldsymbol{\Lambda}$ the number of the neurocontroller corresponding to the given mode, such that its criterial function (3.44) will be for this $\boldsymbol{\lambda} \in \boldsymbol{\Lambda}$ the smallest in comparison with the criterial functions of the remaining neurocontrollers that are part of the ENC.

The problem of the optimal choice of parameters (3.55) for neurocontrollers Ω_i, $I = 1, \ldots, N$, included in the ENC $\boldsymbol{\Omega}$ has the optimal distribution problem (3.54) as a subproblem. It consists in the selection of parameters $\mathbf{W}^{(i)}$ and $\mathbf{V}^{(i)}$ of neurocontrollers Ω_i, $i = 1, \ldots, N$, included in the ENC $\boldsymbol{\Omega}$, in such a way as to minimize the value of the ENC optimality criterion (3.50), (3.52) or (3.51), (3.53), depending on the type of the corresponding application task. We assume that the number of neurocontrollers N in the ENC $\boldsymbol{\Omega}$ is fixed from any considerations external to the problem to be solved.

In the problem of general optimization (3.56) for the ENC $\mathbf{\Omega}$ we remove the condition of fixing the number N of neural controllers Ω_i, $i = 1, \dots, N$. It is possible, in such a case, to vary (select) also the number of neurocontrollers in the ENC minimizing the value of the optimality criterion (3.50), (3.52) or (3.51), (3.53). Obviously, the problem of optimizing ENC parameters (3.55) and, consequently, the optimal distribution problem (3.54) are included in the general optimization problem (3.56) as subtasks.

The solution of the optimal distribution problem (3.54) allows the best divide (in the sense of the criterion (3.50), (3.52) or (3.51), (3.53)) the external set $\mathbf{\Lambda}$ of the considered MDS into the specialization domain $\mathbf{D}_i \subset \mathbf{\Lambda}$, $i = 1, \dots, N$, specifying where it is best to use each of the neurocontrollers Ω_i, $i \in \{1, \dots, N\}$. By varying the parameters of the neurocontrollers in the ENC and solving the optimal distribution problem each time, it is possible to reduce the value of the criterion (3.50), (3.52) or (3.51), (3.53), which evaluates the efficiency of the ENC $\mathbf{\Omega}$ on the external set $\mathbf{\Lambda}$ as a whole. The removal of the restriction on the number of neurocontrollers in the ENC provides, in general, a further improvement in the ENC effectiveness value. In the general case, for the same MDS with a fixed external set $\mathbf{\Lambda}$, the following relation holds:

$$F^{(1)}(\mathbf{\Lambda}) \geqslant F^{(2)}(\mathbf{\Lambda}) \geqslant F^{(3)}(\mathbf{\Lambda}),$$

where $F^{(1)}(\mathbf{\Lambda})$, $F^{(2)}(\mathbf{\Lambda})$, and $F^{(3)}(\mathbf{\Lambda})$ are the values of the optimality criteria (3.50), (3.52) or (3.51), (3.53) obtained by solving the optimal distribution problem (3.54), the parameter optimization problem (3.55), or the general optimization problem (3.56) for a given MDS.

Generally speaking, the required dependence of the controller coefficients on the parameters of the regime description can be approximated with the help of a neural network at once to the entire region of the modes of the MDS operation (that is, on its whole external set). However, here it is necessary to take into account the "price" which will have to be paid for such a decision. For modern and, especially, advanced aircraft with high performances, the required dependence is multidimensional and has a very complicated character, which, also, can be considerably complicated if the aircraft requires the implementation of various types of behavior corresponding to different classes of problems solved by the aircraft. As a result, the synthesized neural network cannot satisfy the designers of the control system due to, for example, a too high network dimension, which makes it difficult to implement this network using aircraft onboard tools, or even makes such an implementation impossible, and also significantly complicates the solution of the problem of training the ANN. Besides, the larger the dimension of the ANN, the longer the time of its response to a given input signal when the network is implemented using serial or limited-parallel hardware, which are the dominant variants now.

It is to this kind of situation that the approach under consideration is oriented, according to which the problem of decomposing one ANN (and, correspondingly, one neurocontroller) into a set of mutually coordinated neurocontrollers, implemented as an ensemble of ANNs, is solved. We have shown how to perform this decomposition optimally within the framework of three classes (levels) of task optimization of ENCs.

We have described here the formation of the optimal ENC concerning the conservative approach to the use of ANNs in control problems. However, this approach is equally suitable after a small adaptation for the radical approach to neurocontrol for multimode dynamic systems, and, consequently, also for a compromise approach to solve this problem.

Moreover, if we slightly reformulate the considered approach, it can also be interpreted as an approach to the decomposition of ANNs, oriented to solving problems under uncertainty conditions, that is, as an approach to replacing one "large" network with a mutually agreed aggregate (ensemble) of "smaller" networks, and it is possible to implement this decomposition in an optimal way. In the general case, such an ensemble can be inhomogeneous, containing an ANN of different architectures, which in principle is an additional source of increasing the efficiency of solving complex applied problems.

It should be noted that, in fact, the ENC is a neural network implementation and the extension of the well-known Gain Scheduling approach [95–97], which is widely used to solve a variety of applied problems.

3.4.2.4 A Formation Example of an Ensemble of Neural Controllers for a Simple Multimode Dynamical System

Let us illustrate the application of the main provisions outlined above, on a synthesis example of the optimal ENC for a simple aperiodic controlled object (plant) [98], described by

$$\dot{x} = -\frac{1}{\tau(\lambda)}x + u, \; t \in [t_0, \infty). \tag{3.57}$$

Here

$$\tau(\lambda) = c_0 + c_1\lambda + c_2\lambda^2, \; \lambda \in [\lambda_0, \lambda_k]. \tag{3.58}$$

As the control law for the plant (3.57), we take

$$u = -kx, \; k_- \leqslant k \leqslant k_+. \tag{3.59}$$

The controller implementing the control law (3.59) must maintain the state x of the controlled object in a neighborhood of zero, i.e., as the desired (reference) object motion (3.57) we assume

$$x_e(t) \equiv 0, \; u_e(t) \equiv 0, \; \forall t \in [t_0, \infty). \tag{3.60}$$

The quality criterion (performance index, functional) J for the MDS (3.57)–(3.60) can be written in the form

$$J(\lambda, k) = \frac{1}{2}\int_{t_0}^{\infty} (ax(\lambda)^2 + bu(k)^2)dt, \tag{3.61}$$

where $a = 1/x^2_{max}$, $b = 1/u^2_{max}$, based on the additional requirement to keep x^2 below the specified value $x^2_{max} = \text{const}$ by using the u^2 control not exceeding $u^2_{max} = \text{const}$.

In the problem under consideration, the external set $\mathbf{\Lambda}$ and the domain $\mathbf{K}$ of the values of the regulator parameters are one-dimensional, that is,

$$\begin{aligned} \mathbf{\Lambda} &= [\lambda_0, \lambda_k] \subset \mathcal{R}^1, \\ \mathbf{K} &= [k_-, k_+] \subset \mathcal{R}^1. \end{aligned} \tag{3.62}$$

The criterial function $f(\lambda, k)$ is written, according to (3.48), in the form

$$f(\lambda, k) = J(\lambda, k) - J(\lambda, k^*). \tag{3.63}$$

For an arbitrary admissible pair (λ, k), $\lambda \in \mathbf{\Lambda}$, $k \in \mathbf{K}$, the expression for $J(\lambda, k)$ applied to the system (3.28)–(3.33) takes the form

$$J(\lambda, k) = \frac{(a^2 + bk^2)x_0^2}{4(1/\tau(\lambda) + k)}. \tag{3.64}$$

Since the function (3.64) is convex $\forall x \in \mathbf{X}$, we can put $x_0 = x_{max}$.

According to [98], the value of the functional $J(\lambda, k^*)$ for some arbitrary $\lambda \in \mathbf{\Lambda}$ can be obtained by knowing the expression for $k^*(\lambda)$, i.e.,

$$k^*(\lambda) = \sqrt{\frac{1}{\tau^2(\lambda)} + \frac{a}{b}} - \frac{1}{\tau(\lambda)}. \tag{3.65}$$

In this problem, the controller realizes the control law (3.59), the neurocorrector reproduces the dependence (3.65) of the k^* coefficient adjustment depending on the current value of $\lambda \in \mathbf{\Lambda}$, and collectively, this regulator and neurocorrector are neurocontroller Ω_1, the only one in the ENC $\mathbf{\Omega}$.

As a neurocorrector here we can use an MLP-type network with one or two hidden layers or some RBF network. Due to the triviality of the formation of the corresponding ANN in the case under consideration, the details of this process are omitted.

By the known criterial function (3.63), we can now construct an optimality criterion (3.50) or (3.51) for the neurocontroller (and then we can build the ENC) for the case under consideration. For definiteness, let us use the guaranteeing approach to evaluate the ENC efficiency for the MDS (3.57)–(3.62) (see (3.52)), that is, the required criterion has the form

$$F(\mathbf{\Lambda}) = \max_{\substack{\lambda \in \mathbf{\Lambda}), \\ k=\text{const}}} f(\lambda, k). \qquad (3.66)$$

In this case, for $\mathbf{\Omega} = (\Omega_0, \Omega_1)$, the neurocontroller Ω_0 implements the distribution function $E(\lambda) = 1,\ \forall \lambda \in \mathbf{\Lambda}$.

If the value of the maximum degree of nonoptimality (3.57) for the obtained ENC is greater than allowed by the conditions of the solved application, it is possible to increase the number N of the neurocontrollers $\Omega_i,\ i = 1, \dots, N$, in $\mathbf{\Omega}$, thereby decreasing the value of the index (3.66). For example, let $N = 3$. Then

$$\mathbf{\Omega} = (\Omega_0, \Omega_1, \Omega_2, \Omega_3),$$

$$E(\lambda) = \begin{cases} 1, & \lambda_0 \leqslant \lambda \leqslant \lambda_{12}, \\ 2, & \lambda_{12} < \lambda \leqslant \lambda_{23}, \\ 3, & \lambda_{23} < \lambda \leqslant \lambda_k. \end{cases}$$

To obtain a numerical estimate, we set $a = b = 1$, $c_0 = 1$, $c_1 = 2$, $c_2 = 5$, $x_{max}^2 = 10$, $\lambda_0 = 0$, $\lambda_k = 1$. It can be shown that in the given problem $k_- \approx 0.4$, $k_+ \approx 0.9$. Then in the considered case for the ENC with $N = 1$ the value of the index (3.66) in the problem (3.55) of ENC parameter optimization will be $F^* = F(\mathbf{\Lambda}, k^*) \approx 0.42$, and for $N = 3$ (i.e., with three "working" neurocontrollers and one "switching" in the ENC) $F^* = F(\mathbf{\Lambda}, k^*) \approx 0.28$.

REFERENCES

[1] Billings SA. Nonlinear system identification: NARMAX methods in the time, frequency and spatio-temporal domains. New York, NY: John Wiley & Sons; 2013.
[2] Codrons B. Process modelling for control: A unified framework using standard black-box techniques. London: Springer; 2005.
[3] Narendra KS, Parthasarathy K. Identification and control of dynamic systems using neural networks. IEEE Trans Neural Netw 1990;1(1):4–27.
[4] Chen S, Billings SA. Neural networks for nonlinear dynamic systems modelling and identification. Int J Control 1992;56(2):319–46.
[5] Sjöberg J, Zhang Q, Ljung L, Benveniste A, Deylon B, Glorennec PY, et al. Nonlinear black-box modeling in system identification: A unified overview. Automatica 1995;31(12):1691–724.
[6] Juditsky A, Hjalmarsson H, Benveniste A, Deylon B, Ljung L, Sjöberg J, et al. Nonlinear black-box modeling in system identification: Mathematical foundations. Automatica 1995;31(12):1725–50.
[7] Rivals I, Personnaz L. Black-box modeling with state-space neural networks. In: Zbikowski R, Hint KJ, editors. Neural Adaptive Control Technology. World Scientific; 1996. p. 237–64.
[8] Billings SA, Jamaluddin HB, Chen S. Properties of neural networks with applications to modelling nonlinear dynamic systems. Int J Control 1992;55(1):193–224.
[9] Chen S, Billings SA. Representation of nonlinear systems: The narmax model. Int J Control 1989;49(3):1013–32.
[10] Chen S, Billings SA. Nonlinear system identification using neural networks. Int J Control 1990;51(6):1191–214.
[11] Chen S, Billings SA, Cowan CFN, Grant PM. Practical identification of narmax using radial basis functions. Int J Control 1990;52(6):1327–50.
[12] Kalman RE, Falb PL, Arbib MA. Topics in mathematical system theory. New York, NY: McGraw Hill Book Company; 1969.
[13] Mesarovic MD, Takahara Y. General systems theory: Mathematical foundations. New York, NY: Academic Press; 1975.
[14] Dreyfus G. Neural networks: Methodology and applications. Berlin ao.: Springer; 2005.
[15] Chen S, Wang SS, Harris C. NARX-based nonlinear system identification using orthogonal least squares basis hunting. IEEE Trans Control Syst Technol 2008;16(1):78–84.
[16] Sahoo HK, Dash PK, Rath NP. NARX model based nonlinear dynamic system identification using low complexity neural networks and robust H_∞ filter. Appl Soft Comput 2013;13(7):3324–34.
[17] Hidayat MIP, Berata W. Neural networks with radial basis function and NARX structure for material lifetime assessment application. Adv Mater Res 2011;277:143–50.
[18] Wong CX, Worden K. Generalised NARX shunting neural network modelling of friction. Mech Syst Signal Process 2007;21:553–72.

[19] Potenza R, Dunne JF, Vulli S, Richardson D, King P. Multicylinder engine pressure reconstruction using NARX neural networks and crank kinematics. Int J Eng Res 2017;8:499–518.
[20] Patel A, Dunne JF. NARX neural network modelling of hydraulic suspension dampers for steady-state and variable temperature operation. Veh Syst Dyn: Int J Veh Mech Mobility 2003;40(5):285–328.
[21] Gaya MS, Wahab NA, Sam YM, Samsudin SI, Jamaludin IW. Comparison of NARX neural network and classical modelling approaches. Appl Mech Mater 2014;554:360–5.
[22] Siegelmann HT, Horne BG, Giles CL. Computational capabilities of recurrent NARX neural networks. IEEE Trans Syst Man Cybern, Part B, Cybern 1997;27(2):208–15.
[23] Kao CY, Loh CH. NARX neural networks for nonlinear analysis of structures in frequency domain. J Chin Inst Eng 2008;31(5):791–804.
[24] Pearson PK. Discrete-time dynamic models. New York–Oxford: Oxford University Press; 1999.
[25] Cybenko G. Approximation by superposition of a sigmoidal function. Math Control Signals Syst 1989;2(4):303–14.
[26] Hornik K, Stinchcombe M, White H. Multilayer feedforward networks are universal approximators. Neural Netw 1989;2(5):359–66.
[27] Gorban AN. Generalized approximation theorem and computational capabilities of neural networks. Sib J Numer Math 1998;1(1):11–24 (in Russian).
[28] Haykin S. Neural networks: A comprehensive foundation. 2nd ed.. Upper Saddle River, NJ, USA: Prentice Hall; 1998.
[29] Hagan MT, Demuth HB, Beale MH, De Jesús O. Neural network design. 2nd ed.. PSW Publishing Co.; 2014.
[30] Sandberg IW, Lo JT, Fancourt CL, Principe JC, Katagiri S, Haykin S. Nonlinear dynamical systems: Feedforward neural network perspectives. Wiley; 2001.
[31] Levin AU, Narendra KS. Recursive identification using feedforward neural networks. Int J Control 2013;61(3):533–47.
[32] Thibault J. Feedforward neural networks for the identification of dynamic processes. Chem Eng Commun 1991;105:109–28.
[33] Kuschewski JG, Hui S, Zak SH. Application of feedforward neural networks to dynamical system identification and control. IEEE Trans Control Syst Technol 1993;1(1):37–49.
[34] Ranković VM, Nikolić IZ. Identification of nonlinear models with feedforward neural network and digital recurrent network. FME Trans 2008;36:87–92.
[35] Mironov K, Pongratz M. Applying neural networks for prediction of flying objects trajectory. Vestn UGATU 2013;17(6):33–7.
[36] Malki HA, Karayiannis NB, Balasubramanian M. Short-term electric power load forecasting using feedforward neural networks. Expert Syst 2004;21(3):157–67.
[37] Messai N, Riera B, Zaytoon J. Identification of a class of hybrid dynamic systems with feed-forward neural networks: About the validity of the global model. Nonlinear Anal Hybrid Syst 2008;2:773–85.
[38] Baek S, Park DS, Cho J, Lee YB. A robot endeffector tracking system based on feedforward neural networks. Robot Auton Syst 1999;28:43–52.
[39] Janczak A. Identification of nonlinear systems using neural networks and polynomial models: A block-oriented approach. Berlin, Heidelberg: Springer-Verlag; 2005.
[40] Giri F, Bai EW. Block-oriented nonlinear system identification. Berlin, Heidelberg: Springer-Verlag; 2010.
[41] Janczak A. Comparison of four gradient-learning algorithms for neural network Wiener models. Int J Syst Sci 2003;34(1):21–35.
[42] Ozer S, Zorlu Y, Mete S. System identification application using Hammerstein model. Int J Syst Sci 2016;4(6):597–605.
[43] Sut HT, McAvoy TJ. Integration of multilayer perceptron networks and linear dynamic models: A Hammerstein modeling approach. Ind Eng Chem Res 1993;32:1927–36.
[44] Peng J, Dubay R, Hernandez JM, Abu-Ayyad M. A Wiener neural network-based identification and adaptive generalized predictive control for nonlinear SISO systems. Ind Eng Chem Res 2011;4:7388–97.
[45] Wills A, Schön TB, Ljung L, Ninness B. Identification of Hammerstein–Wiener models. Ind Eng Chem Res 2012;49:70–81.
[46] Peia JS, Smyth AW, Kosmatopoulos EB. Analysis and modification of Volterra/Wiener neural networks for the adaptive identification of non-linear hysteretic dynamic systems. J Sound Vib 2004;275:693–718.
[47] Li S, Li Y. Model predictive control of an intensified continuous reactor using a neural network Wiener model. Neurocomputing 2016;185:93–104.
[48] Ławryńczuk M. Practical nonlinear predictive control algorithms for neural Wiener models. J Process Control 2013;23:696–714.
[49] Tan AH, Godfrey K. Modeling of direction-dependent processes using Wiener models and neural networks with nonlinear output error structure. IEEE Trans Instrum Meas 2004;53(3):744–53.
[50] Michalkiewicz J. Modified Kolmogorov neural network in the identification of Hammerstein and Wiener systems. IEEE Trans Neural Netw Learn Syst 2012;23(4):657–62.
[51] Lin DT, Dayhoff JE, Ligomenides PA. Trajectory production with the adaptive time-delay neural network. Neural Netw 1995;8(3):447–61.

[52] Guh RS, Shiue YR. Fast and accurate recognition of control chart patterns using a time delay neural network. J Chin Inst Ind Eng 2010;27(1):61–79.
[53] Yazdizadeh A, Khorasani K, Patel RV. Identification of a two-link flexible manipulator using adaptive time delay neural networks. IEEE Trans Syst Man Cybern, Part B, Cybern 2010;30(1):165–72.
[54] Juang JG, Chang HH, Chang WB. Intelligent automatic landing system using time delay neural network controller. Appl Artif Intell 2003;17(7):563–81.
[55] Sun Y, Babovic V, Chan ES. Multi-step-ahead model error prediction using time-delay neural networks combined with chaos theory. J Hydrol 2010;395:109–16.
[56] Zhang J, Wang Z, Ding D, Liu X. H_∞ state estimation for discrete-time delayed neural networks with randomly occurring quantizations and missing measurements. Neurocomputing 2015;148:388–96.
[57] Yazdizadeh A, Khorasani K. Adaptive time delay neural network structures for nonlinear system identification. Neurocomputing 2002;77:207–40.
[58] Ren XM, Rad AB. Identification of nonlinear systems with unknown time delay based on time-delay neural networks. IEEE Trans Neural Netw 2007;18(5):1536–41.
[59] Ljung L, Glad T. Modeling of dynamic systems. Englewood Cliffs, NJ: Prentice Hall; 1994.
[60] Arnold VI. Mathematical methods of classical mechanics. 2nd ed.. Graduate texts in mathematics, vol. 60. Berlin: Springer; 1989.
[61] Krasovsky AA, editor. Handbook of automatic control theory. Moscow: Nauka; 1987 (in Russian).
[62] Brumbaugh, R.W. An aircraft model for the AIAA control design challenge, AIAA Guidance, Navigation and Control Conf., New Orleans, LA, 1991. AIAA Paper–91–2631, 12.
[63] Etkin B, Reid LD. Dynamics of flight: Stability and control. 3rd ed.. New York, NY: John Wiley & Sons, Inc.; 2003.
[64] Boiffier JL. The dynamics of flight: The equations. Chichester, England: John Wiley & Sons; 1998.
[65] Cook MV. Flight dynamics principles. Amsterdam: Elsevier; 2007.
[66] Hull DG. Fundamentals of airplane flight mechanics. Berlin: Springer; 2007.
[67] Gill PE, Murray W, Wright MH. Practical optimization. London, New York: Academic Press; 1981.
[68] Varela L, Acuña S. Handbook of optimization theory: Decision analysis and applications. New York: Nova Science Publishers, Inc.; 2011.
[69] Ljung L. System identification: Theory for the user. 2nd ed.. Upper Saddle River, NJ: Prentice Hall; 1999.
[70] Conti M, Turchetti C. Approximation of dynamical systems by continuous-time recurrent approximate identity neural networks. Neural Parallel Sci Comput 1994;2(3):299–320.
[71] Elanayar S, Shin YC. Radial basis function neural network for approximation and estimation of nonlinear stochastic dynamic systems. IEEE Trans Neural Netw 1994;5(4):594–603.
[72] Pal C, Kayaba N, Morishita S, Hagiwara I. Dynamic system identification by neural network: A new fast learning method based on error back propagation. JSME Int J Ser C, Dyn Control Robot Des Manuf 1995;38(4):686–92.
[73] Ilyin VE. Attack aircraft and fighter-bombers. Moscow: Victoria-AST; 1998 (in Russian).
[74] Astolfi A. Nonlinear and adaptive control: Tools and algorithms for the user. London: Imperial College Press; 2006.
[75] Astolfi A, Karagiannis D, Ortega R. Nonlinear and adaptive control with applications. Berlin: Springer; 2008.
[76] Gros C. Complex and adaptive dynamical systems: A primer. Berlin: Springer; 2008.
[77] Ioannou P, Fidan B. Adaptive control tutorial. Philadelphia, PA: SIAM; 2006.
[78] Ioannou P, Sun J. Robust adaptive control. Englewood Cliffs, NJ: Prentice Hall; 1995.
[79] Ioannou P, Sun J. Optimal, predictive, and adaptive control. Englewood Cliffs, NJ: Prentice Hall; 1994.
[80] Sastry S, Bodson M. Adaptive control: Stability, convergence, and robustness. Englewood Cliffs, NJ: Prentice Hall; 1989.
[81] Spooner JT, Maggiore M, Ordóñez R, Passino KM. Stable adaptive control and estimation for nonlinear systems: Neural and fuzzy approximator techniques. New York, NY: John Wiley & Sons, Inc.; 2002.
[82] Tao G. Adaptive control design and analysis. New York, NY: John Wiley & Sons, Inc.; 2003.
[83] Omatu S, Khalid M, Yusof R. Neuro-control and its applications. London: Springer; 1996.
[84] Leondes CT. Control and dynamic systems: Neural network systems techniques and applications. San Diego, London: Academic Press; 1998.
[85] Omidvar O, Elliott DL. Neural systems for control. San Diego, London: Academic Press; 1997.
[86] Nguyen HT, Prasad NR, Walker CL, Walker EA. A first course in fuzzy and neural control. London, New York: Chapman & Hall/CRC; 1997.
[87] Letov AM. Flight dynamics and control. Moscow: Nauka Publishers; 1969 (in Russian).
[88] Piyavsky SA, Brusov VS, Khvilon EA. Optimization of parameters for multipurpose aircraft. Moscow: Mashinostroyeniye Publishers; 1974 (in Russian).
[89] DiGirolamo R. Flight control law synthesis using neural network theory. In: AIAA Guid., Navig. and Control Conf., Hilton Head Island, S.C., Aug. 10–12, 1992: Collect. Techn. Pap. Pt. 1. AIAA–92–4390–CP. Washington (D.C.); 1992. p. 385–94.

[90] Tanaka T, Chuang CH. Scheduling of linear controllers for X-29 by neural network and genetic algorithm. In: AIAA Guidance, Navigation and Control Conf., Baltimore, Md., Aug.7–10, 1995: Collect. Techn. Pap. Pt 2. AIAA–95–3270—CP. Washington (D.C.); 1995. p. 891–900.
[91] Jacobs RA, Jordan MI. Learning piecewise control strategies in a modular neural network architecture. IEEE Trans Syst Man Cybern 1993;23(2):337–45.
[92] Hush DR, Horne BG. Progress in supervised neural networks. IEEE Control Syst 1993;10(1):8–39.
[93] Germeyer YB. Introduction to the theory of operations research. Moscow: Nauka Publishers; 1971 (in Russian).
[94] Brusov VS, Baranov SK. Optimal design of aircraft: A multipurpose approach. Moscow: Mashinostroyeniye Publishers; 1989 (in Russian).
[95] Rotondo D. Advances in gain-scheduling and fault tolerant control techniques. Berlin: Springer; 2018.
[96] Palm R, Driankov D, Hellendoorn H. Model based fuzzy control: Fuzzy gain schedulers and sliding mode fuzzy controllers. Berlin: Springer; 1997.
[97] Bianchi FD De Battista H, Mantz RJ. Wind turbine control systems: Principles, modelling and gain scheduling design. Berlin: Springer; 2007.
[98] Bryson Ay, Ho YC. Applied optimal control. Toronto, London: Blaisdell Publishing Company; 1969.
[99] Morozov NI, Tiumentsev YV, Yakovenko AV. An adjustment of dynamic properties of a controllable object using artificial neural networks. Aerosp MAI J 2002;(1):73–94 (in Russian).

CHAPTER

4

Neural Network Black Box Modeling of Nonlinear Dynamical Systems: Aircraft Controlled Motion

4.1 ANN MODEL OF AIRCRAFT MOTION BASED ON A MULTILAYER NEURAL NETWORK

As we noted in Chapter 1, many adaptive control schemes require the presence of a controlled object model. To obtain such a model, one needs to solve the classical problem of dynamical systems identification [1]. As experience shows, one of the most effective approaches to solving this problem for nonlinear systems is based on the use of ANNs [2–4]. Neural network modeling allows us to build reasonably accurate and computationally efficient models.

4.1.1 The General Structure of the ANN Model of Aircraft Motion Based on a Multilayer Neural Network

An ANN is an algorithmically universal mathematical model [5–7]. This fact is the basis of the computational efficiency of ANN models. It allows us to represent any nonlinear mapping $\varphi : \mathcal{R}^n \to \mathcal{R}^m$ as some ANN model with any predefined accuracy.[1]

The ANN model design problem for the nonlinear controlled dynamical system is treated further as a problem of a neural network approximation of the initial mathematical model of the aircraft motion, defined in one way or another, more often in the form of a system of differential equations. A structural diagram of the neural network identification process for the controlled system that corresponds to this problem is presented in Fig. 4.1.

The error signal ε that directs the learning of the ANN model is taken to be the squared difference between the outputs of the controlled object y_p and the neural network model y_m for the control signal u. The trained ANN model implements a recurrent relation that allows us to compute the value of the output $\widehat{y}$ at time instant t_{i+1} given the values of $\widehat{y}$ and u at some previous time instants.

We use the Nonlinear Auto-Regressive network with eXternal inputs (NARX) as a model of a dynamical system because it conforms to the nature of the considered problem of flight

[1] That is, any nonlinear mapping of the n-dimensional input vector to the m-dimensional output vector.

Neural Network Modeling and Identification of Dynamical Systems
https://doi.org/10.1016/B978-0-12-815254-6.00014-9

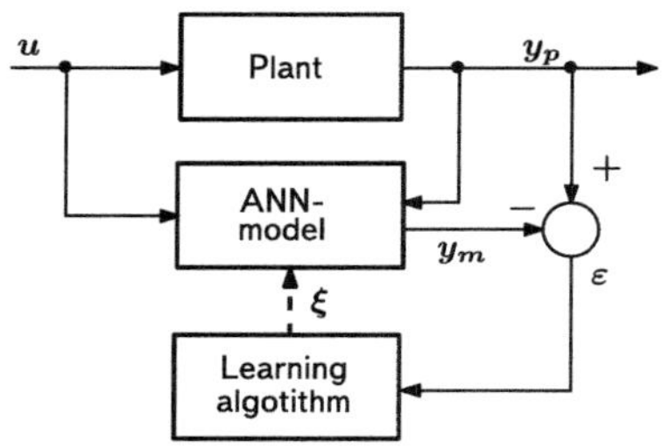

FIGURE 4.1 Scheme of neural network identification of the controlled object. Here u is the control; y_p is the output of the controlled object (plant); y_m is the output of the ANN model for the plant; ε is the difference between the outputs of the plant and ANN model (error signal); ξ is the adjusting action.

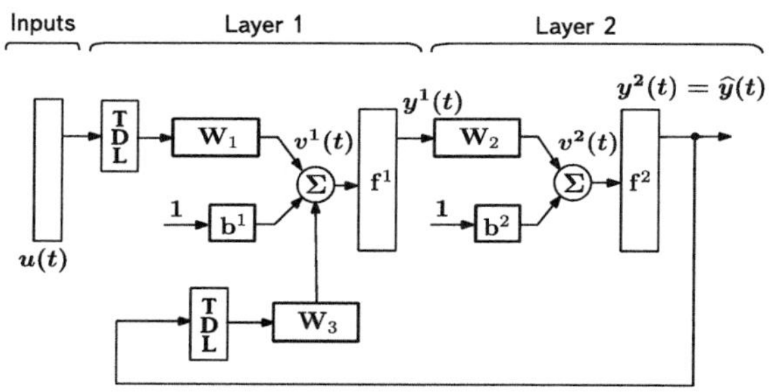

FIGURE 4.2 Structural diagram of the neural network NARX model of the controlled object. Here TDL is time delay line; W_1 is the matrix of the synaptic weights of the connections between the input and the first processing layer of the ANN; W_2 and W_3 are the matrices of the synaptic weights of the connections between the ANN processing layers; b^1 and b^2 are the sets of biases of the ANN layer; f^1 and f^2 are the sets of activation functions of the ANN layer; Σ are sets of summation units of the ANN layer; $v^1(t)$ and $v^2(t)$ are sets of scalar outputs of summation units; $y^1(t)$ and $y^2(t)$ are sets of scalar outputs of activation functions; $u(t)$ is the input signal; $\hat{y}(t)$ is the output of the ANN model.

control (see Fig. 4.2). It is a recurrent dynamic layered ANN model with delay elements (TDL is time delay line) at the inputs of the network and with feedback connections from output to input layers.

The NARX model implements a dynamic mapping described by a difference equation of the following form:

$$\hat{y}(t) = f(\hat{y}(t-1), \hat{y}(t-2), \ldots, \hat{y}(t-N_y), \\ u(t-1), u(t-2), \ldots, u(t-N_u)), \quad (4.1)$$

where the value of the output signal $\hat{y}(t)$ at a given time instant t is computed using the output values $\hat{y}(t-1), \hat{y}(t-2), \ldots, \hat{y}(t-N_y)$ of this signal for the sequence of the preceding time instants, as well as the values of the input (control) signal $u(t-1), u(t-2), \ldots, u(t-N_u)$, external to the NARX model. In the general case, the length of the time window for outputs and controls may not coincide, i.e., $N_y \neq N_u$.

A convenient way to implement the NARX model is to use a multilayer feedforward network of the MLP type for an approximate representation of the $f(\cdot)$ mapping in the relation (4.1), as well as delay lines (TDL elements) $\hat{y}(t-1), \hat{y}(t-2), \ldots, \hat{y}(t-N_y)$ and $u(t-1), u(t-2), \ldots, u(t-N_u)$. The specific form of the neural network implementation of the NARX model, which we can use to simulate the motion of the aircraft, is shown in Fig. 4.2. We can see that this NARX model is a two-layer network, with the nonlinear (sigmoid) activation functions of the hidden layer neurons and the linear activation function of the output layer neurons.

The learning process of the NARX model, in this case, can be constructed in one of two ways.

In the first method (the parallel architecture, Fig. 3.1A), the output of the NARX model can be treated as the estimate $\hat{y}(t)$ of the output for the simulated nonlinear system. This estimate is fed back through the TDL element to the input of the NARX model in order to predict the next $\hat{y}(t+1)$ output of the system.

In the second method (the series-parallel architecture, Fig. 3.1B) we take into account the fact that the supervised learning of the neural network NARX model is carried out. This fact means that information is available not only about the inputs of the model $u(t)$ but also about the values $y(t)$ of the system outputs that correspond to these input values. Hence, these values of the outputs $y(t)$ can be fed to the input of the NARX model instead of their estimates $\widehat{y}$, as was the case in the previous method. This approach has two main advantages. First of all, the accuracy of the obtained NARX model is increased. Second, it becomes possible to use for its training the usual static method of error backpropagation, whereas for learning the NARX model with a purely parallel architecture, we require to use some form of the dynamic error backpropagation method.

4.1.2 Learning of the Neural Network Model of Aircraft Motion in Batch Mode

The ANN model is trained in the standard way [5,6]: training is treated as an optimization problem, namely, the minimization problem for the error $e = y - \widehat{y}$. The objective function is the sum of squares of errors for the entire training sample,

$$E(w) = \frac{1}{2} e^T(w) e(w), \qquad e = [e_1, e_2, \ldots, e_N]^T,$$

where $e(w) = y - \hat{y}(w)$, w is the M-dimensional vector of configurable network parameters, and N is the sample length.

We perform the minimization of the objective function $E(w)$ with respect to the vector w using the Levenberg–Marquardt method. The adjustment of the vector w at each optimization step is as follows:

$$w_{n+1} = w_n + (J^T J + \mu E)^{-1} J^T e,$$

where E is the identity matrix and $J = J(w_n)$ is the Jacobi matrix, i.e., an $(N \times M)$ matrix whose ith row is a vector obtained by transposing the gradient of the function e_i.

The most time-consuming element of the training process is the computation of the Jacobian at each step. This operation is performed using the error backpropagation algorithm [5], which takes up most of the time spent learning the model.

4.1.3 Learning of the Neural Network Model of Aircraft Motion in Real-Time Mode

ANN models discussed in this chapter use sigmoid activation functions for hidden layer neurons. Such global activation functions provide the ANN model with good generalization properties. However, modification of any tunable parameter changes the behavior of the network throughout the entire input domain. This fact means that the adaptation of the network to new data might lead to a decrease of the model accuracy on the previous data. Thus, to take into account the incoming measurements, the ANN models of this type should be trained on a very large sample, which is not reasonable from a computational point of view.

To overcome this problem (that is, to perform adaptation not only for the current measurements, but for some sliding time window), we can use the recursive least-squares method (RLSM), which can be considered as a particular case of the Kalman filter (KF) for estimation of constant parameters. However, KFs and RLSMs are directly applicable only for systems whose observations are linear with respect to the estimated parameters, while the neural network observation equation is nonlinear. Therefore, in order to use the KF, the observation equation must be linearized. In particular, statistical linearization can be used for this purpose.

Application of this approach to the ANN modeling is described in detail in [5]. Again we can see that, just as in the case of batch training of the ANN model, the Jacobian J_k computation is the most time-consuming operation of the whole procedure.

To obtain the model with the required accuracy, the training data are taken to be a sequence of values on a certain sliding observation window,

$$\widehat{y}_k = [\widehat{y}_{i-l}, \widehat{y}_{i-l+1}, \ldots, \widehat{y}_i]^T,$$

where l is the length of the sliding window, the index i refers to the time instant (sampling step), and the index k indicates the estimate number.

To save time, we do not perform the parameter estimation at each sampling step, but at each tenth step (the sampling step is 0.01 sec, and the network parameters are updated in 0.1 sec). The computational experiment shows that such a coarsening is quite acceptable since it does not significantly affect the accuracy of the model.

4.2 PERFORMANCE EVALUATION FOR ANN MODELS OF AIRCRAFT MOTION BASED ON MULTILAYER NEURAL NETWORKS

The performance evaluation of the ANN model under consideration we carry out concerning, as an example, the angular longitudinal motion of the aircraft, which was described using traditional mathematical models for flight dynamics [8–13].

Since the problems of synthesis and analysis of adaptive control algorithms have to be solved for aircraft of various types, we considered two versions of this model. In one version, the relationship between the angle of attack α and the thrust of the engine T_{cr}, which is typical for hypersonic vehicles, was taken into consideration. In another case, as applied to the F-16 maneuverable aircraft, this relationship was not taken into account as it is not characteristic.

The first of the considered models ("single-channel") uses an implicit relationship between the values of the variables α and T_{cr}. It is given through the value of the coefficient $m_z(\alpha, T_{cr})$; additional changes in the effect of thrust on the angle of attack and of the angle of attack on the thrust are not taken into account in this model, thrust control is also not introduced, and control is used on a single channel. This channel provides changing the value $\delta_{e,act}$, which is the command signal for the elevator actuator. This model has the following form [8–13]:

$$\begin{aligned}
\dot{\alpha} &= q - \frac{\bar{q}S}{mV}C_L(\alpha, \delta_e) + \frac{g}{V}\cos\theta, \\
\dot{q} &= \frac{\bar{q}S\bar{c}}{I_z}C_m(\alpha, q, \delta_e), \\
T^2\ddot{\delta}_e &= -2T\zeta\dot{\delta}_e - \delta_e + \delta_{e,act},
\end{aligned} \tag{4.2}$$

where α is the angle of attack, deg; θ is angle of pitch, deg; q is the angular velocity of the pitch, deg/sec; δ_e is the deflection angle of the elevator (controlled stabilizer), deg; C_L is the lift coefficient; C_m is pitch moment coefficient; m is mass of the aircraft, kg; V is the airspeed, m/sec; $\bar{q} = \rho V^2/2$ is the dynamic air pressure, kg·m^{-1} sec^{-2}; ρ is air density, kg/m^3; g is the acceleration of gravity, m/sec^2; S is the wing area, m^2; $\bar{c}$ is mean aerodynamic chord of wing, m; I_z is the moment of inertia of the aircraft relative to the lateral axis, kg·m^2; the dimensionless coefficients C_L and C_m are nonlinear functions of their arguments; T, ζ are the time constant and the relative damping coefficient of the actuator; $\delta_{e_{act}}$ is the command signal to the actuator of the all-turn controllable stabilizer (limited to ±25 deg). In the model (4.2), the variables α, q, δ_e, and $\dot{\delta}_e$ are the states of the controlled object, the variable $\delta_{e,act}$ is the control.

The second motion model ("two-channel"), which was used only for hypersonic research vehicles X-43 and NASP, is a version of the model (4.2), expanded by including the thrust control channel and the explicit relationship between angle of attack and engine thrust, in addition to the implicit one, mentioned above. Thus, the engine thrust control via the command signal δ_{th} is introduced in this model in addition to the command signal $\delta_{e,\,act}$. This model has the form

$$\begin{aligned}
\dot{\alpha} &= q - \frac{\bar{q}S}{mV}C_L(\alpha, \delta_e) + \frac{T_{cr}\sin\alpha}{mV} + \frac{g}{V}\cos\theta, \\
\dot{q} &= \frac{\bar{q}S\bar{c}}{I_z}C_m(\alpha, q, \delta_e) + \frac{T_{cr}\,h_T}{I_z}, \\
T^2\ddot{\delta}_e &= -2T\zeta\dot{\delta}_e - \delta_e + \delta_{e,\,act}, \\
\dot{T}_{cr} &= \omega_{eng}\,(T_{ref}(\delta_{th}) - T_{cr}), \\
n_x &= -\frac{\bar{q}S}{mg}C_D(\alpha, \delta_e) + \frac{T_{cr}\cos\alpha}{mg}.
\end{aligned} \tag{4.3}$$

Here $T_{ref} = T_{ref}(\delta_{th})$ is the specified thrust level (linear function), T_{cr} is the current thrust level, ω_{eng} is the frequency of the aperiodic link, which describes the dynamics of the engine (here $\omega_{eng} = 1$ was assumed). The thrust moment arm is assumed to be equal to $h_T = 0.5$ m; it is calculated relative to the center of mass of the aircraft in the vertical plane, so the change δ_{th} causes a change in the angle of attack. In the model (4.3), the variables α, q, δ_e, $\dot{\delta}_e$, and T_{cr} are the states of the controlled object, $\delta_{e_{act}}$ and δ_{th} are its controls, and n_x is the tangential load factor, i.e., load factor along the velocity vector of the aircraft.

A series of computational experiments has been performed to evaluate the properties of the ANN model under consideration and its suitability for modeling of the aircraft motion. To demonstrate the efficiency of adaptive control under various conditions, aircraft of essentially different classes were chosen as examples: a maneuverable aircraft F-16 [14,15], a heavy hypersonic aircraft (one of the options [16–18], which were considered by NASA within the framework of the National AeroSpace Plane [NASP] program, aimed at creating a single-stage aerospace plane with a horizontal launch, putting a payload on the orbit of the artificial Earth satellite, and horizontal landing), the hypersonic research vehicle X-43 [19–25], and UAV "003" and X-04 micro and minidimensions, respectively [26]. The values of the corresponding parameters and characteristics for these aircraft, required for modeling as source data, were taken from the papers [14,15] for the F-16 aircraft, [21–25] for the hypersonic research vehicle X-43, [16–18] for the NASP research vehicle, and [26] for the UAVs.

For each of the abovementioned aircraft, a computational experiment was performed in order to evaluate the performance of the corresponding ANN models. Some of the results of these experiments are shown in Fig. 4.3 for F-16 and Figs. 4.4 and 4.5 for the UAVs.

In Figs. 4.3A, 4.4A, and 4.5A, examples of training samples used for learning of the ANN models are shown. We can see that for the formation of each of the samples very active work is carried out by the longitudinal motion control element (elevons for UAV, the controlled stabilizer for F-16), expressed in the frequent change in the value of command signal $\delta_{e,\,act}$ of the command control signal at significant differences between its neighboring values (this command signal was formed at random). The purpose of using this method for synthesis of a training set is to provide the broadest possible variety of states of the modeled system (to cover, as much as possible, evenly and densely the entire state space of the system), as well as the highest possible variety of time derivative values that reflect the dynamics of the simulated system.

Since we consider the optimal tracking control problem for the angle of attack, we evaluate the accuracy of the designed model by comparing the actual trajectory of this variable for the controlled object described by the system of differential equations (4.2) or (4.3) with the trajectory predicted by the ANN model. We estimate the accuracy of the model by the error e_α, computed as the difference between the angles of attack for the controlled object and the ANN model at the same time instant.

We can see from these examples that the proposed approach makes it possible to build reasonably accurate ANN models (e_α values lie within the range $\pm(0.5 \div 0.7)$ deg). However, in

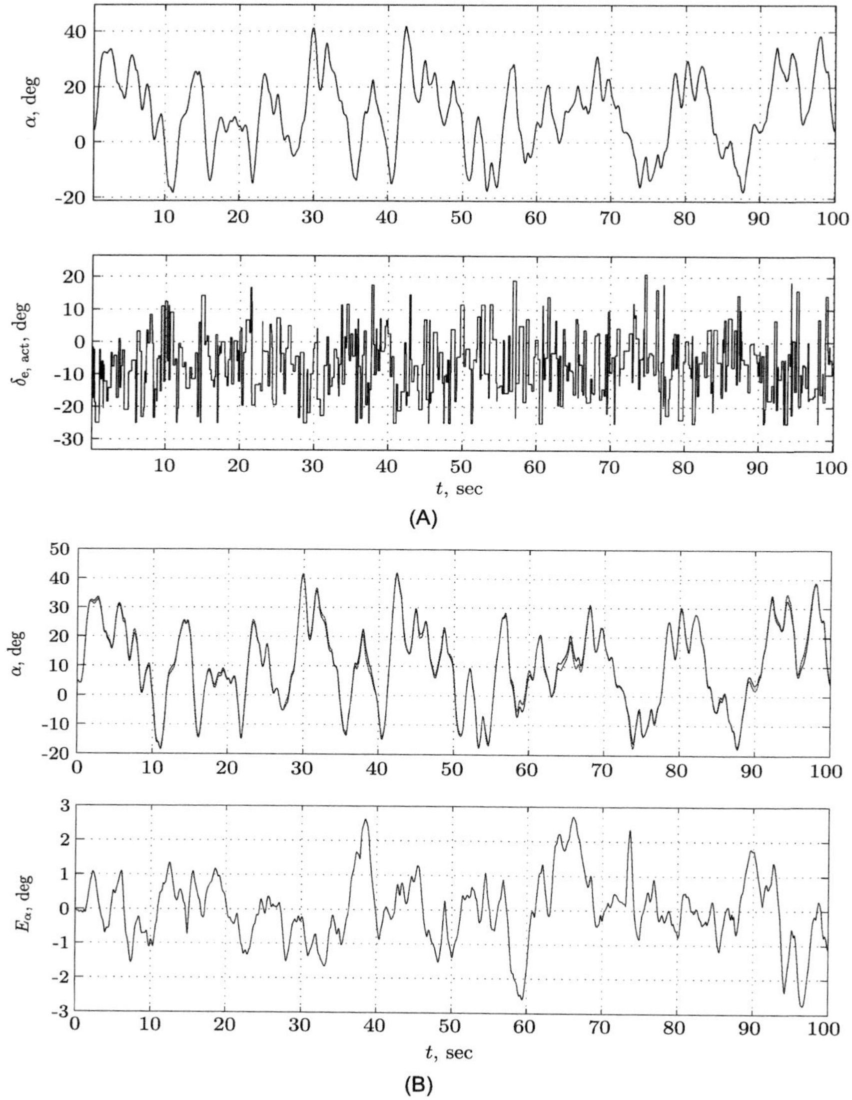

FIGURE 4.3 Neural network model design for F-16 aircraft, flight regime with airspeed $V_i = 500$ km/h. (A) Training sample for the ANN model. (B) Efficiency test for a closed-loop ANN model. Here α is the angle of attack, deg; E_α is the tracking error (difference between angle of attack values for the object and the ANN model), deg; $\delta_{e,\,act}$ is command signal for the actuator of the elevator; t is time, sec; the solid line is the output of the object; the dotted line is the output of the ANN model.

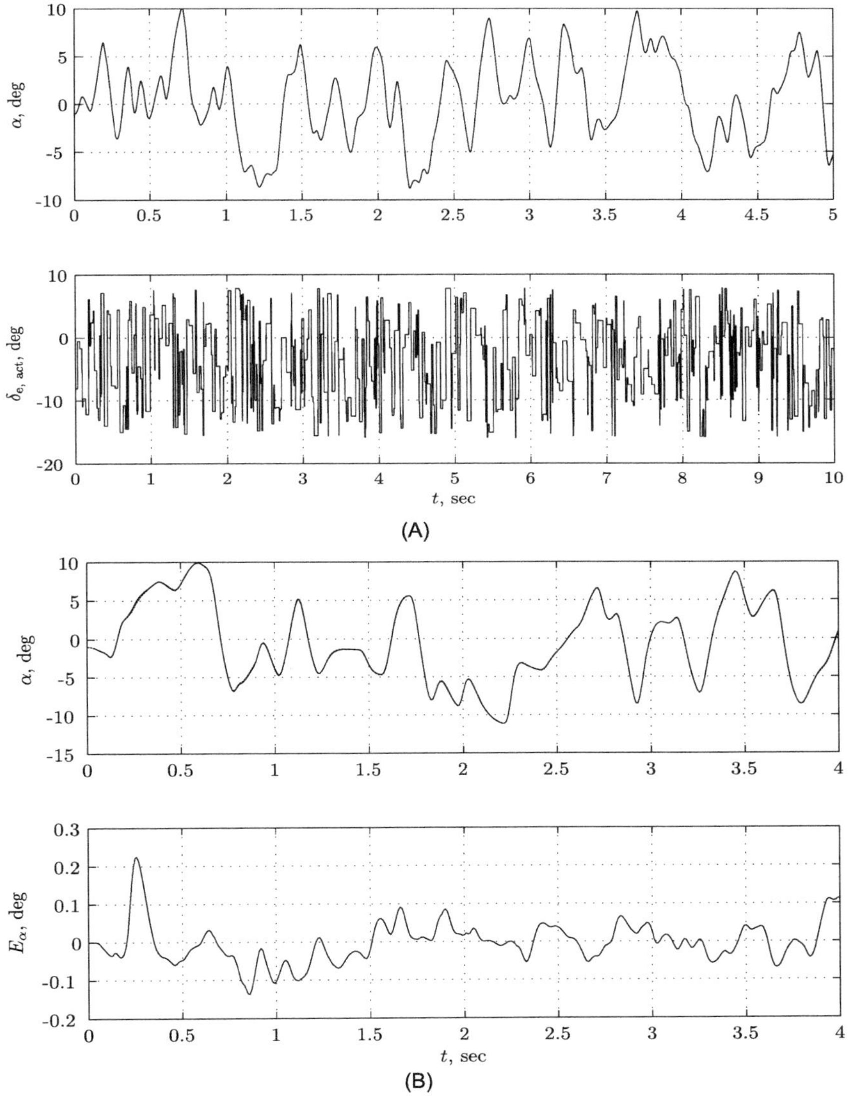

FIGURE 4.4 Neural network model design for micro-UAV "003," flight regime with airspeed $V = 30$ km/h. (A) Training sample for the ANN model. (B) Efficiency test for a closed-loop ANN model. Here α is the angle of attack, deg; E_α is error (difference between angle of attack values for the object and the ANN model), deg; $\delta_{e,\,act}$ is command signal for the actuator of the elevons; t is time, sec; the solid line is the output of the object; the dotted line is the output of the ANN model.

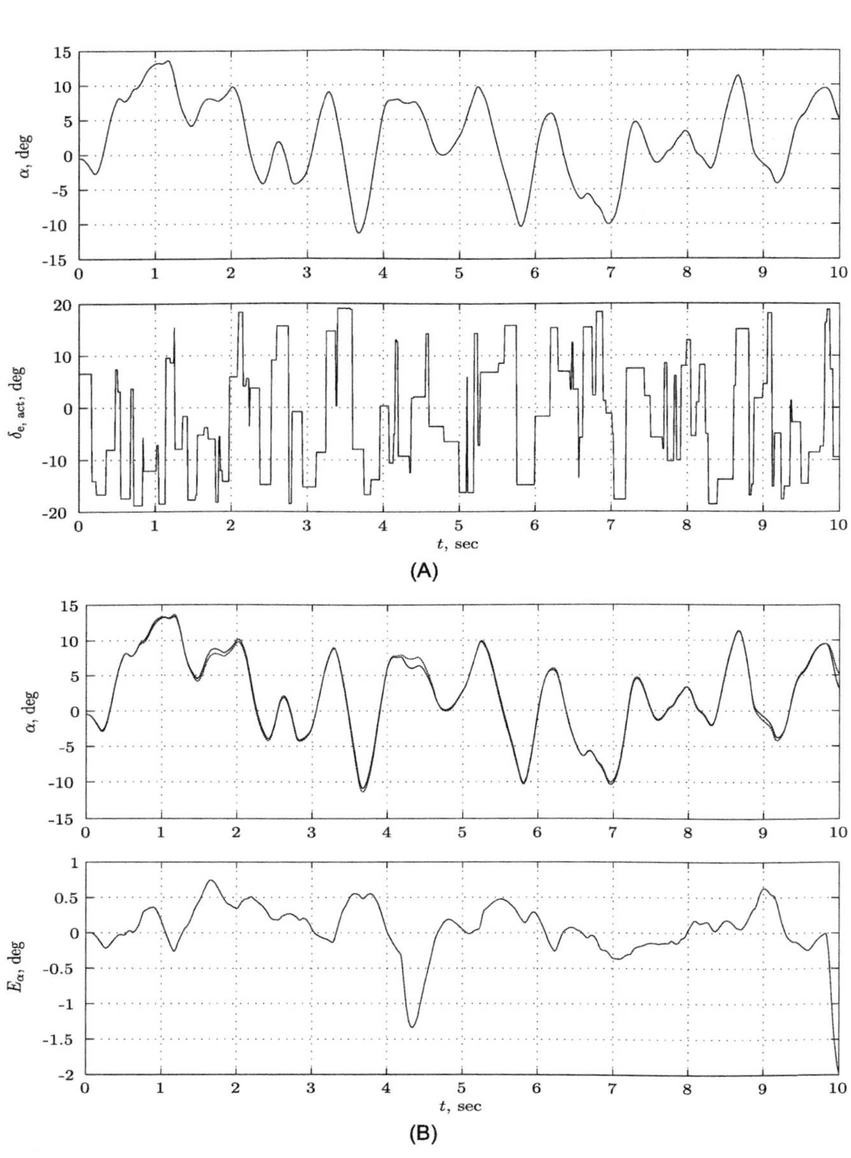

FIGURE 4.5 Neural network model design for mini-UAV X-04, flight regime with airspeed $V = 70$ km/h. (A) Training sample for the ANN model. (B) Efficiency test for a closed-loop ANN model. Here α is the angle of attack, deg; E_α is the tracking error (difference between angle of attack values for the object and the ANN model), deg; $\delta_{e,\,act}$ is command signal for the actuator of the elevons; t is time, sec; the solid line is the output of the object; the dotted line is the output of the ANN model.

some cases, the accuracy degenerates, which in turn leads to poor adaptation performance of the synthesized neurocontroller. Ways to overcome these difficulties are discussed in Chapters 5 and 6.

4.3 APPLICATION OF ANN MODELS TO ADAPTIVE CONTROL PROBLEMS FOR NONLINEAR DYNAMICAL SYSTEMS OPERATING UNDER UNCERTAINTY CONDITIONS

4.3.1 The Demand for Adaptive Systems

One of the most important classes of dynamical systems is aircraft of various types. As already noted in Chapter 1, it is crucial that we provide for control of the motion of modern and advanced aircraft under conditions of significant and varied uncertainties in the values of their parameters and characteristics, flight regimes, and environmental influences. Besides, during flight, various emergencies can arise, in particular, equipment failures and structural damage, the consequences of which in most cases can be overcome by an appropriate reconfiguration of the aircraft control system.

The presence of significant and diverse uncertainties is one of the most severe factors that complicate the solution of all the three classical problems (analysis, synthesis, identification) for dynamical systems and, in particular, for aircraft. The problem is that the current situation can change dramatically, significantly, and unpredictably due to the uncertainties. The controlled system under such conditions must be able to adapt quickly to changes in the situation to ensure the normal operation of the system.

As already noted in Chapter 1, we consider the system to be adaptive if it can quickly adapt to a changing situation by modifying some of its elements. We assume that usually these modifications are applied to the control laws implemented by the control system and to the controlled object model. Modifications of these systems may affect the corresponding parameter values as well as the structure of the control laws and/or system models.

The analysis of the adaptive control algorithms concerning aircraft motion is carried out in the following subsections regarding such basic types of adaptive systems as the model reference adaptive control (MRAC) and model predictive control (MPC).

Another option within the framework of the discussed approach to the control of nonlinear dynamical systems is considered in [27,28]. In this case, we solve the problem of obtaining the specified characteristics of the controllability of the aircraft through automation. The question that is solved by ANN tools is high-precision control over the entire range of flight modes. Synthesis and testing of neural network–based control algorithms are performed using the full nonlinear mathematical model of a maneuverable aircraft through three control channels for the current flight mode. In the structure of the system, the internal contour of the control of angular velocities is distinguished, which is formed by the Inverse Dynamics method, which is based on the feedback linearization [29]. At the same time, using the feedback transformation, the controlled object is reduced to an equivalent linear form, after which the control is selected so that the object moves along a predetermined desired trajectory. The external contour of control of the angle of attack in this system contains a PI controller.

The problem of improving the accuracy of the control system should be considered provided that the aerodynamic and other characteristics of the aircraft are nonlinear and are characterized by a high level of uncertainty. Including, as it was noted earlier, one of the types of uncertainties can be interpreted as the failure of aircraft equipment and damage to its structure, leading to a change in the dynamic properties of the aircraft and/or impeding its piloting. An effective way to solve this problem is to use adaptive control laws that fill in insufficient and/or inaccurate data about an object in the course of its operation.

All these adaptive control schemes require the presence of a controlled object model. As was shown in Chapter 2, neural network implementation of these schemes, which have high computational efficiency, requires the model of the controlled object to be represented as a neural network as well.

Adaptive control schemes are often based on the use of some reference model that specifies the desired behavior of the system under consideration. The studies of which the results are described in this chapter are based on this variant. Adaptive control systems that follow this approach perform the modification of parameters $\theta_c(t)$ for the regulator according to the algorithm implemented by the adaptation law. This modification is based directly on the tracking error value $\varepsilon(t) = y_m(t) - y(t)$, where $y_m(t)$ is the output of the reference model and $y(t)$ is output of the plant (controlled object).

Hands-on experience in the application of the abovementioned adaptive control schemes demonstrates that the choice of reference model parameters has a fundamental influence on the nature of the results obtained. An incorrect choice of these parameters can lead to the control system becoming inoperative. At the same time, if the reference model parameters are selected adequately, it is possible to get a control system that solves the tasks assigned to it well. We present the results of the analysis of the influence of the reference model parameters on the efficiency of the synthesized control system in Section 4.3.2.4.

The studied adaptive control schemes are essentially based on the use of the ANN model of the controlled object as a source of information about the behavior of this object. Since, due to the approximate nature of the ANN model, the real values of the variables describing the motion of the object inevitably differ from those obtained as the outputs of the ANN model, an error appears that decreases the quality of control. We propose one possible approach to decrease this error in Section 4.3.2.3. This approach treats the inaccuracy of the ANN model as a disturbance effect on the system that leads to a deviation of the trajectory of the real object from the reference trajectory. We attempt to reduce the impact of this effect by introducing a compensating circuit into the system.

4.3.2 Model Reference Adaptive Control

4.3.2.1 General Scheme of Model Reference Adaptive Control

In the case of MRAC problems, we implement the controller as a neural network (neurocontroller) using ANN of the NARX type. Training of the neurocontroller is carried out in such a way that the output of the controlled system follows the reference model output as close as possible. A neural network model of the object is required to implement the learning process of the neurocontroller.

Neural network implementation of the MRAC scheme (Fig. 4.6) involves two neural network modules: the controller network (neurocontroller) and the plant model (ANN model). As the first step, we solve the identification problem for the controlled object, and then we use the obtained ANN model to train the neurocontroller, which should provide for the most accurate possible tracking of the reference model output.

The neurocontroller is a two-layer network fed with the reference input signal $r(t)$, the controlled object output $y_p(t)$, and, in some cases, the neurocontroller output $\widetilde{u}(t)$ at previous time steps (this connection is not shown in the diagram) through the time delay lines (TDL elements).

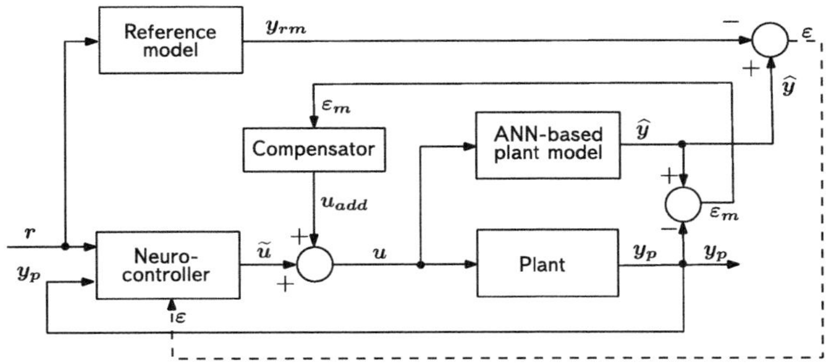

FIGURE 4.6 The neural network–based model reference adaptive control (MRAC) scheme. Here $\widetilde{u}$ is control signal at the output of the neurocontroller, u_{add} is additional control from the compensator, u is resulting control, y_p is output of the controlled object (plant), $\widehat{y}$ is output of the neural network model of the plant; y_{rm} is output of the reference model; ε is the difference between the outputs of the plant and the reference model; ε_m is the difference between the outputs of the plant and the ANN model; r is the reference signal.

The ANN model of the controlled object, whose structure is shown in Fig. 4.2, is fed with the control signal from the neurocontroller, and also with the output of the controlled object through the time delay lines. See Fig. 4.7.

4.3.2.2 *Neurocontroller Synthesis for the Model Reference Adaptive Control*

The equation of the neurocontroller has the following form (for static controllers):

$$u_k = f(r_k, r_{k-1}, \ldots, r_{k-d}, y_k, y_{k-1}, \ldots, y_{k-d}), \tag{4.4}$$

where y is the plant output, r is the reference signal.

By analogy with the model reference control scheme for linear systems, the equation of the neurocontroller should look somewhat different, i.e.,

$$u_k = f(r_k, u_{k-1}, \ldots, u_{k-d}, y_k, y_{k-1}, \ldots, y_{k-d}). \tag{4.5}$$

However, the simulation shows that these two implementations provide similar results, but the former learns a little faster. Therefore, we adopt the static version (4.4) of the neurocontroller as the main one.

The use of the MRAC scheme requires to determine in one way or another the appropriate reference model reflecting the developer's views of what the "good" behavior of this system looks like so that the neurocontroller would attempt to make the controlled system follow this behavior as close as possible.

We can define the reference model in various ways. In this chapter, we build the reference model combining an oscillating-type unit with sufficiently high damping in combination with an aperiodic-type unit.

In the case the motion of an aircraft is described by (4.2), the reference model is defined as follows:

$$\begin{aligned} \dot{x}_1 &= x_2, \\ \dot{x}_2 &= x_3, \\ \dot{x}_3 &= \omega_{act}\,(-x_3 - 2\,\omega_{rm}\,\zeta_{rm}\,x_2 + \omega_{rm}^2\,(r - x_1)). \end{aligned} \tag{4.6}$$

Here $\omega_{act} = 40$, $\omega_{rm} = 3$, $\zeta_{rm} = 0.8$. The state vector is $x = [\alpha_{rm}, \dot{\alpha}_{rm}, \varphi_{act}]$ in this case.

Another version of the reference model, similar to (4.6), is also a third-order linear system that is defined by a transfer function of the fol-

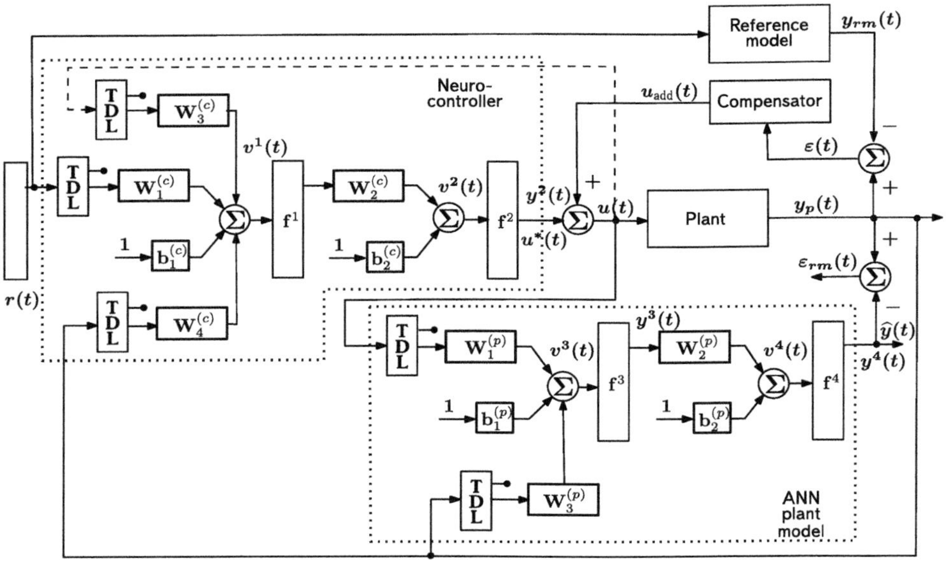

FIGURE 4.7 Structural diagram of the neural network implementation of model reference adaptive control (MRAC). TDL is time delay line; $W_1^{(c)}$ is the matrix of the synaptic weights of the connections between the input and the first processing layer of the ANN; $W_i^{(c)}$, $W_j^{(p)}$, $i = 2, 3, 4$, $j = 1, 2, 3$ are the matrices of the synaptic weights of the connections between the ANN processing layers; $b_1^{(c)}, b_2^{(c)}, b_1^{(p)}, b_2^{(p)}$ are the sets of biases of the ANN layers; $f^1, \ldots, f^4$ are the sets of activation functions of the ANN layers; Σ are sets of summation units of the ANN layers; $v^{(i)}(t)$, $i = 1, \ldots 4$ are sets of scalar outputs of summation units; $y^{(j)}(t)$, $j = 1, 2, 3$ are sets of scalar outputs of activation functions; $r(t)$ is the reference signal; $y_p(t)$ is the output of the plant; $\widehat{y}(t)$ is the output of the ANN model; $y_{rm}(t)$ is the output of the reference model; $u_*(t)$ is the control produced by the neurocontroller; $u_{add}(t)$ is an additional control from the compensator; $u(t)$ is the control used as the input of the plant; $\varepsilon(t) = y_p(t) - y_{rm}(t)$ is the difference between the outputs of the plant and the reference model.

lowing form:

$$W_\alpha = \frac{\omega_{rm}^2}{((1/\omega_{act})p + 1)(p^2 + 2\omega_{rm}\zeta_{rm}p + \omega_{rm}^2)}. \tag{4.7}$$

Fig. 4.8 demonstrates the desired behavior of the controlled object given by the reference model (4.7).[2] The behavior of the reference model (4.6), as shown by the computational experiment, closely resembles the behavior of the model (4.7).

If the motion of the aircraft is described by (4.3), then in addition to the system (4.6), representing the angle of attack channel, we have to add the reference model for the tangential g-load channel, i.e.,

$$\begin{aligned} \dot{x}_1 &= x_2, \\ \dot{x}_2 &= -2\,\omega_{rm}\,\zeta_{rm}\,x_2 + \omega_{rm}^2\,(r - x_1), \end{aligned} \tag{4.8}$$

or, in the transfer function form,

$$W_{n_x} = \frac{\omega_{rm}^2}{(p^2 + 2\omega_{rm}\zeta_{rm}p + \omega_{rm}^2)}. \tag{4.9}$$

[2] We can see characteristics of this model in the time and frequency domain in Fig. 4.8.

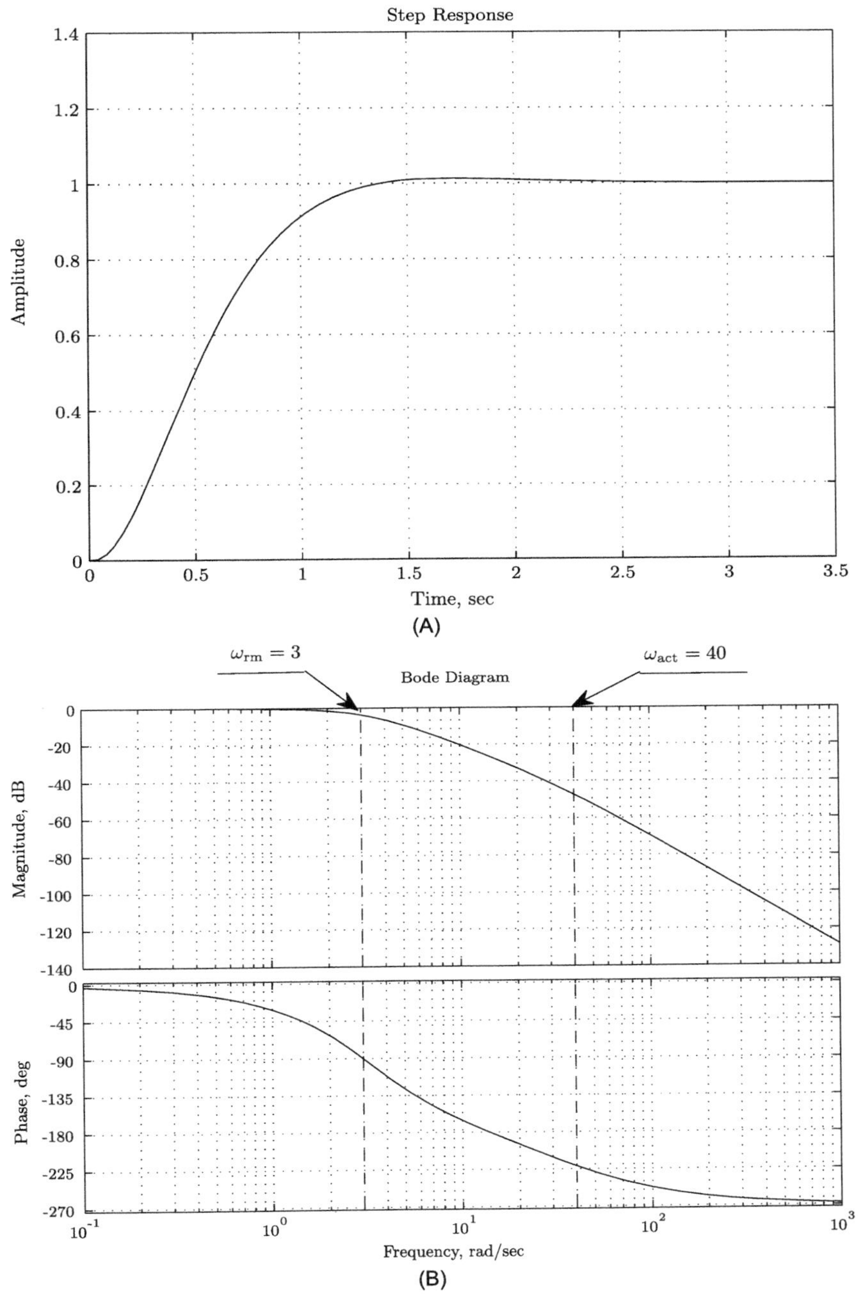

FIGURE 4.8 Characteristics of the reference model of the form (4.7). (A) The transient response of the model. (B) Frequency characteristics of the model.

In this model, $\omega_{rm} = 1$, $\zeta_{rm} = 0.9$, the vector of the state variables $x = [n_{x_a rm}, \dot{n}_{x_a rm}]$, r is the reference signal.

The neurocontroller is configured to minimize the error $y_{rm} - \widehat{y}$, i.e., to approximate the behavior of the reference model with the response of the plant model coupled with the controller. For a good ANN model, this means minimizing the "real" error $y_{rm} - y$ to a certain level.

Although the neurocontroller is static, it works as part of a dynamical system, so we need to configure it as a part of the whole recurrent network. This configurable network consists of two subnets (the neurocontroller itself and the closed-loop object model), closed by the external feedback loop. During the configuration, the parameters of the model subnet do not change, i.e., the ANN model only serves to close the external feedback loop and represent the entire system in a neural network form (to estimate the sensitivities of the outputs of the controlled object to the parameters of the neurocontroller).

In the batch mode, such a network can be trained using the same Levenberg–Marquardt method. However, it requires the computation of dynamic derivatives; hence to compute the Jacobian, we have to either apply the backpropagation through time or the real-time recurrent learning method. A recurrent form of the network presents additional difficulties in the process of ANN learning: the larger the sample, the higher the chance that the learning process will get stuck in some of the local minima. Such chances increase with the length of the sample with catastrophic speed. Therefore, we divide the entire sample into segments.

To configure parameters uniquely, we require the closed-loop network with the controller to start from the reference trajectory on each segment, since the neurocontroller cannot affect the initial conditions.

Thus it is necessary to consider the following factors:

1. Learning the network on small segments (less than 500–1000 points) leads to the situation when the network learns only this particular segment, forgetting about all the others.
2. Learning the network on large segments always leads to a bad local minimum.
3. Learning the network on medium-sized segments also leads to bad local minimum; however, the rotation of these segments allows circumventing this problem to some extent.

For these reasons, it is necessary to use medium-sized segments, to perform training with three to seven epochs for each of them, and to loop over the segments several times, and finally to consolidate the segments to improve the training performance. As a result, the learning process of the ANN becomes very computationally intensive (up to several hours, depending on the implementation details).

According to the previous considerations, it is advantageous to use the sequential training mode mentioned in the last section to train the neurocontroller in the batch mode (i.e., for its pretraining); the only difference is that we need to use dynamic backpropagation to compute the Jacobian.

In this case, the Kalman filter acts as the "stapler" of the individual segments into one data array. Moreover, the segments can be chosen to be small (30–100 points, which saves considerable computational time), so long as the dynamics of the controlled object is reflected on this interval. Although in general sequential methods achieve lower accuracy, it is more important to circumvent the problem of local minima and to decrease the training time.

Thus, the procedure of the neurocontroller configuration is as follows:

1. Set the initial conditions on the reference trajectory. Usually, a first few points of the segment are assigned to the initial conditions.
2. Simulation of the coupled network on this segment (prediction of the behavior of the controlled object with the current parameters of the neural controller), estimation of the error of the reference model tracking, compu-

tation of the Jacobian of this error concerning the parameters of the neurocontroller.
3. Adjustment of parameters on the current segment using the Kalman filter equations.

In real time, the neurocontroller is learned according to the same scheme, but with some differences:

1. Adjacent segments comprise a sliding window (usually 50 points for 0.5 sec). However, the parameters are not updated at each simulation step (0.01 sec), but every 0.1 sec.
2. The ANN model is trained simultaneously; therefore the model subnet parameters change.

It should be noted that the neurocontroller learns to control not the object itself, but its model, so if the ANN model does not have the required accuracy, then the control performance will be unsatisfactory.

The model cannot be ideally accurate since the neural network approach provides only approximate solutions. Therefore, with the help of such a "pure" approach it is impossible to achieve precise control (accurate tracking of the reference model output).

We show this result in Fig. 4.9. For comparison, the same figure shows the performance of the neurocontroller with the object to which it was trained (ANN model). We can see that the accuracy of the operation of the neurocontroller with the real object is somewhat reduced, which indicates that there is a deviation in the behavior of the real object from the behavior of its ANN model. We discuss a way to improve the performance of the neurocontroller in this situation in the next section.

4.3.2.3 *Compensating Loop in the Model Reference Adaptive Control*

We can interpret errors introduced by the neural network model as additional perturbations leading to a deviation of the trajectory of the controlled object from the reference trajectory. To reduce the tracking error, we can use the compensator (additional simple feedback controller). This simple feedback controller does not depend on the model predictions; thus it is more robust to perturbations irrespective of their nature. This kind of compensator integrates very well into the MRAC system.

In the simplest case, the compensator (PD compensator) implements an additional feedback control law of the following form [30] (see also Figs. 4.6 and 4.14):

$$\delta_{e,\,add} = K_p\, e + K_d\, \dot{e}, \tag{4.10}$$

where $e = y_{rm} - y$ is the tracking error y_{rm} for the reference model. In the control system, the compensator is used in discrete time; hence $\dot{e}$ is estimated via finite differences.

Despite its simplicity, the compensating circuit reduces the tracking error by about one order of magnitude. We can compare the effect of the compensator using the data given for the case of hypersonic aircraft in Fig. 4.10 (compensator is used) and in Fig. 4.11 (compensator is not used). Again, we demonstrate here the performance of the neurocontroller with the real plant and with the ANN model of this plant.

We can also use an integral compensator, which is a filter of the form [30]

$$W_{comp} = \frac{W_{rm}^{-1}}{(\tau p + 1)^{n-m} - 1}, \tag{4.11}$$

where n is the order of the numerator of the transfer function of the reference model; m is the order of its denominator; τ is an arbitrary constant, a manually adjustable parameter of the compensation loop (inversely proportional to the gain).

The use of an integral compensator allows us to get rid of a steady state error, completely suppressing the constant disturbances.

However, in an unsteady mode, the integral compensator achieves a similar performance to the PD compensator, and since the steady-state regimes for the systems of the considered class

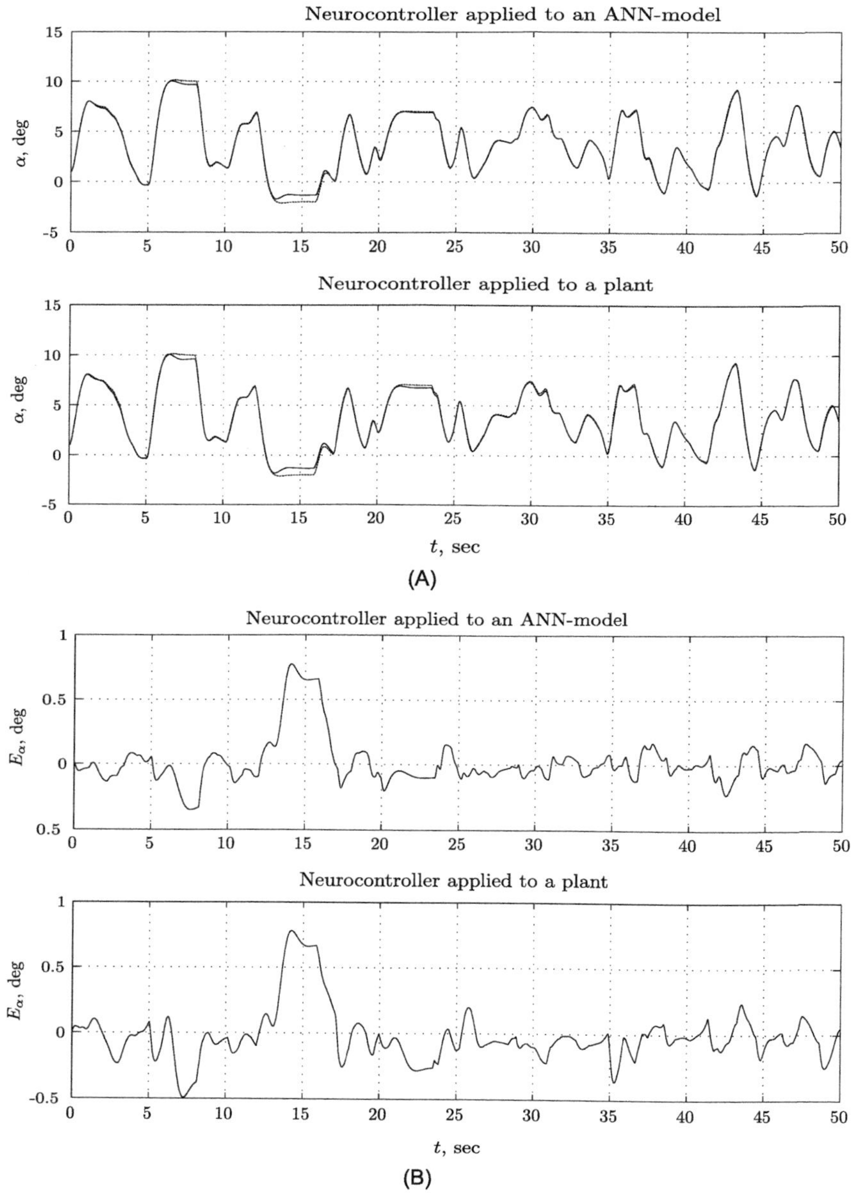

FIGURE 4.9 Performance characteristics of the neurocontroller used with a real controlled object and with its ANN model (hypersonic research vehicle X-43, flight regime $M = 6$). (A) Performance comparison. (B) The reference signal tracking error values. Here $E_\alpha = \alpha - \alpha_{rm}$ is tracking error (the difference between the angle of attack values for the controlled object and reference model); the solid line is for the real plant; the dotted line is for the ANN model.

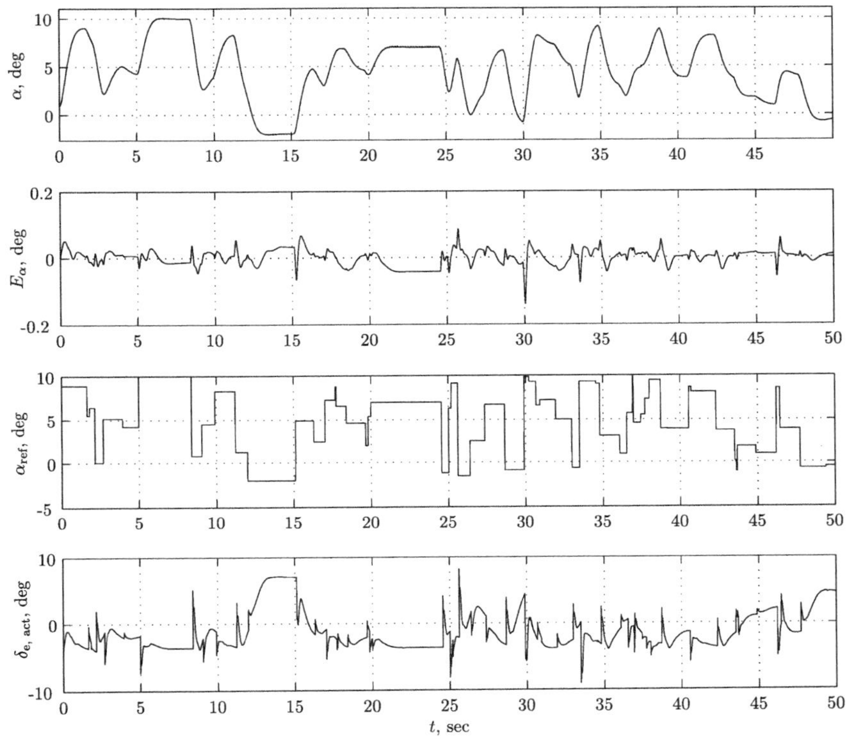

FIGURE 4.10 The results of a computational experiment for model reference adaptive control system with a compensator (hypersonic research vehicle X-43, flight regime $M = 6$). Here α is angle of attack, deg; E_α is the angle of attack tracking error, deg; α_{ref} is the angle of attack reference signal, deg; $\delta_{e,\,act}$ is command signal for the elevator drive, deg; t is time, sec; the solid line is the real plant; the dotted line is the ANN model.

are not typical, it seems reasonable to utilize a simpler PD compensator for these systems.

4.3.2.4 Estimation of the Influence of the Reference Model Parameters on the Efficiency of the Synthesized Control System

As we noted above, the desired behavior of the considered system is defined by a reference model. The computational experiments show that it is expedient to choose the reference model in such a way that the character of the desired behavior determined by it would be close enough to the behavior of the real controlled object. If this condition is not satisfied, then the controller, trying to minimize a difference between the behavior of the system and the reference model, will produce too large values of the command signals for the control actuators, which can lead to a significant deterioration of the control performance.

Taking these considerations into account, in Section 4.3.2.2 a reference model was introduced for the considered adaptive control schemes, which is a serial connection of the aperiodic-type and oscillatory-type units, i.e.,

$$W_{rm} = \frac{\omega_{rm}^2}{(T_{pf}\,p + 1)(p^2 + 2\zeta_{rm}\omega_{rm}\,p + \omega_{rm}^2)}.$$

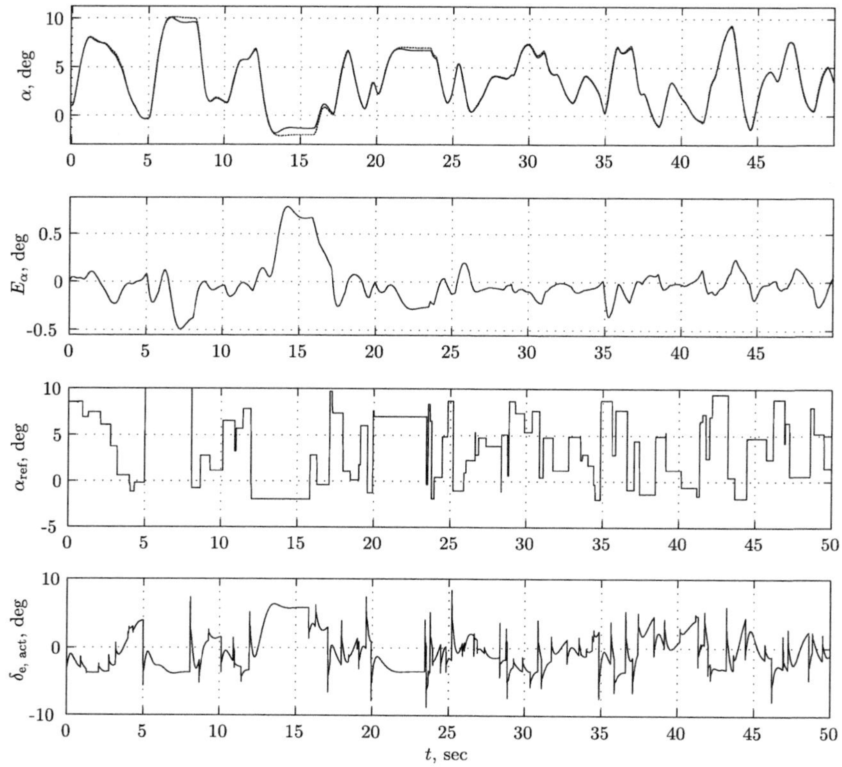

FIGURE 4.11 The results of a computational experiment for model reference adaptive control system without a compensator (hypersonic research vehicle X-43, flight regime $M = 6$). Here α is angle of attack, deg; E_α is the angle of attack tracking error, deg; α_{ref} is the angle of attack reference signal, deg; $\delta_{e,\ act}$ is command signal for the elevator drive, deg; t is time, sec; the solid line is the real plant; the dotted line is the ANN model.

The parameters of this reference model are the eigenfrequency ω_{rm}, the coefficient of relative damping ζ_{rm}, and the time constant of the "prefilter" T_{pf}. Formally the aperiodic-type unit in the reference model is not a prefilter. However, it performs some of its functions, namely the smoothing of the abrupt input signals for the actuator. This reference model structure is selected based on the following considerations:

1. Transient processes in angular longitudinal motion are oscillatory (the system of nonlinear differential equations describing the longitudinal angular motion of the aircraft is of order two). For this reason the oscillatory-type unit is a basis of the reference model.
2. The controlled object is an airplane with an actuator. Therefore a purely oscillatory reference model is unattainable for the controller since following it requires large input signals for the actuator. An aperiodic-type unit is introduced into the reference model to avoid this. This unit plays the role of a prefilter.
3. Reference model parameters are selected based on its feasibility for the controlled object coupled with a control system, and also by taking into account the required range of deflection for the controls.

The value of the relative damping coefficient ζ_{rm} in the reference model was experimentally chosen to be 0.8 to achieve a compromise between the absence of overshoot and the duration of the transient process implemented by the reference model. The prefilter time constant T_{pf} was assumed to be 0.05 sec.

The primary parameter of the reference model is the frequency ω_{rm}, which determines the desired speed of the system as a whole. From the physical point of view, it is possible to increase the reaction rate only using a larger deflection of the controls in the transient mode.

It is well known that the desired speed of the system is selected as a compromise between the speed of the system's reaction and the speed of deflection of the control surfaces. It was demonstrated in a series of computational experiments applied to the nonlinear system under consideration.

As already noted, the desired control performance in the considered adaptive control schemes is set using the reference model. We demonstrate the influence of the desired response speed on the required deflection of the control surfaces and the load on the actuator using the MRAC-type adaptive control scheme. The results of the corresponding computational experiments are shown in Figs. 4.12, A.1–A.3.

Based on the results of these experiments, it is evident that the larger the eigenfrequency of the reference model, the greater the tracking error during the transient process: it increases from ± 0.1 deg for $\omega_{rm} = 1.5$ /sec to ± 2.1 deg for $\omega_{rm} = 4$ /sec. This is due to the fact that the reference model has a third order in accordance with the above considerations, while the controlled object (aircraft + actuator) is a fourth-order oscillatory system. It should be noted that the actuator frequency was set as $\omega_{pf} = 20$ Hz according to the data for the object under consideration. The neural network model of the controlled object by virtue of the procedure for its formation includes an approximate implicit description of the actuator (and therefore the controller also takes into account the actuator approximately), which is adequate until the frequencies of the specified motion (reference model) and the actuator are far apart and the speed of deflection of the elevons does not exceed a certain value. In this formulation, the influence of the actuator reduces only to the appearance of a certain delay in the initial part of the transient process. We can see also the increasing of elevon deflections if we require some high regulation speed. Experiments, the results of which are given in Figs. 4.12, A.1–A.3, were executed for the hypersonic research vehicle. The effectiveness of elevons in the function of the elevator in this aircraft is low. Besides, we need significant deflections of the controls for aircraft balancing at different angles of attack. Thus, it seems reasonable to limit the frequency of the reference model to $\omega_{rm} = 2$ Hz, in order to leave some margin for the deflection of the control surfaces.

The obtained results also show that in some cases the specified actuator speed limitation is achieved (± 60 deg/sec), which affects the transient process (Fig. A.3) and can lead to instability in the case of a significant deviation from the desired motion. To reduce the used actuator speed, it is possible to increase the time constant T_{pf} of the reference model, but this can lead to an increase in the delay in the system, while (with a given actuator) the larger the frequency of the reference model ω_{rm}, the more part of the transient duration will be affected by the actuator.

4.3.2.5 Model Reference Adaptive Control Applied to the Aircraft Angular Motion

A series of computational experiments was carried out to evaluate the properties of the designed MRAC system. Some of the results of these experiments were already presented above. In particular, in Fig. 4.9, as applied to the hypersonic research vehicle X-43 (flight at $M = 6$), the influence of the accuracy of the ANN model on the performance of the obtained ANN mode was shown.

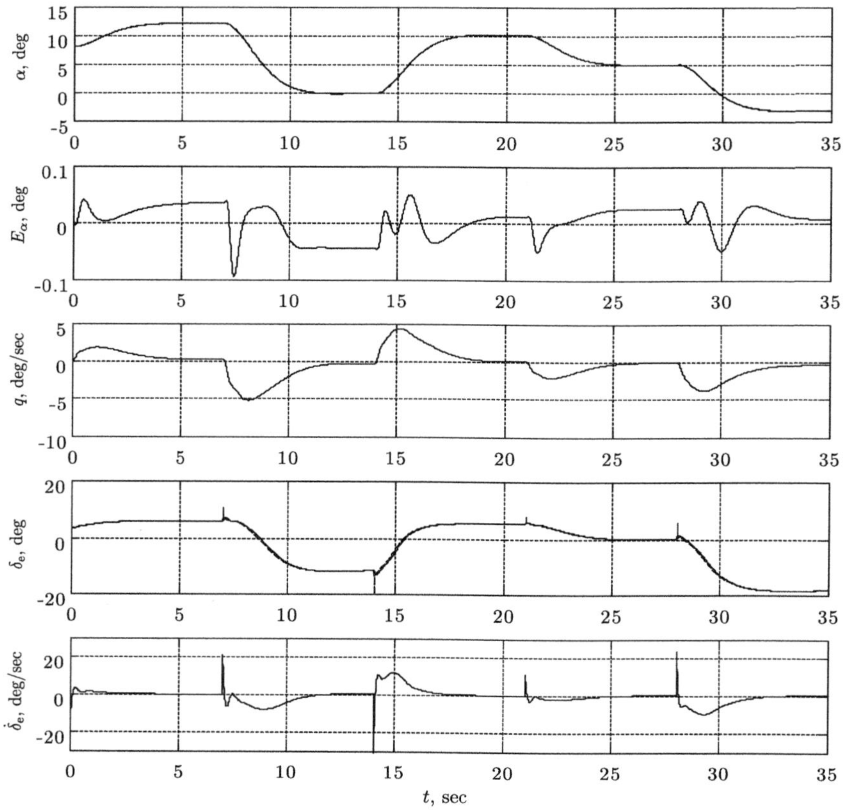

FIGURE 4.12 The results of a computational experiment for an MRAC-type control system applied to the hypersonic research vehicle X-43 for estimating the influence of the natural frequency ω_{rm} of the reference model ($\omega_{rm} = 1.5$; stepwise reference signal on the angle of attack; flight mode $M = 6$, $H = 30$ km). Here α is angle of attack, deg; E_α is tracking error for the given angle of attack, deg; q is the pitch angular velocity, deg/sec; the solid line in the δ_e subgraph is the command signal of the actuator ($\delta_{e,act}$), the dotted line is the deflection angle of the elevons (δ_e); $\dot{\delta}_e = d\delta_e/dt$ is the angular velocity of the deflection of the elevons, deg/sec; t is time, sec.

The results presented in Figs. A.4–A.7 apply to the same aircraft and demonstrate the effect of introducing a compensating loop in the MRAC system. In this case Fig. A.7 shows the value of the α_{ref} reference signal in this experiment, as well as the value of the $\delta_{e,act}$ command signal for the elevator actuator required for realizing this reference signal.

Additional data are shown in Figs. A.8–A.12. In particular, Fig. A.8 shows how the accuracy of the ANN model affects the characteristics of the control system with the reference model and the compensator. We can see that, despite the imperfection of the ANN model used, the quality of control remains very high (tracking error values lie in the range from −0.2 deg to +0.2 deg), although lower than in the case of more accurate ANN models. These facts demonstrate the merit of the compensator, without which in this case the error values become unacceptably large.

The data presented in Figs. A.9, A.10, and A.11 demonstrate the operation of the MRAC system with the compensator under the influence of the traditionally used stepwise reference signals. Fig. A.12 shows the operation of such a system for a sequence of stepwise reference signals, following the intervals during which the previous disturbance has already been canceled, i.e., these stepwise effects can be considered isolated. This kind of verification of the dynamic properties of a controlled system remains essential for the systems of the class considered in this chapter since it makes it possible to observe visually and evaluate the nature of the response of the controlled system when some disturbance effect occurs.

However, in the modern practice of the nonlinear system testing, instead of isolated stepwise disturbances, which we use traditionally for linear systems, we have to use more complex input (reference) signals that allow testing in a much more strict mode. Namely, the input signal is generated, which often and significantly changes in magnitude, so the control system has to start reacting to some ith disturbance even before the transient process associated with the reaction to the $(i-1)$th (and possibly also to the $(i-2)$th, $(i-3)$th, ...) disturbance has completed. In other words, with this approach to testing, the system is required to react regularly not to any single disturbance, but to their combination ("mixture"), the components of which are at different stages of the completion of the transient processes generated by them, and this combination is changing randomly. Here, the control system has to work in much more difficult conditions than with the traditional stepwise effect, but this approach is better suited to the nature of the problems arising in the control of nonlinear systems operating under uncertainty. For example, atmospheric turbulence, affecting the aircraft, will not wait until the control system has fulfilled its previous impact.

For this reason, in the examples of testing adaptive control algorithms considered below, a difficult reference signal is used. This input signal usually changes frequently and significantly.

The angle of attack reference signal, supposed to be tracked by the control system, was constructed by the same rules as the actuator command signal during the training set generation for the ANN model. Namely, a random sequence of stepwise values of the reference angle of attack was generated, with frequent and significant differences between adjacent values of the elements of the sequence. This approach is applied in order to provide the broadest possible variety of states of the simulated system (in order to cover the entire state space of the system as uniformly and densely as possible), as well as the highest possible variety of changes in the adjacent states (in order to reflect the dynamics of the object as accurately as possible in the control algorithm implemented by the neurocontroller). See Fig. 4.13.

Figs. A.13–A.16 show the operation of a control system with a reference model and a compensator for the case when the F-16 aircraft was considered as a controlled object and two emergency situations happened that led to a shift of the center of mass of the aircraft, as well as to a decrease of the effectiveness of its longitudinal control.

Fig. A.17 shows how an adaptive control system with reference model and compensator deals with the influence of two consecutive failures, which significantly affect the dynamics of the hypersonic research vehicle X-43. The first of them leads to a displacement of the center of gravity by 10% (at $t = 20$ sec), the second to a 50% decrease in the effectiveness of the longitudinal motion control (for $t = 50$ sec). We can see that the adaptation scheme provides operation with a small error ($E_\alpha \approx \pm 0.05$ deg) until the first failure situation occurs. Adaptation to the change in the dynamics of the object caused by this situation occurs quickly enough (approximately 1.2–1.5 sec). The error has now become larger (up to the moment of occurrence of the second emergency situation), but it fits, basically, in the range $E_\alpha \approx \pm 0.2$ deg; the stability of the system operation is preserved. Af-

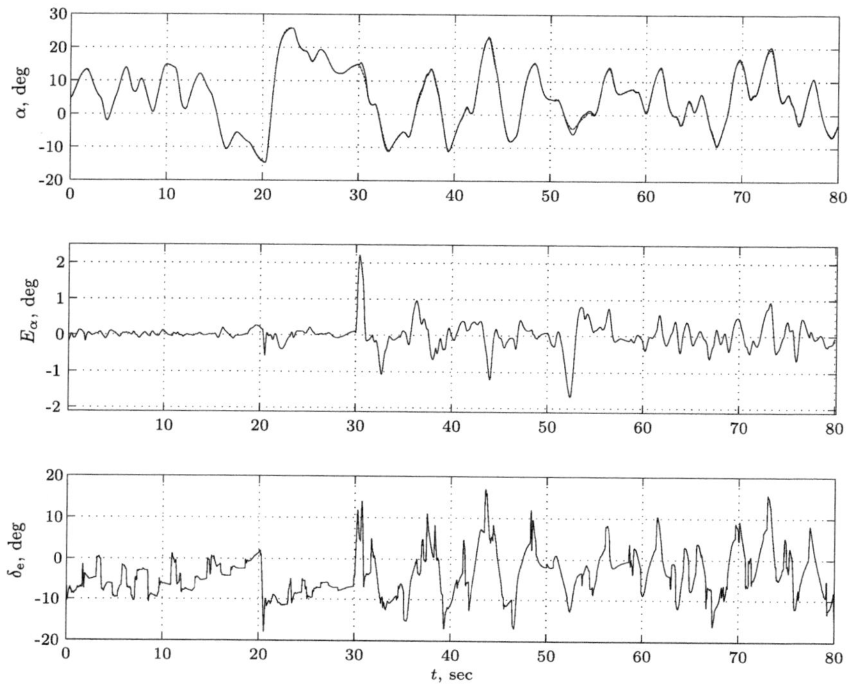

FIGURE 4.13 The results of a computational experiment for the MRAC-type control system with compensator (F-16 aircraft, flight mode with a test speed $V_{ind} = 600$ km/h). Adaptation to the change in the dynamics of the controlled object: the shift of the centering by 10% back ($t = 30$ sec); 50% decrease in the effectiveness of the control ($t = 50$ sec). Here α is angle of attack, deg; E_α is tracking error for the reference angle of attack, deg; δ_e is the deflection angle of the stabilizer, deg; t is time, sec.

ter the second failure, the stability is preserved, but the error values become quite significant ($E_\alpha \approx \pm 0.5$ deg).

In Figs. A.18–A.21 we demonstrate the results of similar computational experiments for the NASP hypersonic aircraft.

In the computational experiments considered above, we have used the model of the aircraft motion of the form (4.2) with a single control variable $\delta_{e,\,act}$. The relation between the angle of attack α and the thrust T_{cr} in this model was introduced through the values of the pitching moment coefficient $C_m(\alpha, T_{cr})$. Additional effects from the thrust on the angle of attack and the angle of attack on the thrust were not taken into account; thrust control was also not introduced.

Another series of computational experiments was performed to assess the significance of the factors excluded from consideration in the tests described above. This additional series was also carried out for the hypersonic research vehicle X-43, which cruised at Mach number $M = 6$.

In this series of experiments, the motion model has the form (4.3), i.e., the interaction of the angle of attack and thrust is taken into account. Besides, we introduce the engine thrust control δ_{th} in addition to control for the angle of attack (command signal $\delta_{e,\,act}$) in order to counteract errors in the angle of attack and tangential g-load.

Various combinations of the reference signals used for both channels were considered, as well as the different versions of control schemes (with or without compensator in one or both channels). All experiments were performed for the MRAC scheme. These experiments include the following:

1. Constant reference g-load $n_x = 0$, random reference angle of attack. Compensators included for both channels.
2. Constant reference g-load $n_x = 0$, random reference angle of attack, there is no compensator in the g-load channel.
3. Constant reference angle of attack (2 degrees), random reference g-load, compensators included for both channels.
4. Constant reference angle of attack (2 degrees), random reference g-load, no compensator in the g-load channel.
5. Both reference signals are random stepwise. Compensators included for both channels.
6. Both reference signals are random stepwise. There is no compensator in the g-load channel.
7. Both reference signals are random stepwise. There are no compensators in all channels.

The results of the computational experiments for the abovementioned conditions are shown in Figs. A.22–A.35. Since for each of the variants, there are eight graphs, they are divided into two parts in four graphs for each of the variants. The first graph in each pair shows the behavior of the object, the reference model, and control signals, while the second graph shows tracking errors and the reference signals.

The results obtained in this series of experiments allow us to draw the following conclusions. The accuracy of the angle of attack tracking when using the extended motion model (4.3) is somewhat reduced.

In the case when the compensator is used for both channels, the error value ranges from $-(0.02 \div 0.04)$ to $+(0.01 \div 0.02)$ by n_{x_a} and from $-(0.05 \div 0.20)$ to $+(0.7 \div 1.1)$ with respect to α. In this case, the lowest error values are obtained by maintaining the regime with $\alpha = \text{const} = 2$ deg, and the largest ones, when the reference signals on both channels are random, with a sharply, often and significantly varying amplitude. Such behavior is not usual for an aircraft of the class under consideration in its normal flight regimes. It will rather be adequate to some abnormal situation, in which the parameters of the flight of the aircraft are changed abruptly and often. But even in this rather complicated situation, the control system has performed quite successfully.

In the case when the compensator was used only for the angle of attack channel, and not for the g-load channel, the quality of control deteriorated somewhat. Namely, the magnitude of the error in this case lies in the range from $-(0.06 \div 0.10)$ to $+(0.05 \div 0.08)$ in n_{x_a} and from $-(0.12 \div 0.60)$ to $+(0.4 \div 1.2)$ with respect to α. It can be seen that the relative error has increased. In particular, the error has increased to a greater extent for g-loading rather than for the angle of attack. However, the absolute errors remain perfectly acceptable, i.e., the adaptation algorithm works quite efficiently, despite the complicated conditions.

The absence of compensators in both channels is more significant. The error value in this case lies in the range from -0.10 to $+0.08$ in n_{x_a} and from -1.2 to $+2.1$ in α. Thus, the role of the compensator in the considered adaptive control scheme is quite significant, but it is not critical.

We can estimate the effect of the change of the angle of attack on the longitudinal trajectory motion due to the relation between the angle of attack and the engine thrust from the dynamics and boundaries of the change in the g-load values n_{x_a}.

From the results presented in Figs. A.22–A.35, we can see that the g-load in the case of using compensators in both channels lies in the range from $-(0.03 \div 0.15)$ up to $+(0.01 \div 0.15)$, in the absence of a compensator in the g-load channel in the range from $-(0.10 \div 0.18)$ to $+(0.10 \div 0.17)$, and in the absence of compensators in both channels from -0.18 to $+0.19$. Thus, the positive effect of the compensators in the control system of this class is also visible here, but the overall effect of changes in the angle of attack on the longitudinal trajectory motion can be considered insignificant. For this reason, it is perfectly acceptable to use a simpler single-channel model (4.2) instead of the two-channel model (4.3) for the estimation of the adaptation algorithms efficiency. Of course, the final evaluation of these algorithms should be performed using the complete model of the aircraft motion.

Figs. A.36–A.39 show the simulation results for one more class of aircraft, namely, micro-UAVs and mini-UAVs. Figs. A.36 and A.37 demonstrate the operation of the adaptive control system under normal operation conditions, and Figs. A.38 and A.39 show how the MRAC system with the compensator copes with the effect of two successive failures that significantly affect the dynamics of the object. The first of them leads to a displacement of the centering by 10% back (at 5 sec for micro-UAV "003" and 10 sec for mini-UAV X-04), the second to a decrease by 50% of efficiency of the longitudinal motion control (at 10 sec for micro-UAVs and 20 sec for mini-UAVs). We can see that the adaptation scheme provides operation with a minor error (as a rule, $E_\alpha \approx \pm 0.05°$) until the first failure situation occurs. Adaptation to the change in the dynamics of the object caused by this situation occurs quickly enough (approximately 1.2–1.5 sec). The error has now become larger (before the appearance of the second fault situation), but it fits, basically, in the range $E_\alpha \approx \pm 0.2°$; the stability of the system's operation is preserved. After the second failure, the stability is preserved, but the error values become quite significant (mostly, their values lie in the range $E_\alpha \approx \pm 0.5°$).

The simulation results presented in this section serve as clear evidence that in most cases, the adaptive neural network control system, whose structure is represented in Fig. 4.6, successfully performs its tasks. It allows to take into account the relationship between the angle of attack and the thrust of the aircraft engine. Also, in the case of a failure situation it performs the reconfiguration of the motion control algorithm and allows us to rapidly cancel the effects of equipment failures and structural damage of the vehicle.

4.3.3 Model Predictive Control

4.3.3.1 *General Scheme of the Model Predictive Control*

The problem of control with the predictive model (MPC) uses the object model, which predicts its future behavior, as well as an optimization algorithm to select the control action that provides the best values of the predicted characteristics of the system.

Control with the predictive model is based on the sliding horizon method, according to which the ANN model predicts the output of the controlled object after a predetermined time interval (forecast horizon). The obtained forecast results are used by the algorithm of numerical optimization to find the control value u, which minimizes the following control quality criterion on the given forecast horizon:

$$J = \sum_{j=N_1}^{N_2} (y_r(t+j) - y_m(t+j))^2 + \rho \sum_{j=1}^{N_u} (u'(t+j-1) - u'(t+j-2))^2.$$

Here N_1, N_2, and N_u are numerical parameters that determine the forecast horizon within which the values of the tracking error and the increments of the control signal are estimated. The values of y_r and y_m are the desired output of the controlled object and the output of the ANN model, respectively, u' are trial control actions, and ρ is the weighting factor, which determines the relative share of the contribution of control variations to the overall value of the efficiency criterion J.

In addition to the forecast horizon, the second important parameter in the MPC scheme is the value of the control horizon, i.e., the values of

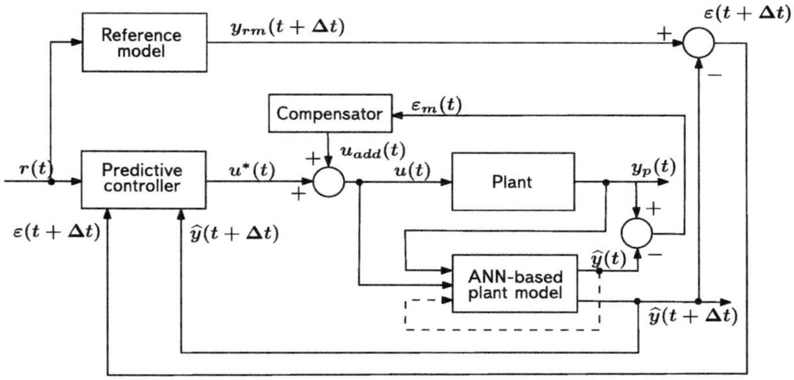

FIGURE 4.14 Adaptive model predictive control (MPC) scheme. Here r is a reference signal; y_p is the plant output; $\hat{y}$ is an output of the ANN model; y_{rm} is the reference model output; $\tilde{u}$ is the control signal generating with predictive controller based on optimization algorithm; u_{add} is additional control signal generated with the compensator; u is combined control input acting on the plant; ε_m is a difference between outputs of the plant and the ANN-based model.

the time intervals through which the optimization algorithm outputs a control signal whose value does not change until the next control horizon is reached.

In general, the horizons of control and forecast do not coincide. As shown by the computational experiment, the ratio of these horizons largely determines the stability of the MPC scheme. According to the obtained experimental data, it is expedient to select the control horizon much less than the forecast horizon. In particular, in the experiments, the results of which are presented below, the control horizon was adopted as the minimum (equal to one time step Δt) and further within the forecast horizon the control was considered constant. Due to this choice, calculations at each step of control generation are simplified, and the stability of the optimization algorithm in the MPC scheme is improved.

We adopted the forecast horizon in the experiments described below as 30 time steps (0.3 sec), based on the following considerations. The optimization algorithm in the MPC scheme tries to minimize the predicted deviation from the reference trajectory. We assume that in the presence of initial deviations, the predicted trajectory should converge to the reference one by the end of the prediction interval. This fact means that the smaller the forecast horizon, the larger the control effect applied to the object will have to be to reduce the expected deviation from the reference model to zero. Thus, the forecast horizon determines the value of the effective gain by the tracking error in the MPC scheme: the smaller the forecast horizon, the higher this gain. For this reason, the minimum horizon of the forecast is limited by considerations of stability of the dynamical system, since if a specific threshold value of this coefficient is exceeded, the stability of this system is lost. On the other hand, there are limitations on the increase in the forecast horizon due to the computational complexity of generating the control signal, the accuracy of the tracking and the approximate nature of the forecast itself, related to the persistence of control on the forecast horizon. In the course of numerical experiments, a compromise solution was found, according to which in the problem to be solved for the advanced hypersonic research vehicle, the forecast horizon should be 30 time steps.

The general adaptive control scheme with the predictive model is shown in Fig. 4.14.

4.3.3.2 *Neurocontroller Synthesis for the Model Predictive Control*

In adaptive MPC schemes, only one multilayer neural network is used, which is the ANN model of the object. The controller is represented here by the optimization algorithm. The structural scheme of the system with the predictive model is illustrated in Fig. 4.14.

The criterion of quality for this system is the root-mean-square error in the forecast interval (5–7 steps), i.e., it does not take into account the increase in control. We have

$$\begin{aligned} E(u) &= \frac{1}{2} e^T e, \quad e(u) = y_{rm} - \hat{y}(u), \\ e &= [e_{k+2}, e_{k+3}, \dots, e_{k+T}]^T, \\ u &= [u_{k+1}, u_{k+2}, \dots, u_{k+T-1}]^T, \end{aligned} \tag{4.12}$$

where T is the length of the forecast interval.

The absence of a control term in the criterion makes it possible to do just one iteration at each step by the Gauss–Newton method.

Thus, here the minimization of the deviation of the behavior of the ANN model from the reference model is performed not by the parameters of the neurocontroller (they are not present here at all), but directly on the control over the forecast interval. To apply effective optimization methods, we need to calculate the dynamic control-related Jacobian, i.e.,

$$J_u = \begin{bmatrix} \frac{\partial e_{k+2}}{\partial u_{k+1}} & \frac{\partial e_{k+2}}{\partial u_{k+2}} & \cdots & \frac{\partial e_{k+2}}{\partial u_{k+T-1}} \\ \frac{\partial e_{k+3}}{\partial u_{k+1}} & \frac{\partial e_{k+3}}{\partial u_{k+2}} & \cdots & \frac{\partial e_{k+3}}{\partial u_{k+T-1}} \\ \cdots & \cdots & \cdots & \cdots \\ \frac{\partial e_{k+T}}{\partial u_{k+1}} & \frac{\partial e_{k+T}}{\partial u_{k+2}} & \cdots & \frac{\partial e_{k+T}}{\partial u_{k+T-1}} \end{bmatrix}. \tag{4.13}$$

This is done using the method of backpropagation of the error in time (BPTT – BackPropagation Through Time) for a closed-loop neural network model.

The Gauss–Newton method is very similar to the Levenberg–Marquardt method; it differs only in that the coefficient μ does not change. This coefficient we choose experimentally for the problem being solved, i.e.,

$$u_{n+1} = u_n + (J_u^T J_u + \lambda E)^{-1} J_u^T e. \tag{4.14}$$

With these facts in mind, the calculation of the control at each step of integration is carried out in the following sequence:

1. Build the desired behavior in the forecast interval. To do this, the reference model is calculated on this interval with a constant reference signal (specifying the signal $r_{k+1} \dots r_{k+T-1} = r_k$: for lack of more sophisticated options, assume the simplest, namely, last available, value of this signal).
2. The forecast of the controlled object behavior according to its ANN model for several steps forward. The initial conditions for the model are the reference trajectory and the previous control values obtained using such a controller.
3. Determination of the error vector, calculation of the Jacobian for the error function by control for each time instant.
4. Adjustment of the control vector by any optimization method (in this case by the Gauss–Newton method). Steps 2–4 represent one iteration of the optimization procedure performed until some reasonable reduction in the forecast error.
5. As the control (in the real object) in the next step (the optimization procedure takes a part of the step), the first control on the forecast interval (i.e., u_{k+1}) is taken. The same value fills the entire vector of the initial control approximation in the next step.

4.3.3.3 Model Predictive Control Applied to the Angular Motion of an Aircraft

The behavior of an aircraft under the control of the MPC system is most similar to its behavior under the control of the MRAC system.

To assess the effectiveness of the adaptive MPC algorithm, we have carried out several series of computational experiments. Initially, the MPC algorithm was tested for a stepwise reference signal concerning the angle of attack. It was required to synthesize the control law for the longitudinal angular motion of the hypersonic research vehicle X-43, which would provide a high accuracy of stabilization of the required angle of attack determined by the input command signal for various combinations of the Mach number and altitude characteristic for a given aircraft. An example of the results from this series is shown in Fig. A.40; an extended set of these results is given in Figs. A.41–A.48. From the results obtained, we can see that the control laws synthesized with the use of adaptation mechanisms with predictive model provide a sufficiently high quality of control. Namely, for all the flight regimes studied, the error of tracking a given angle of attack when it was changed dramatically by up to 12 degrees did not exceed ± 0.27 deg, and in some cases it dropped to ± 0.08 deg. After the transition to the new value of the angle of attack was completed, the value of the tracking error decreased almost in all cases to $\pm(0.01 \div 0.02)$ deg.

The above results also show the nature of the operation of the longitudinal control surface of the hypersonic research vehicle (elevons used as elevator), required for implementing the synthesized control law, by comparing the values of the command signal input to the elevator actuator with the deflection angle of the elevons. In addition, by analyzing the data on the speed of the deflection of the elevons required to implement the obtained control law, it is possible to identify the requirements for the elevator actuator. From the results of computational experiments it follows that the required rate of deviation of the elevons lies in the range of ± 50 deg/sec.

These data can be compared with the results for hypersonic research vehicle X-43 (flight with Mach number $M = 6$), which are shown in Fig. A.49, where the operation of the system is shown in the absence of failures and damage at a more complex (random) nature of the excitation signals. Fig. A.50 demonstrates the operability of the system under consideration in the conditions of two consecutive failures that affect the hypersonic research vehicle dynamics. These failures caused a shift of the centering by 5% back at the time $t = 30$ sec, and then a 30% decrease in the efficiency of the control at $t = 60$ sec.

Similar computational experiments applied to an adaptive control system with a predictive model were performed for a maneuverable F-16 aircraft (the results are shown in Figs. A.51, A.52, and A.53), as well as for the UAV (operation of the MPC control system in the nominal mode); see Figs. A.54 and A.55 and, for the case of failures, see Fig. A.56.

The conclusions that follow from the results of the computational experiments presented in this section are, in general, similar to those that were made for MRAC systems. Namely, in most cases, the adaptive neural network control system, whose structure is represented in Fig. 4.14, successfully copes with its tasks, including emergency situations.

Comparison of the MRAC and MPC schemes does not allow choosing any of them. Each of them has its own positive and negative features. The final decision in favor of any one of these schemes can only be made regarding a particular applied problem, after a fairly extensive series of computational experiments.

4.3.4 Adaptive Control of Angular Aircraft Motion Under Uncertainty Conditions

4.3.4.1 *Influence of Atmospheric Turbulence on the Efficiency of an Adaptive Control System for the Aircraft Longitudinal Motion*

One of the issues traditionally difficult for adaptive control systems is their ability to withstand disturbing external effects of a random nature.

In this section, an attempt is made to evaluate, for the MRAC system, how well a synthesized system can cope with disturbances, including cases when emergencies arise.

As a model of atmospheric turbulence, the well-known Dryden model was used, as described in the MIL-F-8785C standard and implemented in the Simulink modules of the Aerospace Toolbox for the Matlab package. The effect of turbulence is manifested through additional components for the vertical velocity V_z and the pitch angular velocity q.

All results were obtained for the maneuverable F-16 aircraft, for the flight regime $H = 100$ m and $V = 600$ km/h. They are presented in Figs. A.57–A.60.

Two cases are considered:

1. A disturbing effect of $\Delta V_{z,turb}$ in the range ± 10 m/sec, Δq_{turb} in the range ± 0.2 deg/sec (Figs. A.57 and A.58).
2. A disturbing effect of $\Delta V_{z,turb}$ in the range ± 20 m/sec, Δq_{turb} in the range ± 2 deg/sec (Figs. A.59 and A.60).

To evaluate the effect of atmospheric turbulence on the characteristics of the controlled system, for each of the cases considered, two versions were calculated, i.e., with and without turbulence.

In all variants, the behavior of the system was considered with successive emergence of two failure situations introduced in the same way as was done in Section 4.3.2.5. The first of them ($t = 20$ sec) is manifested as a shift of the center by $+10\%$, the second ($t = 50$ sec) as a decrease in longitudinal control efficiency by 50%.

4.3.4.2 *Adaptation to Uncertainties in Source Data*

The presence of uncertainties in the source data leads to the fact that the neural network model and the neurocontroller will be tuned inaccurately and this inaccuracy has a negative effect on the quality of control.

In the computational experiments conducted, inaccurate knowledge of the aircraft dynamics was simulated by the fact that at the beginning of the simulation, an ANN model or neurocontroller tuned to a different flight mode was specified. Thus, the synthesis of the control law was applied to one flight mode (Mach number and altitude of flight), and then the control law began to work quite in other conditions. The correspondence of the flight regime, for which the synthesis of the control law and the regime in which this control law was tested in the computational experiments, is given in Table 4.1.

Two options were considered.

According to the first variant, the adaptation mechanism was activated (its main features were considered in Chapter 1), which allowed for the correction of the control law in relation to the operating conditions in which it appeared.

The second approach was to assess the importance of the introduced adaptation mechanisms by revealing their contribution to the overall task of ensuring the necessary control quality under changing operating conditions. The adaptation mechanisms were disconnected to implement this approach, that is, the control law was not adjusted, and the problem of ensuring the control quality for the system was entirely carried out by the robustness mechanisms, including the compensating circuit introduced into the systems in Section 4.3.2.3.

The results of the computational experiments within the first of the variants listed above are presented in the next two paragraphs. The sec-

TABLE 4.1 Table of correspondence of flight regimes used in the synthesis and testing of the neurocontroller (NC) for the Hypersonic Research Vehicle.

Flight regime, for which the NC was tested	Flight regime, for which the NC was synthesized
M = 5, H = 32 km	M = 7, H = 30 km
M = 5, H = 30 km	M = 7, H = 28 km
M = 6, H = 30 km	M = 5, H = 28 km
M = 7, H = 30 km	M = 5, H = 32 km
M = 7, H = 28 km	M = 6, H = 32 km
M = 7, H = 32 km	M = 5, H = 28 km
M = 6, H = 32 km	M = 7, H = 28 km
M = 6, H = 28 km	M = 7, H = 32 km
M = 5, H = 28 km	M = 7, H = 30 km

ond approach is analyzed in experiments, the results of which are given below.

When considering the first of the two analyzed options, it is actually required to solve for the controlled object the task of real-time identification in order to have reliable information about the nature of the influence of control actions on the behavior of the controlled object.

For the convergence of the real-time identification process, the presence of a test (nontarget) signal at the system input is required for some time (in such a case this is the system adaptation time). This signal can be some additional signal to the actuator to configure the ANN model. However, for nonlinear systems, the condition of proper tuning is also the consideration of as many states as possible from the region in the state space in which the system operates. Therefore, the test signal for the ANN model is introduced not by the control surface deflection, but by the reference signal determined precisely in the state space and passed through the control system. In the neurocontroller, the input signal is directly the reference signal, that is, to set the controller in the case under consideration, the test signal is input as a varying reference signal concerning the angle of attack.

Thus, separate test signals in the form of a deflection of control surfaces (elevons in the case of the hypersonic research vehicle) in the considered approach are not required. This fact, however, does not eliminate the need for disturbing external influences on the controlled object, through which we reveal the information on the reactions of a given object to control actions. This information is the basis for the adaptive adjustment of the control laws used. We used the reference signal for the angle of attack as one of the types of such influences. As shown above, the primary requirement for this signal is that it has to provide the complete coverage of the state space of the system under study. Besides, it is necessary to take into account the dynamics of the system by varying the rate of transitions between individual states. A large number of computational experiments demonstrated the efficiency of this approach.

In all computational experiments, the results of which we give in the next two sections for the MRAC and MPC schemes, the control system operation was set in the same way. The first 20 sec some disturbing reference signal was fed to the input of the control system, which is necessary for solving the identification problem. For the next 20 seconds, we are testing the system. For this testing, we use some sequence of stepwise effects, which are separated in time from one another, so that the transition process from one such disturbance can complete before the moment of submitting the next one.

MODEL REFERENCE ADAPTIVE CONTROL

In this subsection, in Figs. A.61–A.69 the data of the computational experiments on the estimation of the influence of the source data accuracy on the control characteristics for the case of the MRAC system are given.

As noted above, the results of the computational experiments shown in Figs. A.61–A.69 demonstrate the operation of the control system in two time intervals with a duration of 20 sec each: in the first one the adaptation of the control law is carried out by feeding a disturbing influence to the input of the control system (a highly changing signal for the angle of attack); in the second interval with the same duration, the system is tested using a sequence of stepwise input signals that are separate in time so that the disturbed motion caused by the applied signal can be decayed until the next signal is applied. If this condition is met, then we can assume that the system in one session is tested by a set of independent stepwise input signals, differing in magnitude.

The value of the natural frequency of the reference model in these experiments was chosen, for the reasons stated in Section 4.3.2.4, equal to 2 /sec.

The analysis of the data shown in Figs. A.61–A.69 allows us to draw the following conclusions. The quality of the control, estimated by the error of tracking the given angle of attack, reaches in most cases acceptable values within 3–5 sec from the process of adjusting the control law to the changed flight conditions. Thus, the time interval of adaptation of the control system with a length of 20 sec is clearly redundant; it can be substantially reduced, as a rule, to several seconds.

The considered mechanisms of adaptation provide sufficiently high accuracy of the control: the tracking error of the prescribed angle of attack does not exceed, as a rule, values in the range from ± 0.25 deg up to ± 0.45 deg, and this is true for both the MRAC and the MPC scheme. In steady-state regimes during testing (when the disturbance from the stepwise disturbance was attenuated), the error in tracking the angle of attack became practically zero. At the same time, the initial error (at that moment in time when the adaptation mechanism was activated) in some cases reached ± 1 deg. As we can see from Table 4.1, the gap between the flight regimes at which the control law was synthesized and tested was large enough: up to 2 units in Mach number and up to 4 km in flight altitude. Nevertheless, the adaptation mechanisms (both in the MPC scheme and in the MRAC scheme) in practically all cases successfully coped with the task of adjusting the control law to the changed flight conditions, restoring the accuracy of tracking the prescribed angle of attack.

MODEL PREDICTIVE CONTROL

The results of computational experiments on the estimation of the influence for inaccuracy in the initial model concerning the adaptive MPC scheme are shown in Figs. A.70–A.78.

4.3.4.3 *Estimation of the Importance of the Adaptation Mechanisms in the Problem of Controlling the Angular Motion of an Aircraft*

The neural network control system in the variants studied in this chapter consists of two parts: the proper neural network and an additional compensating loop. These two parts are independent; each of them performs their specific functions. The neural network part is adaptive, and its task is to provide the desired dynamics of the system. The compensating element reduces errors arising from inaccuracy of the ANN model and provides a robustness property for the system.

As shown in Section 4.3.2.3, the presence of the compensating element is fundamentally essential for the MRAC and MPC adaptive control schemes. In these schemes, the basic idea is that we use an ANN model as the source of information about the behavior of the controlled object when forming the values of the regulator parameters. Due to the approximate nature of the ANN model, the results obtained with its help inevitably differ from the real values of the variables describing the motion of the object. The approach to compensating such an error was proposed in Section 4.3.2.3. According to this approach, we interpret an inaccuracy of the ANN model as some disturbance acting on the system which deviates the trajectory of the real object from the reference trajectory. We try to reduce this deviation embedding a compensating loop (compensator) into the system. As shown in Section 4.3.2.3, the embedding of the compensating loop into the adaptive system increases its robustness, without which the adaptation algorithm may not meet the adjustment of the control law in that time window within which the situation must not have time to grow from an emergency to a catastrophe. In a number of cases, changes in the dynamics of the controlled object and/or the conditions of its operation are not significant. In these cases, its robustness will be sufficient for the normal operation of the system, and it is possible not to use the adaptation mechanism. For this reason, questions arise about the importance of the mechanism of adaptation. We need to know what this mechanism provides in comparison with the version of the system without adaptation, but with robustness improved by using a compensating loop.

The experiments were carried out to assess the contribution of adaptability and robustness to the behavior of the system. We deactivate the adaptation mechanisms in these experiments. As in the previous case (analysis of the influence of uncertainties in the source data on the properties of the system being synthesized), the system was initially set up incorrectly. In the previous series of experiments described in Section 4.3.4.2, the control system was adapted to the changed conditions for the operation of the controlled object. For this purpose, in the experiments whose results are shown in Figs. A.61–A.69 (MRAC scheme) and Figs. A.70–A.78 (MPC scheme), we set up for 20 sec the disturbing reference signal required to adjust the control law, and then the test stepwise signal for estimating the tuning quality. In a series of experiments, the results of which are presented below, the control system continued to operate with this "incorrect" setting, which was obtained initially, without performing the adaptive adjustment, only due to the robustness properties (see Figs. A.79–A.81 for the MRAC system and Figs. A.82–A.84 for the MPC system).

In the conducted experiments, the results of which are presented in Figs. A.79–A.84, the system operation time was divided, as in the previous case, into two segments of 20 sec each. The nature of the control action concerning the angle of attack is the same as in the last case, i.e., the first 20 sec command signal is a random sequence of frequently changing values of the angle of attack. With such a signal, the disturbance from the input signal does not have enough time to decay, as the next signal acts on the object. Such a complicated signal makes it possible to estimate the cumulative effect of several disturbances whose responses intersect in time, i.e., it allows us to test the control system in fairly harsh conditions. Unlike the previous case, the adjustment of the control law was not carried out, that is, the adaptation mechanism was disabled.

In the second time interval, the control system is tested using a more traditional input signal, which makes it possible to evaluate the behavior of the system with isolated stepwise disturbances.

The results demonstrate that the inaccuracy of the adjusting is not critical for the stability of the system, but causes significant deviations from the reference trajectory. In a steady state, these deviations can be reduced to zero using an integral compensator, but in transient modes, this cannot be handled, and the only way to improve accuracy is to activate adaptation mechanisms to reduce the error below some prescribed level.

As we can see from the results presented in Figs. A.79–A.84, the lack of mechanisms of adaptation leads to a sharp deterioration in the quality of control. Namely, the error of tracking the reference signal with respect to the angle of attack lies now within ±2°, and in some cases even within ±4°.

Thus, the adaptation mechanisms allow to reduce the tracking error in a much larger region of state space (in the case of a nonlinear system) and to expand the frequency band in which the tracking error does not exceed a specific predetermined value. That is, endowing the system with adaptive properties allows it to deal with a much broader class of parametric uncertainties in controlled objects.

We can draw some conclusions from the results obtained in Section 4.3.4. The methods of adaptive-robust modeling and control in variants with a reference and predictive model are the powerful and promising tools that allow solving the problems of fault-tolerant aircraft motion control under uncertainty conditions.

REFERENCES

[1] Ljung L. System identification: Theory for the user. 2nd ed. Upper Saddle River, NJ: Prentice Hall; 1999.
[2] Narendra KS, Parthasarathy K. Identification and control of dynamic systems using neural networks. IEEE Trans Neural Netw 1990;1(1):4–27.
[3] Chen S, Billings SA. Neural networks for nonlinear dynamic systems modelling and identification. Int J Control 1992;56(2):319–46.
[4] Heister F, Müller R. An approach for the identification of nonlinear, dynamic processes with Kalman-filter-trained recurrent neural structures. Research report series, Report No. 193, University of Würzburg, Institute of Computer Science; April 1999.
[5] Haykin S. Neural networks: A comprehensive foundation. 2nd ed. Upper Saddle River, NJ, USA: Prentice Hall; 1998.
[6] Hagan MT, Demuth HB, Beale MH, De Jesús O. Neural network design. 2nd ed. PSW Publishing Co.; 2014.
[7] Gorban AN. Generalized approximation theorem and computational capabilities of neural networks. Sib J Numer Math 1998;1(1):11–24 (in Russian).
[8] Etkin B, Reid LD. Dynamics of flight: Stability and control. 3rd ed. New York, NY: John Wiley & Sons, Inc.; 2003.
[9] Boiffier JL. The dynamics of flight: The equations. Chichester, England: John Wiley & Sons; 1998.
[10] Roskam J. Airplane flight dynamics and automatic flight control. Part I. Lawrence, KS: DAR Corporation; 1995.
[11] Roskam J. Airplane flight dynamics and automatic flight control. Part II. Lawrence, KS: DAR Corporation; 1998.
[12] Cook MV. Flight dynamics principles. Amsterdam: Elsevier; 2007.
[13] Hull DG. Fundamentals of airplane flight mechanics. Berlin: Springer; 2007.
[14] Nguyen LT, Ogburn ME, Gilbert WP, Kibler KS, Brown PW, Deal PL. Simulator study of stall/post-stall characteristics of a fighter airplane with relaxed longitudinal static stability. NASA TP-1538, Dec. 1979.
[15] Sonneveld L. Nonlinear F-16 model description. The Netherlands: Control & Simulation Division, Delft University of Technology; June 2006.
[16] Shaughnessy JD, Pinckney SZ, et al. Hypersonic vehicle simulation model: Winged-cone configuration. NASA–TM–102610, November 1990.
[17] Boyden RP, Dress DA, Fox CH. Subsonic static and dynamic stability characteristics of the test technique demonstrator NASP configuration. In: 31st Aerospace Sciences Meeting & Exhibit, January 11–14, 1993, Reno, NV, AIAA-93-0519.
[18] Boyden RP, Dress DA, Fox CH, Huffman JK, Cruz CI. Subsonic static and dynamic stability characteristics of a NASP configuration. J Aircr 1994;31(4):879–85.
[19] Davidson J. et al., Flight control laws for NASA's Hyper-X research vehicle. AIAA–99–4124.
[20] Engelund WC. Holland SD, et al., Propulsion system airframe integration issues and aerodynamic database development for the Hyper-X flight research vehicle. ISOABE–99–7215.
[21] Engelund WC, Holland SD, Cockrell CE, Bittner RD, Aerodynamic database development for the Hyper-X airframe integrated scramjet propulsion experiments. In: AIAA 18th Applied Aerodynamics Conference, August 14–17, 2000, Denver, Colorado. AIAA 2000–4006.
[22] Holland SD, Woods WC, Engelund WC. Hyper-X research vehicle experimental aerodynamics test program overview. J Spacecr Rockets 2001;38(6):828–35.
[23] Morelli, E.A. Derry, S.D. Smith, M.S. Aerodynamic parameter estimation for the X-43A (Hyper-X) from flight data. In: AIAA Atmospheric Flight Mechanics Conference and Exhibit, August 15–18, 2005. San Francisco, CA. AIAA 2005-5921.
[24] Davis MC, White JT. X-43A flight-test-determined aerodynamic force and moment characteristics at Mach 7.0. J Spacecr Rockets 2008;45(3):472–84.
[25] Morelli EA. Flight test experiment design for characterizing stability and control of hypersonic vehicles. J Guid Control Dyn 2009;32(3):949–59.

[26] Brusov VS, Petruchik VP, Morozov NI. Aerodynamics and flight dynamics of small unmanned aerial vehicles. Moscow: MAI-Print; 2010 (in Russian).
[27] Kondratiev AI, Tyumentsev YV. Application of neural networks for synthesizing flight control algorithms. I. Neural network inverse dynamics method for aircraft flight control. Russ Aeronaut (IzVUZ) 2013;56(2):23–30.
[28] Kondratiev AI, Tyumentsev YV. Application of neural networks for synthesizing flight control algorithms. II. Adaptive tuning of neural network control law. Russ Aeronaut (IzVUZ) 2013;56(3):34–9.
[29] Khalil HK. Nonlinear systems. 3rd ed. Upper Saddle River, NJ: Prentice Hall; 2002.
[30] Ioannou P, Sun J. Robust adaptive control. Englewood Cliffs, NJ: Prentice Hall; 1995.

CHAPTER

5

Semiempirical Neural Network Models of Controlled Dynamical Systems

5.1 SEMIEMPIRICAL ANN-BASED APPROACH TO MODELING OF DYNAMICAL SYSTEMS

Theoretical ("white box") modeling approach relies on the knowledge of some fundamental relationships (such as the laws of mechanics, thermodynamics, etc.), as well as the knowledge of the simulated system structure. Theoretical models might lack the required accuracy due to incomplete and inaccurate knowledge of the properties of the simulated system and environment in which it operates. Moreover, such models are unable to adapt to changes in the simulated system properties.

Empirical ("black box") modeling approach, described in Chapters 3 and 4, relies only on experimental data for the behavior of the simulated system. This approach has its benefits, and it is the only possible option in cases when there is no a priori knowledge of the nature of the system being modeled, of its operational mechanisms, and of essential aspects of its behavior. However, the results presented in this chapter show that empirical ANN-based models of dynamical systems have severe limitations on the complexity level of the problems being solved. In order to overcome these limitations, we need to reduce the number of parameters of the ANN model, while preserving its flexibility at the same time.

So, both traditional theoretical and empirical modeling approaches have certain flaws. Manually designed theoretical models often lack the required accuracy, because it is difficult to take all the factors into account. Moreover, such models are not suited for real-time adaptation. Hence, any changes in a simulated system or its operating environment lead to a decrease in model accuracy. On the other hand, empirical models require the acquisition and preprocessing of an extensive amount of experimental data. Also, a poor choice of the family of empirical models will likely result in a nonparsimonious model with low generalization ability due to overfitting. We propose a hybrid semiempirical ("gray box") modeling approach that utilizes both theoretical domain-specific knowledge and experimental data of system behavior [1–3].

In this book, we assume that the mentioned domain-specific knowledge about the object of modeling is presented in the form of ODEs. There is also an extension of this approach to the case of subject knowledge in the form of DAEs [4–6]. This approach can be extended for the case when the object of modeling is described by partial differential equations (PDEs), but we do not consider this variant in our book.

Neural Network Modeling and Identification of Dynamical Systems
https://doi.org/10.1016/B978-0-12-815254-6.00015-0

Another approach to semiempirical modeling of dynamical systems is suggested in [7–15]. It is also based on the combination of theoretical knowledge about the object of modeling and experimental data on its behavior. This interesting and promising approach allows us to deal with objects that are described in the traditional version of both ODE and PDE cases.

To describe our techniques, let us consider, at first, the semiempirical approach to function approximation. Recall that in the case of a purely empirical approach, the only assumption we make about the function to be approximated is its continuity. With this approach, we pick a parametric family of functions which possesses the universal approximation property (i.e., a set of functions everywhere dense in the space of continuous functions on compact subsets of Euclidean space), such as the family of layered feedforward neural networks. Finally, we search this parametric family for the best approximant according to some predefined objective function via numerical optimization methods. In the case of a semiempirical approach, we assume that some additional a priori knowledge is available for the unknown function apart from the continuity. This knowledge is used to reduce the search space, that is, to select a more specific parametric family of functions which simplifies the optimization of an objective function. In this way, we regularize the original problem and reduce the number of free parameters for the model, while preserving its accuracy at the same time. Thus, the semiempirical model is a parametric function family, whose elements are the compositions of:

- specific nonparametric functions which reflect the exact knowledge of some dependencies;
- functions that belong to a parametric family with specific properties (weighted linear combinations, trigonometric polynomials, etc.) which reflect the knowledge of general features of some dependencies;
- functions that belong to a general parametric family, everywhere dense in the space of continuous functions on compact subsets of Euclidean space (feedforward neural networks, polynomials, etc.) which reflect the absence of any a priori knowledge of some dependencies.

Available experimental data is utilized as well for the purpose of tuning the model parameters, adjusting its structure to improve the accuracy and performing the adaptation in case the unknown function is nonstationary.

In the general case, semiempirical models of this kind need not be universal approximators of continuous functions. However, they allow us to approximate functions of the specific type defined by the model structure (which, in turn, is given by theoretical knowledge) up to any predefined accuracy, as shown in the following theorem.

Theorem 1. *Let m be a positive integer. Suppose that $\mathcal{X}_i$ is a compact subset of $\mathbb{R}^{n_{x_i}}$, and $\mathcal{Y}_i$ are compact subsets of $\mathbb{R}^{n_{y_i}}$ for $i = 1, \ldots, m$. Also suppose that $\mathcal{Z}$ is a subset of $\mathbb{R}^{n_z}$. Let $\mathcal{F}_i$ be a subspace of the continuous vector-valued functions space from $\mathcal{X}_i$ to $\mathcal{Y}_i$, and let $\hat{\mathcal{F}}_i$ be a set of vector-valued functions, everywhere dense in $\mathcal{F}_i$. Finally, let $\mathcal{G}$ be a subspace of Lipschitz continuous vector-valued functions from $\mathcal{Y}_1 \times \cdots \times \mathcal{Y}_m$ to $\mathcal{Z}$, and let $\hat{\mathcal{G}}$ be a set of vector-valued functions, everywhere dense in $\mathcal{G}$. Then the set of vector-valued functions $\hat{\mathcal{H}} = \left\{\hat{\mathbf{g}}(\hat{\mathbf{f}}_1(\mathbf{x}_1), \ldots, \hat{\mathbf{f}}_m(\mathbf{x}_m)) \mid \mathbf{x}_i \in \mathcal{X}_i, \hat{\mathbf{f}}_i \in \hat{\mathcal{F}}_i, \hat{\mathbf{g}} \in \hat{\mathcal{G}}\right\}$ is everywhere dense in the space $\mathcal{H} = \{\mathbf{g}(\mathbf{f}_1(\mathbf{x}_1), \ldots, \mathbf{f}_m(\mathbf{x}_m)) \mid \mathbf{x}_i \in \mathcal{X}_i, \mathbf{f}_i \in \mathcal{F}_i, \mathbf{g} \in \mathcal{G}\}$.*

Proof. Since the sets of vector-valued functions $\hat{\mathcal{F}}_i$ are everywhere dense in the respective spaces $\mathcal{F}_i$ for any vector-valued function $\mathbf{f}_i \in \mathcal{F}_i$ and any positive real ε_i there exists a vector-valued function $\hat{\mathbf{f}}_i \in \hat{\mathcal{F}}_i$ such that

$$\left\|\mathbf{f}_i(\mathbf{x}_i) - \hat{\mathbf{f}}_i(\mathbf{x}_i)\right\| < \varepsilon_i, \quad \forall \mathbf{x}_i \in \mathcal{X}_i.$$

In the same manner, for any vector-valued function $\mathbf{g} \in \mathcal{G}$ and any positive real ε there exists a vector-valued function $\hat{\mathbf{g}} \in \hat{\mathcal{G}}$ such that

$$\left\|\mathbf{g}(\mathbf{y}_1, \ldots, \mathbf{y}_m) - \hat{\mathbf{g}}(\mathbf{y}_1, \ldots, \mathbf{y}_m)\right\| < \varepsilon, \quad \forall \mathbf{y}_i \in \mathcal{Y}_i, \ i = 1, \ldots, m.$$

According to the assumption of the theorem, all functions $\mathbf{g} \in \mathcal{G}$ satisfy the Lipschitz condition

$$\left\|\mathbf{g}(\mathbf{y}_1, \ldots, \mathbf{y}_i, \ldots, \mathbf{y}_m) - \mathbf{g}(\mathbf{y}_1, \ldots, \hat{\mathbf{y}}_i, \ldots, \mathbf{y}_m)\right\| \leqslant M_i \left\|\mathbf{y}_i - \hat{\mathbf{y}}_i\right\|$$

for some nonnegative real constant M_i, referred to as a Lipschitz constant. By applying the triangle inequality, we obtain

$$\begin{aligned}
&\left\|\mathbf{g}(\mathbf{f}_1(\mathbf{x}_1), \ldots, \mathbf{f}_m(\mathbf{x}_m)) - \hat{\mathbf{g}}(\hat{\mathbf{f}}_1(\mathbf{x}_1), \ldots, \hat{\mathbf{f}}_m(\mathbf{x}_m))\right\| \\
&\leqslant \left\|\mathbf{g}(\mathbf{f}_1(\mathbf{x}_1), \ldots, \mathbf{f}_m(\mathbf{x}_m)) - \mathbf{g}(\hat{\mathbf{f}}_1(\mathbf{x}_1), \ldots, \hat{\mathbf{f}}_m(\mathbf{x}_m))\right\| \\
&+ \left\|\hat{\mathbf{g}}(\hat{\mathbf{f}}_1(\mathbf{x}_1), \ldots, \hat{\mathbf{f}}_m(\mathbf{x}_m)) - \hat{\mathbf{g}}(\hat{\mathbf{f}}_1(\mathbf{x}_1), \ldots, \hat{\mathbf{f}}_m(\mathbf{x}_m))\right\| \\
&\leqslant \left\|\mathbf{g}(\mathbf{f}_1(\mathbf{x}_1), \ldots, \mathbf{f}_m(\mathbf{x}_m)) \right. \\
&\quad \left. - \mathbf{g}(\hat{\mathbf{f}}_1(\mathbf{x}_1), \mathbf{f}_2(\mathbf{x}_2), \ldots, \mathbf{f}_m(\mathbf{x}_m))\right\| \\
&+ \left\|\mathbf{g}(\hat{\mathbf{f}}_1(\mathbf{x}_1), \mathbf{f}_2(\mathbf{x}_2), \ldots, \mathbf{f}_m(\mathbf{x}_m)) \right. \\
&\quad \left. - \mathbf{g}(\hat{\mathbf{f}}_1(\mathbf{x}_1), \hat{\mathbf{f}}_2(\mathbf{x}_2), \mathbf{f}_3(\mathbf{x}_3), \ldots, \mathbf{f}_m(\mathbf{x}_m))\right\| \\
&+ \cdots + \\
&+ \left\|\mathbf{g}(\hat{\mathbf{f}}_1(\mathbf{x}_1), \ldots, \hat{\mathbf{f}}_{m-1}(\mathbf{x}_{m-1}), \mathbf{f}_m(\mathbf{x}_m)) - \mathbf{g}(\hat{\mathbf{f}}_1(\mathbf{x}_1), \right. \\
&\quad \left. \ldots, \hat{\mathbf{f}}_m(\mathbf{x}_m))\right\| \\
&+ \left\|\hat{\mathbf{g}}(\hat{\mathbf{f}}_1(\mathbf{x}_1), \ldots, \hat{\mathbf{f}}_m(\mathbf{x}_m)) - \hat{\mathbf{g}}(\hat{\mathbf{f}}_1(\mathbf{x}_1), \ldots, \hat{\mathbf{f}}_m(\mathbf{x}_m))\right\| \\
&< \sum_{i=1}^{m} M_i \left\|\mathbf{f}_i(\mathbf{x}_i) - \hat{\mathbf{f}}_i(\mathbf{x}_i)\right\| + \varepsilon < \sum_{i=1}^{m} M_i \varepsilon_i + \varepsilon. \quad \square
\end{aligned}$$

Nevertheless, under certain conditions the universal approximation property holds for semiempirical models.

Theorem 2. *Let $\mathcal{X}$ be a compact subset of $\mathbb{R}^{n_x}$, let $\mathcal{F}$ be a space of continuous vector-valued functions from $\mathcal{X}$ to $\mathbb{R}^{n_y}$, and let $\hat{\mathcal{F}}$ be a set of vector-valued functions, everywhere dense in $\mathcal{F}$. Also, let $\mathcal{G}$ be a subspace of Lipschitz continuous vector-valued functions from $\mathbb{R}^{n_y}$ to $\mathbb{R}^{n_y}$ with the continuous inverse, and let $\hat{\mathcal{G}}$ be a set of vector-valued functions, everywhere dense in $\mathcal{G}$. Then the set of vector-valued functions $\hat{\mathcal{H}} = \left\{\hat{\mathbf{g}}(\hat{\mathbf{f}}(\mathbf{x})) \mid \mathbf{x} \in \mathcal{X}, \hat{\mathbf{f}} \in \hat{\mathcal{F}}, \hat{\mathbf{g}} \in \hat{\mathcal{G}}\right\}$ is everywhere dense in $\mathcal{F}$.*

Proof. According to the assumption of this Theorem, for any vector-valued function $\mathbf{g} \in \mathcal{G}$ there exists a continuous inverse $\mathbf{g}^{-1}$. Hence, for any continuous vector-valued function $\mathbf{f} \in \mathcal{F}$ the composition $\mathbf{g}^{-1} \circ \mathbf{f}$ is also continuous. This implies that there exists $\hat{\mathbf{f}} \in \hat{\mathcal{F}}$ such that

$$\left\|\hat{\mathbf{f}}(\mathbf{x}) - \mathbf{g}^{-1}(\mathbf{f}(\mathbf{x}))\right\| < \varepsilon_1, \quad \forall \mathbf{x} \in \mathcal{X}.$$

Since the vector-valued function $\mathbf{g}$ is Lipschitz continuous, the following condition holds:

$$\begin{aligned}
\left\|\mathbf{g}(\hat{\mathbf{f}}(\mathbf{x})) - \mathbf{f}(\mathbf{x})\right\| &= \left\|\mathbf{g}(\hat{\mathbf{f}}(\mathbf{x})) - \mathbf{g}(\mathbf{g}^{-1}(\mathbf{f}(\mathbf{x})))\right\| \\
&\leqslant M \left\|\hat{\mathbf{f}}(\mathbf{x}) - \mathbf{g}^{-1}(\mathbf{f}(\mathbf{x}))\right\| < M\varepsilon_1,
\end{aligned}$$

for some nonnegative real Lipschitz constant M. Moreover, there exists a vector-valued function $\hat{\mathbf{g}} \in \hat{\mathcal{G}}$ such that

$$\left\|\hat{\mathbf{g}}(\mathbf{y}) - \mathbf{g}(\mathbf{y})\right\| < \varepsilon_2, \quad \forall \mathbf{y} \in \mathbb{R}^{n_y}.$$

Finally, we apply the triangle inequality and obtain

$$\begin{aligned}
&\left\|\hat{\mathbf{g}}(\hat{\mathbf{f}}(\mathbf{x})) - \mathbf{f}(\mathbf{x})\right\| \\
&\leqslant \left\|\hat{\mathbf{g}}(\hat{\mathbf{f}}(\mathbf{x})) - \mathbf{g}(\hat{\mathbf{f}}(\mathbf{x}))\right\| + \left\|\mathbf{g}(\hat{\mathbf{f}}(\mathbf{x})) - \mathbf{f}(\mathbf{x})\right\| \\
&< \varepsilon_2 + M\varepsilon_1. \qquad \square
\end{aligned}$$

In the case of a dynamical system modeling problem, theoretical knowledge of the plant may be used to select the meaningful state variables (i.e., state variables interpretable in domain-specific terms) in contrast to black box models, such as Elman recurrent networks with abstract context units. Therefore, the initial values of these state variables might be estimated during the same experiment carried out to collect the training data set via additional measurements or calibration procedures. As mentioned in Chapter 2, the initial values of state variables may also be subject to optimization along with the model parameters; however, even in this case such estimates might serve as a good initial guess. Thus, the experimental data set required for the semiempirical model training includes the estimates of state variables at the initial time instant. Moreover, this data set also contains the values of sampling time instants t. The experimental data set has the following form:

$$\left\{\left\langle\left\langle \tilde{\mathbf{x}}^{(p)}(0), \mathbf{u}^{(p)}(0), \tilde{\mathbf{y}}^{(p)}(0)\right\rangle, \left\{\left\langle t_k^{(p)}, \mathbf{u}^{(p)}(t_k^{(p)}), \tilde{\mathbf{y}}^{(p)}(t_k^{(p)})\right\rangle\right\}_{k=1}^{K^{(p)}}\right\rangle\right\}_{p=1}^{P}. \tag{5.1}$$

Here, we assume that $t_0^{(p)} \equiv 0$.

Since the state variables of semiempirical models are interpretable in domain-specific terms, it is usually possible to utilize additional theoretical knowledge of the internal structure of the simulated object. This knowledge might allow us to determine the specific form of some relationships between state variables, controls, and observable outputs with sufficient accuracy. These accurately known relationships are to be embedded in mappings $\hat{\mathbf{f}}$ and $\hat{\mathbf{g}}$, just as in the case of function approximation. For example, in the problem of aircraft motion modeling, discussed in Chapter 6, the relationships between state variables and observable outputs are known theoretically with sufficient accuracy; hence the function $\hat{\mathbf{g}}$ from the observation equation exactly matches the theoretical model counterpart $\mathbf{g}$ and lacks any parameters to be tuned. The function $\hat{\mathbf{f}}$ from the state equation in these problems also includes some known dependencies borrowed from the theoretical model function $\mathbf{f}$, as well as the neural network–based approximations for unknown dependencies.

We consider the following family of continuous time state space semiempirical models:

$$\begin{aligned} \frac{d\hat{\mathbf{x}}(t,\mathbf{w})}{dt} &= \hat{\mathbf{f}}(\hat{\mathbf{x}}(t,\mathbf{w}), \mathbf{u}(t), \mathbf{w}), \\ \hat{\mathbf{y}}(t,\mathbf{w}) &= \hat{\mathbf{g}}(\hat{\mathbf{x}}(t,\mathbf{w}), \mathbf{w}), \end{aligned} \tag{5.2}$$

where $\hat{\mathbf{x}}\colon [0,\bar{t}] \times \mathbb{R}^{n_w} \to \mathbb{R}^{n_x}$ is an estimate of the state space trajectory, $\hat{\mathbf{y}}\colon [0,\bar{t}] \times \mathbb{R}^{n_w} \to \mathbb{R}^{n_y}$ is an estimate of the observable outputs trajectory, and $\hat{\mathbf{f}}\colon \mathbb{R}^{n_x} \times \mathbb{R}^{n_u} \times \mathbb{R}^{n_w} \to \mathbb{R}^{n_x}$ and $\hat{\mathbf{g}}\colon \mathbb{R}^{n_x} \times \mathbb{R}^{n_w} \to \mathbb{R}^{n_y}$ are parametric families of functions, such as layered feedforward neural networks or semiempirical neural network–based function approximators.

The following theorem describes the capabilities of this family of models.

Theorem 3. *Let $\mathcal{U}$ be a compact subset of $\mathbb{R}^{n_u}$, let $\mathcal{X}$ be a compact subset of $\mathbb{R}^{n_x}$, and let $\mathcal{Y}$ be a subset of $\mathbb{R}^{n_y}$. Also, let $\mathcal{F}$ be a subspace of the space of continuous vector-valued functions from $\mathcal{X} \times \mathcal{U}$ to $\mathbb{R}^{n_x}$, locally Lipschitz continuous with respect to all of its arguments, and let $\hat{\mathcal{F}}$ be a set of vector-valued functions, everywhere dense in $\mathcal{F}$. Similarly, let $\mathcal{G}$ be a subspace of continuous vector-valued functions from $\mathcal{X}$ to $\mathbb{R}^{n_y}$, locally Lipschitz continuous with respect to all of its arguments, and let $\hat{\mathcal{G}}$ be a set of vector-valued functions, everywhere dense in $\mathcal{G}$. Then, for all vector-valued functions $\mathbf{f} \in \mathcal{F}$ and $\mathbf{g} \in \mathcal{G}$ and all positive real $\bar{t}$ and ε there exist a positive real δ and vector-valued functions $\hat{\mathbf{f}} \in \hat{\mathcal{F}}$, $\hat{\mathbf{g}} \in \hat{\mathcal{G}}$ such that for any $\mathbf{x}^s \in \mathcal{X}$, any $\tilde{\mathbf{x}}^s$ contained in the δ-neighborhood of $\mathbf{x}^s$, any $\tilde{t} \in (0,\bar{t}]$, and any measurable, locally integrable function $\mathbf{u}\colon [0,\tilde{t}] \to \mathcal{U}$ such that the solution $\mathbf{x}\colon [0,\tilde{t}] \to \mathcal{X}$ of the initial value problem for the system of ODEs with the right hand side $\mathbf{f}^{\mathbf{u}}(t,\mathbf{x}(t)) \equiv \mathbf{f}(\mathbf{x}(t),\mathbf{u}(t))$ and initial condition $\mathbf{x}(0) = \mathbf{x}^s$ exists on the whole segment $[0,\tilde{t}]$ and is contained in $\mathcal{X}$ along with the closure of its ε-neighborhood, the following conditions hold:*

$$\begin{aligned} \|\mathbf{x}(t) - \hat{\mathbf{x}}(t)\| &< \varepsilon \\ \|\mathbf{y}(t) - \hat{\mathbf{y}}(t)\| &< \varepsilon \end{aligned} \quad \forall t \in [0,\tilde{t}],$$

where $\hat{\mathbf{x}}\colon [0,\bar{t}] \to \mathbb{R}^{n_x}$ *is a solution of the initial value problem for the system of ODEs with right hand side* $\hat{\mathbf{f}}^{\mathbf{u}}(t,\hat{\mathbf{x}}(t)) \equiv \hat{\mathbf{f}}(\hat{\mathbf{x}}(t),\mathbf{u}(t))$ *and initial condition* $\hat{\mathbf{x}}(0) = \tilde{\mathbf{x}}^s$; $\mathbf{y}(t) = \mathbf{g}(\mathbf{x}(t))$; $\hat{\mathbf{y}}(t) = \hat{\mathbf{g}}(\hat{\mathbf{x}}(t))$.

Proof. For any vector-valued functions $\mathbf{f} \in \mathcal{F}$ and $\hat{\mathbf{f}} \in \hat{\mathcal{F}}$, the vector-valued functions $\mathbf{f}^{\mathbf{u}}$ and $\hat{\mathbf{f}}^{\mathbf{u}}$ satisfy all conditions of the theorem on existence and uniqueness of solutions to initial value problems (see Theorem 54 in [16]), due to the following considerations:

- the vector-valued function $\mathbf{u}(t)$ is measurable, while vector-valued functions $\mathbf{f}(\mathbf{x},\mathbf{u})$ and $\hat{\mathbf{f}}(\hat{\mathbf{x}},\mathbf{u})$ are continuous with respect to $\mathbf{u}$; therefore the compositions $\mathbf{f}^{\mathbf{u}}(t,\mathbf{x}^*)$ and $\hat{\mathbf{f}}^{\mathbf{u}}(t,\hat{\mathbf{x}}^*)$ are measurable with respect to t for all $\hat{\mathbf{x}}^*$;
- the vector-valued function $\mathbf{u}(t)$ is locally integrable, while vector-valued functions $\mathbf{f}(\mathbf{x},\mathbf{u})$ and $\hat{\mathbf{f}}(\hat{\mathbf{x}},\mathbf{u})$ are locally Lipschitz continuous with respect to $\mathbf{u}$; therefore the compositions $\mathbf{f}^{\mathbf{u}}(t,\mathbf{x}^*)$ and $\hat{\mathbf{f}}^{\mathbf{u}}(t,\hat{\mathbf{x}}^*)$ are locally integrable with respect to t for all $\hat{\mathbf{x}}^*$;
- vector-valued functions $\mathbf{f}(\mathbf{x},\mathbf{u})$ and $\hat{\mathbf{f}}(\hat{\mathbf{x}},\mathbf{u})$ are continuous and locally Lipschitz continuous with respect to $\mathbf{x}$; therefore the vector-valued functions $\mathbf{f}^{\mathbf{u}}(t^*,\mathbf{x})$ and $\hat{\mathbf{f}}^{\mathbf{u}}(t^*,\hat{\mathbf{x}})$ are also continuous and locally Lipschitz continuous with respect to $\mathbf{x}$ for all t^*.

Thus, there exists the unique solution $\mathbf{x}$ of the initial value problem for the system of ODEs with right hand side $\mathbf{f}^{\mathbf{u}}$ and initial condition $\mathbf{x}(0) = \mathbf{x}^s$, as well as the unique solution $\hat{\mathbf{x}}$ of the similar problem with right hand side $\hat{\mathbf{f}}^{\mathbf{u}}$ and initial condition $\hat{\mathbf{x}}(0) = \tilde{\mathbf{x}}^s$. According to the assumption of this theorem, the solution $\mathbf{x}$ exists on the whole segment $[0,\bar{t}]$ and is contained in $\mathcal{X}$ along with the closure of its ε-neighborhood.

Since the set of vector-valued functions $\hat{\mathcal{F}}$ is everywhere dense in the space $\mathcal{F}$, for any vector-valued function $\mathbf{f} \in \mathcal{F}$ and any positive real ε^f there exists a vector-valued function $\hat{\mathbf{f}} \in \hat{\mathcal{F}}$ such that

$$\|\mathbf{f}(\mathbf{x}^*,\mathbf{u}^*) - \hat{\mathbf{f}}(\mathbf{x}^*,\mathbf{u}^*)\| < \varepsilon^f \quad \forall \mathbf{x}^* \in \mathcal{X} \ \forall \mathbf{u}^* \in \mathcal{U}.$$

Next, we estimate the upper bound on the approximation error for the right hand side of the system of ODEs, evaluated at each point of the true state space trajectory $\mathbf{x}(t)$ on the whole time segment. We have

$$\begin{aligned}
&\sup_{t\in[0,\bar{t}]} \left\| \int_0^t \mathbf{f}^{\mathbf{u}}(s,\mathbf{x}(s)) - \hat{\mathbf{f}}^{\mathbf{u}}(s,\mathbf{x}(s))\,ds \right\| \\
&\quad \leqslant \sup_{t\in[0,\bar{t}]} \int_0^t \left\| \mathbf{f}^{\mathbf{u}}(s,\mathbf{x}(s)) - \hat{\mathbf{f}}^{\mathbf{u}}(s,\mathbf{x}(s)) \right\| ds \\
&\quad = \int_0^{\bar{t}} \left\| \mathbf{f}^{\mathbf{u}}(s,\mathbf{x}(s)) - \hat{\mathbf{f}}^{\mathbf{u}}(s,\mathbf{x}(s)) \right\| ds < \int_0^{\bar{t}} \varepsilon^f\,ds \\
&\quad = \bar{t}\varepsilon^f.
\end{aligned}$$

According to the assumption of this theorem, the initial condition $\tilde{\mathbf{x}}^s$ is contained in the δ-neighborhood of $\mathbf{x}^s$, i.e.,

$$\|\mathbf{x}^s - \tilde{\mathbf{x}}^s\| < \delta.$$

Also, we have assumed that all vector-valued functions $\hat{\mathbf{f}} \in \hat{\mathcal{F}}$ and $\hat{\mathbf{g}} \in \hat{\mathcal{G}}$ satisfy the Lipschitz condition, i.e.,

$$\begin{aligned}
\left\|\hat{\mathbf{f}}(\mathbf{x}',\mathbf{u}^*) - \hat{\mathbf{f}}(\mathbf{x}'',\mathbf{u}^*)\right\| &\leqslant M^f \left\|\mathbf{x}' - \mathbf{x}''\right\|, \ \forall \mathbf{u}^* \in \mathcal{U}, \\
\left\|\hat{\mathbf{g}}(\mathbf{x}') - \hat{\mathbf{g}}(\mathbf{x}'')\right\| &\leqslant M^g \left\|\mathbf{x}' - \mathbf{x}''\right\|, \ \forall \mathbf{x}',\mathbf{x}'' \in \mathcal{X},
\end{aligned}$$

for some nonnegative real Lipschitz constants M^f and M^g. Let us also assume that

$$\begin{aligned}
\delta &= \frac{\varepsilon}{3M^g} e^{-M^f \bar{t}}, \\
\varepsilon^f &= \frac{\varepsilon}{3M^g \bar{t}} e^{-M^f \bar{t}}.
\end{aligned}$$

Note that since $\tilde{t} < \bar{t}$, the following inequalities hold:

$$\frac{\varepsilon}{3M^g} e^{-M^f \bar{t}} < \frac{\varepsilon}{3M^g} e^{-M^f \tilde{t}},$$
$$\frac{\varepsilon}{3M^g \bar{t}} e^{-M^f \bar{t}} < \frac{\varepsilon}{3M^g \tilde{t}} e^{-M^f \tilde{t}}.$$

Hence, the well-posedness theorem for initial value problems (see Theorem 55 in [16]) implies that on the whole segment $[0, \tilde{t}]$ there exists a unique solution $\hat{\mathbf{x}}$ of the initial value problem for the system of ODEs with the right hand side $\hat{\mathbf{f}}^{\mathbf{u}}$ and initial condition $\hat{\mathbf{x}}(0) = \hat{\mathbf{x}}^s$. It also implies that this solution is uniformly close to the solution of the original initial value problem, i.e.,

$$\|\mathbf{x}(t) - \hat{\mathbf{x}}(t)\| \leqslant \frac{2\varepsilon}{3M^g} < \varepsilon \ \forall t \in [0, \tilde{t}].$$

According to the assumption of this theorem, the solution $\mathbf{x}$ exists on the whole segment $[0, \tilde{t}]$ and is contained in $\mathcal{X}$ along with the closure of its ε-neighborhood; therefore the solution $\hat{\mathbf{x}}$ is contained in $\mathcal{X}$.

Since the set of vector-valued functions $\hat{\mathcal{G}}$ is everywhere dense in the space $\mathcal{G}$, for any vector-valued function $\mathbf{g} \in \mathcal{G}$ and any positive real ε^g there exists a vector-valued function $\hat{\mathbf{g}} \in \hat{\mathcal{G}}$ such that

$$\|\mathbf{g}(\mathbf{x}^*) - \hat{\mathbf{g}}(\mathbf{x}^*)\| < \varepsilon^g \ \forall \mathbf{x}^* \in \mathcal{X}.$$

If we assume that $\varepsilon^g = \frac{\varepsilon}{3}$, then we obtain the following inequality:

$$\begin{aligned}
&\|\mathbf{y}(t) - \hat{\mathbf{y}}(t)\| \\
&\quad = \|\mathbf{g}(\mathbf{x}(t)) - \hat{\mathbf{g}}(\hat{\mathbf{x}}(t))\| \\
&\quad \leqslant \|\mathbf{g}(\mathbf{x}(t)) - \hat{\mathbf{g}}(\mathbf{x}(t))\| + \|\hat{\mathbf{g}}(\mathbf{x}(t)) - \hat{\mathbf{g}}(\hat{\mathbf{x}}(t))\| \\
&\quad < \varepsilon^g + M^g \|\mathbf{x}(t) - \hat{\mathbf{x}}(t)\| \\
&\quad < \frac{\varepsilon}{3} + M^g \frac{2\varepsilon}{3M^g} = \varepsilon, \ t \in [0, \tilde{t}]. \qquad \square
\end{aligned}$$

Similar results for the case of purely empirical recurrent neural networks were given in [17].

The initial value problem for the system of ODEs defined by the model (5.2) can be solved by various numerical methods; consequently, theoretical domain-specific knowledge as well as the knowledge of the properties of these numerical methods provides useful insights as to which method should be applied to each specific problem. Meanwhile, this knowledge is completely ignored by purely empirical discrete time state space models of the form (2.13), because in these models the state at the next time instant is given directly by the black box parametric family of functions $\mathbf{F}$ which performs the functions of both the ODE right hand side evaluation and the numerical ODE solver. Thus, the family of functions $\mathbf{F}$ is implicitly forced to learn the numerical method for each problem, which leads to unnecessary complication of the model. The semiempirical approach not only overcomes this problem, but also provides additional flexibility, because it allows to use the experimental data set with variable time step for the model training and testing; it also allows to use different numerical ODE solvers for different trajectories from the data set; finally, the trained model may be used in combination with another numerical ODE solver and with any time step. One example of successful application of this approach is the Runge–Kutta Neural Network (RKNN) model, based on the explicit fourth-order Runge–Kutta method [18].

5.2 SEMIEMPIRICAL ANN-BASED MODEL DESIGN PROCESS

The semiempirical ANN model design procedure consists of the following stages:

1. Design of the original theoretical state space continuous time model for the system.
2. Acquisition of experimental data of system behavior (5.1).
3. Evaluation of theoretical model accuracy using the available experimental data.

4. Approximation of the initial theoretical continuous time model with the corresponding discrete time model.
5. Transformation of the discrete time model to the neural network form.
6. Training of the discrete time neural network–based model.
7. Evaluation of the trained model accuracy.
8. Structural adjustment of neural network–based model.

In order to demonstrate this procedure, we consider the following simple dynamical system:

$$\begin{aligned} \frac{dx_1(t)}{dt} &= -(x_1(t) + 2x_2(t))^2 + u(t), \\ \frac{dx_2(t)}{dt} &= 8.322109 \sin x_1(t) + 1.135 x_2(t), \\ y(t) &= x_2(t). \end{aligned} \tag{5.3}$$

In the first step, we need to design an initial theoretical continuous time model of the system. As mentioned before, in all cases we assume the control u to be known exactly. Let us assume that we have perfect theoretical knowledge of the behavior for x_1, so the theoretical model includes the first equation from (5.3). Assume also that we have imperfect knowledge of the behavior for x_2. We imitate this partial knowledge by including the modified second equation from (5.3) in the theoretical model. Finally, assume that we have exact theoretical knowledge of the relationship between state variables and the measured output y. Thus, we obtain the following continuous time theoretical model:

$$\begin{aligned} \frac{d\hat{x}_1(t)}{dt} &= -(\hat{x}_1(t) + 2\hat{x}_2(t))^2 + u(t), \\ \frac{d\hat{x}_2(t)}{dt} &= 8.32 \hat{x}_1(t), \\ \hat{y}(t) &= \hat{x}_2(t). \end{aligned} \tag{5.4}$$

The experimental data set for this simple problem consists of a single trajectory on time segment $t \in [0, 100]$ with constant sampling period $\Delta t = 0.025$. Initial conditions are known exactly, i.e., $\tilde{x}_1(0) = x_1(0) = 0$ and $\tilde{x}_2(0) = x_2(0) = 0$. The control signal $u(t_k)$ is a sequence of steps with random amplitude. Measured output values y are corrupted by additive white Gaussian noise η, i.e., $\tilde{y}(t_k) = y(t_k) + \eta(t_k)$. Standard deviation of the noise η equals $\sigma = 0.01$. The root-mean-square error (RMSE) is used as a measure of model quality. In the best possible case, when the ANN model exactly matches the unknown dynamical system, the modeling error would match the additive measurement noise. Thus, we can use the measurement noise standard deviation as a target value for the root-mean-square modeling error.

The next step is to evaluate the theoretical model accuracy using the available experimental data set. Hence, we need to numerically solve the corresponding initial value problem defined by the system of ODEs (5.4) together with the initial condition and control function given by the experimental data set. In this example, we will compute the initial value problem solution using the first-order explicit one-step Euler method, as well as the fourth-order explicit multistep Adams–Bashforth method [19]. Computational experiments show that the RMSE of the theoretical model predictions equals 0.13947 for the Euler method and 0.071429 for the Adams–Bashforth method, which is much larger than the target value of 0.01. In order to increase the model accuracy we will perform its adjustment using the available experimental data.

Now we need to transform the initial continuous time theoretical model in the form of differential equations to an approximate discrete time model in the form of difference equations. A well-studied topic of numerical methods for ODEs [19] presents a solid algorithmic foundation for the solution of this problem. We transform the system (5.4) to discrete time using the abovementioned explicit finite difference methods: the first-order Euler method and the fourth-order Adams–Bashforth method. Thus, we obtain the following discrete time counterparts of the initial theoretical model:

$$\begin{aligned}\bar{x}_1(t_{k+1}) &= \bar{x}_1(t_k) \\ &\quad + \Delta t\left[-(\bar{x}_1(t_k) + 2\bar{x}_2(t_k))^2 + u(t_k)\right], \\ \bar{x}_2(t_{k+1}) &= \bar{x}_2(t_k) + \Delta t\,[8.32\bar{x}_1(t_k)], \\ \bar{y}(t_k) &= \bar{x}_2(t_k)\end{aligned} \tag{5.5}$$

for the Euler method and

$$\begin{aligned}r_1(t_k) &= -(\bar{x}_1(t_k) + 2\bar{x}_2(t_k))^2 + u(t_k), \\ r_2(t_k) &= 8.32\bar{x}_1(t_k), \\ \bar{x}_1(t_{k+4}) &= \bar{x}_1(t_{k+3}) + \frac{\Delta t}{24}\big[55r_1(t_{k+3}) - 59r_1(t_{k+2}) \\ &\quad + 37r_1(t_{k+1}) - 9r_1(t_k)\big], \\ \bar{x}_2(t_{k+4}) &= \bar{x}_2(t_{k+3}) + \frac{\Delta t}{24}\big[55r_2(t_{k+3}) - 59r_2(t_{k+2}) \\ &\quad + 37r_2(t_{k+1}) - 9r_2(t_k)\big], \\ \bar{y}(t_k) &= \bar{x}_2(t_k)\end{aligned} \tag{5.6}$$

for the Adams–Bashforth method.

The theoretical model discretization is followed by the transformation of the resulting difference equations into a neural network form. To that end, elements of the difference equations are reinterpreted as the respective elements of the neural network–based model. As a result, we obtain a recurrent neural network that is in one-to-one correspondence with the original system of difference equations. However, different continuous time models lead to different difference equations that, in turn, lead to different recurrent neural network–based models. In general, these neural network–based models may have a highly diverse and complicated architecture. A direct approach to the training problem for such neural networks would in general require the development of different training algorithms, suited for the corresponding recurrent neural network architectures. In order to unify the training approach and avoid the necessity to modify the training algorithm for each specific recurrent neural network, we perform the transformation of the recurrent neural network–based models to the unique canonical form. In [2,20,21], authors propose an algorithm that allows to transform a discrete time dynamical system model defined by a bond graph to a corresponding state space model of the form (2.13), called the canonical form of the model. The resulting canonical form of the recurrent neural network consists of a feedforward layered neural network with unit delay feedback loop that connects some outputs of this network with some of its inputs (Fig. 5.1). In Fig. 5.1 and in the following, we use the notation z^{-1} for the unit delay element.

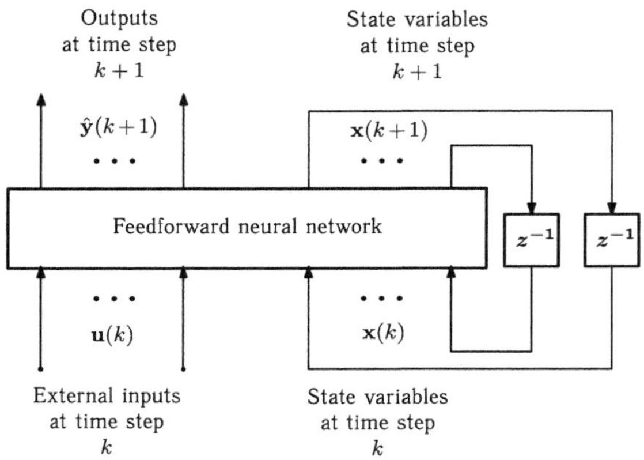

FIGURE 5.1 Canonical form of the recurrent neural network (From G. Dreyfus. Neural networks: Methodology and applications, Springer-Verlag, 2005).

Note that the abovementioned difficulties arise if we consider the recurrent neural networks derived from arbitrary discrete time models. If we consider instead only the networks derived from the discrete time approximations of systems of ODEs in the normal form provided by explicit one-step finite difference methods, then the resulting recurrent neural network architecture immediately has the canonical form (2.13). Also, if we apply an explicit multistep finite difference method, the only transformation required to put the model in the canonical form

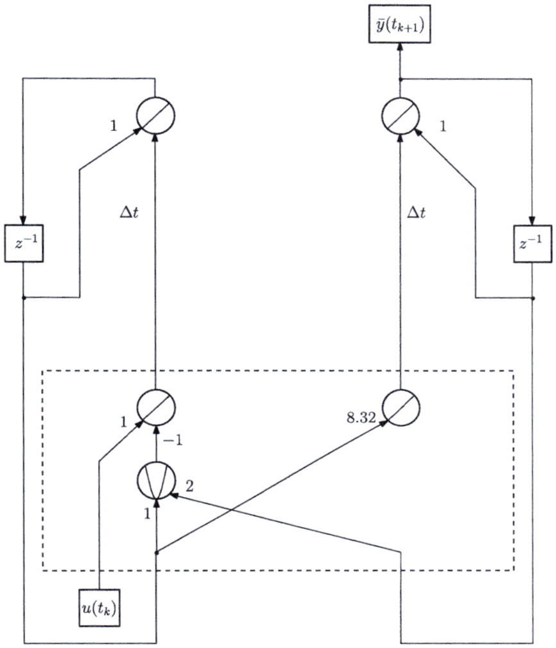

FIGURE 5.2 Canonical form of the initial theoretical model (5.5) discretized by the explicit Euler method.

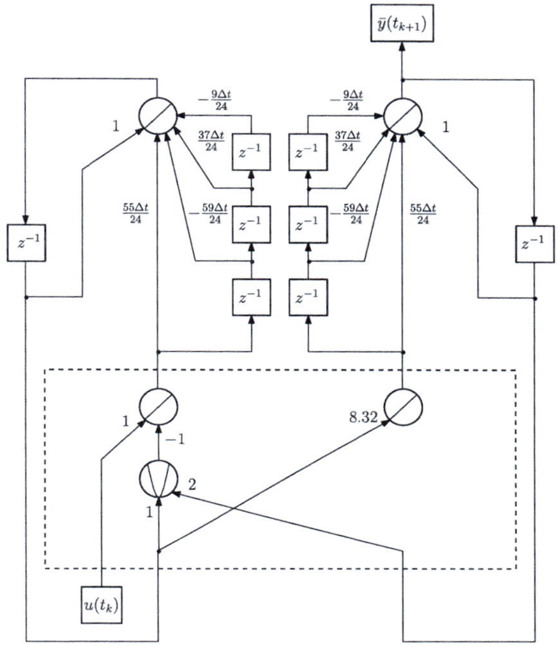

FIGURE 5.3 Canonical form of the initial theoretical model (5.6) discretized by the explicit Adams–Bashforth method.

is the inclusion of additional time delay elements for the ODE right hand side values at previous time instants. Canonical form of the recurrent neural network derived from the discretization of initial theoretical model (5.4) by the Euler method (5.5) and the Adams–Bashforth method (5.6) is shown in Figs. 5.2 and 5.3, respectively.

The following notation is used in Figs. 5.2–5.9: square-shaped elements represent external inputs (control variables), measured outputs, and time delay elements; round-shaped elements represent the neurons; neuron activation functions are depicted schematically within the round-shaped elements (for instance, in this example we use linear, quadratic, and sigmoidal functions); arrows between the elements represent the corresponding input and output connections; arrows without the input element represent neuron biases; values of the biases and connection weights are labeled along the arrows; thick arrows represent the connections with varying weights.

Transition from a system of finite difference equations to a neural network model representation allows us to explicitly preserve the localization of function, inherent to the original continuous time model as well as the corresponding discrete time approximation. This feature of semiempirical models allows us to "freeze" specific parts of the model, i.e., to prohibit the modification of both its structure and the values of its parameters during the training phase. This approach is justified for the models that include some relationships derived from accurate theoretical domain-specific knowledge, that is, there is little doubt concerning these parts of the model. Other parts of the model that might be doubtful (i.e., these parts are suspected to be the main reason of the low accuracy of the

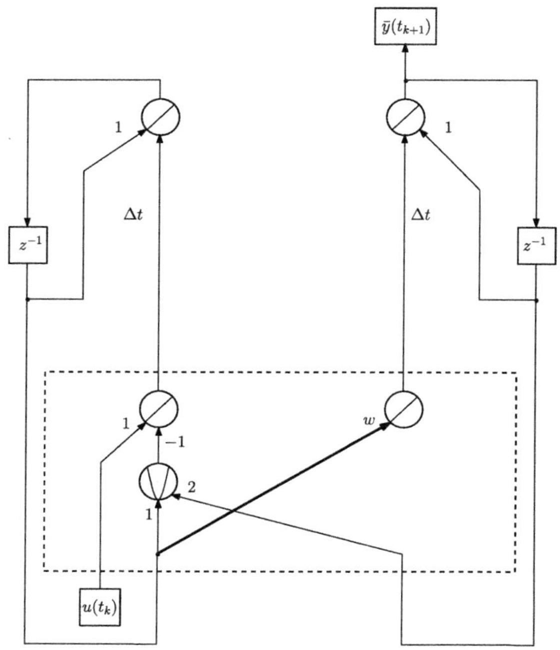

FIGURE 5.4 Canonical form of the semiempirical model (Euler method) with tunable weight.

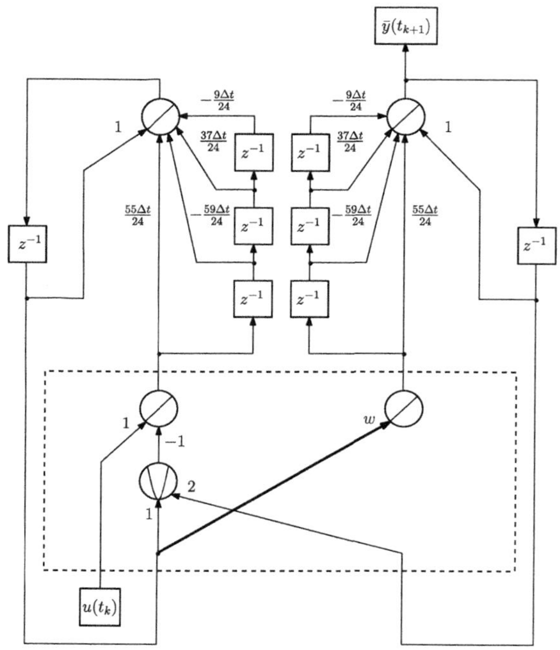

FIGURE 5.5 Canonical form of the semiempirical model (Adams–Bashforth method) with tunable weight.

model), are subject to modification, both parametrical and structural. Thus, the training procedure of semiempirical neural network–based models usually implies the modification of only specific parts of the model, while keeping the other parts unchanged.

Further analysis of the possible reasons for the low accuracy of the initial theoretical model allows us to propose the following hypotheses. As already mentioned, the first equation of the initial theoretical model accurately describes the behavior of $x_1(t)$. However, we lack confidence in the accuracy of the second equation of the model, (5.4); this equation might be the reason for the low accuracy. We propose the following list of possible reasons for this problem:

- the value 8.32 for the parameter in the second equation might be inaccurate, i.e., we should try to adjust this parameter value;
- linear dependence of the right hand side of the second equation on the variable x_1 might be inadequate, i.e., if the adjustment of the parameter value mentioned above does not allow for an accurate solution, we should try to use the nonlinear dependence;
- the right hand side of the second equation might be lacking the dependence on x_2, i.e., if the previous attempts have failed, we should incorporate the dependence on x_2 in the second equation.

In order to test the first of the three hypotheses concerning the reasons for low accuracy of the initial theoretical model, we replace the numerical constant 8.32 in the second equation of (5.4) with a single tunable parameter, namely, the weight w (see Figs. 5.4 and 5.5), and perform the training procedure for the corresponding neural network–based model afterwards.

The initial guess for the parameter w value is taken to be $w^{(0)} = 8.32$. Training is performed using the Levenberg–Marquardt optimization method. The Jacobian of the error function is computed using the RTRL algorithm. Tuning the parameter w this way provides a minor improvement to the model accuracy; the prediction error has decreased from 0.13947 to 0.13593 for the model discretized by the Euler method and from 0.07143 to 0.07104 for the model discretized by the Adams–Bashforth method.

Based on these results, we can conclude that it is impossible to achieve the required model accuracy by keeping the original model structure and tuning values of its parameters; hence we need to modify the model structure itself. The exact form of these structural modifications is suggested by the abovementioned hypotheses concerning the possible reasons for the unsatisfactory accuracy of the model. Structural modification is performed on a modular basis: only specific parts of the model are subject to this modification, while other parts remain unchanged. These parts of the model may be viewed as separate modules which interact with the other parts of the model via the corresponding input and output connections. Implementation details for each module may differ; however, in general, this has no effect on the rest of the model. Structural adjustment of these modules also has a local nature, since they directly affect only the respective parts of the whole model. This modularity feature combined with a step-by-step structural adjustment process significantly simplifies the overall model design procedure.

The second hypothesis from the list above suggests that we should replace the linear dependence of the right hand side of the second equation from (5.4) on the variable x_1 with a nonlinear one. This structural modification is performed by replacing a corresponding single neuron with a linear activation function by a layered feedforward neural network with one hidden layer that includes 10 neurons with a hyperbolic tangent activation function (a suitable number of neurons was selected experimentally). A structural diagram of the corresponding neural network model for the case of the explicit Euler difference method is shown in Fig. 5.6. Modifications of the neural network model for the case of the explicit Adams–Bashforth difference method follow the same pattern.

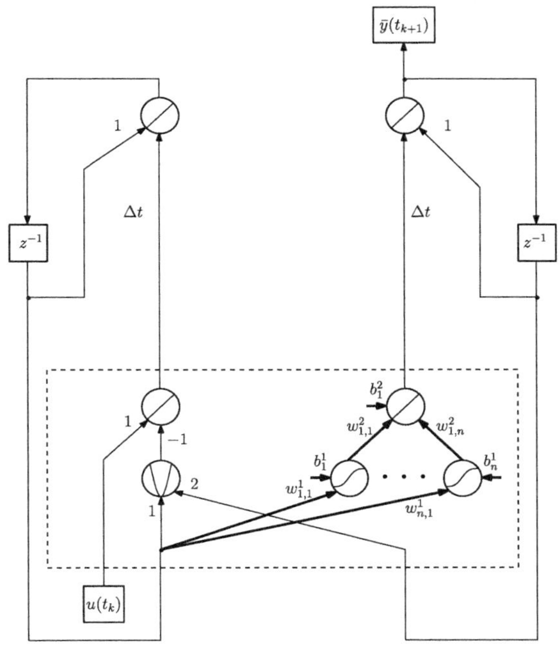

FIGURE 5.6 Canonical form of the semiempirical model (Euler method) with nonlinear dependence on x_1.

These models are also trained using the Levenberg–Marquardt optimization method. Evaluation of the trained models shows that although the introduction of a nonlinear dependence on x_1 into the second equation of (5.4) provides some improvement, the model accuracy still remains unsatisfactory: the prediction error has decreased from 0.13593 to 0.12604 for the model discretized by the Euler method and from 0.07104 to 0.03883 for the model discretized by the Adams–Bashforth method. This means

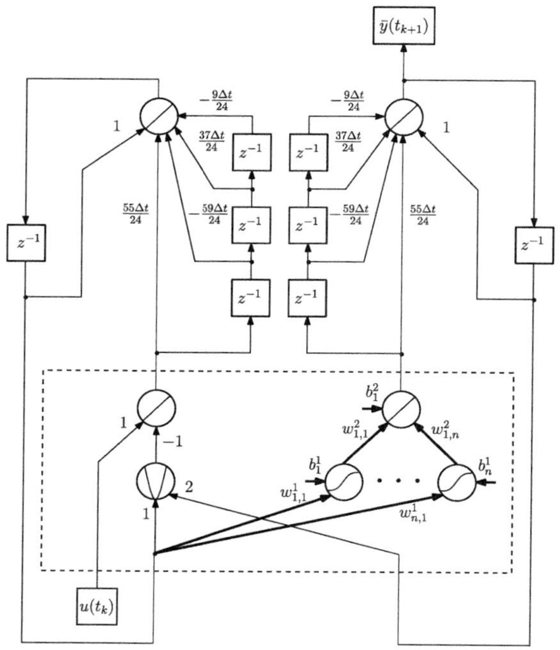

FIGURE 5.7 Canonical form of the semiempirical model (Adams–Bashforth method) with nonlinear dependence on x_1.

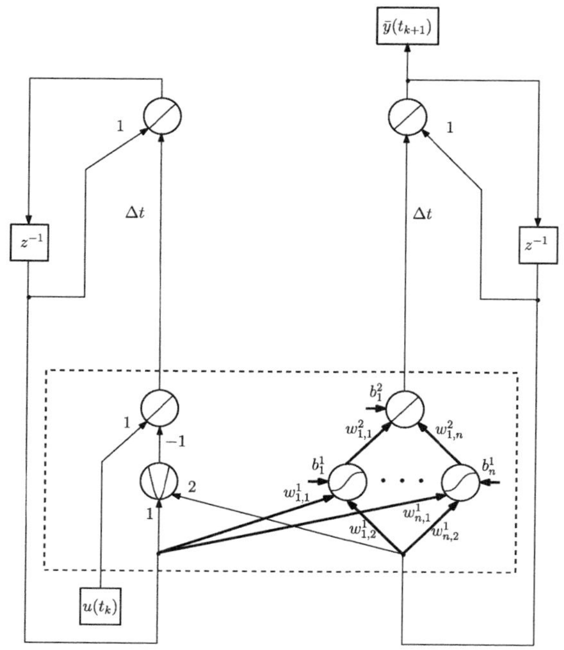

FIGURE 5.8 Canonical form of the semiempirical model (Euler method) with additional dependence on x_2.

that the inclusion of nonlinear dependence of the right hand side of the second equation from (5.4) on x_1 is not enough to achieve the required accuracy. Therefore we should proceed to the third hypothesis and include the dependence on x_2 as well. This dependence is included by the appropriate structural modification of the neural network model, namely, by adding the connections from the element that represents x_2 to the hidden layer neurons of the layered feedforward neural network module. A structural diagram of the corresponding neural network model for the case of the explicit Euler difference method is shown in Fig. 5.8. Modification of the neural network model for the case of the explicit Adams–Bashforth difference method is performed in a similar way.

Introduction of the additional dependence on x_2 along with the dependence on x_1 into the second equation of (5.4) has allowed us to achieve the required model accuracy. Evaluation of the trained model gives a prediction error of 0.01394 for the model discretized by the Euler method and 0.01219 for the model discretized by the Adams–Bashforth method.

Next, we will compare the semiempirical model results with results achieved by the purely empirical NARX model. In our computational experiments the best accuracy (of 0.02821) was achieved by a NARX model with three hidden layer neurons and five delayed feedback inputs. These results provide clear evidence of the superiority of semiempirical models over purely empirical ones in terms of the generalization ability: even in the case of a first-order Euler method the semiempirical model achieves a prediction error of 0.01394, as compared to an error of 0.02821 for the NARX model; in the case of a fourth-order Adams–Bashforth method the ac-

TABLE 5.1 Model prediction error for each stage of model design.

	Theory	ANN-1	ANN-2	ANN-3	Empir
Euler	0.13947	0.13593	0.12604	0.01394	–
Adams–Bashforth	0.07143	0.07104	0.03883	0.01219	–
NARX	–	–	–	–	0.02821

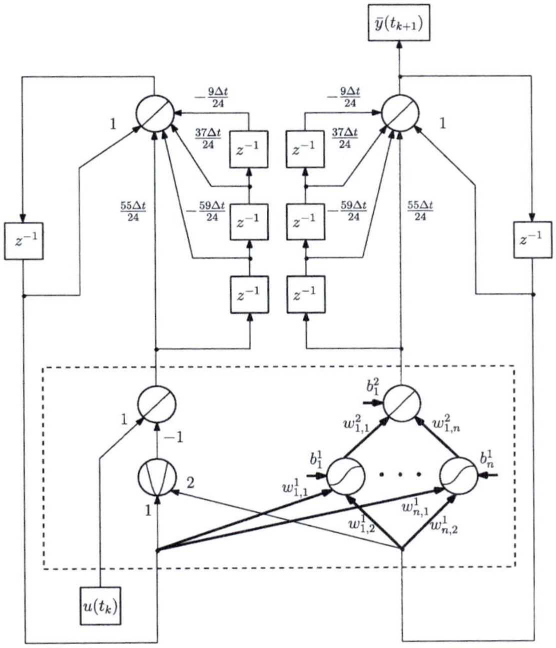

FIGURE 5.9 Canonical form of the semiempirical model (Adams–Bashforth method) with additional dependence on x_2.

curacy of the semiempirical model is even better – the prediction error equals 0.01219, hence we might expect even better results for the more sophisticated numerical methods.

Computational experimental results for each modeling stage are presented in Table 5.1. The following abbreviations are used in this table: "Theory" signifies results for the initial theoretical model (5.4); ANN-1, ANN-2, ANN-3 denote results for the semiempirical model after the first, second, and third modification stages; "Empir" signifies the best results for the empirical NARX model of the system.

5.3 SEMIEMPIRICAL ANN-BASED MODEL DERIVATIVES COMPUTATION

Semiempirical neural network–based models of the form (5.2) are continuous time models, in contrast to discrete time purely empirical models, described in Chapter 2; hence the training methods for these models are also formulated in continuous time. Despite the fact that the actual implementation of these algorithms requires the appropriate finite difference approximations of the ODEs, the continuous time algorithm versions provide an additional flexibility in the choice of the most suitable finite difference method. The total error function $\bar{E}\colon \mathbb{R}^{n_w} \to \mathbb{R}$ evaluated on the training set of the form (5.1) is a sum of errors $E^{(p)}\colon \mathbb{R}^{n_w} \to \mathbb{R}$ for its individual trajectories:

$$\bar{E}(\mathbf{w}) = \sum_{p=1}^{P} E^{(p)}(\mathbf{w}). \tag{5.7}$$

The individual errors $E^{(p)}$ have the following form:

$$E^{(p)}(\mathbf{w}) = \int_{0}^{\bar{t}^{(p)}} e(\tilde{\mathbf{y}}^{(p)}(t), \hat{\mathbf{x}}^{(p)}(t, \mathbf{w}), \mathbf{w})\, dt, \tag{5.8}$$

where $\bar{t}^{(p)} = t_{K^{(p)}}$ are the time segment durations, $\tilde{\mathbf{y}}^{(p)}(t)$ are the target values of observable outputs, $\hat{\mathbf{x}}^{(p)}(t, \mathbf{w})$ are the model states, and $e^{(p)}\colon \mathbb{R}^{n_y} \times \mathbb{R}^{n_x} \times \mathbb{R}^{n_w} \to \mathbb{R}$ represent the model prediction error at time instant t. Under the usual assumptions of the additive Gaussian white measurement noise, it seems reasonable to utilize the instantaneous error function e of the following form:

$$e(\tilde{\mathbf{y}}, \hat{\mathbf{x}}, \mathbf{w}) = \frac{1}{2} \left(\tilde{\mathbf{y}} - \hat{\mathbf{g}}(\hat{\mathbf{x}}, \mathbf{w})\right)^T \boldsymbol{\Omega} \left(\tilde{\mathbf{y}} - \hat{\mathbf{g}}(\hat{\mathbf{x}}, \mathbf{w})\right), \tag{5.9}$$

where $\boldsymbol{\Omega} = \operatorname{diag}(\omega_1, \ldots, \omega_{n_y})$ is the diagonal matrix of error weights, usually taken inversely proportional to the corresponding variances of measurement noise.

The error function (5.7) has to be minimized with respect to the semiempirical neural network–based model parameters $\mathbf{w} \in \mathbb{R}^{n_w}$. The minimization of the error function may be performed using various numerical optimization methods mentioned before, such as the Levenberg–Marquardt method. Hence, we assume the error function to be twice continuously differentiable with respect to all tunable parameters.

Since the error function (5.7) is a summation of errors for each individual trajectory, its gradient as well as its Hessian may also be represented as summations of gradients and Hessians of these errors, i.e.,

$$\frac{\partial \bar{E}(\mathbf{w})}{\partial \mathbf{w}} = \sum_{p=1}^{P} \frac{\partial E^{(p)}(\mathbf{w})}{\partial \mathbf{w}}, \tag{5.10}$$

$$\frac{\partial^2 \bar{E}(\mathbf{w})}{\partial \mathbf{w}^2} = \sum_{p=1}^{P} \frac{\partial^2 E^{(p)}(\mathbf{w})}{\partial \mathbf{w}^2}. \tag{5.11}$$

Thus, all the algorithms presented below describe the computation of derivatives for these individual errors, and the trajectory index p is omitted.

Just like in the discrete time case, there are two methods for computation of error function derivatives: one method operates forward-in-time, and the other operates backward-in-time. We begin with a description of a continuous time counterpart of a forward-in-time RTRL algorithm.

The gradient of the individual trajectory error function (5.8) equals

$$\frac{\partial E(\mathbf{w})}{\partial \mathbf{w}} = \int_0^{\bar{t}} \frac{\partial e(\tilde{\mathbf{y}}(t), \hat{\mathbf{x}}(t, \mathbf{w}), \mathbf{w})}{\partial \mathbf{w}} + \frac{\partial \hat{\mathbf{x}}(t, \mathbf{w})}{\partial \mathbf{w}}^T \frac{\partial e(\tilde{\mathbf{y}}(t), \hat{\mathbf{x}}(t, \mathbf{w}), \mathbf{w})}{\partial \hat{\mathbf{x}}} dt, \tag{5.12}$$

and its Hessian equals

$$\begin{aligned}\frac{\partial^2 E(\mathbf{w})}{\partial \mathbf{w}^2} = \int_0^{\bar{t}} & \frac{\partial^2 e(\tilde{\mathbf{y}}(t), \hat{\mathbf{x}}(t, \mathbf{w}), \mathbf{w})}{\partial \mathbf{w}^2} \\ & + \frac{\partial^2 e(\tilde{\mathbf{y}}(t), \hat{\mathbf{x}}(t, \mathbf{w}), \mathbf{w})}{\partial \mathbf{w} \partial \hat{\mathbf{x}}} \frac{\partial \hat{\mathbf{x}}(t, \mathbf{w})}{\partial \mathbf{w}} \\ & + \frac{\partial \hat{\mathbf{x}}(t, \mathbf{w})}{\partial \mathbf{w}}^T \frac{\partial^2 e(\tilde{\mathbf{y}}(t), \hat{\mathbf{x}}(t, \mathbf{w}), \mathbf{w})}{\partial \hat{\mathbf{x}} \partial \mathbf{w}} \\ & + \frac{\partial \hat{\mathbf{x}}(t, \mathbf{w})}{\partial \mathbf{w}}^T \frac{\partial^2 e(\tilde{\mathbf{y}}(t), \hat{\mathbf{x}}(t, \mathbf{w}), \mathbf{w})}{\partial \hat{\mathbf{x}}^2} \frac{\partial \hat{\mathbf{x}}(t, \mathbf{w})}{\partial \mathbf{w}} \\ & + \sum_{i=1}^{n_x} \frac{\partial e(\tilde{\mathbf{y}}(t), \hat{\mathbf{x}}(t, \mathbf{w}), \mathbf{w})}{\partial \hat{\mathbf{x}}_i} \frac{\partial^2 \hat{\mathbf{x}}_i(t, \mathbf{w})}{\partial \mathbf{w}^2} dt.\end{aligned} \tag{5.13}$$

First-order derivatives of the instantaneous error function (5.9) have the form

$$\begin{aligned}\frac{\partial e(\tilde{\mathbf{y}}, \hat{\mathbf{x}}, \mathbf{w})}{\partial \mathbf{w}} &= -\frac{\partial \hat{\mathbf{g}}(\hat{\mathbf{x}}, \mathbf{w})}{\partial \mathbf{w}}^T \boldsymbol{\Omega} \left(\tilde{\mathbf{y}} - \hat{\mathbf{g}}(\hat{\mathbf{x}}, \mathbf{w})\right), \\ \frac{\partial e(\tilde{\mathbf{y}}, \hat{\mathbf{x}}, \mathbf{w})}{\partial \hat{\mathbf{x}}} &= -\frac{\partial \hat{\mathbf{g}}(\hat{\mathbf{x}}, \mathbf{w})}{\partial \hat{\mathbf{x}}}^T \boldsymbol{\Omega} \left(\tilde{\mathbf{y}} - \hat{\mathbf{g}}(\hat{\mathbf{x}}, \mathbf{w})\right),\end{aligned} \tag{5.14}$$

and its second-order derivatives are

$$\frac{\partial^2 e(\tilde{\mathbf{y}}, \hat{\mathbf{x}}, \mathbf{w})}{\partial \mathbf{w}^2} = \frac{\partial \hat{\mathbf{g}}(\hat{\mathbf{x}}, \mathbf{w})}{\partial \mathbf{w}}^T \boldsymbol{\Omega} \frac{\partial \hat{\mathbf{g}}(\hat{\mathbf{x}}, \mathbf{w})}{\partial \mathbf{w}} - \sum_{i=1}^{n_y} \frac{\partial^2 \hat{\mathbf{g}}_i(\hat{\mathbf{x}}, \mathbf{w})}{\partial \mathbf{w}^2} \omega_i \left(\tilde{\mathbf{y}}_i - \hat{\mathbf{g}}_i(\hat{\mathbf{x}}, \mathbf{w})\right),$$

$$\frac{\partial^2 e(\tilde{\mathbf{y}}, \hat{\mathbf{x}}, \mathbf{w})}{\partial \mathbf{w} \partial \hat{\mathbf{x}}} = \frac{\partial \hat{\mathbf{g}}(\hat{\mathbf{x}}, \mathbf{w})}{\partial \mathbf{w}}^T \boldsymbol{\Omega} \frac{\partial \hat{\mathbf{g}}(\hat{\mathbf{x}}, \mathbf{w})}{\partial \hat{\mathbf{x}}} - \sum_{i=1}^{n_y} \frac{\partial^2 \hat{\mathbf{g}}_i(\hat{\mathbf{x}}, \mathbf{w})}{\partial \mathbf{w} \partial \hat{\mathbf{x}}} \omega_i \left(\tilde{\mathbf{y}}_i - \hat{\mathbf{g}}_i(\hat{\mathbf{x}}, \mathbf{w})\right),$$

$$\frac{\partial^2 e(\tilde{\mathbf{y}}, \hat{\mathbf{x}}, \mathbf{w})}{\partial \hat{\mathbf{x}}^2} = \frac{\partial \hat{\mathbf{g}}(\hat{\mathbf{x}}, \mathbf{w})}{\partial \hat{\mathbf{x}}}^T \boldsymbol{\Omega} \frac{\partial \hat{\mathbf{g}}(\hat{\mathbf{x}}, \mathbf{w})}{\partial \hat{\mathbf{x}}} - \sum_{i=1}^{n_y} \frac{\partial^2 \hat{\mathbf{g}}_i(\hat{\mathbf{x}}, \mathbf{w})}{\partial \hat{\mathbf{x}}^2} \omega_i \left(\tilde{\mathbf{y}}_i - \hat{\mathbf{g}}_i(\hat{\mathbf{x}}, \mathbf{w})\right). \tag{5.15}$$

Recall that according to Schwarz's theorem on equality of mixed partials,

$$\frac{\partial^2 e(\tilde{\mathbf{y}}, \hat{\mathbf{x}}, \mathbf{w})}{\partial \hat{\mathbf{x}} \partial \mathbf{w}} = \frac{\partial^2 e(\tilde{\mathbf{y}}, \hat{\mathbf{x}}, \mathbf{w})}{\partial \mathbf{w} \partial \hat{\mathbf{x}}}^T.$$

Also, the individual trajectory error function gradient (5.12) depends on first-order sensitivities $\frac{\partial \hat{\mathbf{x}}(t,\mathbf{w})}{\partial \mathbf{w}}$, and the Hessian (5.13) additionally depends on second-order sensitivities

$$\left\{ \frac{\partial^2 \hat{\mathbf{x}}_i(t, \mathbf{w})}{\partial \mathbf{w}^2} \right\}_{i=1}^{n_x}.$$

Differentiating the initial value problem for the semiempirical model (5.2) with respect to parameters $\mathbf{w}$ yields the initial value problem for the first-order sensitivities, i.e.,

$$\frac{\partial \hat{\mathbf{x}}(0, \mathbf{w})}{\partial \mathbf{w}} = \mathbf{0},$$
$$\frac{d}{dt} \frac{\partial \hat{\mathbf{x}}(t, \mathbf{w})}{\partial \mathbf{w}} = \frac{\partial \hat{\mathbf{f}}(\hat{\mathbf{x}}(t, \mathbf{w}), \mathbf{u}(t), \mathbf{w})}{\partial \mathbf{w}} + \frac{\partial \hat{\mathbf{f}}(\hat{\mathbf{x}}(t, \mathbf{w}), \mathbf{u}(t), \mathbf{w})}{\partial \hat{\mathbf{x}}} \frac{\partial \hat{\mathbf{x}}(t, \mathbf{w})}{\partial \mathbf{w}}. \tag{5.16}$$

Similarly, the second-order sensitivities are represented by the solution of the following initial value problem for $i = 1, \ldots, n_x$:

$$\frac{\partial^2 \hat{\mathbf{x}}_i(0, \mathbf{w})}{\partial \mathbf{w}^2} = \mathbf{0},$$
$$\begin{aligned}\frac{d}{dt} \frac{\partial^2 \hat{\mathbf{x}}_i(t, \mathbf{w})}{\partial \mathbf{w}^2} &= \frac{\partial^2 \hat{\mathbf{f}}_i(\hat{\mathbf{x}}(t, \mathbf{w}), \mathbf{u}(t), \mathbf{w})}{\partial \mathbf{w}^2} + \frac{\partial^2 \hat{\mathbf{f}}_i(\hat{\mathbf{x}}(t, \mathbf{w}), \mathbf{u}(t), \mathbf{w})}{\partial \mathbf{w} \partial \hat{\mathbf{x}}} \frac{\partial \hat{\mathbf{x}}(t, \mathbf{w})}{\partial \mathbf{w}} \\ &+ \frac{\partial \hat{\mathbf{x}}(t, \mathbf{w})}{\partial \mathbf{w}}^T \frac{\partial^2 \hat{\mathbf{f}}_i(\hat{\mathbf{x}}(t, \mathbf{w}), \mathbf{u}(t), \mathbf{w})}{\partial \hat{\mathbf{x}} \partial \mathbf{w}} \\ &+ \frac{\partial \hat{\mathbf{x}}(t, \mathbf{w})}{\partial \mathbf{w}}^T \frac{\partial^2 \hat{\mathbf{f}}_i(\hat{\mathbf{x}}(t, \mathbf{w}), \mathbf{u}(t), \mathbf{w})}{\partial \hat{\mathbf{x}}^2} \frac{\partial \hat{\mathbf{x}}(t, \mathbf{w})}{\partial \mathbf{w}} \\ &+ \sum_{j=1}^{n_x} \frac{\partial \hat{\mathbf{f}}_i(\hat{\mathbf{x}}(t, \mathbf{w}), \mathbf{u}(t), \mathbf{w})}{\partial \hat{\mathbf{x}}_j} \frac{\partial^2 \hat{\mathbf{x}}_j(t, \mathbf{w})}{\partial \mathbf{w}^2}.\end{aligned} \tag{5.17}$$

A computationally cheaper Gauss–Newton Hessian approximation may be obtained by discarding the second-order terms in (5.13), i.e.,

$$\begin{aligned}\frac{\partial^2 E(\mathbf{w})}{\partial \mathbf{w}^2} \approx \int_0^{\bar{t}} &\frac{\partial^2 e(\tilde{\mathbf{y}}(t), \hat{\mathbf{x}}(t, \mathbf{w}), \mathbf{w})}{\partial \mathbf{w}^2} + \frac{\partial^2 e(\tilde{\mathbf{y}}(t), \hat{\mathbf{x}}(t, \mathbf{w}), \mathbf{w})}{\partial \mathbf{w} \partial \hat{\mathbf{x}}} \frac{\partial \hat{\mathbf{x}}(t, \mathbf{w})}{\partial \mathbf{w}} \\ &+ \frac{\partial \hat{\mathbf{x}}(t, \mathbf{w})}{\partial \mathbf{w}}^T \frac{\partial^2 e(\tilde{\mathbf{y}}(t), \hat{\mathbf{x}}(t, \mathbf{w}), \mathbf{w})}{\partial \hat{\mathbf{x}} \partial \mathbf{w}} \\ &+ \frac{\partial \hat{\mathbf{x}}(t, \mathbf{w})}{\partial \mathbf{w}}^T \frac{\partial^2 e(\tilde{\mathbf{y}}(t), \hat{\mathbf{x}}(t, \mathbf{w}), \mathbf{w})}{\partial \hat{\mathbf{x}}^2} \frac{\partial \hat{\mathbf{x}}(t, \mathbf{w})}{\partial \mathbf{w}} dt,\end{aligned} \tag{5.18}$$

as well as in (5.15), i.e.,

$$\begin{aligned}
\frac{\partial^2 e(\tilde{\mathbf{y}}, \hat{\mathbf{x}}, \mathbf{w})}{\partial \mathbf{w}^2} &\approx \frac{\partial \hat{\mathbf{g}}(\hat{\mathbf{x}}, \mathbf{w})}{\partial \mathbf{w}}^T \boldsymbol{\Omega} \frac{\partial \hat{\mathbf{g}}(\hat{\mathbf{x}}, \mathbf{w})}{\partial \mathbf{w}}, \\
\frac{\partial^2 e(\tilde{\mathbf{y}}, \hat{\mathbf{x}}, \mathbf{w})}{\partial \mathbf{w} \partial \hat{\mathbf{x}}} &\approx \frac{\partial \hat{\mathbf{g}}(\hat{\mathbf{x}}, \mathbf{w})}{\partial \mathbf{w}}^T \boldsymbol{\Omega} \frac{\partial \hat{\mathbf{g}}(\hat{\mathbf{x}}, \mathbf{w})}{\partial \hat{\mathbf{x}}}, \\
\frac{\partial^2 e(\tilde{\mathbf{y}}, \hat{\mathbf{x}}, \mathbf{w})}{\partial \hat{\mathbf{x}}^2} &\approx \frac{\partial \hat{\mathbf{g}}(\hat{\mathbf{x}}, \mathbf{w})}{\partial \hat{\mathbf{x}}}^T \boldsymbol{\Omega} \frac{\partial \hat{\mathbf{g}}(\hat{\mathbf{x}}, \mathbf{w})}{\partial \hat{\mathbf{x}}}.
\end{aligned} \tag{5.19}$$

Now, we describe a continuous time counterpart of a backward-in-time BPTT algorithm. This method is based on the application of *adjoint sensitivity analysis* [22–24]. First, we treat the set of ODEs that describe the semiempirical model dynamics (5.2) as a set of constraints. Next, we include them into the original individual trajectory error function along with the Lagrange multipliers $\boldsymbol{\lambda}$ to obtain the following Lagrange function:

$$\begin{aligned}
L(\mathbf{w}) = \int_0^{\bar{t}} & e(\tilde{\mathbf{y}}(t), \hat{\mathbf{x}}(t, \mathbf{w}), \mathbf{w}) \\
& - \boldsymbol{\lambda}(t, \mathbf{w})^T \left[\frac{d\hat{\mathbf{x}}(t, \mathbf{w})}{dt} - \hat{\mathbf{f}}(\hat{\mathbf{x}}(t, \mathbf{w}), \mathbf{u}(t), \mathbf{w}) \right] dt.
\end{aligned} \tag{5.20}$$

Note that since the state variable predictions $\hat{\mathbf{x}}$ are in fact computed by solving the initial value problem for the ODE system (5.2), these newly imposed constraints are satisfied on the whole time segment, and the second term of (5.20) is identically zero. Hence, the gradient and Hessian of the individual trajectory error function are identical to the gradient and Hessian of the Lagrange function. The Lagrange function gradient equals

$$\begin{aligned}
\frac{\partial L(\mathbf{w})}{\partial \mathbf{w}} = \int_0^{\bar{t}} & \frac{\partial e(\tilde{\mathbf{y}}(t), \hat{\mathbf{x}}(t, \mathbf{w}), \mathbf{w})}{\partial \mathbf{w}} \\
& + \frac{\partial \hat{\mathbf{x}}(t, \mathbf{w})}{\partial \mathbf{w}}^T \frac{\partial e(\tilde{\mathbf{y}}(t), \hat{\mathbf{x}}(t, \mathbf{w}), \mathbf{w})}{\partial \hat{\mathbf{x}}} \\
& - \left[\frac{d}{dt} \frac{\partial \hat{\mathbf{x}}(t, \mathbf{w})}{\partial \mathbf{w}} - \frac{\partial \hat{\mathbf{f}}(\hat{\mathbf{x}}(t, \mathbf{w}), \mathbf{u}(t), \mathbf{w})}{\partial \mathbf{w}} \right. \\
& \left. - \frac{\partial \hat{\mathbf{f}}(\hat{\mathbf{x}}(t, \mathbf{w}), \mathbf{u}(t), \mathbf{w})}{\partial \hat{\mathbf{x}}} \frac{\partial \hat{\mathbf{x}}(t, \mathbf{w})}{\partial \mathbf{w}} \right]^T \boldsymbol{\lambda}(t, \mathbf{w}) \\
& - \frac{\partial \boldsymbol{\lambda}(t, \mathbf{w})}{\partial \mathbf{w}}^T \left[\frac{d\hat{\mathbf{x}}(t, \mathbf{w})}{dt} - \hat{\mathbf{f}}(\hat{\mathbf{x}}(t, \mathbf{w}), \mathbf{u}(t), \mathbf{w}) \right] dt.
\end{aligned} \tag{5.21}$$

Integrating by parts, we obtain

$$\begin{aligned}
\int_0^{\bar{t}} & \frac{d}{dt} \frac{\partial \hat{\mathbf{x}}(t, \mathbf{w})}{\partial \mathbf{w}}^T \boldsymbol{\lambda}(t, \mathbf{w}) \, dt \\
&= \frac{\partial \hat{\mathbf{x}}(\bar{t}, \mathbf{w})}{\partial \mathbf{w}}^T \boldsymbol{\lambda}(\bar{t}, \mathbf{w}) - \frac{\partial \hat{\mathbf{x}}(0, \mathbf{w})}{\partial \mathbf{w}}^T \boldsymbol{\lambda}(0, \mathbf{w}) \\
&\quad - \int_0^{\bar{t}} \frac{\partial \hat{\mathbf{x}}(t, \mathbf{w})}{\partial \mathbf{w}}^T \frac{d\boldsymbol{\lambda}(t, \mathbf{w})}{dt} \, dt.
\end{aligned} \tag{5.22}$$

If we substitute (5.22) into (5.21) and exclude the term $\left[\frac{d\hat{\mathbf{x}}(t, \mathbf{w})}{dt} - \hat{\mathbf{f}}(\hat{\mathbf{x}}(t, \mathbf{w}), \mathbf{u}(t), \mathbf{w})\right]$, which is identically zero, we get

$$\begin{aligned}
\frac{\partial L(\mathbf{w})}{\partial \mathbf{w}} = \int_0^{\bar{t}} & \frac{\partial e(\tilde{\mathbf{y}}(t), \hat{\mathbf{x}}(t, \mathbf{w}), \mathbf{w})}{\partial \mathbf{w}} \\
& + \frac{\partial \hat{\mathbf{f}}(\hat{\mathbf{x}}(t, \mathbf{w}), \mathbf{u}(t), \mathbf{w})}{\partial \mathbf{w}}^T \boldsymbol{\lambda}(t, \mathbf{w}) \\
& + \frac{\partial \hat{\mathbf{x}}(t, \mathbf{w})}{\partial \mathbf{w}}^T \left[\frac{d\boldsymbol{\lambda}(t, \mathbf{w})}{dt} \right. \\
& + \frac{\partial \hat{\mathbf{f}}(\hat{\mathbf{x}}(t, \mathbf{w}), \mathbf{u}(t), \mathbf{w})}{\partial \hat{\mathbf{x}}}^T \boldsymbol{\lambda}(t, \mathbf{w}) \\
& \left. + \frac{\partial e(\tilde{\mathbf{y}}(t), \hat{\mathbf{x}}(t, \mathbf{w}), \mathbf{w})}{\partial \hat{\mathbf{x}}} \right] dt \\
& + \frac{\partial \hat{\mathbf{x}}(0, \mathbf{w})}{\partial \mathbf{w}}^T \boldsymbol{\lambda}(0, \mathbf{w}) - \frac{\partial \hat{\mathbf{x}}(\bar{t}, \mathbf{w})}{\partial \mathbf{w}}^T \boldsymbol{\lambda}(\bar{t}, \mathbf{w}).
\end{aligned} \tag{5.23}$$

Again, since the artificial constraints are satisfied on the whole time segment, the Lagrange multipliers are completely arbitrary. In order to simplify (5.23), we define them as

$$\begin{aligned}\boldsymbol{\lambda}(\bar{t}, \mathbf{w}) &= \mathbf{0},\\ \frac{d\boldsymbol{\lambda}(t, \mathbf{w})}{dt} &= -\frac{\partial \hat{\mathbf{f}}(\hat{\mathbf{x}}(t, \mathbf{w}), \mathbf{u}(t), \mathbf{w})}{\partial \hat{\mathbf{x}}}^T \boldsymbol{\lambda}(t, \mathbf{w}) \\ &\quad - \frac{\partial e(\tilde{\mathbf{y}}(t), \hat{\mathbf{x}}(t, \mathbf{w}), \mathbf{w})}{\partial \hat{\mathbf{x}}}.\end{aligned} \tag{5.24}$$

The ODEs in (5.24) are called the adjoint equations. Note that this initial value problem is to be solved in reverse time. Further simplification of (5.23) comes from the fact that the initial value of state variables does not depend on parameters $\mathbf{w}$; hence $\frac{\partial \hat{\mathbf{x}}(0, \mathbf{w})}{\partial \mathbf{w}} \equiv \mathbf{0}$. Finally, we have

$$\begin{aligned}\frac{\partial E(\mathbf{w})}{\partial \mathbf{w}} &= \frac{\partial L(\mathbf{w})}{\partial \mathbf{w}} \\ &= \int_0^{\bar{t}} \frac{\partial e(\tilde{\mathbf{y}}(t), \hat{\mathbf{x}}(t, \mathbf{w}), \mathbf{w})}{\partial \mathbf{w}} \\ &\quad + \frac{\partial \hat{\mathbf{f}}(\hat{\mathbf{x}}(t, \mathbf{w}), \mathbf{u}(t), \mathbf{w})}{\partial \mathbf{w}}^T \boldsymbol{\lambda}(t, \mathbf{w})\, dt.\end{aligned} \tag{5.25}$$

Then, the Hessian of the Lagrange function equals

$$\begin{aligned}\frac{\partial^2 E(\mathbf{w})}{\partial \mathbf{w}^2} &= \frac{\partial^2 L(\mathbf{w})}{\partial \mathbf{w}^2} \\ &= \int_0^{\bar{t}} \frac{\partial^2 e(\tilde{\mathbf{y}}(t), \hat{\mathbf{x}}(t, \mathbf{w}), \mathbf{w})}{\partial \mathbf{w}^2} \\ &\quad + \frac{\partial \hat{\mathbf{f}}(\hat{\mathbf{x}}(t, \mathbf{w}), \mathbf{u}(t), \mathbf{w})}{\partial \mathbf{w}}^T \frac{\partial \boldsymbol{\lambda}(t, \mathbf{w})}{\partial \mathbf{w}} \\ &\quad + \sum_{i=1}^{n_x} \lambda_i(t, \mathbf{w}) \frac{\partial^2 \hat{\mathbf{f}}_i(\hat{\mathbf{x}}(t, \mathbf{w}), \mathbf{u}(t), \mathbf{w})}{\partial \mathbf{w}^2} \\ &\quad + \left[\frac{\partial^2 e(\tilde{\mathbf{y}}(t), \hat{\mathbf{x}}(t, \mathbf{w}), \mathbf{w})}{\partial \mathbf{w} \partial \hat{\mathbf{x}}} \right. \\ &\quad \left. + \sum_{i=1}^{n_x} \lambda_i(t, \mathbf{w}) \frac{\partial^2 \hat{\mathbf{f}}_i(\hat{\mathbf{x}}(t, \mathbf{w}), \mathbf{u}(t), \mathbf{w})}{\partial \mathbf{w} \partial \hat{\mathbf{x}}} \right] \frac{\partial \hat{\mathbf{x}}(t, \mathbf{w})}{\partial \mathbf{w}}\, dt.\end{aligned} \tag{5.26}$$

Differentiating the initial value problem for the first-order sensitivities (5.24) with respect to parameters $\mathbf{w}$ yields the initial value problem for the second-order sensitivities, i.e.,

$$\begin{aligned}\frac{\partial \boldsymbol{\lambda}(\bar{t}, \mathbf{w})}{\partial \mathbf{w}} &= \mathbf{0},\\ \frac{d}{dt}\frac{\partial \boldsymbol{\lambda}(t, \mathbf{w})}{\partial \mathbf{w}} &= -\frac{\partial^2 e(\tilde{\mathbf{y}}(t), \hat{\mathbf{x}}(t, \mathbf{w}), \mathbf{w})}{\partial \hat{\mathbf{x}} \partial \mathbf{w}} \\ &\quad - \frac{\partial \hat{\mathbf{f}}(\hat{\mathbf{x}}(t, \mathbf{w}), \mathbf{u}(t), \mathbf{w})}{\partial \hat{\mathbf{x}}}^T \frac{\partial \boldsymbol{\lambda}(t, \mathbf{w})}{\partial \mathbf{w}} \\ &\quad - \sum_{i=1}^{n_x} \lambda_i(t, \mathbf{w}) \frac{\partial^2 \hat{\mathbf{f}}_i(\hat{\mathbf{x}}(t, \mathbf{w}), \mathbf{u}(t), \mathbf{w})}{\partial \hat{\mathbf{x}} \partial \mathbf{w}} \\ &\quad - \left[\frac{\partial^2 e(\tilde{\mathbf{y}}(t), \hat{\mathbf{x}}(t, \mathbf{w}), \mathbf{w})}{\partial \hat{\mathbf{x}}^2} \right. \\ &\quad \left. + \sum_{i=1}^{n_x} \lambda_i(t, \mathbf{w}) \frac{\partial^2 \hat{\mathbf{f}}_i(\hat{\mathbf{x}}(t, \mathbf{w}), \mathbf{u}(t), \mathbf{w})}{\partial \hat{\mathbf{x}}^2} \right] \frac{\partial \hat{\mathbf{x}}(t, \mathbf{w})}{\partial \mathbf{w}}.\end{aligned} \tag{5.27}$$

Again, the initial value problem (5.27) is to be solved in reverse time. Thus, in order to compute the gradient and Hessian of the Lagrange function, we first need to solve the initial value problems for (5.2) and (5.16) forward-in-time, next we need to solve the initial value problems for adjoint equations (5.24) and (5.27) in reverse time, and finally we need to integrate (5.25) and (5.26).

In the above equations, expressions for derivatives of vector-valued functions $\hat{\mathbf{f}}$ and $\hat{\mathbf{g}}$ with respect to state variables $\hat{\mathbf{x}}$ as well as parameters $\mathbf{w}$ are assumed to be known. In the case these vector-valued functions are represented by layered feedforward neural networks, the corresponding derivatives are presented in Chapter 2. Otherwise, if they are represented by more general semiempirical models, their derivatives at required points may be computed by utilizing the *algorithmic differentiation* technique [25]. Note that the Hessian-vector products of the form

$$\sum_{i=1}^{n_x} \boldsymbol{\lambda}_i(t, \mathbf{w}) \frac{\partial^2 \hat{\mathbf{f}}_i(\hat{\mathbf{x}}(t, \mathbf{w}), \mathbf{u}(t), \mathbf{w})}{\partial \mathbf{w}^2}$$

can be computed significantly faster by a combination of forward and reverse sweeps, in contrast to a straightforward approach that relies on evaluation of a full second derivative tensor

$$\left\{ \frac{\partial^2 \hat{\mathbf{f}}_i(\hat{\mathbf{x}}(t, \mathbf{w}), \mathbf{u}(t), \mathbf{w})}{\partial \mathbf{w}^2} \right\}_{i=1}^{n_x}$$

by forward sweeps. The abovementioned algorithmic differentiation methods are implemented in a number of software packages, such as CppAD [26] and ADOL-C [27].

The continuous time versions of RTRL and BPTT algorithms share the same advantages and disadvantages as their discrete time counterparts. Namely, the backward-in-time method is more algorithmically efficient since it does not require the computation of second-order sensitivities

$$\left\{ \frac{\partial^2 \hat{\mathbf{x}}_i(t, \mathbf{w})}{\partial \mathbf{w}^2} \right\}_{i=1}^{n_x}$$

for all state variables, but only the computation of first-order sensitivities

$$\frac{\partial \boldsymbol{\lambda}(t, \mathbf{w})}{\partial \mathbf{w}}$$

for Lagrange multipliers. For instance, in the case of an aircraft longitudinal angular motion modeling problem, the backward-in-time algorithm speeds up the computation of the error function Hessian for a semiempirical neural network–based model with 472 weights by a factor of 11.8 as compared to a forward-in-time algorithm. This speedup further increases as the number of parameters n_w grows. On the other hand, the backward-in-time method requires additional memory for storage of $\hat{\mathbf{x}}(t, \mathbf{w})$ and

$$\frac{\partial \hat{\mathbf{x}}(t, \mathbf{w})}{\partial \mathbf{w}}.$$

Moreover, this method is only applicable when the whole training set is available beforehand, which is not the case with real-time adaptation problems.

The error function value is estimated via numerical integration of the definite integral in Eq. (5.8). Note that the target values of observable outputs are usually available only at some discrete time instants; thus, if the numerical integration procedure requires the integrand evaluation at other time instants, we need to perform interpolation or approximation of the corresponding vector-valued function of time $\tilde{\mathbf{y}}(t)$. Estimates of the state variable values $\hat{\mathbf{x}}(t, \mathbf{w})$ given by the model are computed by numerical solution of the initial value problem for the system of ODEs (5.2). In a special case, when all the time steps $(t_{k+1}^{(p)} - t_k^{(p)})$ are equal and the definite integral value is approximated by the right Riemann sum with the same time step, the error function value estimate given by (5.8) is equivalent to the usual sum of the squared residuals error function. Also note that all these computations for each trajectory may be performed in parallel.

Despite the fact that the abovementioned equations for the error function as well as its derivatives are exact, the application of numerical methods for the solution of initial value problems, estimation of definite integrals, and interpolation of observable output target values introduces some error to the computed values. The size of this error depends on the specific numerical methods used, as well as the magnitude of integration time steps and the sampling period of measurements. In the rest of this section, we analyze the asymptotic behavior of these errors for the forward-in-time method with respect to the magnitude of time steps, assuming the lack of irreducible errors (i.e., the lack of measurement noise which implies $\mathbf{y}(t) = \tilde{\mathbf{y}}(t)$).

Theorem 4. *Let $\mathcal{U}$ be a compact subset of $\mathbb{R}^{n_u}$, let $\mathcal{X}$ be a compact subset of $\mathbb{R}^{n_x}$, and let $\mathcal{Y}$ be a subset of $\mathbb{R}^{n_y}$. Assume that an unknown dynamical system has the form*

$$\frac{d\mathbf{x}(t)}{dt} = \mathbf{f}(\mathbf{x}(t), \mathbf{u}(t)),$$
$$\mathbf{y}(t) = \mathbf{g}(\mathbf{x}(t)),$$

where the vector-valued functions $\mathbf{f}\colon \mathcal{X} \times \mathcal{U} \to \mathbb{R}^{n_x}$ and $\mathbf{g}\colon \mathcal{X} \to \mathcal{Y}$ have r continuous derivatives with respect to all of their arguments for some positive integer r.

Assume that we are given a finite set of control signals $\{\mathbf{u}^{(p)}\}_{p=1}^{P}$, such that $\mathbf{u}^{(p)}\colon [0, \bar{t}^{(p)}] \to \mathcal{U}$ are piecewise-defined vector-valued functions of time that have r continuous derivatives on each subdomain. Suppose that $\mathbf{x}^{(p)}\colon [0, \bar{t}^{(p)}] \to \mathcal{X}$ and $\mathbf{y}^{(p)}\colon [0, \bar{t}^{(p)}] \to \mathcal{Y}$ represent states and outputs of an unknown dynamical system for the corresponding controls $\mathbf{u}^{(p)}$.

Let the finite data set of the form (5.1) be given. Assume that this data set contains the exact values for controls $\mathbf{u}^{(p)}(t_k^{(p)})$, initial states $\tilde{\mathbf{x}}^{(p)}(0) = \mathbf{x}^{(p)}(0)$, and observable outputs $\tilde{\mathbf{y}}^{(p)}(t_k^{(p)}) = \mathbf{y}^{(p)}(t_k^{(p)})$ over the whole time segment $t_{K^{(p)}}^{(p)} = \bar{t}^{(p)}$. Denote the maximum time step $\overline{\Delta t} = \max\limits_{p=1,\dots,P} \max\limits_{k=1,\dots,K^{(p)}} \left(t_k^{(p)} - t_{k-1}^{(p)}\right)$.

Let $\mathcal{W}$ be a compact subset of $\mathbb{R}^{n_w}$. Suppose that the parametric families of vector-valued functions $\hat{\mathbf{f}}\colon \mathcal{X} \times \mathcal{U} \times \mathcal{W} \to \mathbb{R}^{n_x}$ and $\hat{\mathbf{g}}\colon \mathcal{X} \times \mathcal{W} \to \mathcal{Y}$ which define the semiempirical state space continuous time model of the form (5.2) have $r+2$ continuous derivatives with respect to all of their arguments, including the parameters $\mathbf{w} \in \mathcal{W}$.

Let $\hat{\mathbf{x}}^{(p)}\colon [0, \bar{t}^{(p)}] \to \mathcal{X}$ and $\hat{\mathbf{y}}^{(p)}\colon [0, \bar{t}^{(p)}] \to \mathcal{Y}$ represent states and outputs of a semiempirical model given initial conditions $\hat{\mathbf{x}}^{(p)}(0) = \tilde{\mathbf{x}}^{(p)}(0)$. Suppose that for the parameter values $\mathbf{w}$, for each initial condition $\tilde{\mathbf{x}}^{(p)}(0)$ and control $\mathbf{u}^{(p)}$, the corresponding initial value problem solution $\hat{\mathbf{x}}^{(p)}$ exists on the whole time segment $[0, \bar{t}^{(p)}]$ and is contained in $\mathcal{X}$ along with the closure of its ε-neighborhood for some positive real ε.

Suppose that the estimates of error function values $\bar{E}(\mathbf{w})$ (5.7) as well as the values of its derivatives $\frac{\partial \bar{E}(\mathbf{w})}{\partial \mathbf{w}}$ (5.10), $\frac{\partial^2 \bar{E}(\mathbf{w})}{\partial \mathbf{w}^2}$ (5.11) with respect to parameters $\mathbf{w}$ are computed numerically, using the following procedure. The initial value problems for systems of ODEs (5.2), (5.16), (5.17) are solved by an explicit one-step method which has the order of accuracy r_1 if the true solution has $r_1 + 1$ continuous time derivatives; time steps of the solver are selected in such a way that all the points of the control discontinuity match the nodes. The target values of observable outputs $\tilde{\mathbf{y}}^{(p)}(t_k^{(p)})$ are interpolated with respect to the time variable using the method which has the order of accuracy r_2 if the true functions have $r_2 + 1$ continuous time derivatives. Values of definite integrals (5.8), (5.12), (5.13) are estimated by a method which linearly depends on the integrand values at some nodes and has the order of accuracy r_3 if the integrand has r_3 continuous derivatives with respect to the variable of integration; the numerical integration step does not exceed $\overline{\Delta t}$. Finally, suppose that the following inequalities hold: $r_1 \leqslant r$, $r_2 < r$, $r_3 \leqslant r$.

Then there exists a maximal time step size $\Delta t = \Delta t(\varepsilon) \leqslant \overline{\Delta t}$, such that if time step sizes used by numerical solvers for initial value problems (5.2), (5.16), (5.17) do not exceed Δt, then the estimates of the error function and its derivatives with respect to the parameters have the asymptotic error $O(\overline{\Delta t}^{\min\{r_1-1, r_2-1, r_3\}})$.

Proof. Since the total error function (5.7) is a summation of errors for each individual trajectory (5.8) over a finite set of trajectories, its estimate has the same order of accuracy as the estimates for individual trajectory errors. The same argument applies to the total gradient (5.10) and the Hessian (5.11). Thus, we analyze the accuracy of individual trajectory estimates and omit the corresponding index for convenience.

Nodes used for the computation of definite integrals are denoted as τ_m, $m = 0, \dots, M$ ($\tau_m \in [0, \bar{t}]$). In order to evaluate the integrands, we need the corresponding initial value problem solutions at time instants τ_m; therefore we require the numerical solver to visit these nodes. Only the errors of initial value problem solution estimates at these nodes influence the total error of the definite integral estimates; hence the other time instants are not treated explicitly in this analysis.

Since the vector-valued functions $\mathbf{u}$, $\mathbf{f}$, and $\hat{\mathbf{f}}$ satisfy the conditions of the theorem on existence and uniqueness of solutions to initial value problems (see Theorem 54 in [16]), there exist unique solutions to the corresponding initial value problems: $\mathbf{x}$ and $\hat{\mathbf{x}}$. Moreover, according to the assumptions of this theorem, these solutions exist on the whole segment $[0, \bar{t}]$ and are contained in $\mathcal{X}$. Since both the control $\mathbf{u}$ and the evolution function $\mathbf{f}$ have r continuous derivatives, the state space trajectory $\mathbf{x}$ of unknown dynamical system has $r+1$ continuous derivatives with respect to time. The observation function $\mathbf{g}$ has r continuous derivatives, which implies that the output $\mathbf{y}$ also has r continuous derivatives. Similarly, since the control has r continuous derivatives and the evolution function $\hat{\mathbf{f}}$ has $r+2$ continuous derivatives, the state space trajectory $\hat{\mathbf{x}}$ of the semiempirical model has $r+1$ continuous derivatives with respect to time. Two additional continuous derivatives of the vector-valued function $\hat{\mathbf{f}}$ also provide the r continuous derivatives for right hand sides of (5.16) and (5.17); hence the corresponding sensitivities $\frac{\partial \hat{\mathbf{x}}}{\partial \mathbf{w}}$ and $\frac{\partial^2 \hat{\mathbf{x}}}{\partial \mathbf{w}^2}$ have $r+1$ continuous derivatives with respect to time. Also, the observation function $\hat{\mathbf{g}}$ has $r+2$ continuous derivatives, which implies that e has $r+2$ continuous derivatives. Thus, all the integrands in (5.8), (5.12), (5.13) also have r continuous derivatives. Since r is strictly positive, the integrands are at least continuous, and therefore the corresponding definite integrals do exist.

Since the vector-valued functions $\hat{\mathbf{x}}$, $\frac{\partial \hat{\mathbf{x}}}{\partial \mathbf{w}}$, and $\frac{\partial^2 \hat{\mathbf{x}}}{\partial \mathbf{w}^2}$ have $r+1$ continuous derivatives with respect to time, and since $r_1 \leqslant r$, the numerical solutions to the initial value problems given by the explicit one-step r_1th-order method have the global truncation error of the form $O(\Delta t^{r_1})$. Thus, we obtain the following inequalities for the discrete time approximations of these solutions:

$$\begin{aligned}
\left\| \hat{\mathbf{x}}(\tau_m, \mathbf{w}) - \check{\mathbf{x}}(\tau_m, \mathbf{w}) \right\| &= O(\Delta t^{r_1}), \\
\left\| \frac{\partial \hat{\mathbf{x}}(\tau_m, \mathbf{w})}{\partial \mathbf{w}} - \frac{\partial \check{\mathbf{x}}(\tau_m, \mathbf{w})}{\partial \mathbf{w}} \right\| &= O(\Delta t^{r_1}), \\
\left\| \frac{\partial^2 \hat{\mathbf{x}}_i(\tau_m, \mathbf{w})}{\partial \mathbf{w}^2} - \frac{\partial^2 \check{\mathbf{x}}_i(\tau_m, \mathbf{w})}{\partial \mathbf{w}^2} \right\| &= O(\Delta t^{r_1}), \\
& i = 1, \dots, n_x.
\end{aligned} \tag{5.28}$$

Note that we assume all the matrix norms to be Frobenius norms.

Hence, for all positive real ε there exists such a small time step size $\Delta t = \Delta t(\varepsilon) \leqslant \overline{\Delta t}$ that $\left\| \hat{\mathbf{x}}(\tau_m, \mathbf{w}) - \check{\mathbf{x}}(\tau_m, \mathbf{w}) \right\| \leqslant \varepsilon$. According to the assumption of this theorem, solutions $\hat{\mathbf{x}}$ are contained in the compact set $\mathcal{X}$ along with the closure of their ε-neighborhood; therefore the approximate solutions $\check{\mathbf{x}}$ are contained in $\mathcal{X}$. Initial conditions of the initial value problems (5.16) and (5.17) are identical to zero. Meanwhile, the corresponding solutions $\frac{\partial \hat{\mathbf{x}}}{\partial \mathbf{w}}$ and $\frac{\partial^2 \hat{\mathbf{x}}}{\partial \mathbf{w}^2}$ are represented by (at least) continuous vector-valued functions of time, defined on compact time segments. Therefore, the abovementioned solutions are contained in some compact subsets of Euclidean space. The vector-valued function $\mathbf{g}$ is the continuous function of states; therefore values of $\mathbf{y}$ are contained in some compact set. According to the assumption of the absence of measurement noise, we have $\tilde{\mathbf{y}}(t) = \mathbf{y}(t)$, i.e., the values of $\tilde{\mathbf{y}}$ belong to some compact set. The vector-valued function $\hat{\mathbf{g}}$ and its two derivatives with respect to parameters $\mathbf{w}$ are the continuous functions of the states and the parameter values (note that the parameters also belong to the compact set $\mathcal{W}$). Hence, the values of

$$e, \quad \frac{\partial e}{\partial \mathbf{w}}, \quad \frac{\partial e}{\partial \hat{\mathbf{x}}}, \quad \frac{\partial^2 e}{\partial \mathbf{w}^2}, \quad \frac{\partial^2 e}{\partial \mathbf{w} \partial \hat{\mathbf{x}}}, \quad \frac{\partial^2 e}{\partial \hat{\mathbf{x}}^2}$$

are contained in some compact set.

The model outputs as well as their derivatives depend on the values of model states; hence their estimates are influenced by the global truncation error of numerical solutions to corresponding initial value problems. Since the vector-valued function $\hat{\mathbf{g}}$ has $r+2$ continuous derivatives, it satisfies the Lipschitz condition along with its two derivatives on the compact set $\mathcal{X}$. Using (5.28), we obtain

$$\begin{aligned}
&\left\|\hat{\mathbf{g}}(\hat{\mathbf{x}}(\tau_m, \mathbf{w}), \mathbf{w}) - \hat{\mathbf{g}}(\check{\mathbf{x}}(\tau_m, \mathbf{w}), \mathbf{w})\right\| \\
&\quad = O(\Delta t^{r_1}), \\
&\left\|\frac{\partial \hat{\mathbf{g}}(\hat{\mathbf{x}}(\tau_m, \mathbf{w}), \mathbf{w})}{\partial \mathbf{w}} - \frac{\partial \hat{\mathbf{g}}(\check{\mathbf{x}}(\tau_m, \mathbf{w}), \mathbf{w})}{\partial \mathbf{w}}\right\| \\
&\quad = O(\Delta t^{r_1}), \\
&\left\|\frac{\partial \hat{\mathbf{g}}(\hat{\mathbf{x}}(\tau_m, \mathbf{w}), \mathbf{w})}{\partial \hat{\mathbf{x}}} - \frac{\partial \hat{\mathbf{g}}(\check{\mathbf{x}}(\tau_m, \mathbf{w}), \mathbf{w})}{\partial \check{\mathbf{x}}}\right\| \\
&\quad = O(\Delta t^{r_1}), \\
&\left\|\frac{\partial^2 \hat{\mathbf{g}}_i(\hat{\mathbf{x}}(\tau_m, \mathbf{w}), \mathbf{w})}{\partial \mathbf{w}^2} - \frac{\partial^2 \hat{\mathbf{g}}_i(\check{\mathbf{x}}(\tau_m, \mathbf{w}), \mathbf{w})}{\partial \mathbf{w}^2}\right\| \\
&\quad = O(\Delta t^{r_1}), \quad i = 1, \ldots, n_x, \\
&\left\|\frac{\partial^2 \hat{\mathbf{g}}_i(\hat{\mathbf{x}}(\tau_m, \mathbf{w}), \mathbf{w})}{\partial \mathbf{w} \partial \hat{\mathbf{x}}} - \frac{\partial^2 \hat{\mathbf{g}}_i(\check{\mathbf{x}}(\tau_m, \mathbf{w}), \mathbf{w})}{\partial \mathbf{w} \partial \check{\mathbf{x}}}\right\| \\
&\quad = O(\Delta t^{r_1}), \quad i = 1, \ldots, n_x, \\
&\left\|\frac{\partial^2 \hat{\mathbf{g}}_i(\hat{\mathbf{x}}(\tau_m, \mathbf{w}), \mathbf{w})}{\partial \hat{\mathbf{x}}^2} - \frac{\partial^2 \hat{\mathbf{g}}_i(\check{\mathbf{x}}(\tau_m, \mathbf{w}), \mathbf{w})}{\partial \check{\mathbf{x}}^2}\right\| \\
&\quad = O(\Delta t^{r_1}), \quad i = 1, \ldots, n_x.
\end{aligned} \tag{5.29}$$

According to assumptions on the interpolation method for target values of observable outputs with respect to the time variable, and due to the fact that these vector-valued functions have r continuous derivatives ($r < r_2$), the interpolation error satisfies

$$\|\tilde{\mathbf{y}}(t) - \check{\mathbf{y}}(t)\| = O(\overline{\Delta t}^{r_2}), \quad t \in [0, \bar{t}]. \tag{5.30}$$

By the triangle inequality we obtain

$$\begin{aligned}
&\left\|\left(\tilde{\mathbf{y}}(\tau_m) - \hat{\mathbf{g}}(\hat{\mathbf{x}}(\tau_m, \mathbf{w}), \mathbf{w})\right)\right. \\
&\quad \left. - \left(\check{\mathbf{y}}(\tau_m) - \hat{\mathbf{g}}(\check{\mathbf{x}}(\tau_m, \mathbf{w}), \mathbf{w})\right)\right\| \\
&\quad = \left\|\left(\tilde{\mathbf{y}}(\tau_m) - \check{\mathbf{y}}(\tau_m)\right)\right. \\
&\qquad \left. + \left(\hat{\mathbf{g}}(\check{\mathbf{x}}(\tau_m, \mathbf{w}), \mathbf{w}) - \hat{\mathbf{g}}(\hat{\mathbf{x}}(\tau_m, \mathbf{w}), \mathbf{w})\right)\right\| \\
&\quad \leqslant \left\|\tilde{\mathbf{y}}(\tau_m) - \check{\mathbf{y}}(\tau_m)\right\| \\
&\qquad + \left\|\hat{\mathbf{g}}(\check{\mathbf{x}}(\tau_m, \mathbf{w}), \mathbf{w}) - \hat{\mathbf{g}}(\hat{\mathbf{x}}(\tau_m, \mathbf{w}), \mathbf{w})\right\| \\
&\quad = O(\overline{\Delta t}^{r_2}) + O(\Delta t^{r_1}) = O(\overline{\Delta t}^{\min\{r_1, r_2\}}),
\end{aligned} \tag{5.31}$$

thus, the error estimate has the order of accuracy $\min\{r_1, r_2\}$.

Using (5.29) and (5.31) and the fact that the Frobenius norm is submultiplicative and considering that all these quantities are bounded, we obtain

$$\begin{aligned}
&\left|e(\tilde{\mathbf{y}}(\tau_m), \hat{\mathbf{x}}(\tau_m, \mathbf{w}), \mathbf{w})\right. \\
&\quad \left. - e(\check{\mathbf{y}}(\tau_m), \check{\mathbf{x}}(\tau_m, \mathbf{w}), \mathbf{w})\right| = O(\overline{\Delta t}^{\min\{r_1, r_2\}}), \\
&\left\|\frac{\partial e(\tilde{\mathbf{y}}(\tau_m), \hat{\mathbf{x}}(\tau_m, \mathbf{w}), \mathbf{w})}{\partial \mathbf{w}}\right. \\
&\quad \left. - \frac{\partial e(\check{\mathbf{y}}(\tau_m), \check{\mathbf{x}}(\tau_m, \mathbf{w}), \mathbf{w})}{\partial \mathbf{w}}\right\| = O(\overline{\Delta t}^{\min\{r_1, r_2\}}), \\
&\left\|\frac{\partial e(\tilde{\mathbf{y}}(\tau_m), \hat{\mathbf{x}}(\tau_m, \mathbf{w}), \mathbf{w})}{\partial \hat{\mathbf{x}}}\right. \\
&\quad \left. - \frac{\partial e(\check{\mathbf{y}}(\tau_m), \check{\mathbf{x}}(\tau_m, \mathbf{w}), \mathbf{w})}{\partial \check{\mathbf{x}}}\right\| = O(\overline{\Delta t}^{\min\{r_1, r_2\}}), \\
&\left\|\frac{\partial^2 e(\tilde{\mathbf{y}}(\tau_m), \hat{\mathbf{x}}(\tau_m, \mathbf{w}), \mathbf{w})}{\partial \mathbf{w}^2}\right. \\
&\quad \left. - \frac{\partial^2 e(\check{\mathbf{y}}(\tau_m), \check{\mathbf{x}}(\tau_m, \mathbf{w}), \mathbf{w})}{\partial \mathbf{w}^2}\right\| = O(\overline{\Delta t}^{\min\{r_1, r_2\}}), \\
&\left\|\frac{\partial^2 e(\tilde{\mathbf{y}}(\tau_m), \hat{\mathbf{x}}(\tau_m, \mathbf{w}), \mathbf{w})}{\partial \mathbf{w} \partial \hat{\mathbf{x}}}\right. \\
&\quad \left. - \frac{\partial^2 e(\check{\mathbf{y}}(\tau_m), \check{\mathbf{x}}(\tau_m, \mathbf{w}), \mathbf{w})}{\partial \mathbf{w} \partial \check{\mathbf{x}}}\right\| = O(\overline{\Delta t}^{\min\{r_1, r_2\}}),
\end{aligned}$$

$$\left\| \frac{\partial^2 e(\tilde{\mathbf{y}}(\tau_m), \hat{\mathbf{x}}(\tau_m, \mathbf{w}), \mathbf{w})}{\partial \hat{\mathbf{x}}^2} - \frac{\partial^2 e(\check{\mathbf{y}}(\tau_m), \check{\mathbf{x}}(\tau_m, \mathbf{w}), \mathbf{w})}{\partial \check{\mathbf{x}}^2} \right\| = O(\overline{\Delta t}^{\min\{r_1, r_2\}}). \tag{5.32}$$

Next, we analyze the errors of estimates for integrands in equations (5.8), (5.12), and (5.13). Considering that all the terms are bounded and that the Frobenius norm is submultiplicative and taking into account (5.28) and (5.32), we get the same order of accuracy $\min\{r_1, r_2\}$.

Denote the maximum integration step as $\overline{\Delta\tau} = \max\limits_{m=1,\dots,M} (\tau_m - \tau_{m-1})$. Since the true integrands have r continuous derivatives with respect to the variable of integration, and since $r_3 \leqslant r$, the estimates of definite integrals given by the r_3th-order numerical integration method would have the global error of the form $O(\overline{\Delta\tau}^{r_3})$ in case the true integrand values were used for the computations. According to the assumption of this theorem, the numerical integration method linearly depends on the values of the integrand. Denote the weights of this linear combination by ω^I_m. Thus, we have

$$\left\| \int_0^{\bar{t}} e(\tilde{\mathbf{y}}(t), \hat{\mathbf{x}}(t, \mathbf{w}), \mathbf{w})\, dt - \sum_{m=0}^{M} \omega^I_m e(\tilde{\mathbf{y}}(\tau_m), \hat{\mathbf{x}}(\tau_m, \mathbf{w}), \mathbf{w}) \right\| = O(\overline{\Delta\tau}^{r_3}), \tag{5.33}$$

and also

$$\begin{aligned} &\left\| \sum_{m=0}^{M} \omega^I_m e(\tilde{\mathbf{y}}(\tau_m), \hat{\mathbf{x}}(\tau_m, \mathbf{w}), \mathbf{w}) - \sum_{m=0}^{M} \omega^I_m e(\check{\mathbf{y}}(\tau_m), \check{\mathbf{x}}(\tau_m, \mathbf{w}), \mathbf{w}) \right\| \\ &\leqslant \sum_{m=0}^{M} \left|\omega^I_m\right| \left\| e(\tilde{\mathbf{y}}(\tau_m), \hat{\mathbf{x}}(\tau_m, \mathbf{w}), \mathbf{w}) - e(\check{\mathbf{y}}(\tau_m), \check{\mathbf{x}}(\tau_m, \mathbf{w}), \mathbf{w}) \right\| \\ &\leqslant (M+1) \max_{m=0,\dots,M} \left|\omega^I_m\right| \left\| e(\tilde{\mathbf{y}}(\tau_m), \hat{\mathbf{x}}(\tau_m, \mathbf{w}), \mathbf{w}) - e(\check{\mathbf{y}}(\tau_m), \check{\mathbf{x}}(\tau_m, \mathbf{w}), \mathbf{w}) \right\| \\ &= O(\overline{\Delta t}^{\min\{r_1, r_2\}-1}). \end{aligned} \tag{5.34}$$

By the triangle inequality, from (5.33) and (5.34) we obtain

$$\begin{aligned} &\left\| \int_0^{\bar{t}} e(\tilde{\mathbf{y}}(t), \hat{\mathbf{x}}(t, \mathbf{w}), \mathbf{w})\, dt - \sum_{m=0}^{M} \omega^I_m e(\check{\mathbf{y}}(\tau_m), \check{\mathbf{x}}(\tau_m, \mathbf{w}), \mathbf{w}) \right\| \\ &\leqslant \left\| \int_0^{\bar{t}} e(\tilde{\mathbf{y}}(t), \hat{\mathbf{x}}(t, \mathbf{w}), \mathbf{w})\, dt - \sum_{m=0}^{M} \omega^I_m e(\tilde{\mathbf{y}}(\tau_m), \hat{\mathbf{x}}(\tau_m, \mathbf{w}), \mathbf{w}) \right\| \\ &+ \left\| \sum_{m=0}^{M} \omega^I_m e(\tilde{\mathbf{y}}(\tau_m), \hat{\mathbf{x}}(\tau_m, \mathbf{w}), \mathbf{w}) - \sum_{m=0}^{M} \omega^I_m e(\check{\mathbf{y}}(\tau_m), \check{\mathbf{x}}(\tau_m, \mathbf{w}), \mathbf{w}) \right\| \\ &= O(\overline{\Delta\tau}^{r_3}) + O(\overline{\Delta t}^{\min\{r_1, r_2\}-1}) \\ &= O(\overline{\Delta t}^{\min\{r_1-1, r_2-1, r_3\}}). \end{aligned} \tag{5.35}$$

Using the same argument for definite integrals in (5.12) and (5.13), we obtain the final result for individual trajectories, i.e.,

$$\begin{aligned} \left| E(\mathbf{w}) - \check{E}(\mathbf{w}) \right| &= O(\overline{\Delta t}^{\min\{r_1-1, r_2-1, r_3\}}), \\ \left\| \frac{\partial E(\mathbf{w})}{\partial \mathbf{w}} - \frac{\partial \check{E}(\mathbf{w})}{\partial \mathbf{w}} \right\| &= O(\overline{\Delta t}^{\min\{r_1-1, r_2-1, r_3\}}), \\ \left\| \frac{\partial^2 E(\mathbf{w})}{\partial \mathbf{w}^2} - \frac{\partial^2 \check{E}(\mathbf{w})}{\partial \mathbf{w}^2} \right\| &= O(\overline{\Delta t}^{\min\{r_1-1, r_2-1, r_3\}}). \end{aligned} \tag{5.36}$$

□

To illustrate this theoretical result, let us consider the asymptotic error for some specific combination of numerical methods. For instance, the initial value problem may be solved using the family of explicit one-step Runge–Kutta methods, i.e.,

$$\mathbf{x}(t_{k+1}) = \mathbf{x}(t_k) + \Delta t \sum_{i=1}^{s} b_i \mathbf{r}^{k,i},$$
$$\mathbf{r}^{k,i} = \mathbf{f}^{\mathbf{u}} \left(t_k + c_i \Delta t, \mathbf{x}(t_k) + \Delta t \sum_{j=1}^{i-1} a_{i,j} \mathbf{r}^{k,j} \right), \tag{5.37}$$

where s is the number of stages, c_i are the coefficients that define the locations of intermediate nodes, $a_{i,j}$ and b_i are the corresponding weights, and $\mathbf{f}^{\mathbf{u}}(t, \mathbf{x}(t)) \equiv \mathbf{f}(\mathbf{x}(t), \mathbf{u}(t))$. The fourth-order explicit Runge–Kutta method has the following form:

$$\begin{aligned}
\mathbf{r}^{k,1} &= \mathbf{f}^{\mathbf{u}}(t_k, \mathbf{x}(t_k)), \\
\mathbf{r}^{k,2} &= \mathbf{f}^{\mathbf{u}}(t_k + \frac{\Delta t}{2}, \mathbf{x}(t_k) + \frac{\Delta t}{2}\mathbf{r}^{k,1}), \\
\mathbf{r}^{k,3} &= \mathbf{f}^{\mathbf{u}}(t_k + \frac{\Delta t}{2}, \mathbf{x}(t_k) + \frac{\Delta t}{2}\mathbf{r}^{k,2}), \\
\mathbf{r}^{k,4} &= \mathbf{f}^{\mathbf{u}}(t_k + \Delta t, \mathbf{x}(t_k) + \Delta t \mathbf{r}^{k,3}), \\
\mathbf{x}(t_{k+1}) &= \mathbf{x}(t_k) + \frac{\Delta t}{6}\left(\mathbf{r}^{k,1} + 2\mathbf{r}^{k,2} + 2\mathbf{r}^{k,3} + \mathbf{r}^{k,4}\right).
\end{aligned} \tag{5.38}$$

Interpolation may be performed using splines; in particular, the cubic splines with not-a-knot end conditions provide the fourth order of accuracy. Definite integrals can be computed using the family of composite Newton–Cotes rules, which approximate the integral on each subsegment $[a, b]$:

$$\int_a^b f(x)\,dx \approx \sum_{m=0}^{M} \omega_m^I f(x_m), \tag{5.39}$$

where $x_m = a + m\frac{b-a}{M}$, and ω_m^I are the corresponding weights. For example, the trapezoidal rule

$$\int_a^b f(x)\,dx \approx \frac{b-a}{2}(f(a) + f(b)) \tag{5.40}$$

has the second order of accuracy. Thus, the estimates of the error function and its derivatives provided by the combination of the explicit fourth-order Runge–Kutta method, cubic spline interpolation, and composite trapezoidal rule have the asymptotic error

$$O(\Delta t^{\min\{4-1, 4-1, 2\}}) = O(\Delta t^2).$$

5.4 HOMOTOPY CONTINUATION TRAINING METHOD FOR SEMIEMPIRICAL ANN-BASED MODELS

We have already discussed some reasons for difficulties of training recurrent neural networks in Chapter 2. These difficulties include the vanishing and the exploding gradients problem, bifurcations in the recurrent neural networks, and the presence of spurious valleys in the error function landscape. Thus, traditional gradient-based methods often fail to find a sufficiently good solution unless the initial guess for parameter values lies very close to it.

But what if we consider a problem of finding such an initial guess which lies close to a good solution? We might further assume that this initial guess itself is an exact solution to another optimization problem which closely resembles the original problem. Following this logic, we can construct a sequence of optimization problems, such that: the first problem is trivial to solve; each subsequent problem resembles the previous one so that their solutions lie close to each other; and the sequence converges to the original, difficult optimization problem. In the limit, for infinitesimal perturbations of optimization problems, we expect to have a continuous curve of solutions.

We can see these considerations as an informal explanation of the ideas behind the homotopy continuation method [28,29] for the solution of nonlinear equation systems

$$\mathbf{F}(\mathbf{w}) = \mathbf{0}, \tag{5.41}$$

where $\mathbf{F}\colon \mathbb{R}^{n_w} \to \mathbb{R}^{n_w}$ is a smooth vector-valued function. First, we choose a smooth vector-valued function $\mathbf{G}\colon \mathbb{R}^{n_w} \to \mathbb{R}^{n_w}$ such that the system of equations $\mathbf{G}(\mathbf{w}) = \mathbf{0}$ is easy to solve. For instance, we might construct a system of linear equations with the unique solution. Next, we introduce a homotopy between functions $\mathbf{G}$ and $\mathbf{F}$, that is, a continuous mapping $\mathbf{H}\colon [0,1] \times \mathbb{R}^{n_w} \to \mathbb{R}^{n_w}$ such that

$$\mathbf{H}(0, \mathbf{w}) = \mathbf{G}(\mathbf{w}), \quad \mathbf{H}(1, \mathbf{w}) = \mathbf{F}(\mathbf{w}). \tag{5.42}$$

We denote the set of solutions of $\mathbf{H}(\tau, \mathbf{w}) = \mathbf{0}$ by Γ. Under certain conditions the set Γ contains a continuous curve $\gamma \subset [0,1] \times \mathbb{R}^{n_w}$, which connects some solution of a simple equation system $\mathbf{G}(\mathbf{w}) = 0$ with a solution of an original, difficult equation system $\mathbf{F}(\mathbf{w}) = 0$ (see Fig. 5.10). In this case, it is possible to numerically trace this curve starting with a simple equation system solution and to find a solution of the difficult equation system. There exist variations of this approach which utilize a vector of additional parameters $\boldsymbol{\tau}$ instead of a scalar τ; however, they are not considered in this book.

At first, we discuss the existence conditions for the abovementioned solution curve. For convenience, we restate the following standard definition.

Definition 1. A vector $\mathbf{v} \in \mathbb{R}^n$ is called a regular value of a differentiable vector-valued function $\mathbf{f}\colon \mathbb{R}^m \to \mathbb{R}^n$ $(n \leqslant m)$, if at every point $\mathbf{u} \in \mathbf{f}^{-1}(\mathbf{v})$ the Jacobian of $\mathbf{f}$ has full rank n. Otherwise, the vector is called a singular value of a vector-valued function.

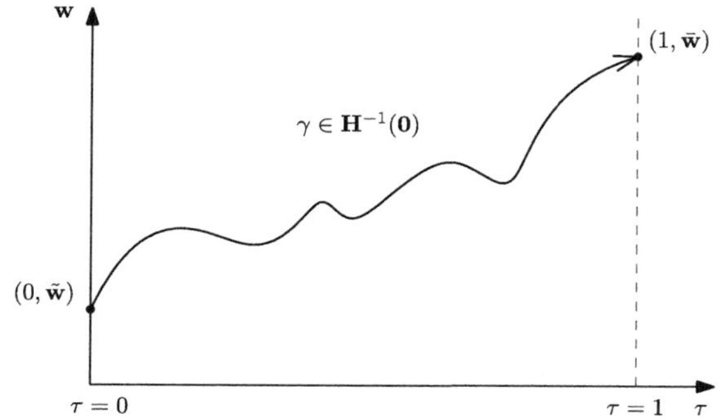

FIGURE 5.10 Example of a continuous solution curve for $\mathbf{H}(\tau, \mathbf{w}) = 0$.

By the implicit function theorem, if the zero vector $\mathbf{0} \in \mathbb{R}^{n_w}$ is a regular value of the smooth map $\mathbf{H}$ defined as above, then the set of solutions $\Gamma = \mathbf{H}^{-1}(\mathbf{0})$ is a smooth one-dimensional manifold in $[0,1] \times \mathbb{R}^{n_w}$. Thus, we need to make sure that the zero vector is a regular value of $\mathbf{H}$.

In order to do that, we adopt a globally convergent probability-one homotopy approach [30,31], which relies on the parametrized Sard theorem [30].

Theorem 5. *Let $\mathcal{V}$ be an open subset of $\mathbb{R}^q$ and let $\mathcal{U}$ be an open subset of $\mathbb{R}^m$. Assume that the vector-valued function $\mathbf{f}\colon \mathcal{V} \times \mathcal{U} \to \mathbb{R}^p$ is C^r-smooth with $r > \max\{0, m - p\}$. If a zero vector $\mathbf{0} \in \mathbb{R}^p$ is a regular value of $\mathbf{f}$, then for almost all (with respect to Lebesgue measure) $\mathbf{a} \in \mathcal{V}$ it is also a regular value of a vector-valued function $\mathbf{f}_{\mathbf{a}}(\cdot) = \mathbf{f}(\mathbf{a}, \cdot)$.*

In particular, if we include n_w additional parameters $\mathbf{a} \in \mathbb{R}^{n_w}$ into the homotopy $\mathbf{H}$ to obtain $\mathbf{H}\colon \mathbb{R}^{n_w} \times [0,1] \times \mathbb{R}^{n_w} \to \mathbb{R}^{n_w}$, and we also make sure that $\mathbf{H}$ is C^2-smooth and it has a zero vector as a regular value, then for almost all values of $\mathbf{a}$, a zero vector will also be a regular value of $\mathbf{H}_{\mathbf{a}}(\tau, \mathbf{w}) \triangleq \mathbf{H}(\mathbf{a}, \tau, \mathbf{w})$. A simple way to achieve this guarantee is to utilize the following convex homotopy:

$$\mathbf{H}(\mathbf{a}, \tau, \mathbf{w}) = (1 - \tau)(\mathbf{w} - \mathbf{a}) + \tau \mathbf{F}(\mathbf{w}). \tag{5.43}$$

It is due to these additional parameters $\mathbf{a}$ that the Jacobian of $\mathbf{H}$ will have full rank for all $\tau \in [0, 1)$.

The following theorem [31] offers theoretical justification for the probability-one homotopy methods.

Theorem 6. *Let* $\mathbf{H}\colon \mathbb{R}^{n_w} \times [0, 1) \times \mathbb{R}^{n_w} \to \mathbb{R}^{n_w}$ *be a* C^2*-smooth vector-valued function, and let* $\mathbf{H_a}\colon [0, 1) \times \mathbb{R}^{n_w} \to \mathbb{R}^{n_w}$ *be a vector-valued function which satisfies* $\mathbf{H_a}(\tau, \mathbf{w}) \equiv \mathbf{H}(\mathbf{a}, \tau, \mathbf{w})$*. Assume that the zero vector* $\mathbf{0} \in \mathbb{R}^{n_w}$ *is a regular value of* $\mathbf{H}$*. Finally, assume that for each value of additional parameters* $\mathbf{a} \in \mathbb{R}^{n_w}$ *the equation system* $\mathbf{H_a}(0, \mathbf{w}) = \mathbf{0}$ *has a unique solution* $\tilde{\mathbf{w}}$*. Then, for almost all* $\mathbf{a} \in \mathbb{R}^{n_w}$ *there exists a* C^2*-smooth curve* $\gamma \subset [0, 1) \times \mathbb{R}^{n_w}$*, emanating from* $(0, \tilde{\mathbf{w}})$ *and satisfying* $\mathbf{H_a}(\tau, \mathbf{w}) = 0$ $\forall (\tau, \mathbf{w}) \in \gamma$*. Also, if the curve* γ *is bounded, then it has an accumulation point* $(1, \bar{\mathbf{w}})$ *for some* $\bar{\mathbf{w}} \in \mathbb{R}^{n_w}$*. Moreover, if the Jacobian of* $\mathbf{H_a}$ *at the point* $(1, \bar{\mathbf{w}})$ *has full rank, then the curve* γ *has a finite arc length.*

Since the zero vector is a regular value, the Jacobian of $\mathbf{H}$ has full rank at all points of the curve γ, therefore this curve has no self-intersections or intersections with the other solution curves of $\mathbf{H_a}$. Also, since the equation system $\mathbf{H_a}(0, \mathbf{w}) = \mathbf{0}$ has a unique solution, the curve γ cannot return to cross the hyperplane $\tau = 0$. The convex homotopy (5.43) satisfies all the conditions of Theorem 6 for any C^2-smooth function $\mathbf{F}$ with $\tilde{\mathbf{w}} \equiv \mathbf{a}$. In order to guarantee the boundedness of γ, we may require that the equation system $\mathbf{H_a}(\tau, \mathbf{w}) = \mathbf{0}$ does not have solutions at infinity. This can be achieved by means of regularization.

Although this method is designed for solving systems of nonlinear equations, it can be applied to optimization problems as well. In order to do that, we replace the error function minimization problem $\bar{E}(\mathbf{w}) \to \min\limits_{\mathbf{w}}$ with a problem of finding a stationary point $\frac{\partial \bar{E}(\mathbf{w})}{\partial \mathbf{w}} = \mathbf{0}$, i.e., with a system of nonlinear equations. We should mention that these equations represent only the necessary conditions for a local minimum of the error function. These conditions are not sufficient, unless the error function is pseudoconvex. Hence, the solution of this system $\mathbf{w}^*$ should be additionally verified: for example, if the Hessian $\frac{\partial^2 \bar{E}(\mathbf{w}^*)}{\partial \mathbf{w}^2}$ has full rank and all of its eigenvalues are positive, then the solution is a local minimum. Also note that we have two possibilities: either we transform an optimization problem to a system of equations and then construct a homotopy for this system, or we construct a homotopy for the error function and then differentiate it to obtain a homotopy for the equation system.

The homotopy continuation approach has been previously applied to a feedforward neural network training problem. Some authors [32, 33] have applied the convex homotopy (5.43) as well as the so-called "global" homotopy of the form

$$\mathbf{H}(\mathbf{a}, \tau, \mathbf{w}) = \mathbf{F}(\mathbf{w}) - (1 - \tau)\mathbf{F}(\mathbf{a}) \tag{5.44}$$

to the sum of the squared errors objective function. Gorse, Shepherd, and Taylor [34] have suggested a homotopy that scales the training set target outputs from their mean value at $\tau = 0$ to the original value at $\tau = 1$. Coetzee [35] has proposed a "natural" homotopy that transforms neuron activation functions from linear to nonlinear ones ($\boldsymbol{\varphi}(\tau, \mathbf{n}) = (1 - \tau)\mathbf{n} + \tau \operatorname{th} \mathbf{n}$), thereby deforming the problem from linear to nonlinear regression. Coetzee has also suggested the use of regularization in order to keep the solution curve γ bounded. Authors of [32,35] have also studied the homotopy continuation methods which allow for a search of multiple solutions to the problem. However, in this book we are only concerned with a single solution search.

However, these homotopies are less efficient for a recurrent neural network training problem because the individual trajectory error function (5.8) sensitivity to parameters $\mathbf{w}$ grows exponentially over time. Thus, even for a moderate prediction horizon $\bar{t}$, the error function landscape becomes quite complicated. For instance, if we utilize a convex homotopy (5.43)) and fail to choose a good initial guess $\mathbf{w}^{(0)}$ for parameters, then the error function growth will be very rapid even for the slightest increase of τ. To overcome this issue, we propose the following homotopy for the error function which controls the prediction horizon value:

$$\bar{E}(\mathbf{a}, \tau, \mathbf{w}) = (1-\tau)\frac{\|\mathbf{w}-\mathbf{a}\|^2}{2} + \sum_{p=1}^{P} E^{(p)}(\tau, \mathbf{w}), \tag{5.45}$$

$$E^{(p)}(\tau, \mathbf{w}) = \int_{0}^{\tau\bar{t}^{(p)}} e^{(p)}(\tilde{\mathbf{y}}^{(p)}(t), \hat{\mathbf{x}}^{(p)}(t, \mathbf{w}), \mathbf{w})\, dt. \tag{5.46}$$

Thus, for $\tau = 0$ the error function has a unique stationary point – the global minimum $\mathbf{w} = \mathbf{a}$. The prediction horizon for each trajectory of the training set grows linearly with the parameter τ, so that for $\tau = 1$ the individual trajectory error function (5.46) is identical to (5.8).

The corresponding total error function gradient homotopy has the form

$$\begin{aligned} \mathbf{H}(\mathbf{a}, \tau, \mathbf{w}) &= \frac{\partial \bar{E}(\mathbf{a}, \tau, \mathbf{w})}{\partial \mathbf{w}} \\ &= (1-\tau)(\mathbf{w}-\mathbf{a}) + \sum_{p=1}^{P} \frac{\partial E^{(p)}(\tau, \mathbf{w})}{\partial \mathbf{w}}. \end{aligned} \tag{5.47}$$

As described in the previous section, the individual trajectory error function gradient can be computed either by taking a forward-in-time or a backward-in-time approach. In fact, the corresponding expressions (5.12) and (5.25) remain almost the same for $\frac{\partial E(\tau, \mathbf{w})}{\partial \mathbf{w}}$ and the only difference is the upper limit of integration, which needs to be changed from $\bar{t}$ to $\tau\bar{t}$.

Derivatives of the total error function gradient homotopy (5.47) with respect to τ and $\mathbf{w}$ are as follows:

$$\begin{aligned} \frac{\partial \mathbf{H}(\mathbf{a}, \tau, \mathbf{w})}{\partial \tau} &= \frac{\partial^2 \bar{E}(\mathbf{a}, \tau, \mathbf{w})}{\partial \mathbf{w}\partial \tau} \\ &= -(\mathbf{w}-\mathbf{a}) + \sum_{p=1}^{P} \frac{\partial^2 E^{(p)}(\tau, \mathbf{w})}{\partial \mathbf{w}\partial \tau}, \\ \frac{\partial \mathbf{H}(\mathbf{a}, \tau, \mathbf{w})}{\partial \mathbf{w}} &= \frac{\partial^2 \bar{E}(\mathbf{a}, \tau, \mathbf{w})}{\partial \mathbf{w}^2} \\ &= (1-\tau)\mathbf{I} + \sum_{p=1}^{P} \frac{\partial^2 E^{(p)}(\tau, \mathbf{w})}{\partial \mathbf{w}^2}. \end{aligned} \tag{5.48}$$

Again, the individual trajectory error function Hessian expressions (5.13) and (5.26) can be adapted to compute $\frac{\partial^2 E(\tau, \mathbf{w})}{\partial \mathbf{w}^2}$ by replacing the upper limit of integration from $\bar{t}$ to $\tau\bar{t}$. In order to derive the expression for $\frac{\partial^2 E(\tau, \mathbf{w})}{\partial \mathbf{w}\partial \tau}$, we apply the Leibniz integral rule, which gives us

$$\begin{aligned} \frac{\partial^2 E(\tau, \mathbf{w})}{\partial \mathbf{w}\partial \tau} = \bar{t}\Bigg[& \frac{\partial e(\tilde{\mathbf{y}}(\tau\bar{t}), \hat{\mathbf{x}}(\tau\bar{t}, \mathbf{w}), \mathbf{w})}{\partial \mathbf{w}} \\ & + \frac{\partial \hat{\mathbf{x}}(\tau\bar{t}, \mathbf{w})}{\partial \mathbf{w}}^T \frac{\partial e(\tilde{\mathbf{y}}(\tau\bar{t}), \hat{\mathbf{x}}(\tau\bar{t}, \mathbf{w}), \mathbf{w})}{\partial \hat{\mathbf{x}}} \Bigg] \end{aligned} \tag{5.49}$$

in the case of the forward-in-time method and

$$\begin{aligned} \frac{\partial^2 E(\tau, \mathbf{w})}{\partial \mathbf{w}\partial \tau} = \bar{t}\Bigg[& \frac{\partial e(\tilde{\mathbf{y}}(\tau\bar{t}), \hat{\mathbf{x}}(\tau\bar{t}, \mathbf{w}), \mathbf{w})}{\partial \mathbf{w}} \\ & + \frac{\partial \hat{\mathbf{f}}(\hat{\mathbf{x}}(\tau\bar{t}, \mathbf{w}), \mathbf{u}(\tau\bar{t}), \mathbf{w})}{\partial \mathbf{w}}^T \boldsymbol{\lambda}(\tau\bar{t}, \mathbf{w}) \Bigg] \end{aligned} \tag{5.50}$$

in the case of the backward-in-time method.

Now, we discuss the numerical methods that allow us to trace the solution curve $\gamma \subset [0, 1] \times \mathbb{R}^{n_w}$. If we parametrize the curve γ with respect to some parameter $s \in \mathbb{R}$, so that $\gamma(s) = \begin{pmatrix} \tau(s) \\ \mathbf{w}(s) \end{pmatrix}$, and then differentiate the equation system $\mathbf{H}_{\mathbf{a}}(\tau(s), \mathbf{w}(s)) = \mathbf{0}$ with respect to the parameter s, we obtain the following system of differential equations:

$$\frac{\partial \mathbf{H_a}(\tau(s), \mathbf{w}(s))}{\partial \tau}\frac{d\tau(s)}{ds} + \frac{\partial \mathbf{H_a}(\tau(s), \mathbf{w}(s))}{\partial \mathbf{w}}\frac{d\mathbf{w}(s)}{ds} = \mathbf{0}. \quad (5.51)$$

If we introduce an additional constraint of the form

$$\frac{d\tau(s)}{ds}^2 + \frac{d\mathbf{w}(s)}{ds}^T\frac{d\mathbf{w}(s)}{ds} = 1, \quad (5.52)$$

then the parameter s will represent the arc length of γ. Thus, we can trace $\gamma(s)$ by solving the initial value problem

$$\begin{pmatrix} \frac{\partial \mathbf{H_a}(\tau, \mathbf{w})}{\partial \tau} & \frac{\partial \mathbf{H_a}(\tau, \mathbf{w})}{\partial \mathbf{w}} \\ \frac{d\tau(s)}{ds} & \frac{d\mathbf{w}(s)}{ds}^T \end{pmatrix} \begin{pmatrix} \frac{d\tau(s)}{ds} \\ \frac{d\mathbf{w}(s)}{ds} \end{pmatrix} = \begin{pmatrix} \mathbf{0} \\ 1 \end{pmatrix},$$
$$\tau(0) = 0, \ \mathbf{w}(0) = \mathbf{a}. \quad (5.53)$$

As shown in [29], the arc length parametrization of curve γ is optimal in the sense that the associated system of linear equations has the smallest possible condition number.

The initial value problem can be solved by various methods, both explicit and implicit. Note that although the global truncation error of the initial value problem solution inevitably accumulates as we trace the curve, we can significantly reduce it by applying the iterative corrector process which converges to the solution curve γ. This correction procedure is based on the fact that each point of γ satisfies the equation system $\mathbf{H_a}(\tau, \mathbf{w}) = \mathbf{0}$. Hence, given a point $(\tilde{\tau}, \tilde{\mathbf{w}})$ which lies in the neighborhood of γ, we can find a closest point of γ by solving the following optimization problem:

$$\min_{\tau, \mathbf{w}} \left\{ (\tilde{\tau} - \tau)^2 + \|\tilde{\mathbf{w}} - \mathbf{w}\|^2 \mid \mathbf{H_a}(\tau, \mathbf{w}) = \mathbf{0} \right\}. \quad (5.54)$$

We need to mention that the numerical continuation method described above requires the evaluation of the error function Hessian (5.48) at each step, which incurs a significant computational burden. The quasi-Newton methods allow for a faster estimation of the error function Hessian, but the accuracy of these estimates might be insufficient. Unfortunately, the Gauss–Newton approximation cannot be utilized, because it assumes the positive semidefiniteness of the Hessian. However, under the additional assumption that the error function Hessian $\frac{\partial \mathbf{H_a}(\tau, \mathbf{w})}{\partial \mathbf{w}}$ has full rank at all points of the solution curve γ, the following properties hold. First, all the eigenvalues of the Hessian $\frac{\partial \mathbf{H_a}(\tau, \mathbf{w})}{\partial \mathbf{w}}$ never change their sign along the γ curve. Since all the eigenvalues are positive at $(0, \mathbf{a})$, they remain positive at all points of γ (see [36]). This means that all points of γ, including the solution of the original problem $(1, \mathbf{w}^*)$, actually represent the local minima of the error function for each fixed τ. Thus, the iterative corrector process may be implemented as a minimization process for the error function with respect to $\mathbf{w}$, while keeping τ fixed. Also, the efficient Gauss–Newton Hessian approximation may be utilized. Finally, the parameter τ monotonically increases along the curve γ (i.e., the curve has no turning points with respect to τ). Therefore, the solution curve may be parametrized with τ instead of arc length s. In this case, the homotopy continuation is performed by solving the initial value problem for Davidenko's system of ODEs, i.e.,

$$\mathbf{w}(0) = \mathbf{a},$$
$$\frac{d\mathbf{w}}{d\tau} = -\frac{\partial \mathbf{H_a}(\tau, \mathbf{w})}{\partial \mathbf{w}}^{-1}\frac{\partial \mathbf{H_a}(\tau, \mathbf{w})}{\partial \tau}. \quad (5.55)$$

This simple version of a homotopy continuation training algorithm is summarized below (see Algorithm 1). The iterative corrector process is implemented as a Levenberg–Marquardt method for minimization of the error

Algorithm 1 Simple homotopy continuation training algorithm for semiempirical ANN model (5.2).

Require: $\underline{\delta}, \bar{\delta}, \Delta\tau^{\min}, \Delta\tau$

$\mathbf{a} \sim U(\mathcal{W})$
$\mathbf{w} \leftarrow \mathbf{a}$
$\tau \leftarrow 0$
while $\tau < 1$ **and** $\Delta\tau > \Delta\tau^{\min}$ **do**
    $\tilde{\tau} \leftarrow \min\{\tau + \Delta\tau, 1\}$
    $\tilde{\mathbf{w}} \leftarrow \mathbf{LM}(E, \mathbf{a}, \mathbf{w}, \tau)$
    if $\|\tilde{\mathbf{w}} - \mathbf{w}\| < \bar{\delta}$ **then**
        $\mathbf{w} \leftarrow \tilde{\mathbf{w}}$
        $\tau \leftarrow \tilde{\tau}$
        if $\|\tilde{\mathbf{w}} - \mathbf{w}\| < \underline{\delta}$ **then**
            $\Delta\tau \leftarrow 2\Delta\tau$
        end if
    else
        $\Delta\tau \leftarrow \frac{1}{2}\Delta\tau$
    end if
end while

function (5.45) with respect to parameters $\mathbf{w}$, while keeping τ fixed. It uses the current parameter values as initial guess. The Levenberg–Marquardt method is denoted by `LM` in the algorithm description. The continuation algorithm also involves some form of step length adaptation, whereby if the norm of model parameters change exceeds $\bar{\delta}$, the predictor step length $\Delta\tau$ is decreased and the corrector step is reevaluated. Conversely, if the norm of model parameters change does not exceed $\underline{\delta}$, the step length is increased. The initial guess for parameter values $\mathbf{a}$ is picked at random.

Note that a conceptually similar approach of solving a series of problems with increasing prediction horizon was suggested in [37–40], and it has proved to be highly successful. Results of computational experiments with this algorithm for training of semiempirical ANN-based models of a maneuverable F-16 aircraft motion are presented in Chapter 6.

5.5 OPTIMAL DESIGN OF EXPERIMENTS FOR SEMIEMPIRICAL ANN-BASED MODELS

The indirect approach to acquisition of training data sets for ANN-based models of dynamical systems, described in Section 2.4.3, can also benefit from theoretical knowledge of the simulated system. Recall that we need to design a set of reference maneuvers that maximize the resulting training set representativeness. Such a set of maneuvers might be designed manually by an expert in the specific domain, although this procedure is quite time consuming and the results tend to be suboptimal. Methods for automation of this procedure constitute the subject of study for optimal design of experiments [41]. Classical theory of optimal design of experiments is mostly dedicated to linear regression models. Extensions of this theory to active selection of most informative training examples for feedforward neural networks were suggested in [42,43]. More recently, these results were extended to active selection of controls that provide most informative training examples for recurrent neural networks [44]; however, the primary focus is on the greedy optimization with one-step-ahead prediction horizon. All of the abovementioned methods alternate between the following three steps: search for the most informative training examples to include in the data set guided by the current model estimate; acquisition of the selected training examples; retraining or adaptation of the model using the new training set. Since this approach relies on the specific form of the model and involves model training after inclusion of each training example, it is better suited for online adaptation of the existing model rather than the design of a new model from scratch.

In this section we discuss an approach to the optimal design of reference maneuvers for semiempirical neural network–based models of controlled dynamical systems in off-line setting according to the optimality criterion that does not depend on the form of the network and also allows to account for the constraints on the control and state variables. We assume that the total number of reference maneuvers P and their durations $\bar{t}^{(p)}$ are given. Thus, we need to find the optimal set of reference maneuvers of the following form:

$$\left\{\left\langle \bar{\mathbf{x}}^{(p)}(0), \bar{\mathbf{u}}^{(p)} \right\rangle\right\}_{p=1}^{P}, \tag{5.56}$$

where $\bar{\mathbf{x}}^{(p)}(0) \in \mathcal{X}$ and $\bar{\mathbf{u}}^{(p)}: [0, \bar{t}^{(p)}] \to \mathcal{U}$ are the initial state and the control signal of the pth reference trajectory, respectively.

The corresponding set of true reference trajectories has the form $\left\{\mathbf{x}^{(p)}\right\}_{p=1}^{P}$, where $\mathbf{x}^{(p)}: [0, \bar{t}^{(p)}] \to \mathcal{X}$ satisfy $\frac{d\mathbf{x}^{(p)}(t)}{dt} = \mathbf{f}(\mathbf{x}^{(p)}(t), \mathbf{u}^{(p)}(t))$ and $\mathbf{x}^{(p)}(0) = \bar{\mathbf{x}}^{(p)}(0)$. However, the true function $\mathbf{f}$ is unknown. Moreover, we do not have an experimental data set to build an empirical or a semiempirical model yet. Therefore, we utilize a theoretical model of the system $\hat{\mathbf{f}}$ to obtain a set of predicted reference trajectories $\left\{\hat{\mathbf{x}}^{(p)}\right\}_{p=1}^{P}$. As already mentioned, this theoretical model might be a crude approximation to an unknown system. Fortunately, the accuracy requirements for the problem of reference maneuvers design are not too strict: the associated reference trajectories are required only to reach some area of interest within the state space.

Now, we need to define the optimality criterion for a set of predicted reference trajectories. We treat each point $\left\langle \bar{\mathbf{u}}^{(p)}(t), \hat{\mathbf{x}}^{(p)}(t) \right\rangle$ as a sample point of $(n_u + n_x)$-dimensional random vector $\boldsymbol{\xi}$ and assume that we want it to have a uniform distribution on $\mathcal{U} \times \mathcal{X}$. Since the random vector uniformly distributed on compact set $\mathcal{U} \times \mathcal{X}$ achieves the maximum value of differential entropy among all continuous distributions supported in $\mathcal{U} \times \mathcal{X}$, it seems reasonable to utilize differential entropy as the optimality criterion for the set of predicted reference trajectories, i.e.,

$$h(\boldsymbol{\xi}) = -\int_{\mathcal{U} \times \mathcal{X}} \mathbf{p}(\mathbf{z}) \ln \mathbf{p}(\mathbf{z}) d\mathbf{z}, \tag{5.57}$$

where $\mathbf{p}$ is the probability density function of $\boldsymbol{\xi}$. This criterion was proposed and analyzed in [45, 46].

Since the probability density function $\mathbf{p}$ is unknown, we cannot compute differential entropy using (5.57). Instead, we estimate it from a sample $\mathcal{Z} = \{\mathbf{z}_i\}_{i=1}^{n_z}$ using the Kozachenko–Leonenko method [47], i.e.,

$$\begin{aligned} \hat{h}(\boldsymbol{\xi}) &= \frac{n_u + n_x}{n_z} \sum_{i=1}^{n_z} \ln \rho_i + \ln(n_z - 1) \\ &\quad + \ln\left(\frac{\pi^{\frac{n_u+n_x}{2}}}{\Gamma\left(\frac{n_u+n_x}{2} + 1\right)}\right) + \gamma, \\ \rho_i &= \min_{\substack{j=1,\dots,n_z \\ j \neq i}} \rho\left(\mathbf{z}_i, \mathbf{z}_j\right), \end{aligned} \tag{5.58}$$

where ρ_i is the distance between the ith sample point and its nearest neighbor according to some metric ρ, Γ is the gamma function, and γ is the Euler–Mascheroni constant. The sample $\mathcal{Z}$ is obtained by numerical solution of a set of initial value problems for the predicted reference trajectories $\hat{\mathbf{x}}^{(p)}$, using the theoretical model $\hat{\mathbf{f}}$ and the set of reference maneuvers (5.56).

Then, the problem of optimal design of reference maneuvers may be viewed as an optimal control problem, i.e.,

$$\begin{aligned} &\underset{\left\{\left\langle \bar{\mathbf{x}}^{(p)}(0), \bar{\mathbf{u}}^{(p)} \right\rangle\right\}_{p=1}^{P}}{\text{minimize}} && -\hat{h} \\ &\text{subject to} && \bar{\mathbf{u}}^{(p)}(t) \in \mathcal{U},\ t \in [0, \bar{t}^{(p)}], \\ &&& p = 1, \dots, P, \\ &&& \hat{\mathbf{x}}^{(p)}(t) \in \mathcal{X},\ t \in [0, \bar{t}^{(p)}], \\ &&& p = 1, \dots, P. \end{aligned} \tag{5.59}$$

In order to solve it numerically, we parametrize a set of reference maneuvers (5.56) by a finite-dimensional parameter vector $\boldsymbol{\theta} \in \mathbb{R}^{n_\theta}$, i.e.,

$$\boldsymbol{\theta} = \begin{pmatrix} \boldsymbol{\theta}^{(1)} \\ \vdots \\ \boldsymbol{\theta}^{(P)} \end{pmatrix}, \qquad \bar{\mathbf{x}}^{(p)}(0) = \begin{pmatrix} \boldsymbol{\theta}_1^{(p)} \\ \vdots \\ \boldsymbol{\theta}_{n_x}^{(p)} \end{pmatrix},$$
$$\bar{\mathbf{u}}^{(p)}(t) = \begin{pmatrix} \boldsymbol{\theta}_{n_x+kn_u+1}^{(p)} \\ \vdots \\ \boldsymbol{\theta}_{n_x+(k+1)n_u}^{(p)} \end{pmatrix}, \tag{5.60}$$
$$t \in \left[\Delta t^{(p)} k, \Delta t^{(p)}(k+1)\right),$$
$$k = 0, \dots, K^{(p)} - 1,$$

where $\Delta t^{(p)} = \frac{\bar{t}^{(p)}}{K^{(p)}}$ and $K^{(p)} \in \mathbb{N}$ is given. Thus, each control signal $\bar{\mathbf{u}}^{(p)}$ is a piecewise constant function of time defined on segments of duration $\Delta t^{(p)}$ and parametrized by the corresponding set of step values. The total number of parameters equals $n_\theta = n_x P + n_u \sum_{i=1}^{P} K^{(p)}$.

The resulting nonlinear inequality-constrained optimization problem can be replaced with a series of unconstrained problems using the penalty function method. We also adopt a homotopy continuation method that gradually increases the prediction horizon for each trajectory in a similar fashion to the algorithm described in Section 5.4.

Since the objective function is discontinuous, the optimization can be performed only by means of zero-order algorithms. Numerical experiments evidence that the particle swarm method [48–51] is not well suited for this problem, because the system becomes ill-conditioned for long prediction horizons. Hence, the Covariance Matrix Adaptation Evolution Strategy (CMA-ES) method [52–55] seems to be more appropriate for the task.

The CMA-ES algorithm is a stochastic local optimization method for nonlinear nonconvex functions of real variables. This algorithm belongs to a class of evolution strategy algorithms; therefore the iterative search procedure relies on the mutation, selection, and recombination mechanisms. Mutation of the vector of real variables amounts to the addition of a realization of a random vector drawn from the multivariate normal distribution with zero mean and the covariance matrix $\mathbf{C} \in \mathbb{R}^{n_\theta \times n_\theta}$. Thus, the current value of the vector of parameters can be viewed as a mean vector for this normal distribution. Mutation is used to obtain $\lambda \geqslant 2$ candidate parameter vectors (population). Then, the selection and recombination take place: the new value for the parameter vector is a weighted linear combination of $\mu \in [1, \lambda]$ best individuals (i.e., candidate solutions with the lowest objective function values) which maximizes their likelihood. Obviously, the values of the elements of covariation matrix $\mathbf{C}$ (also called strategy parameters) have a significant impact on the effectiveness of the algorithm. However, the values for strategy parameters leading to the efficient search steps are unknown a priori and usually tend to change during the search. Therefore, it is necessary to provide some form of adaptation for the strategy parameters during the search (hence the name Covariance Matrix Adaptation) in order to maximize the likelihood of successful search steps. The covariance matrix adaptation is performed incrementally, i.e., it is based not only on the current population, but also on the search history that is stored in the vector $\mathbf{p}_c \in \mathbb{R}^{n_\theta}$, referred to as the search path. Similarly, the search path $\mathbf{p}_\sigma \in \mathbb{R}^{n_\theta}$ is used for adaptation of the step length σ. The Active CMA-ES algorithm extends the basic algorithm by incorporating the information of the most unsuccessful search steps (with negative weights) to the covariance matrix adaptation step. Note that in the case of a convex quadratic objective function, the covariance matrix adaptation leads to the matrix proportional to the inverse Hessian matrix, just like the quasi-Newton methods. This optimization algorithm is scale and rotation invariant. Its convergence has not been proved for

Algorithm 2 Active CMA-ES.

Require: $E: \mathbb{R}^{n_\theta} \to \mathbb{R}$ ▷ objective function to be minimized
Require: $\boldsymbol{\theta}^+ \in \mathbb{R}^{n_\theta}$ ▷ initial guess for parameter vector
Require: $\sigma \in \mathbb{R}_{>0}$ ▷ initial step length
Require: $\lambda \geqslant 2$ ▷ population size
$\mu \leftarrow \left\lfloor \frac{\lambda}{4} \right\rfloor$ ▷ number of individuals subject to recombination
$c_c \leftarrow \frac{4}{n_\theta+4}$ ▷ learning rate for search path $\mathbf{p}_c$
$c_\sigma \leftarrow c_c$ ▷ learning rate for search path $\mathbf{p}_\sigma$
$d_\sigma \leftarrow 1 + \frac{1}{c_\sigma}$ ▷ damping factor for step length
$c_{\text{cov}} \leftarrow \frac{2}{(n_\theta+\sqrt{2})^2}$ ▷ learning rate for $\mathbf{C}$ based on the search history
$c_\mu \leftarrow \frac{4\mu-2}{(n_\theta+12)^2+4\mu}$ ▷ learning rate for $\mathbf{C}$ based on the current population
$\mathbf{C} \leftarrow \mathbf{I}$ ▷ initial guess for covariance matrix
$\mathbf{p}_\sigma = \mathbf{0}$ ▷ initial value for search path
$\mathbf{p}_c = \mathbf{0}$
$\chi_{n_\theta} \leftarrow \sqrt{2}\frac{\Gamma\left(\frac{n_\theta+1}{2}\right)}{\Gamma\left(\frac{n_\theta}{2}\right)}$ ▷ $M\left[\|N(\mathbf{0},\mathbf{I})\|\right]$
repeat
  $\mathbf{C} = \mathbf{BD}\,(\mathbf{BD})^T$ ▷ eigendecomposition of the covariance matrix
  for $i = 1, \ldots, \lambda$ **do**
    $\boldsymbol{\zeta}_i \sim N(\mathbf{0},\mathbf{I})$
    $\boldsymbol{v}_i \leftarrow \boldsymbol{\theta}^+ + \sigma \mathbf{BD}\boldsymbol{\zeta}_i$ ▷ $\boldsymbol{v}_i \sim N(\boldsymbol{\theta}^+, \sigma^2\mathbf{C})$
    $E_i \leftarrow E(\boldsymbol{v}_i)$
  end for
  $\boldsymbol{\zeta}_{1,\ldots,\lambda} \leftarrow \text{argsort}(E_{1,\ldots,\lambda})$ ▷ sort $\boldsymbol{\zeta}_i$ according to objective function values $E(\boldsymbol{v}_i)$
  $\bar{\boldsymbol{\zeta}} \leftarrow \frac{1}{\mu}\sum_{i=1}^{\mu} \boldsymbol{\zeta}_i$
  $\boldsymbol{\theta}^- \leftarrow \boldsymbol{\theta}^+$
  $\boldsymbol{\theta}^+ \leftarrow \boldsymbol{\theta}^- + \sigma \mathbf{BD}\bar{\boldsymbol{\zeta}}$
  $\mathbf{p}_\sigma \leftarrow (1-c_\sigma)\mathbf{p}_\sigma + \sqrt{\mu c_\sigma(2-c_\sigma)}\mathbf{B}\bar{\boldsymbol{\zeta}}$
  $\mathbf{p}_c \leftarrow (1-c_c)\mathbf{p}_c + \sqrt{\mu c_c(2-c_c)}\mathbf{BD}\bar{\boldsymbol{\zeta}}$
  $\mathbf{C} \leftarrow (1-c_{\text{cov}})\mathbf{C} + c_{\text{cov}}\mathbf{p}_c\mathbf{p}_c^T + c_\mu \mathbf{BD}\left(\frac{1}{\mu}\sum_{i=1}^{\mu}\boldsymbol{\zeta}_i\boldsymbol{\zeta}_i^T - \frac{1}{\mu}\sum_{i=\lambda-\mu+1}^{\lambda}\boldsymbol{\zeta}_i\boldsymbol{\zeta}_i^T\right)(\mathbf{BD})^T$
  $\sigma \leftarrow \sigma \exp\frac{\|\mathbf{p}_\sigma\|-\chi_{n_\theta}}{d_\sigma \chi_{n_\theta}}$
until $\|\boldsymbol{\theta}^+ - \boldsymbol{\theta}^-\| > \varepsilon$

the general case, but experimentally confirmed for many real-world problems. In the following we present a pseudocode for the basic version of the Active CMA-ES algorithm (see Algorithm 2).

Another important aspect of the effective training set is the choice of weights for contributions of individual training examples to the error function (so-called error weights). Note that the situation when the values of inputs for $\bar{n}$ training examples are located in a small area of the input space has a similar effect as the assignment of the weight $\bar{n}$ to some mean example from this area. Thus, the nonuniform distribution of training examples in $\mathcal{U} \times \mathcal{X}$ might lead to higher model accuracy in some areas of the input space at the expense of the lower accuracy in other areas. In order to avoid this effect, we need to perform appropriate weighting for individual training examples. In particular, the weight of a training example might be taken inversely proportional to the number of training examples in its neighborhood of fixed radius.

An efficient software implementation of the algorithms described in this section should take advantage of a special data structure for the storage of training set points in order to provide fast operations of the nearest neighbor search as well as the search for all neighbors in a fixed radius. One reasonable candidate is a k-d tree structure that allows for approximate nearest neighbor searches, such as the one implemented in the FLANN library [56, 57].

REFERENCES

[1] Bohlin T. Practical grey-box process identification: Theory and applications. London: Springer-Verlag; 2006.
[2] Dreyfus G. Neural networks: Methodology and applications. Berlin ao.: Springer; 2005.
[3] Oussar Y, Dreyfus G. How to be a gray box: Dynamic semi-physical modeling. Neural Netw 2001;14(9):1161–72.
[4] Kozlov DS, Tiumentsev YV. Neural network based semi-empirical models for dynamical systems described by differential-algebraic equations. Opt Memory Neural Netw (Inf Opt) 2015;24(4):279–87.
[5] Kozlov DS, Tiumentsev YV. Neural network based semi-empirical models for dynamical systems represented by differential-algebraic equations of index 2. Proc Comput Sci 2018;123:252–7.
[6] Kozlov DS, Tiumentsev YV. Neural network based semi-empirical models of 3d-motion of hypersonic vehicle. In: Kryzhanovsky B, et al., editors. Advances in neural computation, machine learning, and cognitive research. Studies in computational intelligence, vol. 799. Berlin: Springer Nature; 2019. p. 196–201.
[7] Tarkhov DA, Vasilyev AN. New neural network technique to the numerical solution of mathematical physics problems. I: Simple problems. Opt Memory Neural Netw (Inf Opt) 2005;14(1):59–72.
[8] Tarkhov DA, Vasilyev AN. New neural network technique to the numerical solution of mathematical physics problems. II: Complicated and nonstandard problems. Opt Memory Neural Netw (Inf Opt) 2005;14(2):97–122.
[9] Kainov NU, Tarkhov DA, Shemyakina TA. Application of neural network modeling to identification and prediction problems in ecology data analysis for metallurgy and welding industry. Nonlinear Phenom Complex Syst 2014;17(1):57–63.
[10] Vasilyev AN, Tarkhov DA. Mathematical models of complex systems on the basis of artificial neural networks. Nonlinear Phenom Complex Syst 2014;17(3):327–35.
[11] Budkina EM, Kuznetsov EB, Lazovskaya TV, Tarkhov DA, Shemyakina TA, Vasilyev AN. Neural network approach to intricate problems solving for ordinary differential equations. Opt Memory Neural Netw (Inf Opt) 2017;26(2):96–109.
[12] Lazovskaya TN, Tarkhov DA, Vasilyev AN. Parametric neural network modeling in engineering. Recent Patents Eng 2017;11(1):10–5.
[13] Lazovskaya TN, Tarkhov DA, Vasilyev AN. Multilayer solution of heat equation. Stud Comput Intell 2018;736:17–22.
[14] Vasilyev AN, Tarkhov DA, Tereshin VA, Berminova MS, Galyautdinova AR. Semi-empirical neural network model of real thread sagging. Stud Comput Intell 2018;736:138–46.
[15] Antonov V, Tarkhov D, Vasilyev A. Unified approach to constructing the neural network models of real objects. Part 1. Math Models Methods Appl Sci 2018;48(18):1–8.
[16] Sontag ED. Mathematical control theory: Deterministic finite dimensional systems. 2nd ed. New York, NY, USA: Springer-Verlag New York, Inc.. ISBN 0-387-98489-5, 1998.
[17] Sontag ED. Neural nets as systems models and controllers. In: Proc. Seventh Yale Workshop on Adaptive and Learning Systems; 1992. p. 73–9.
[18] Wang YJ, Lin CT. Runge–Kutta neural network for identification of dynamical systems in high accuracy. IEEE Trans Neural Netw 1998;9(2):294–307.
[19] Scott LR. Numerical analysis. Princeton and Oxford: Princeton University Press; 2011.
[20] Dreyfus G, Idan Y. The canonical form of nonlinear discrete-time models. Neural Comput 1998;10:133–64.
[21] Nerrand O, Roussel-Ragot P, Personnaz L, Dreyfus G. Neural networks and non-linear adaptive filtering: Unifying concepts and new algorithms. Neural Comput 1993;5(2):165–97.

[22] Pearlmutter BA. Learning state space trajectories in recurrent neural networks. In: International 1989 Joint Conference on Neural Networks, vol. 2; 1989. p. 365–72.
[23] Sato MA. A real time learning algorithm for recurrent analog neural networks. Biol Cybern 1990;62(3):237–41.
[24] Özyurt DB, Barton PI. Cheap second order directional derivatives of stiff ODE embedded functionals. SIAM J Sci Comput 2005;26(5):1725–43.
[25] Griewank A, Walther A. Evaluating derivatives: Principles and techniques of algorithmic differentiation. 2nd ed. Philadelphia, PA, USA: Society for Industrial and Applied Mathematics. ISBN 0898716594, 2008.
[26] CppAD, a package for differentiation of C++ algorithms. https://www.coin-or.org/CppAD/.
[27] Walther A, Griewank A. Getting started with ADOL-C. In: Naumann U, Schenk O, editors. Combinatorial scientific computing. Chapman-Hall CRC computational science; 2012. p. 181–202. Chap. 7.
[28] Allgower E, Georg K. Introduction to numerical continuation methods. Philadelphia, PA, USA: Society for Industrial and Applied Mathematics. ISBN 089871544X, 2003.
[29] Shalashilin VI, Kuznetsov EB. Parametric continuation and optimal parametrization in applied mathematics and mechanics. Dordrecht, Boston, London: Kluwer Academic Publishers; 2003.
[30] Chow SN, Mallet-Paret J, Yorke JA. Finding zeros of maps: Homotopy methods that are constructive with probability one. Math Comput 1978;32:887–99.
[31] Watson LT. Theory of globally convergent probability-one homotopies for nonlinear programming. SIAM J Optim 2000;11(3):761–80.
[32] Chow J, Udpa L, Udpa SS. Homotopy continuation methods for neural networks. In: IEEE International Symposium on Circuits and Systems, vol. 5; 1991. p. 2483–6.
[33] Lendl M, Unbehauen R, Luo FL. A homotopy method for training neural networks. Signal Process 1998;64(3):359–70.
[34] Gorse D, Shepherd AJ, Taylor JG. The new era in supervised learning. Neural Netw 1997;10(2):343–52.
[35] Coetzee FM. Homotopy approaches for the analysis and solution of neural network and other nonlinear systems of equations. Ph.D. thesis, Carnegie Mellon University; 1995.
[36] Allgower EL, Georg K. Numerical path following. In: Techniques of scientific computing (Part 2). Handbook of numerical analysis, vol. 5. Elsevier; 1997. p. 3–207.
[37] Elman JL. Learning and development in neural networks: the importance of starting small. Cognition 1993;48(1):71–99.
[38] Ludik J, Cloete I. Incremental increased complexity training. In: Proc. ESANN 1994, 2nd European Sym. on Artif. Neural Netw.; 1994. p. 161–5.
[39] Suykens JAK, Vandewalle J. Learning a simple recurrent neural state space model to behave like Chua's double scroll. IEEE Trans Circuits Syst I, Fundam Theory Appl 1995;42(8):499–502.
[40] Bengio Y, Louradour J, Collobert R, Weston J. Curriculum learning. In: Proceedings of the 26th Annual International Conference on Machine Learning, ICML '09. New York, NY, USA: ACM. ISBN 978-1-60558-516-1, 2009. p. 41–8.
[41] Fedorov VV. Theory of optimal experiments. New York: Academic Press; 1972.
[42] MacKay DJC. Information-based objective functions for active data selection. Neural Comput 1992;4(4):590–604.
[43] Cohn DA. Neural network exploration using optimal experiment design. Neural Netw 1996;9(6):1071–83.
[44] Póczos B, Lörincz A. Identification of recurrent neural networks by Bayesian interrogation techniques. J Mach Learn Res 2009;10:515–54.
[45] Shewry MC, Wynn HP. Maximum entropy sampling. J Appl Stat 1987;14(2):165–70.
[46] Wynn HP. Maximum entropy sampling and general equivalence theory. In: Di Bucchianico A, Läuter H, Wynn HP, editors. mODa 7 — Advances in model-oriented design and analysis. Heidelberg: Physica-Verlag HD; 2004. p. 211–8.
[47] Kozachenko L, Leonenko N. Sample estimate of the entropy of a random vector. Probl Inf Transm 1987;23:95–101.
[48] Kennedy J, Eberhart R. Particle swarm optimization. In: Proceedings of ICNN'95 – IEEE International Conference on Neural Networks, vol. 4. ISBN 0-7803-2768-3, 1995. p. 1942–8.
[49] van den Bergh F, Engelbrecht A. A new locally convergent particle swarm optimiser. In: IEEE International Conference on Systems, Man and Cybernetics, vol. 3; 2002. p. 6.
[50] Peer ES, van den Bergh F, Engelbrecht AP. Using neighbourhoods with the guaranteed convergence PSO. In: Proceedings of the SIS '03 – IEEE Swarm Intelligence Symposium; 2003. p. 235–42.
[51] Clerc M. Particle swarm optimization. Newport Beach, CA, USA: ISTE. ISBN 9781905209040, 2010.
[52] Hansen N, Ostermeier A. Adapting arbitrary normal mutation distributions in evolution strategies: The covariance matrix adaptation. In: Proceedings of the IEEE Conference on Evolutionary Computation. ISBN 0-7803-2902-3, 1996. p. 312–7.
[53] Hansen N, Ostermeier A. Completely derandomized self-adaptation in evolution strategies. Evol Comput 2001;9(2):159–95.
[54] Jastrebski G, Arnold D. Improving evolution strategies through active covariance matrix adaptation. Evol Comput 2006:2814–21.

[55] Hansen N. The CMA evolution strategy: A comparing review. In: Towards a new evolutionary computation. Studies in fuzziness and soft computing, vol. 192. 2007. p. 75–102.

[56] Friedman JH, Bentley JL, Finkel RA. An algorithm for finding best matches in logarithmic expected time. ACM Trans Math Softw 1977;3(3):209–26. http://doi.acm.org/10.1145/355744.355745.

[57] Muja M, Lowe DG. Scalable nearest neighbor algorithms for high dimensional data. IEEE Trans Pattern Anal Mach Intell 2014;36(11):2227–40.

CHAPTER

6

Neural Network Semiempirical Modeling of Aircraft Motion

6.1 THE PROBLEM OF MOTION MODELING AND IDENTIFICATION OF AIRCRAFT AERODYNAMIC CHARACTERISTICS

In the process of forming aircraft motion models, we need to solve a problem, which is very significant for practice. Namely, the initial theoretical model of the object contains, as a rule, elements that we cannot determine with the required accuracy without involving experimental data on the behavior of the modeled object due to the lack of knowledge about this object. For an aircraft, these are, most often, the nonlinear dependencies of the aerodynamic forces and moments on the parameters characterizing its motion. The reconstruction of the form of such dependencies from available experimental data (for example, based on flight tests results for aircraft) is a traditional system identification task. The approach proposed in the book as part of the formation process for semiempirical ANN models provides the restoration of unknown (or insufficiently known) dependencies that are included in these models. We proposed this approach to solving this problem in [1,2]. It is based on the use of semiempirical ANN models of nonlinear controlled dynamical systems introduced in [3–6].

The typical problem of system identification in aviation is based on the use of the motion model for aircraft as a rigid body. Such a model is described by a system of ODEs or DAEs.

In the most common case, the motion model of an aircraft is described by the following ODE system:

$$\begin{aligned} \dot{x} &= f(x,u,t), \quad x=(x_1,\ldots,x_n),\ u=(u_1,\ldots,u_m),\\ y &= h(x,t), \qquad y=(y_1,\ldots,y_p). \end{aligned} \tag{6.1}$$

The right hand sides of the equations of the aircraft motion include, among others, the values of the aerodynamic forces (longitudinal, lateral, and normal, respectively)

$$X = C_x\bar{q}S; \quad Y = C_y\bar{q}S; \quad Z = C_z\bar{q}S; \quad \bar{q} = \frac{\rho V^2}{2}$$

and aerodynamic moments (roll, pitch, and yaw, respectively)

$$L = C_l\bar{q}Sb; \quad M = C_m\bar{q}Sb; \quad N = C_n\bar{q}S\bar{c}.$$

A typical feature of the aircraft motion model is that it is determined up to aerodynamic forces

https://doi.org/10.1016/B978-0-12-815254-6.00016-2

and moments acting on the aircraft, i.e.,

$$X = C_x\bar{q}S;\ Y = C_y\bar{q}S;\ Z = C_z\bar{q}S;$$
$$L = C_l\bar{q}Sb;\ M = C_m\bar{q}S\bar{c};\ N = C_n\bar{q}Sb. \tag{6.2}$$

The dimensionless coefficients of aerodynamic forces and moments are nonlinear functions of several variables; for example, in the typical case

$$\begin{aligned}
C_x &= C_x(\alpha, \beta, \delta_e, q, M_0);\\
C_y &= C_y(\alpha, \beta, \delta_r, \delta_a, p, r, M_0);\\
C_z &= C_z(\alpha, \beta, \delta_e, q, M_0);\\
C_l &= C_l(\alpha, \beta, \delta_r, \delta_a, p, r, M_0);\\
C_m &= C_m(\alpha, \beta, \delta_e, q, M_0);\\
C_n &= C_n(\alpha, \beta, \delta_r, \delta_a, p, r, M_0).
\end{aligned} \tag{6.3}$$

The problem of identification in this case is that it is required to restore the unknown dependencies for the available experimental data for $C_x, C_y, C_z, C_l, C_m, C_n$.

The system of equations of motion for the aircraft (6.1) can be substantially simplified if we accept the assumption of small changes in all parameters relative to some initial (reference) motion. If this condition is satisfied, instead of the source nonlinear motion model, its variant can be used, linearized with respect to some reference motion.

With this approach, which is traditional for flight simulation problems [7–12], the dependencies for the aerodynamic forces and the moments acting on the aircraft in flight are represented as Taylor series in powers of the increments of the parameters relative to the reference flight regime, with terms in the expansions of order not higher than the first; for example, for the coefficient of normal force $C_z = C_z(\alpha, \delta_e)$,

$$\begin{aligned}
C_z &= C_{z0} + \frac{\partial C_z}{\partial \alpha}\Delta\alpha + \frac{\partial C_z}{\partial \delta_e}\Delta\delta_e\\
&= C_{z0} + C_{z\alpha}\Delta\alpha + C_{z\delta_e}\Delta\delta_e.
\end{aligned} \tag{6.4}$$

The recovery object from the experimental data with this approach is the set of partial derivatives $C_{z_\alpha}, C_{z\delta_e}, \dots, C_{m_\alpha}, \dots$.

In some cases, when the assumption of small changes in the parameters is not satisfied, we leave the terms of the second-order expansion in the representations for the forces and moments. In other words, we introduce in these cases nonlinearities into expressions for the coefficients of forces and moments.

In comparison with the traditional approach, which is based on the linearization of the relationships for aerodynamic forces and moments, in the semiempirical approach to the modeling and identification of dynamical systems, we restore the nonlinear functions C_x, C_y, C_z, C_l, C_m, C_n (6.3) as whole objects in the entire range of changing the values of their arguments. In contrast, in the traditional approach, the derivatives $C_{z_\alpha}, C_{z\delta_e}, \dots, C_{m_\alpha}, \dots$ are restored.

6.2 SEMIEMPIRICAL MODELING OF LONGITUDINAL SHORT-PERIOD MOTION FOR A MANEUVERABLE AIRCRAFT

In this chapter, we demonstrate that semiempirical ANN models (gray box models) provide highly efficient solutions to applied problems. We use in this section as the first example the modeling of longitudinal short-period (rotational) motion of a maneuverable aircraft. These models are based on the traditional theoretical model of aircraft motion in the form of a system of ODEs. The semiempirical ANN model designed in this particular example includes two black box module elements. These elements describe the dependence of the normal force and pitch moment on the state variables (angle of attack, pitch angular velocity, and a deflection angle of the controlled stabilizer) which is initially unknown and meant to be extracted from available experimental data for the observed state variables of the dynamical system.

In this example, we consider the theoretical model of the angular longitudinal motion of the aircraft that is traditional for aircraft flight dynamics [13–19]. This model is written as follows:

$$\begin{aligned} \dot{\alpha} &= q - \frac{\bar{q}S}{mV} C_L(\alpha, q, \delta_e) + \frac{g}{V}\cos\theta, \\ \dot{q} &= \frac{\bar{q}S\bar{c}}{I_y} C_m(\alpha, q, \delta_e), \\ T^2\ddot{\delta}_e &= -2T\zeta\dot{\delta}_e - \delta_e + \delta_{e_{act}}, \end{aligned} \tag{6.5}$$

where α is the angle of attack, deg; θ is angle of pitch, deg; q is the angular velocity of the pitch, deg/sec; δ_e is the deflection angle of the controlled stabilizer, deg; C_L is the lift coefficient; C_m is the pitch moment coefficient; m is mass of the aircraft, kg; V is the airspeed, m/sec; $\bar{q} = \rho V^2/2$ is the dynamic pressure, kg·m^{-1} sec^{-2}; ρ is air density, kg/m^3; g is the acceleration due to gravity, m/sec^2; S is the wing area, m^2; $\bar{c}$ is the mean aerodynamic chord of the wing, m; I_y is the moment of inertia of the aircraft relative to the lateral axis, kg·m^2; the dimensionless coefficients C_L and C_m are nonlinear functions of their arguments; T, ζ are the time constant and the relative damping coefficient of the actuator; and $\delta_{e_{act}}$ is the command signal to the actuator of the all-turn controllable stabilizer (limited to ±25 deg). In the model (6.5), the variables α, q, δ_e, and $\dot{\delta}_e$ are the states of the controlled object, the variable $\delta_{e_{act}}$ is the control. Here, $g(H)$ and $\rho(H)$ are the variables describing the state of the environment (gravitational field and atmosphere, respectively), where H is the altitude of the flight; $m, S, \bar{c}, I_z, T, \zeta$ are constant parameters of the simulated object, C_L and C_m are variable parameters of the simulated object.

As an example of a particular simulation object, a maneuverable F-16 aircraft was considered, the source data for which were taken from [20,21]. Computational experiments with the model (6.5) were performed on the time interval $t \in [0; 20]$ sec with a sample step $\Delta t =$ 0.02 sec for the partially observed state vector $y(t) = [\alpha(t); q(t)]^T$, corrupted by additive Gaussian noise with a mean square deviation $\sigma = 0.01$.

As already noted, one of the critical issues arising in the development of empirical and semiempirical ANN models is the problem of the acquisition of a training set that adequately describes the behavior of the modeled system. We get this training data set by developing an appropriate test control signal for the simulated object and evaluating the response of the object to this signal.

Let us analyze the influence of the type of test control signal on the representativeness of the resulting set of experimental data used as a training set for the ANN model. We compare the typical excitations (step, impulse, doublet, and random signal) with the polyharmonic signal designed specially. The comparison is based on the simulation results for various test maneuvers, including a straight-line horizontal flight with a constant speed and a flight with a monotonically increasing angle of attack.

When solving problems of the considered type, one of the most critical tasks is the generation of an informative (representative) data set that characterizes the behavior of the simulated dynamical system over the whole range of the values of the variables describing the dynamical system and the derivatives (rates of change) of these quantities. As shown in Section 2.2.2, the required training data for the generated ANN model can be obtained by application of the specially designed test excitations to the simulated dynamical system. As evidenced by computational experiments, sufficient informativity of the training set for the considered problem is provided only by the random and polyharmonic signals (Fig. 6.1) of all the test signals listed in Section 2.2.2.

We can graphically present the effectiveness of various types of test signals, mentioned above. For this purpose, we will use the coverage diagrams for the range of acceptable values of the

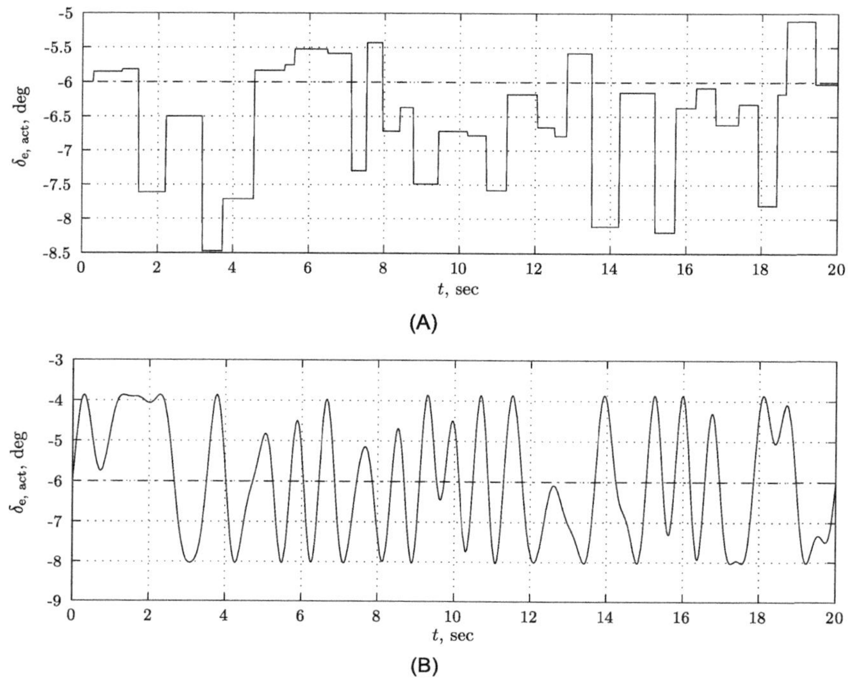

FIGURE 6.1 Test excitation signals used in studying the dynamics of controlled systems. (A) A random signal. (B) A polyharmonic signal. Here $\delta_{e_{act}}$ is the command signal of the elevator (all-moving horizontal tail) actuator; the dash-dotted horizontal line in both subfigures is the elevator deflection providing flight mode with constant altitude and airspeed.

variables and their derivatives describing the modeling object. We build these diagrams using the system response data obtained when an object is affected by a particular test signal. As applied to the problem (6.5), the variables and derivatives include $\alpha, \dot{\alpha}, q, \dot{q}, \delta_e, \dot{\delta}_e, \ddot{\delta}_e$.

Coverage diagrams allow us to compare the representativeness of training sets obtained by the application of various test excitations to the modeled object. The better the training set is, the more dense and uniform is the corresponding coverage of the required range of values for state and control variables. However, the original representation in the seven-dimensional space defined by the indicated list of variables is not suited for visualization. For this purpose, we use two-dimensional representations given by pairwise combinations of variables $(\alpha, \dot{\alpha})$, (α, q), $(\alpha, \dot{q})$, (α, δ_e), $(q, \dot{q})$, (δ_e, q) that are most informative from the point of view of the considered task. As an example, Fig. 6.2 shows coverage diagrams $(\alpha, \dot{\alpha})$ for one of the widely used test signals (doublet) as well as for a polyharmonic signal generated according to the specific procedure (Fig. 6.3). The advantages of a polyharmonic signal according to the density and uniformity of the placement for training examples are obvious.

In Section 2.2.2 we have presented the algorithm for the polyharmonic signal generation that achieves the minimum peak factor value. The coverage diagrams make it possible to visualize the influence of the peak factor on the coverage uniformity for the region of acceptable values for the variables that describe the state of the dynamical system (Fig. 6.4). Be-

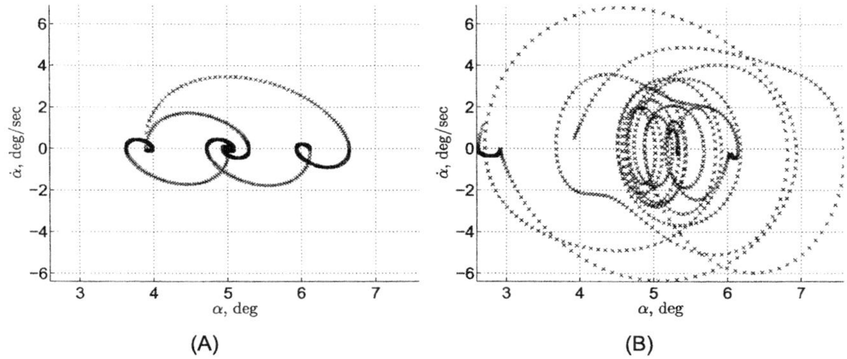

FIGURE 6.2 Coverage diagrams for the training set section $(\alpha, \dot{\alpha})$ for (A) doublet and (B) polyharmonic signals with an equal number (1000) of training examples.

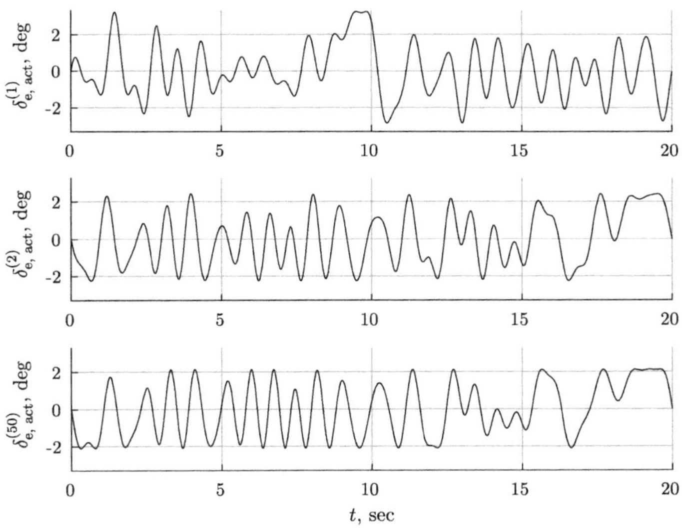

FIGURE 6.3 The process of polyharmonic signal generation (see also Fig. 6.4).

sides, we can also compare the final distribution of training examples with the distribution given by one of the widely used test signals, namely, the doublet. It is evident that the doublet is substantially inferior to the polyharmonic signal regarding informativeness of the corresponding training data set. Similarly, we can see that all the other types of control signals listed above are also inferior to the polyharmonic signal concerning the training set informativeness.

As in the example considered in Section 5.5, we will use the standard deviation of additive noise acting on the output of the system as the target value of the simulation error.

In order to utilize the Matlab Neural Network Toolbox, we represent the ANN model in the form of an LDDN. Neural network learning is

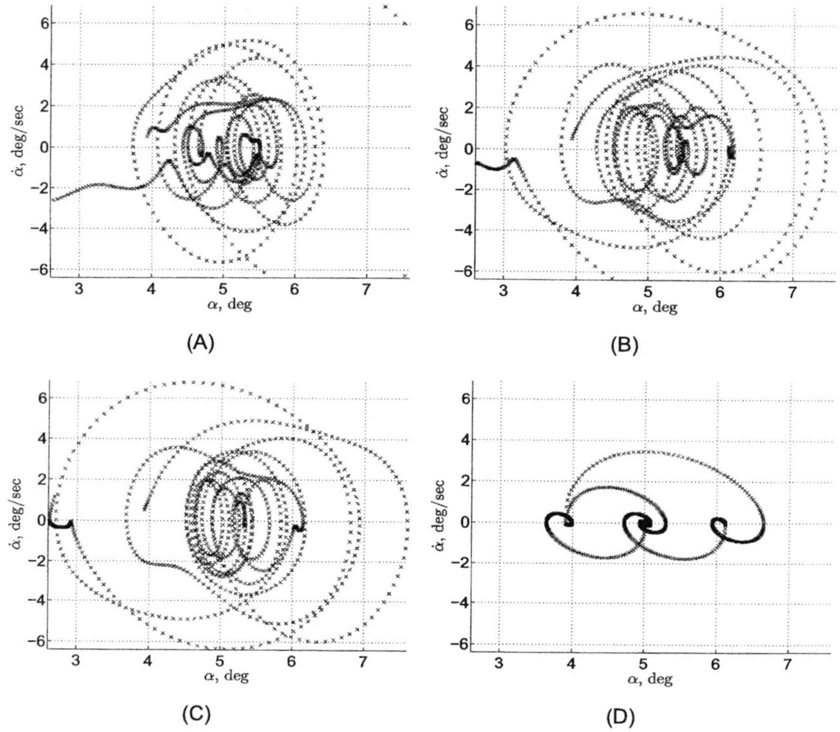

FIGURE 6.4 The change in the shape of the coverage diagrams $(\alpha, \dot{\alpha})$ for the process of the polyharmonic signal generation shown in Fig. 6.3, for iterations (A) 1, (B) 2, and (C) 50 as compared to (D) the doublet signal; the number of examples (1000) is the same everywhere (see also Fig. 6.3).

performed using the Levenberg–Marquardt algorithm for minimization of the mean square error objective function evaluated on the training data set $\{y_i\}$, $i = 1, \dots, N$, that was obtained using the initial theoretical model (6.5). The Jacobi matrix is calculated using the RTRL algorithm [22].

The application of the above semiempirical ANN model generation procedure to the theoretical model (6.5) results in the semiempirical model structure shown in Fig. 6.5 (the discrete time model is obtained using the Euler finite difference scheme). For comparison, a purely empirical NARX model structure for the same modeling problem (6.5) is shown in Fig. 6.6.

An extensive series of computational experiments were performed to compare the efficiency of all the above test signals for two types of test maneuvers: a straight-line horizontal flight with a constant speed ("point mode") and a flight with a monotonically increasing angle of attack ("monotonous mode"). As a typical example, Fig. 6.7 shows how accurately the unknown dependencies are approximated for nonlinear functions $C_L(\alpha, q, \delta_e)$, $C_m(\alpha, q, \delta_e)$. We also evaluate the accuracy of the whole semiempirical ANN model that includes abovementioned approximations for $C_L(\alpha, q, \delta_e)$ and $C_m(\alpha, q, \delta_e)$ by comparing the trajectories predicted by this model with the trajectories given by the original system (6.5). These trajectories are so close

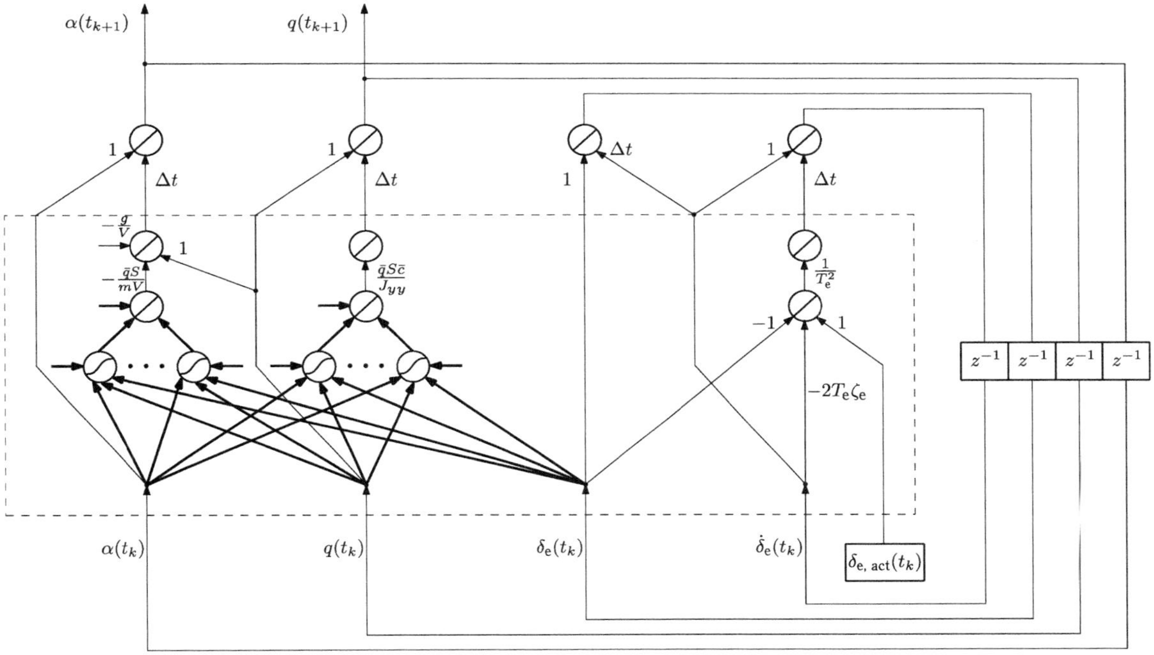

FIGURE 6.5 Semiempirical (gray box) ANN model of the longitudinal angular motion of the aircraft (based on Euler difference scheme).

TABLE 6.1 Simulation error on the training set (polyharmonic signal).

Problem	Point mode		Monotonous mode	
	$RMSE_\alpha$	$RMSE_q$	$RMSE_\alpha$	$RMSE_q$
Adjusting C_L	$1.02 \cdot 10^{-3}$	$1.24 \cdot 10^{-4}$	$1.02 \cdot 10^{-3}$	$1.24 \cdot 10^{-4}$
Learning C_L	$1.02 \cdot 10^{-3}$	$1.23 \cdot 10^{-4}$	$1.02 \cdot 10^{-3}$	$1.24 \cdot 10^{-4}$
Learning C_L, C_m	$1.02 \cdot 10^{-3}$	$1.19 \cdot 10^{-4}$	$1.02 \cdot 10^{-3}$	$1.27 \cdot 10^{-4}$
NARX simulation	$1.85 \cdot 10^{-3}$	$3.12 \cdot 10^{-3}$	$1.12 \cdot 10^{-3}$	$7.36 \cdot 10^{-4}$

TABLE 6.2 Simulation error on the test set (polyharmonic signal).

Problem	Point mode		Monotonous mode	
	$RMSE_\alpha$	$RMSE_q$	$RMSE_\alpha$	$RMSE_q$
Adjusting C_L	$1.02 \cdot 10^{-3}$	$1.59 \cdot 10^{-4}$	$1.02 \cdot 10^{-3}$	$1.17 \cdot 10^{-4}$
Learning C_L	$1.02 \cdot 10^{-3}$	$1.59 \cdot 10^{-4}$	$1.02 \cdot 10^{-3}$	$1.17 \cdot 10^{-4}$
Learning C_L, C_m	$1.02 \cdot 10^{-3}$	$1.32 \cdot 10^{-4}$	$1.02 \cdot 10^{-3}$	$1.59 \cdot 10^{-4}$
NARX simulation	$2.32 \cdot 10^{-2}$	$4.79 \cdot 10^{-2}$	$3.16 \cdot 10^{-2}$	$5.14 \cdot 10^{-2}$

that the curves on the graphs practically coincide. Quantitative estimates of the accuracy for the obtained models are given in Table 6.1 (estimate of the model accuracy on the training set) and Table 6.2 (estimate of the generalization properties for the ANN model), along with the

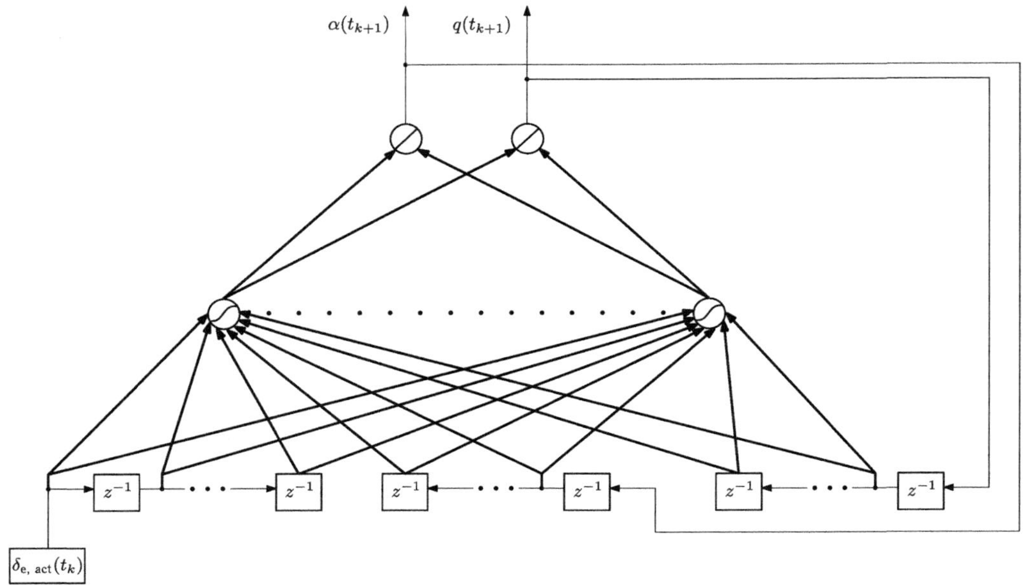

FIGURE 6.6 Empirical (black box) ANN model of the longitudinal angular motion of the aircraft (NARX).

TABLE 6.3 Simulation error on the test set for the semiempirical model and three types of excitation signals.

Problem	Point mode		Monotonous mode	
	$RMSE_\alpha$	$RMSE_q$	$RMSE_\alpha$	$RMSE_q$
Doublet	0.0202	0.0417	8.6723	34.943
Random	0.0041	0.0071	0.0772	0.2382
Polyharmonic	0.0029	0.0076	0.0491	0.1169

results given by purely empirical NARX models.

In Table 6.3 we present a comparison of magnitudes of prediction errors for various kinds of excitation signals used to generate the training data set for the semiempirical model of the angular longitudinal motion of the aircraft. It is evident that results given by the empirical model are much less accurate; for example, in the case of a polyharmonic excitation signal, the NARX model has $\text{RMSE}_\alpha = 1.3293$ deg, $\text{RMSE}_q = 2.7445$ deg/sec.

In Tables 6.1, 6.2, and 6.3 the term "point mode" denotes a straight and level flight at a constant speed while the term "monotone mode" denotes a flight with a monotonically increasing angle of attack. Also, the term "learning" for the aerodynamic coefficients denotes the problem of restoring the corresponding unknown functions "from scratch," i.e., under the assumption that there is no information on the possible values of these coefficients. The term "adjusting" refers to the task of improving the initial approximations of the corresponding coefficients, known, for example, from wind tunnel tests.

As noted above, in a number of cases, we need not only to restore the unknown functions (in this problem, $C_L(\alpha, q, \delta_e)$ and $C_m(\alpha, q, \delta_e)$), but also their derivatives by state variables, for

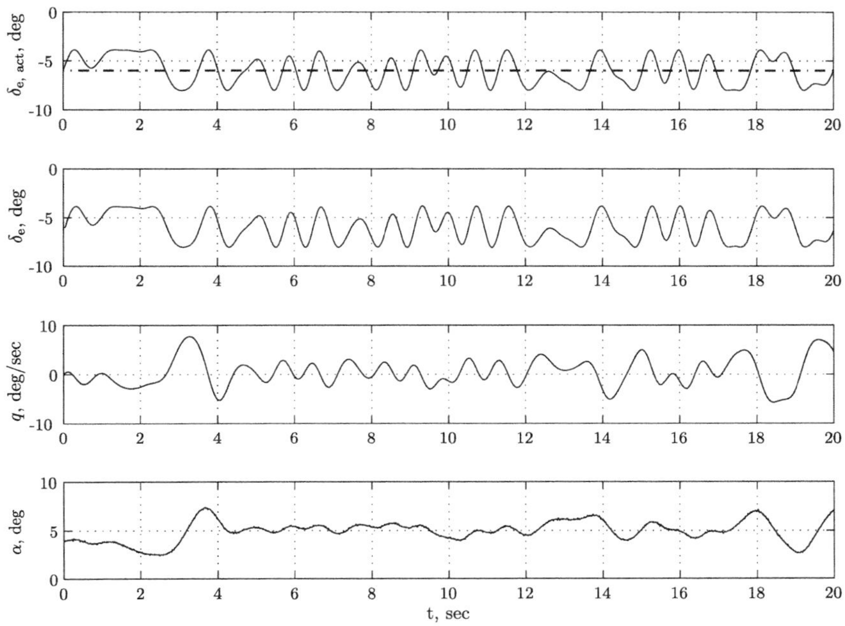

FIGURE 6.7 Accuracy of the estimated dependencies $C_L(\alpha, q, \delta_e)$ and $C_m(\alpha, q, \delta_e)$ based on the ANN model testing results (point mode, identification, and testing with polyharmonic control signal). The values of the outputs for system (6.5) and the ANN model are shown by a solid line and a dashed line, respectively.

example, C_{L_α} and C_{m_α}. When the training of the semiempirical model is completed, it is possible to extract from this model the ANN modules that represent the approximations for the functions C_L and C_m. Then, we can estimate the derivatives of functions C_L and C_m with respect to α, q, and δ_e. We can do this by computing the derivatives of the corresponding outputs of the ANN modules concerning their inputs. We can perform the computation of these derivatives by an algorithm similar to the forward propagation, originally designed to calculate the derivatives of the network outputs with respect to its weights and biases.

Using the chain rule for differentiation, we can express the derivative of the output a_k^m for the kth neuron of the mth (output) layer with respect to the input p_i in terms of the sensitivities $s_{k,j}^{m,l}$, i.e.,

$$\frac{\partial a_k^m}{n_j^l} = \sum_{(j,l)\in IC_i} s_{k,j}^{m,l} \cdot \frac{\partial n_j^l}{p_j}, \qquad s_{k,j}^{m,l} = \frac{\partial a_k^m}{n_j^l},$$

where n_j^l is the weighted input of the jth neuron of the lth layer; IC_i is the set of pairs of indices $\langle j, l\rangle$ defining the number j of a neuron in an lth layer that has a connection with the ith input p_i. In this case, the sensitivities $s_{k,j}^{m,l}$ are computed during execution of the forward propagation algorithm, and the derivatives $(\partial n_j^l/\partial p_j)$ are equal to the weights of the corresponding input connections (for a neuron with a weighted summation as an input mapping).

For example, application of this algorithm gives the following values of the derivatives C_{L_α} and C_{m_α}, corresponding to the point mode ($q = 0$) and the balancing values of the deflection angle of the stabilizer δ_e and the angle of attack α: $C_{L_\alpha} = 0.5$, $C_{m_\alpha} = -0.5$. In a similar way, we can compute derivatives for any other combinations of the values of the arguments for the functions C_L and C_m.

Based on these results, we can conclude that the semiempirical neural network modeling approach, which combines domain-specific knowledge and experience with computational mathematics methods, is a powerful and promising tool potentially suitable for solving complicated problems of describing and analyzing the controlled motion of aircraft. Comparison of the results obtained using the semiempirical approach with the traditional (black box) ANN modeling (NARX-type models) approach shows the definite advantages of semiempirical models.

6.3 SEMIEMPIRICAL MODELING OF AIRCRAFT THREE-AXIS ROTATIONAL MOTION

In the previous section, we have demonstrated the effectiveness of the semiempirical approach to ANN modeling of dynamical systems by applying it to the problem of longitudinal angular motion of the maneuverable aircraft. This task is a relatively simple one, due to its low dimensionality and, more importantly, due to the use of single-channel control (pitch channel, a single control surface is used, namely an all-movable stabilizer). In this section, we solve a much more complicated problem. We will design the ANN model of three-axis rotational motion (with three simultaneously used controls: stabilizer, rudder, and ailerons) and perform the identification for five of the six unknown aerodynamic coefficients.

As in the previous case, the theoretical model for the problem being solved is the corresponding traditional model of aircraft motion, which contains some uncertainty factors. To eliminate the existing uncertainties, we form the semiempirical ANN model, which includes five black box modules that represent the normal and lateral force coefficients, as well as the pitch, yaw, and roll moment coefficients, each of which depends nonlinearly on several parameters of the aircraft motion. These five dependencies need to be extracted (restored) from available experimental data for the observed variables of the dynamical system, i.e., we need to solve the identification problem for the aerodynamic characteristics of the aircraft.

The proposed approach to the identification of aerodynamic characteristics of an aircraft differs substantially from the traditionally accepted way of solving such problems. Namely, the traditional approach [7–11,23–29] relies on the use of a linearized model of the disturbed motion of an aircraft. In this case, the dependencies for the aerodynamic forces and moments acting on the aircraft are represented in the form of the Taylor series expansion, truncated after the first-order terms (in rare cases after the second-order terms). In such a case, we reduce the solution of the identification problem to the reconstruction of the coefficients of the Taylor expansion using the experimental data. In this expansion, the dominant terms are the partial derivatives of the dimensionless coefficients of aerodynamic forces and moments concerning the various parameters of the aircraft motion (C_{z_α}, C_{y_β}, C_{m_α}, C_{m_q}, etc.). In contrast, the semiempirical approach implements the reconstruction of the relations for the coefficients of the forces C_x, C_y, C_z and the moments C_l, C_n, C_m as whole nonlinear dependencies from the corresponding arguments. We perform this reconstruction without resorting to a Taylor series expansion for the aerodynamic coefficients. That is, the functions C_x, C_y, C_z, C_l, C_n, C_m themselves are estimated, and not the coefficients of their series expansion. We represent each of these dependencies as a separate ANN module embedded into the semiempirical model. If the derivatives C_{z_α}, C_{y_β}, C_{m_α}, C_{m_q}, etc. are required for the solution of some problems, for example, for the analysis of the stability characteristics and controllability of an aircraft, they can be easily estimated using the appropriate ANN modules obtained during the generation of a semiempirical ANN model (see also the end of the previous section).

The initial theoretical model of the total angular motion of the aircraft, used for the development of the semiempirical ANN model, is a system of ODEs, traditional for flight dynamics of aircraft [13–19]. This model has the following form:

$$\begin{cases} \dot{p} = (c_1 r + c_2 p)q + c_3 \bar{L} + c_4 \bar{N}, \\ \dot{q} = c_5 pr - c_6(p^2 - r^2) + c_7 \bar{M}, \\ \dot{r} = (c_8 p - c_2 r)q + c_4 \bar{L} + c_9 \bar{N}, \end{cases} \tag{6.6}$$

$$\begin{cases} \dot{\phi} = p + q \tan\theta \sin\phi + r \tan\theta \cos\phi, \\ \dot{\theta} = q \cos\phi - r \sin\phi, \\ \dot{\psi} = q \dfrac{\sin\phi}{\cos\theta} + r \dfrac{\cos\phi}{\cos\theta}, \end{cases} \tag{6.7}$$

$$\begin{cases} \dot{\alpha} = q - (p\cos\alpha + r\sin\alpha)\tan\beta \\ \qquad + \dfrac{1}{mV\cos\beta}(-L + mg_3), \\ \dot{\beta} = p\sin\alpha - r\cos\alpha + \dfrac{1}{mV}(Y + mg_2), \end{cases} \tag{6.8}$$

$$\begin{cases} T_e^2 \ddot{\delta}_e = -2T_e\zeta_e\dot{\delta}_e - \delta_e + \delta_e^{act}, \\ T_a^2 \ddot{\delta}_a = -2T_a\zeta_a\dot{\delta}_a - \delta_a + \delta_a^{act}, \\ T_r^2 \ddot{\delta}_r = -2T_r\zeta_r\dot{\delta}_r - \delta_r + \delta_r^{act}. \end{cases} \tag{6.9}$$

The following notation is used for this model: p, r, q are the roll, yaw, and pitch angular velocities, deg/sec; ϕ, ψ, θ are the roll, yaw, and pitch angles, deg; α, β are the angles of attack and sideslip, deg; δ_e, δ_r, δ_a are the deflection angles of the controlled stabilizer, rudder, and ailerons, deg; $\dot{\delta}_e$, $\dot{\delta}_r$, $\dot{\delta}_a$ are the angular velocities of the deflection of the controlled stabilizer, rudder, and ailerons, deg/sec; V is the airspeed, m/sec; δ_e^{act}, δ_r^{act}, δ_a^{act} are the command signals to the actuators of the controlled stabilizer, rudder, and ailerons, deg; T_e, T_r, T_a are the time constants for the actuators of the controlled stabilizer, rudder, and ailerons, sec; ζ_e, ζ_r, ζ_a are the relative damping coefficients for the actuators of the controlled stabilizer, rudder, and ailerons; D, L, Y are the drag, lift, and side forces; $\bar{L}$, $\bar{M}$, $\bar{N}$ are the roll, pitch, and yaw moments; m is the mass of the aircraft, kg.

The coefficients $c_1, \ldots, c_9$ in (6.6) are defined as follows:

$$\begin{aligned} c_0 &= I_x I_z - I_{xz}^2, \\ c_1 &= [(I_y - I_z)I_z - I_{xz}^2]/c_0, \\ c_2 &= [(I_x - I_y + I_z)I_{xz}]/c_0, \\ c_3 &= I_z/c_0, \\ c_4 &= I_{xz}/c_0, \\ c_5 &= (I_z - I_x)/I_y, \\ c_6 &= I_{xz}/I_y, \\ c_7 &= 1/I_y, \\ c_8 &= [I_x(I_x - I_y) + I_{xz}^2]/c_0, \\ c_9 &= I_x/c_0, \end{aligned}$$

where I_x, I_y, I_z are moments of inertia of the aircraft with respect to the axial, lateral, and normal axes, kg·m^2; I_{xz}, I_{xy}, I_{yz} are centrifugal moments of inertia of the aircraft, kg·m^2.

The aerodynamic forces D, L, Y in (6.7) and the moments $\bar{L}$, $\bar{M}$, $\bar{N}$ in (6.6) are defined by relationships of the following form:

$$\begin{cases} D = -\bar{X}\cos\alpha\cos\beta - \bar{Y}\sin\beta - \bar{Z}\sin\alpha\cos\beta, \\ Y = -\bar{X}\cos\alpha\sin\beta + \bar{Y}\cos\beta - \bar{Z}\sin\alpha\sin\beta, \\ L = \bar{X}\sin\alpha - \bar{Z}\cos\alpha, \end{cases} \tag{6.10}$$

$$\begin{cases} \bar{X} = q_p S C_x(\alpha, \beta, \delta_e, q), \\ \bar{Y} = q_p S C_y(\alpha, \beta, \delta_r, \delta_a, p, r), \\ \bar{Z} = q_p S C_z(\alpha, \beta, \delta_e, q), \end{cases} \tag{6.11}$$

$$\begin{cases} \bar{L} = q_p S b C_l(\alpha, \beta, \delta_e, \delta_r, \delta_a, p, r), \\ \bar{M} = q_p S \bar{c} C_m(\alpha, \beta, \delta_e, q), \\ \bar{N} = q_p S b C_n(\alpha, \beta, \delta_e, \delta_r, \delta_a, p, r). \end{cases} \tag{6.12}$$

The variables g_1, g_2, g_3 required in (6.8) are the projections of the acceleration of gravity on the axes of the wind frame, m/sec^2, i.e.,

$$\begin{cases} g_1 = g(-\sin\theta\cos\alpha\cos\beta + \cos\phi\cos\theta\sin\alpha\cos\beta \\ \qquad + \sin\phi\cos\theta\sin\beta), \\ g_2 = g(\sin\theta\cos\alpha\sin\beta - \cos\phi\cos\theta\sin\alpha\sin\beta \\ \qquad + \sin\phi\cos\theta\cos\beta), \\ g_3 = g(\sin\theta\sin\alpha + \cos\phi\cos\theta\cos\alpha). \end{cases} \tag{6.13}$$

In addition, in Eqs. (6.11), (6.12) we use the following notation: $\bar{X}$, $\bar{Y}$, $\bar{Z}$ are the aerodynamic axial, lateral, and normal forces; S is the area of the aircraft wing, m^2; b, $\bar{c}$ are the wingspan and mean aerodynamic chord of the wing, m; q_p is the dynamic air pressure, kg·m^{-1}sec^{-2}. Also, C_x, C_y, C_z denote the dimensionless coefficients of axial, lateral, and normal forces, and C_l, C_m, C_n denote the dimensionless coefficients of roll, pitch, and yaw moments. All these aerodynamic coefficients are nonlinear functions of their arguments, as listed in (6.11) and (6.12).

It should be noted that the dependencies of the coefficients of aerodynamic forces and, especially, of the aerodynamic moments on their respective arguments are highly nonlinear within the domain of interest, which significantly complicates the process of the aerodynamic characteristics identification for a maneuverable aircraft. As an example, in Fig. 6.8 we show the cross-section of the hypersurface given by the function $C_m = C_m(\alpha, \beta, \delta_e, q)$ at $\delta_e \in \{-25, 0, 25\}$ deg, $q = 0$ deg/sec within the domain $\alpha \in [-10, 45]$ deg, $\beta \in [-30, 30]$ deg.

We consider the maneuverable aircraft F-16 as an example of a simulated object. The source data for it were taken from the report [20], which presents experimental results obtained by wind tunnels tests.

The following particular values of the corresponding variables in (6.6)–(6.13) were adopted for the simulation: the mass of the aircraft $m =$ 9295.44 kg; wing span $b = 9.144$ m; the wing area $S = 27.87$ m^2; the mean aerodynamic chord of the wing is $\bar{c} = 3.45$ m; moments of inertia $I_x = 12874.8$ kg·m^2, $I_y = 75673.6$ kg·m^2, $I_z =$ 85552.1 kg·m^2, $I_{xz} = 1331.4$ kg·m^2, $I_{xy} = I_{yz} =$ 0 kg·m^2; center of gravity place is 5% of the mean aerodynamic chord; time constants of actuators $T_e = T_r = T_a = 0.025$ sec; relative damping coefficients for actuators are $\zeta_e = \zeta_r = \zeta_a = 0.707$.

During the transient processes of the angular motion for the aircraft, the airspeed V and the flight altitude H do not change significantly. Thus, we assume them to be constant and do not include the corresponding equations that describe the translational motion in the model. In the experiments carried out, we used the following constant values: altitude above sea level $H =$ 3000 m; airspeed $V = 147.86$ m/sec. Accordingly, the other variables, which depend only on constants V and H, have the following values: gravitational acceleration $g = 9.8066$ m/sec^2; air density $\rho = 0.8365$ kg/m^3; local speed of sound $a = 328.5763$ m/sec; the free stream Mach number $M_0 = 0.45$; dynamic air pressure $q_p =$ 9143.6389 kg·m^{-1} sec^{-2}.

In the model (6.6)–(6.9), the 14 variables p, q, r, ϕ, θ, ψ, α, β, δ_e, δ_r, δ_a, $\dot{\delta}_e$, $\dot{\delta}_r$, $\dot{\delta}_a$ represent the state of the controlled object, and the other three variables δ_e^{act}, δ_r^{act}, δ_a^{act} represent the controls. The values of the control variables are restricted to the following ranges: $\delta_e^{act} \in [-25, 25]$ deg, $\delta_r^{act} \in [-30, 30]$ deg, $\delta_a^{act} \in [-21.5, 21.5]$ deg for the command signals to the actuators of the controlled stabilizer, rudder, and ailerons, respectively.

During the process of the training set generation, as well as during the testing of the final semiempirical ANN model, the control actions were applied to the aircraft simultaneously along all three channels (elevator, rudder, ailerons). We utilized the polyharmonic excitation signals δ_e^{act}, δ_r^{act}, δ_a^{act} for the training set generation and random excitation signals for the test set generation.

The computational experiments for the model (6.6)–(6.9) were performed on the time interval $t \in [0, 20]$ sec in the ANN model training phase and on the interval $t \in [0, 40]$ sec in the testing phase. In both cases we used the sampling

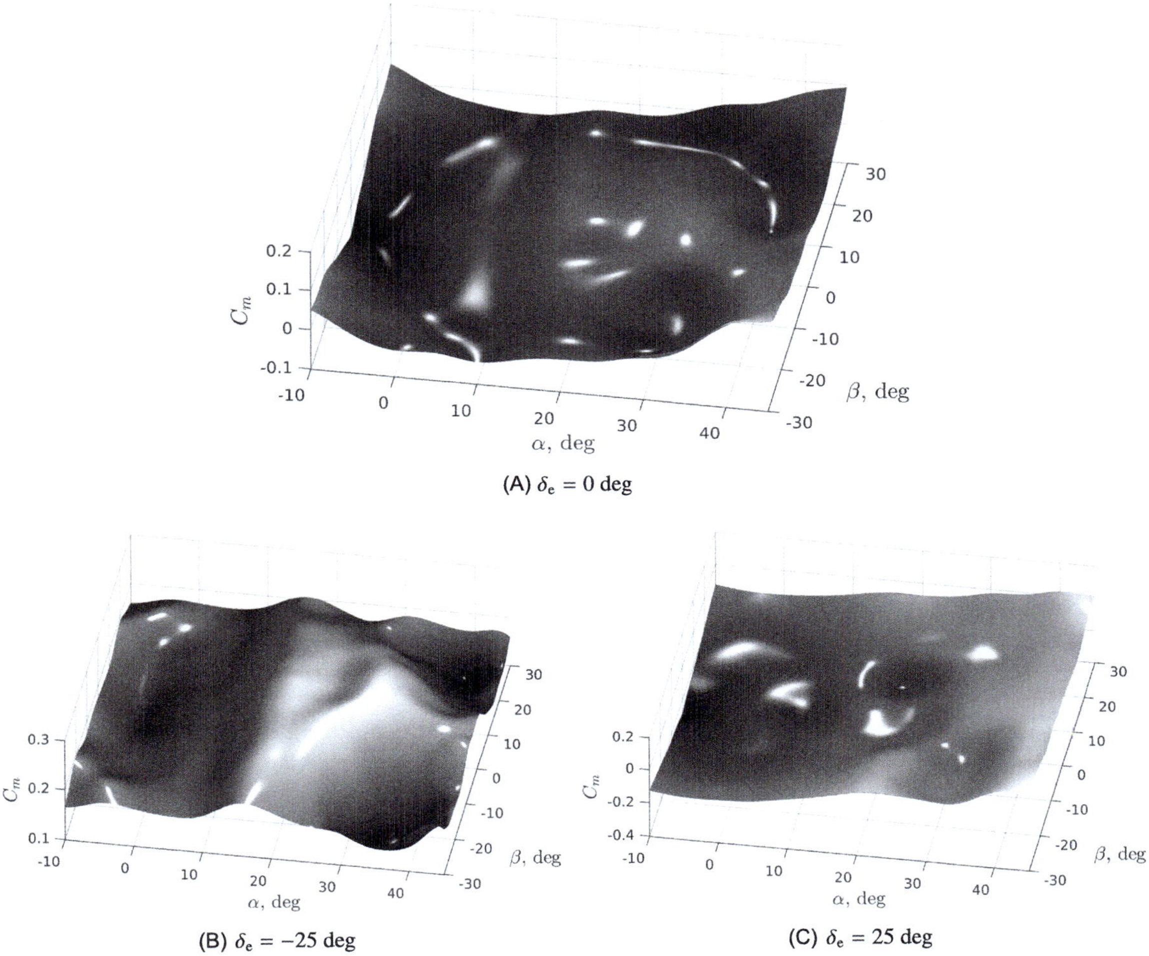

(A) $\delta_e = 0$ deg

(B) $\delta_e = -25$ deg

(C) $\delta_e = 25$ deg

FIGURE 6.8 Cross-sections of the hypersurface $C_m = C_m(\alpha, \beta, \delta_e, q)$ for several values of δ_e at $q = 0$ deg/sec, $V = 150$ m/sec within the domain $\alpha \in [-10, 45]$ deg, $\beta \in [-30, 30]$ deg.

period $\Delta t = 0.02$ sec and a partially observed state vector $y(t) = [\alpha(t); \beta(t); p(t); q(t); r(t)]^T$. The output of the system $y(t)$ is corrupted by an additive Gaussian noise with a standard deviation $\sigma_\alpha = \sigma_\beta = 0.02$ deg, $\sigma_p = 0.1$ deg/sec, $\sigma_q = \sigma_r = 0.05$ deg/sec.

As in the previous example (Section 6.2), we will use the standard deviation of additive noise affecting the output of the system as the target value of the simulation error. We perform LDDN neural network training using the Levenberg–Marquardt algorithm for minimization of the mean square error objective function evaluated on the training data set $\{y_i\}$, $i = 1, \dots, N$, that was obtained using the initial theoretical model (6.6)–(6.9). The Jacobi matrix is computed using the RTRL algorithm [22]. The learning strategy for the ANN model was based on the segmentation of the training set considered in Chapter 5.

The structural diagram of the semiempirical model corresponding to the system (6.6)–(6.9) is quite cumbersome and thus not shown here. This diagram is conceptually similar to the one shown in Fig. 6.5; however, it includes a much larger number of elements and connections between them. Most of these elements correspond to the additional terms in the initial theoretical model and do not contain any unknown tunable parameters. Also, the ANN model of the system (6.6)–(6.9) contains five black box–type ANN modules that represent unknown dependencies for the coefficients of aerodynamic forces and moments (C_y, C_z, C_l, C_n, C_m) to be reconstructed, as compared to only two modules (C_z, C_m) for the system (6.5).

It is important to note that since we consider the problem of aircraft short-period angular motion modeling, we can assume the altitude H and the airspeed V to be constant (these variables do not change significantly during the transient time). This assumption allows us to reduce the initial theoretical model by eliminating the differential equations for translational motion of an aircraft as well as equations that describe the engine dynamics. However, this also leads to the lack of possibility to efficiently control the aircraft velocity using the engine thrust or air brake deflection. Thus, we cannot obtain a representative training set for the axial force coefficient C_x using only the stabilizer, rudder, and ailerons deflections. In order to overcome this problem, we first train the ANN module for C_x directly using the wind tunnel data [20], separately from the whole model. Then, we embed this ANN module into the semiempirical model and "freeze" its parameters (i.e., prohibit their modification during the model training). Finally, we perform training for the semiempirical model to simultaneously approximate the unknown functions C_y, C_z, C_l, C_m, C_n.[1]

If we expand the initial theoretical model (6.6)–(6.9) by adding the equations for the translational motion of the aircraft as well as the equations that describe the engine dynamics, it becomes possible to reconstruct all the six functions $C_x, C_y, C_z, C_l, C_m, C_n$ by training the semiempirical ANN model. This problem is conceptually similar, although the model training is somewhat more time consuming due to the increased dimensionality.

As already noted, to ensure the adequacy of the semiempirical ANN model being created, we require a representative (informative) training set that describes the response of the simulated object to control signals from a given range. These constraints on the values of control signals, in turn, lead to the constraints on the values of the state variables that describe the system. Adequacy of the designed model[2] can only be ensured within the corresponding domain of values for control and state variables, which is formed by the constraints mentioned above.

In the computational experiments, the control variables $\delta_e^{act}, \delta_r^{act}, \delta_a^{act}$ took values within intervals specified in Table 6.4 for both the training phase (polyharmonic control signal) and the testing phase (random control signal). Corresponding intervals for the values of the state variables $p, q, r, \phi, \theta, \psi, \alpha, \beta$ are also included in Table 6.4.

In order to expand these ranges of values for control and state variables up to the full operational area of the simulated system, we have to develop appropriate algorithms for model generation. One of the approaches to solving this problem relies upon the incremental learning methods for the ANN model [30,31]. Under this approach, initially only the core of the model is designed that provides required accuracy within some subspace of the operational area, and then

[1] All the ANN modules, both for the functions C_y, C_z, C_l, C_m, C_n and for the function C_x, are represented by sigmoidal feedforward neural networks with one hidden layer.

[2] That is, the model should generalize well enough to provide the required simulation accuracy for the whole range of its operating modes.

TABLE 6.4 Ranges of variables in the model (6.6)–(6.9).

Variables	Training set		Test set	
	min	max	min	max
α	3.8405	6.3016	3.9286	5.8624
β	−1.9599	1.7605	−0.4966	0.9754
p	−16.0310	18.1922	−10.1901	11.8683
q	−3.0298	3.1572	−1.2555	3.6701
r	−4.6205	4.1017	−0.9682	4.1661
δ_e	−7.2821	−4.7698	−7.2750	−5.0549
$\dot{\delta}_e$	−8.1746	8.0454	−39.4708	36.8069
δ_a	−1.2714	1.2138	−2.0423	1.0921
$\dot{\delta}_a$	−8.6386	8.7046	−56.8323	48.9997
δ_r	−2.5264	1.7844	−1.7308	1.4222
$\dot{\delta}_r$	−20.4249	17.8579	−48.6391	58.5552
ϕ	−22.3955	7.7016	0	59.6928
θ	0	5.3013	−20.8143	3.8094
ψ	−11.9927	0	−0.0099	98.5980
δ_e^{act}	−7.2629	−4.7886	−7.0105	−5.3111
δ_a^{act}	−1.2518	1.1944	−1.4145	0.7694
δ_r^{act}	−2.4772	1.7321	−1.3140	1.0044

TABLE 6.5 Simulation error on the test set for semiempirical model at different learning stages.

Prediction horizon	MSE_α	MSE_β	MSE_p	MSE_r	MSE_q
2	0.1376	0.2100	1.5238	0.4523	0.4517
4	0.1550	0.0870	0.5673	0.2738	0.4069
6	0.1647	0.0663	0.4270	0.2021	0.3973
9	0.1316	0.0183	0.1751	0.0530	0.2931
14	0.0533	0.0109	0.1366	0.0300	0.1116
21	0.0171	0.0080	0.0972	0.0193	0.0399
1000	0.0171	0.0080	0.0972	0.0193	0.0399

the domain of the model is iteratively expanded, while preserving the behavior within the previous subdomain.

This algorithm has been successfully applied to the problem of the aerodynamic coefficients identification for the five unknown coefficients C_y, C_z, C_l, C_m, C_n and the 1000-time step prediction horizon. Computational experiment results for this problem are presented in Table 6.5 and in Figs. 6.9 and 6.10.

Analysis of the obtained simulation results allows us to draw the following conclusions.

The most important characteristic of the generated model is its ability to generalize. For the neural network models, that usually means the ability of the model to ensure the desired accuracy not only for the data used for the model learning, but also for any values of the inputs (in this case, the control and state variables) within the domain of interest. This type of verification is performed on the test data set that covers the abovementioned domain and does not coincide with the training data set.

Successful solution to the modeling and identification problem should ensure, firstly, that the required modeling accuracy is attained throughout the whole domain of interest for the model and, secondly, that the aerodynamic characteristics of the aircraft are approximated to the desired accuracy.

From the results presented in Fig. 6.9 and Table 6.5, we can conclude that the first of these problems is successfully solved. Fig. 6.9 demonstrates that the prediction errors for all of the observed variables are insignificant and that these errors grow very slowly over time, which indicates good generalization properties of the ANN model. Namely, the model does not "fall apart" with a sufficiently large prediction horizon.

Testing was carried out for a prediction horizon of 40 sec, which is a sufficiently long time interval for the problem of aircraft short-period motion modeling. We need to emphasize that the model was tested in rather strict conditions. We can see from Fig. 6.9 that very active work is performed by the control surfaces of the aircraft (controlled stabilizer, rudder, ailerons), expressed in the frequent change in the value of command signals δ_e^{act}, δ_r^{act}, δ_a^{act} for actuators of control surfaces. In this situation, there is a significant difference between adjacent values of the command signals that were randomly generated. The purpose of this method of a test data set generation is to provide a wide variety of states for the simulated system (in order to cover

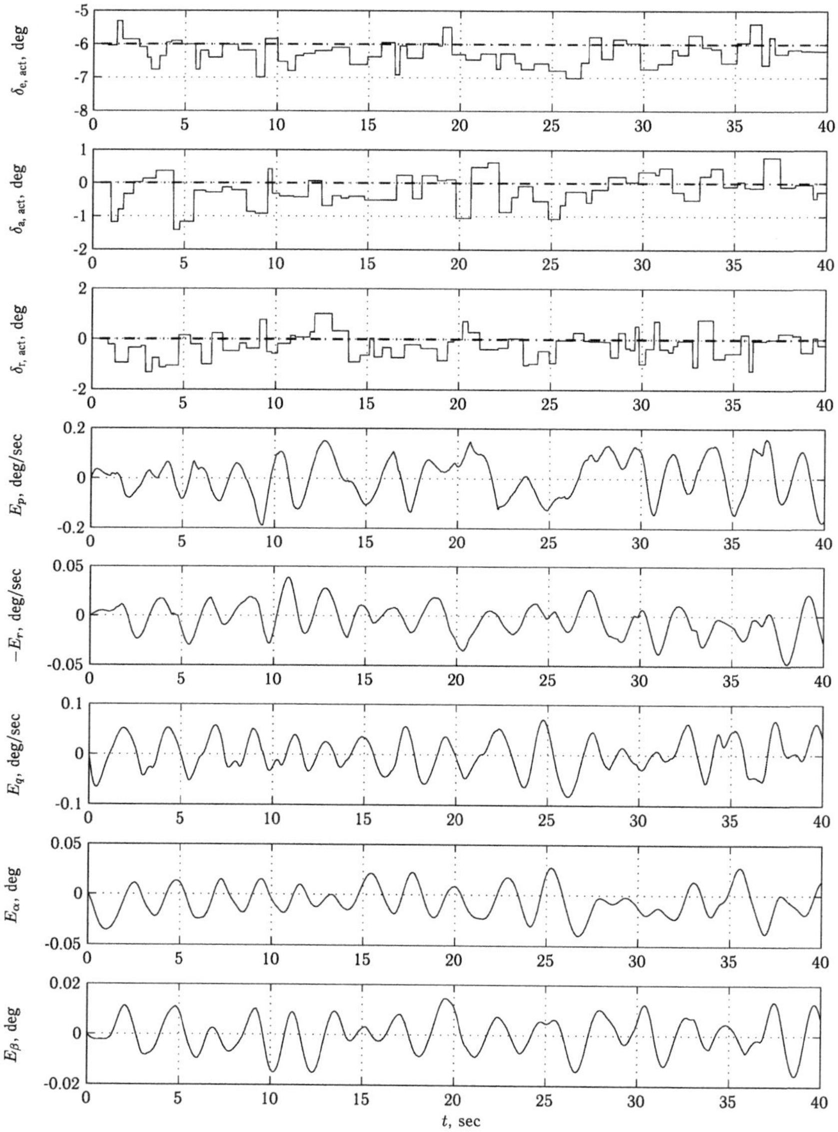

FIGURE 6.9 Evaluation of the generalization ability of the ANN model after the final 1000-step learning phase: E_α, E_β, E_p, E_r, E_q are the prediction errors for the corresponding observable variables; the straight lines on the three upper subgraphs show the values of the control variables corresponding to the test maneuver (From [4], used with permission from Moscow Aviation Institute).

as evenly and densely as possible the entire state space of the system), as well as the variety of changes in the neighboring states in time (in order to authentically reflect in the ANN model the dynamics of the simulated system). An additional complicating factor is that the subsequent

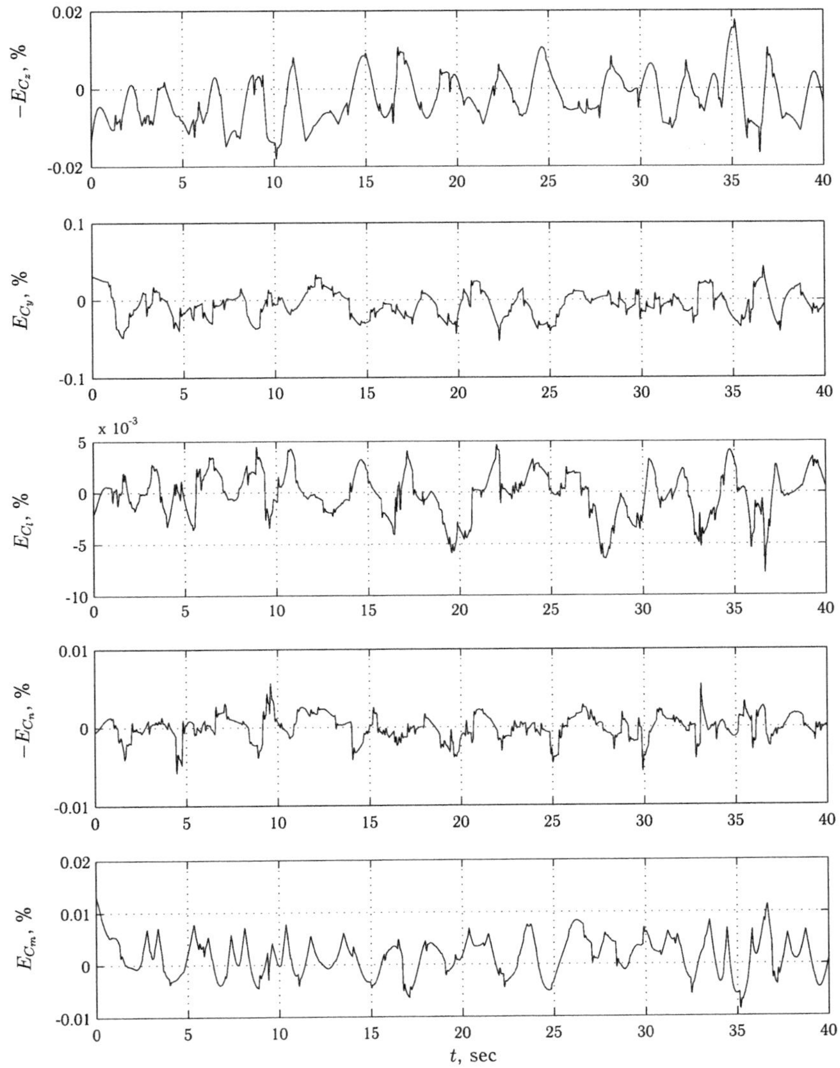

FIGURE 6.10 The values of the reproduction error for the values C_y, C_z, C_l, C_n, C_m according to the reconstructed dependencies for them during the testing of the semiempirical model (refer to the ranges of these values obtained during testing) (From [4], used with permission from Moscow Aviation Institute).

input disturbance affects the aircraft before the transition processes from one or several preceding disturbances have died out.

Fig. 6.9 characterizes the final model after the training procedure has already been completed. Data presented in Table 6.5 allow us to analyze the accuracy dynamics for this model during the training.

The accuracy of the model is determined by how accurately the nonlinear functions that represent the aerodynamic characteristics of the aircraft are reconstructed. The data in Fig. 6.9 characterize the total effect that the errors of these function approximations have on the accuracy of the trajectory predictions given by the model. These results can be regarded as completely satisfactory. However, it is also of interest to analyze how accurately the problem of the aerodynamic characteristics identification has been solved.

To answer this question, we need to extract the ANN modules corresponding to the approximated functions C_y, C_z, C_l, C_n, C_m and then compare the values they yield with the available experimental data [20]. Integral estimates of the accuracy can be obtained, for example, with the RMSE function. In the experiments above we have the following error estimates: $\text{RMSE}_{C_y} = 5.4257 \cdot 10^{-4}$, $\text{RMSE}_{C_z} = 9.2759 \cdot 10^{-4}$, $\text{RMSE}_{C_l} = 2.1496 \cdot 10^{-5}$, $\text{RMSE}_{C_m} = 1.4952 \cdot 10^{-4}$, $\text{RMSE}_{C_n} = 1.3873 \cdot 10^{-5}$. The values of the reproduction error for the functions C_y, C_z, C_l, C_n, C_m for each time instant during the testing of the semiempirical model are shown in Fig. 6.10.

6.4 SEMIEMPIRICAL MODELING OF LONGITUDINAL TRANSLATIONAL AND ANGULAR MOTION FOR A MANEUVERABLE AIRCRAFT

In this section, we consider the problem of the longitudinal motion modeling for a maneuverable aircraft as well as the identification problem for its aerodynamic characteristics, such as the coefficients of aerodynamic axial and normal forces, and the pitching moment. We solve this problem in the same way as the problems in the previous two sections, by using a class of semiempirical dynamic models that combine the possibilities of theoretical and neural network modeling.

In Section 6.3, it was shown that the simultaneous reconstruction of the dependencies for all six aerodynamic forces and moments is impractical due to the difficulties encountered in obtaining the informative data set required to find the axial force coefficient C_x. For this reason, the overall problem of aerodynamic characteristics identification has been divided into two subproblems. The first subproblem, considered in Section 6.3, is the problem of identification of the five coefficients C_y, C_z, C_l, C_n, C_m in the case of three-axis rotational motion. The second subproblem, considered in this section, amounts to the identification of the remaining axial force coefficient C_x in the case of longitudinal motion (both translational and angular). In addition, in [3,4], in order to reduce the computational complexity of the problem being solved, the solution of the modeling and identification problem was not carried out for the full range of possible values of the state variables and the controls for the dynamical system under consideration, but only for its part (on the order of several percent of the range of values of each of the variables). In this section, we extract the dependencies for the coefficients C_x, C_z, C_m on a radically more wide range of possible values of their arguments (for the list of these arguments, see (6.11) and (6.12)).

The identification problem for the axial aerodynamic force $\bar{X}$ as a nonlinear function of the corresponding arguments is traditionally challenging to solve (see, for example, [32,33]). Similarly, the problem of finding the aircraft engine thrust F_T value is difficult [32,33]. We need this value to extract $\bar{X}$ from the total force R_X measured during the flight experiment. The ANN modeling methods seem to be a promising tool for the solution of this problem in the same way as it was for the other aerodynamic characteristics identification. This hypothesis is supported by the theoretical results (see, for example, [34–36]), which show that an artificial neural network has the properties of a universal approximator, i.e., it can represent any mapping of an n-dimensional input into an m-dimensional output up to any preassigned accuracy. One purpose of this section is to verify the validity of this hypothesis concerning a somewhat complicated applied problem. We consider the problem of extracting the dependencies for the coefficients C_x, C_z, C_m from the experimental data for a rather wide range of possible values of their arguments, typical for a maneuverable aircraft. As is well known [22], in order to obtain the successful solution of this problem, it is necessary to provide the ANN learning algorithm with a representative set of data (training examples) with the required level of informativeness. Obtaining such a set of data that adequately describes the behavior of the simulated system is one of the critical issues that arise in the design of ANN models. As was shown in [3–6], this problem is solved by developing the appropriate test control actions for the modeled object and estimating the response of the object to these excitations. Taking into account that in the previous section the problem was solved only for a small fragment of the range of possible values of the state variables and controls that describe the motion of the aircraft, it is also necessary to provide coverage with training examples of the entire region under consideration.

Further in this section, a mathematical model of the longitudinal motion of a maneuverable aircraft is used, which we need to build the corresponding semiempirical ANN model, and also for the generation of a training set. We propose an algorithm for such generation that provides sufficiently uniform and dense coverage by training examples for the region of possible values of state and control variables of a maneuverable aircraft. Next, we form a semiempirical ANN model of the longitudinal controlled aircraft motion, including ANN modules realizing functional dependencies for the coefficients C_x, C_z, and C_m. In the process of learning the obtained ANN model, the identification problem for these coefficients is solved. The corresponding results of computational experiments characterizing the accuracy of the generated ANN model as a whole are given as well as the efficiency of solving the problem of identifying the aerodynamic characteristics of the aircraft.

To solve this problem, we need to develop a mathematical model of the longitudinal motion of the aircraft. In this case we consider a system of nonlinear ODEs traditional for aircraft flight dynamics [13–19], i.e.,

$$\begin{cases} \dot{V} = \dfrac{1}{m} R_x, \\ \dot{\gamma} = \dfrac{1}{mV} R_z, \\ \dot{x}_E = V \cos\gamma, \\ \dot{q} = \dfrac{1}{J_y} \bar{M}, \\ \dot{\theta} = q, \\ \dot{P}_a = \dfrac{1}{\tau_{eng}} (P_c - P_a), \\ T^2 \ddot{\delta}_e = -2T\zeta\dot{\delta}_e - \delta_e + \delta_{e_{act}}. \end{cases} \tag{6.14}$$

In the model (6.14), the following notation is used: V is airspeed, m/sec; γ is the flight path angle, deg; h is the altitude of the flight, m; x_E is the position of the center of mass of the aircraft in the earth-fixed reference frame, m; q is the pitch angular velocity, deg/sec; θ is the pitch angle, deg; δ_e is the deflection angle of the all-turn controllable stabilizer, deg; $\delta_{e_{act}}$ is command signal to the stabilizer actuator, deg; T, ζ are the time constant (sec) and the relative damping factor of the controlled stabilizer actuator; R_x, R_z are the total drag and lift forces, N; $\bar{M}$ is the total pitching moment, N·m; P_c, P_a are command and current value of the relative thrust of the engine, percent; τ_{eng} is the engine time constant, sec; m is the mass of the aircraft, kg; g is the gravitational acceleration, m/sec^2; I_y is the moment of inertia of the aircraft relative to the lateral axis, kg·m^2.

The resulting forces R_x, R_z and the moment $\bar{M}$ in (6.14) are given by the relationships of the form

$$\begin{cases} R_x = F_T(h, M, P_c)\cos\alpha - \bar{q}SC_x(V, \alpha, \phi, q) \\ \qquad - mg\sin\gamma, \\ R_z = F_T(h, M, P_c)\sin\alpha + \bar{q}SC_z(V, \alpha, \phi, q) \\ \qquad - mg\cos\gamma, \\ \bar{M} = q_p S\bar{c}C_m(V, \alpha, \phi, q), \end{cases} \tag{6.15}$$

where α is the angle of attack, deg; V is the airspeed, m/sec; S is the aircraft wing area, m^2; $\bar{c}$ is the mean aerodynamic chord, m; $q_p = (\rho V^2)/2$ is dynamic pressure, N/m^2; M is the Mach number. Here C_x, C_z are the dimensionless coefficients of the axial and normal forces and C_m is the pitching moment coefficient, which are nonlinear functions of their arguments listed in (6.15). In addition, F_T is a nonlinear function describing the dependence of the engine thrust on the altitude H, the Mach number M, and the current value of the relative thrust P_a.

The function $F_T(H, M, P_a)$, which determines the altitude-velocity and throttle characteristics of the engine, was obtained using data from [20, 21]. The mathematical model (6.14) also includes the equation that describes the engine dynamics. It is represented by a differential equation for the actual relative thrust value P_a and depends on the corresponding command value P_c as well as the relative position of the engine throttle δ_{th}. We have

$$P_c^* = \begin{cases} 64.94\,\delta_{th}, & \text{if } 0 \le \delta_{th} \le 0.77, \\ 217.38\,\delta_{th} - 117.38, & \text{if } 0.77 \le \delta_{th} \le 1.0, \\ P_c^*. \end{cases} \tag{6.16}$$

The dependence of the time constant τ_{eng} of the engine on the actual P_a value and the command P_c value of the relative thrust is determined by the following equations:

$$P_c = \begin{cases} 60, & \text{if } P_c^* \ge 50 \text{ and } P_a < 50, \\ 40, & \text{if } P_c^* < 50 \text{ and } P_a \ge 50, \\ P_c^*, & \text{otherwise,} \end{cases} \tag{6.17}$$

$$\frac{1}{\tilde{\tau}_{eng}} = \begin{cases} 1.0, & \text{if } (P_c - P_a) \le 25, \\ 0.1, & \text{if } (P_c - P_a) \ge 25, \\ 1.9 - 0.036(P_c - P_a), & \text{otherwise,} \end{cases} \tag{6.18}$$

$$\frac{1}{\tau_{eng}} = \begin{cases} 5, & \text{if } P_a \ge 50, \\ \frac{1}{\tilde{\tau}_{eng}}, & \text{otherwise.} \end{cases} \tag{6.19}$$

The function $F_T(h, M, P_a)$ from (6.15) is given in [20] as follows:

$$F_T = \begin{cases} T_{idle} + (T_{mil} - T_{idle})(P_a/50), \\ \qquad\qquad \text{if } P_a \le 50, \\ T_{mil} + (T_{max} - T_{mil})((P_a - 50)/50), \\ \qquad\qquad \text{if } P_a \ge 50, \end{cases} \tag{6.20}$$

where T_{idle}, T_{mil}, and T_{max} are the thrust of the engine in the idle, military, and maximum modes, respectively. These quantities are the functions of flight altitude and Mach number, interpolated using the experimental data given by [20] (page 93, Table VI). As an example, these values are $T_{idle} = 111.2$ N, $T_{mil} = 41421.9$ N, and $T_{max} = 74997.0$ N for $H = 3000$ m and $M = 0.4$.

The values of the atmosphere parameters (air density ρ and local speed of sound a) at a given altitude H required by the model (6.14)–(6.20) are estimated using the International Standard Atmosphere (ISA) model. The gravitational acceleration g is assumed to be constant and equal to its value at sea level.

We consider the maneuverable F-16 aircraft as an example of a specific simulated object. The experimental data for this specific aircraft given by wind tunnel tests are presented in [20]. Some additional data are contained in [21].

The following particular values of the corresponding variables in (6.14)–(6.20) were used for the simulation: the mass of the aircraft $m = 9295.44$ kg; the wing area $S = 27.87\ m^2$; the mean aerodynamic chord of the wing $\bar{c} = 3.45$ m; moment of inertia $I_y = 75673.6$ kg·m^2; center of gravity is located at 5 percent of the mean aerodynamic chord; the time constant of the stabilizer actuator $T_\phi = 0.025$ sec; coefficient of relative damping of the stabilizer actuator $\zeta = 0.707$. In these experiments we consider a range of altitudes from 1000 m to 9000 m and Mach numbers from 0.1 to 0.6.

When solving problems of the type in question, one of the most critical tasks is the generation of a representative set of data that presents the behavior of the simulated dynamical system on a sufficiently wide range of values of the variables describing the given object. This task is essential for obtaining a reliable model of such a system, but it has no simple solution. We can collect the required training data for the generated ANN model using the specially organized test excitation signals applied to the simulated dynamical system.

In this section, we propose an automatic procedure for synthesizing control actions that provide sufficiently dense coverage of the region of change for the values of the variables describing the dynamical system. Such technique assumes the availability of some initial theoretical model of the dynamical system. This model may have low accuracy, or for other reasons not satisfy the requirements for final models. However, it can be used to synthesize control signals corresponding to sufficiently diverse trajectories in the state space.

Then, we apply the resulting set of control actions to the simulated object, and the resulting trajectories are used to fill the training set. The test set is generated similarly. The description of this procedure is presented in Algorithm 1.

In addition to the representative training set, we use the weighting of individual examples from the training set to improve the generalization error of the neural network model. This procedure is based on the following considerations: if the arguments of K examples from the training set are located in a small neighborhood, then this situation is equivalent to assigning the weight K to some typical example from this region. Thus, the uneven distribution of examples can lead to a model with high accuracy in some regions of input space and a much lower accuracy in the others. To avoid this, at the end of the procedure for synthesizing the training set, we assign weights to its elements. For each element $\lambda \in \Lambda$, we find the elements $\tilde{\lambda} \in \Lambda$ located in its ε-neighborhood. Then, we assign a weight to each example from Q, which is inversely proportional to the number of neighbors found for this example.

When implementing this algorithm on a computer, one should choose an appropriate data structure for the representation of the sets Λ, $\bar{\Lambda}$, and $\bar{\Lambda}_m$, which would ensure the efficient operations of the nearest neighbor search, the search for neighbors in a given region, and the addition of new items. An example of such structure is a k-dimensional tree, implemented, for example, in the FLANN library [37].

This algorithm was successfully applied to the generation of a training set for a semiempirical model of the longitudinal motion of a maneuverable aircraft. The following ranges for variables were considered:

$$\delta_{e_{act}} \in [-25, 25] \text{ deg}, \quad \delta_e \in [-25, 25] \text{ deg},$$
$$\delta_{th} \in [0, 1], \quad P_c \in [0, 100]\%,$$
$$\theta \in [-90, 90] \text{ deg}, \ q \in [-100, 100] \text{ deg/sec},$$
$$V \in [35, 180] \text{ m/sec}, \ \alpha \in [-20, 90] \text{ deg}.$$

The effectiveness of this algorithm can be estimated using the coverage diagrams [38], for the range of acceptable values of the variables and their derivatives that describe the simulated object, using examples obtained when the test signal is applied to the object. These diagrams make it possible to evaluate the representativeness (informativeness) of training sets obtained by applying various test excitations to the modeled object. The set will be better if it covers the required range of values describing the behavior of the object under consideration more

Algorithm 1 Generation of the test maneuvers.

Require: set of the admissible values for the state variables $\mathcal{X} \subset \mathbb{R}^{n_x}$ and the control variables $\mathcal{U} \subset \mathbb{R}^{n_u}$;

Require: maximum number of maneuvers P, admissible maneuver duration limits $[K_{\min}, K_{\max}]$;

Require: maximum number of candidate trajectory segments Q, minimum admissible quality of a candidate trajectory segment $d_{\min}$, admissible candidate trajectory segment duration limits $[S_{\min}, S_{\max}]$ and number of trials R;

Require: dynamical system model right hand side $\mathbf{f}\colon \mathcal{X} \times \mathcal{U} \to \mathbb{R}^{n_x}$;

Require: metric ρ for comparison of sets of vectors;

Ensure: set of the test maneuvers M that contains pairs $\langle \mathbf{x}_0, \{\mathbf{u}_k\}_{k=1}^{K} \rangle$, where $\mathbf{x}_0 \in \mathcal{X}$ is an initial state, $\mathbf{u}_k \in \mathcal{U}$ is a sequence of controls;

Ensure: set of the function $\mathbf{f}$ argument values A for the selected trajectories, which contains vectors $\mathbf{a} \in \mathbb{R}^{n_x+n_u}$;

$M \leftarrow \varnothing$;

$A \leftarrow \varnothing$;

$p \leftarrow 1$;

while $p \leqslant P$ **and** $S_{\max} > S_{\min}$ **do**

  $r \leftarrow 1$;

  while $r < R$ **do**

    $\bar{A} \leftarrow A$;

    $\mathbf{x}_0^p \sim U(\mathcal{X})$;

    $K^p \leftarrow 0$;

    while $K^p < K_{\max}$ **do**

      $S \leftarrow \min\{S_{\max}, K_{\max} - K^p\}$;

      Generate a set of candidate maneuver segments $\left\{\mathbf{u}_k^{p,q}\right\}_{k=K^p}^{K^p+S-1}, q = 1, \ldots, Q$ within $\mathcal{U}$, for example, a sequence of steps with uniformly distributed amplitudes and frequencies;

      Numerically solve the corresponding initial value problems using the dynamical system model to obtain candidate trajectory segments $\left\{\mathbf{x}_k^{p,q}\right\}_{k=K^p}^{K^p+S}, q = 1, \ldots, Q$;

      $\tilde{A}^{p,q} \leftarrow \left\{(\mathbf{x}_k^{p,q}, \mathbf{u}_k^{p,q})^T\right\}_{k=K^p}^{K^p+S-1}, q = 1, \ldots, Q$;

      Evaluate fitness of each candidate maneuver segment $d^{p,q} = \begin{cases} 0, & \text{if } \exists k \in [K^p+1, K^p+S]\colon \mathbf{x}_k^{p,q} \notin \mathcal{X}, \\ 0, & \text{if maximum eigenvalue of the cov}\left(\tilde{A}^{p,q}\right) \text{ is too small}, q = 1, \ldots, Q; \\ \rho(\bar{A}, \tilde{A}^{p,q}), & \text{otherwise}, \end{cases}$

      Find the best candidate maneuver segment $q^* \leftarrow \underset{q}{\operatorname{argmax}}\, d^{p,q}$ with fitness $d^* \leftarrow \max\limits_q d^{p,q}$;

      if $d^* > d_{\min}$ **then**

        $\mathbf{u}_k^p \leftarrow \mathbf{u}_k^{p,q^*}, k = K^p, \ldots, K^p + S - 1$;

        $\mathbf{x}_k^p \leftarrow \mathbf{x}_k^{p,q^*}, k = K^p + 1, \ldots, K^p + S$;

        $\bar{A} \leftarrow \bar{A} \cup \tilde{A}^{p,q^*}$;

        $K^p \leftarrow K^p + S$;

```
                break;
            else
                r ← r + 1;
            end if
        end while
        if r < R then
            A ← Ā;
            break;
        else
            Decrease S_max by some amount;
        end if
    end while
end while
```

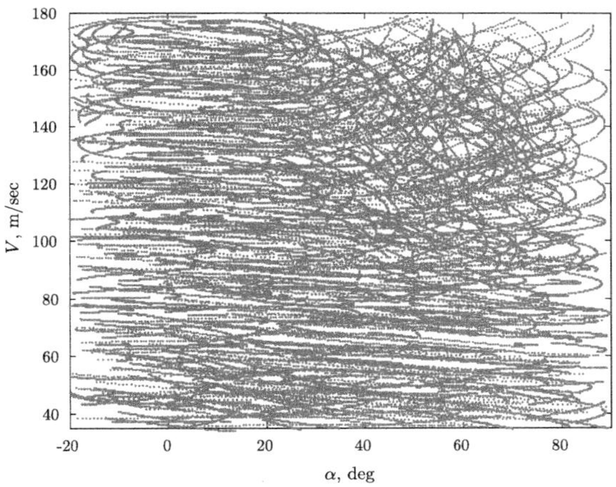

FIGURE 6.11 Coverage diagram (α, V) for the training set (From [38], used with permission from Moscow Aviation Institute).

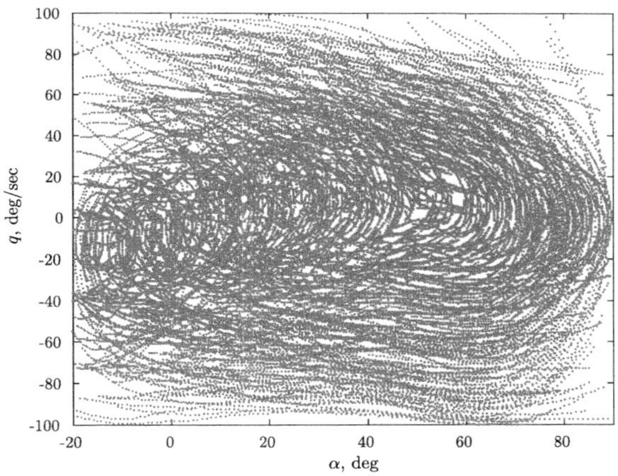

FIGURE 6.12 Coverage diagram (α, q) for the training set (From [38], used with permission from Moscow Aviation Institute).

densely and evenly. Since the original representation is multidimensional, for the sake of clarity, we use its two-dimensional cross-sections. As examples, in Figs. 6.11 and 6.12, diagrams for the two most significant pairs of variables (α, V) and (α, q), respectively, are shown. A single cross-like point represents each of the examples of the training set on the diagram. The total number of examples in each diagram is 70,000.

A general approach to the semiempirical ANN model design for controllable dynamical systems was described in [3–6]. In [1,2] it was applied to the problems of identification of the aircraft aerodynamic characteristics C_y, C_z, C_l, C_m, C_n, for the problem of the three-dimensional rotational motion. In this section, we develop a semiempirical ANN model of longitudinal motion, based on the theoretical model (6.14)–(6.20). This ANN model allows us to approximate the coefficients C_x, C_z, C_m on a vast range of possible values of their arguments.

In the model (6.14)–(6.20), the variables V, h, γ, x_E, q, θ, P_a, δ_e, $\dot{\delta}_e$ are the states of the controlled object, and the variables $\delta_{e_{act}}$ and δ_{th}

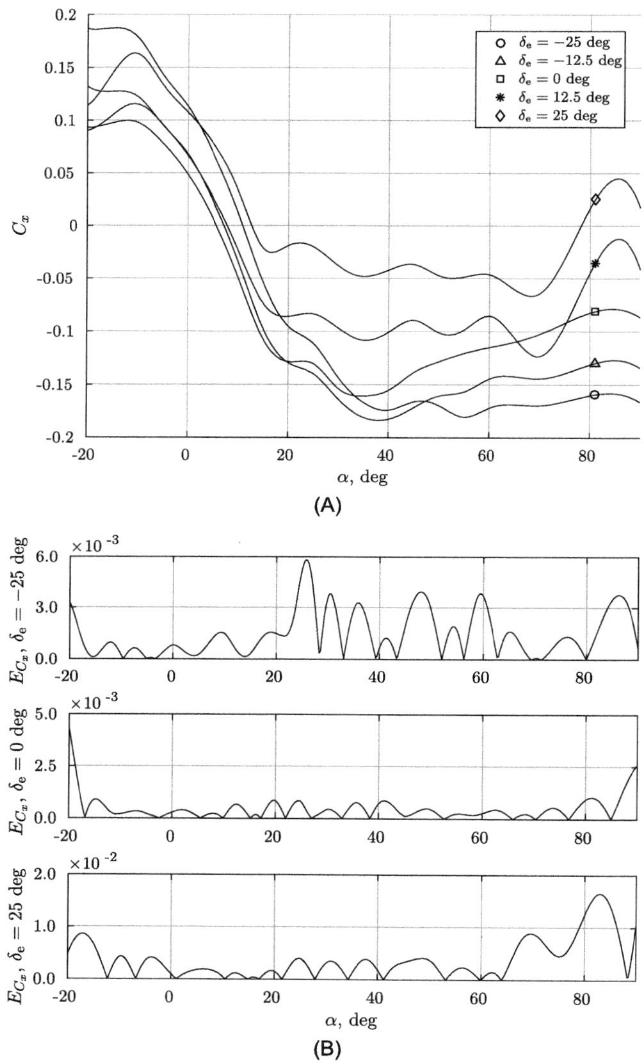

FIGURE 6.13 The coefficient $C_x(\alpha, \phi)$ for various values of ϕ according to (A) the data from [20] and (B) the error of its approximation E_{C_x} for fixed $q = 0$ deg/sec and $V = 150$ m/sec (From [38], used with permission from Moscow Aviation Institute).

are controls. The values of the controls are constrained; $\delta_{e_{act}} \in [-25, 25]$ deg, $\delta_{th} \in [0, 1]$. The training and test sets were generated according to the procedure described in the previous section, with a sampling period of $\Delta t = 0.01$ sec. The vector of state variables is partially observable, $y(t) = [V(t), \theta(t), q(t)]^T$. The output of the system $y(t)$ is affected by additive Gaussian noise with a standard deviation $\sigma_V = 0.01$ m/sec, $\sigma_\theta = 0.01$ deg, $\sigma_q = 0.005$ deg/sec.

The semiempirical ANN model learning is performed by the algorithms described in Chap-

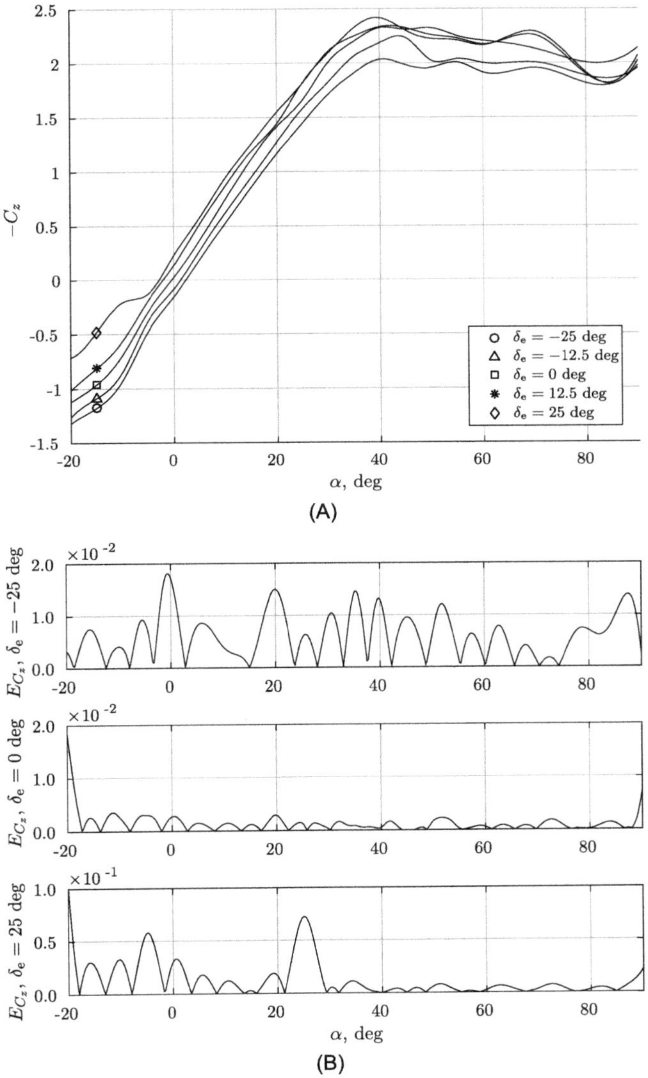

FIGURE 6.14 The coefficient $C_z(\alpha, \phi)$ for various values of ϕ according to (A) the data from [20] and (B) the error of its approximation E_{C_z} for fixed $q = 0$ deg/sec and $V = 150$ m/sec (From [38], used with permission from Moscow Aviation Institute).

ter 2. These algorithms were implemented using the Matlab Neural Network Toolbox and rely on the Levenberg–Marquardt algorithm for minimization of the mean square error objective function for the neural network in the LDDN form. The Jacobi matrix is calculated using the RTRL algorithm [22].

ANN modules for functions C_x, C_z, and C_m are represented by sigmoidal feedforward ANNs. The variables α, δ_e, and q/V are used as

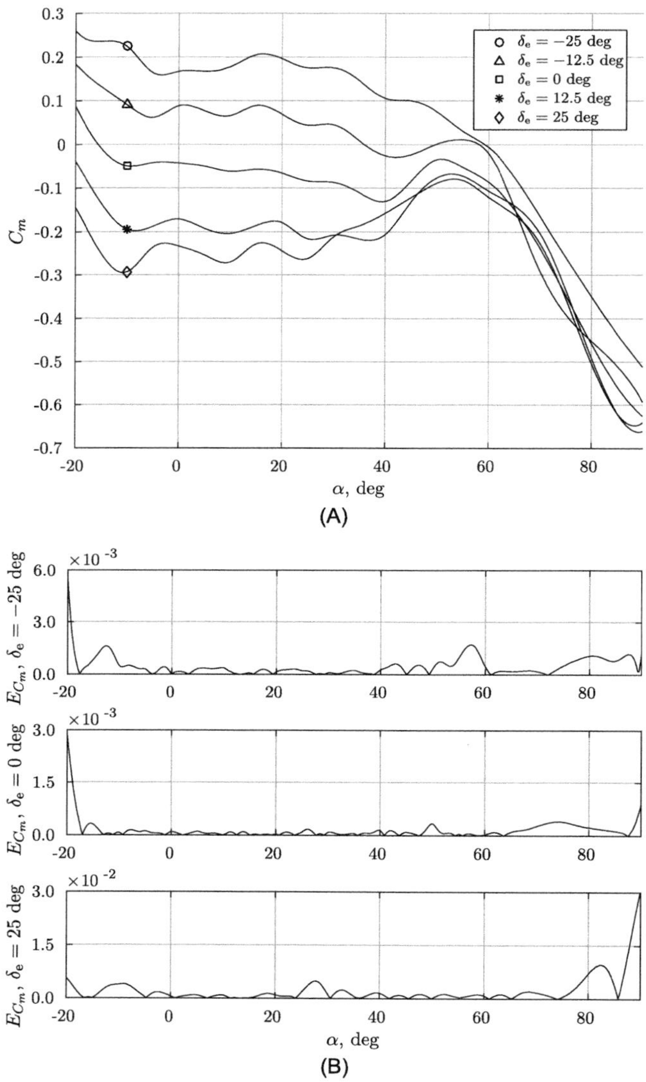

FIGURE 6.15 The coefficient $C_m(\alpha, \phi)$ for various values of ϕ according to (A) the data from [20] and (B) the error of its approximation E_{C_m} for fixed $q = 0$ deg/sec and $V = 150$ m/sec (From [38], used with permission from Moscow Aviation Institute).

inputs of each of the modules. The ANN modules for the functions C_x and C_z have two hidden layers, the first of which includes 10 neurons and the second of which includes 20. The ANN module for the function C_m has three hidden layers, the first of which includes 10 neurons, the second includes 15 neurons, and the third includes 20 neurons.

The simulation errors on the test set for the obtained semiempirical ANN model of the longitudinal motion of the maneuverable aircraft are $\text{RMSE}_V = 0.0026\ \text{m/sec}$, $\text{RMSE}_\alpha = 0.183\ \text{deg}$, $\text{RMSE}_q = 0.0071\ \text{deg/sec}$.

The identification of the accuracy of the aerodynamical coefficients C_x, C_z, and C_m is demonstrated by the data shown in Figs. 6.13, 6.14, and 6.15.

In each of these figures, the upper part of the figure shows the actual values (according to the data from [20,21]) of the required coefficients depending on the angle of attack α and the deflection angle of the controlled stabilizer δ_e. The lower part shows the approximation errors for the corresponding ANN modules. We can see that the accuracy achieved is very high.

The results presented above allow us to draw the following conclusions.

Just as in the case described in the previous section and in [1,2] for the identification problem of the aerodynamic coefficients C_y, C_z, C_l, C_n, C_m, the semiempirical ANN modeling methods successfully provide the solution to the identification problem of the axial force coefficient C_x, given the known engine model. If this model is unavailable, then the solution to the identification problem would provide the approximation for the total axial force coefficient, dependent on the control variable δ_{th} as well. As follows from (6.14) and (6.15), this is quite enough to simulate the motion of the aircraft.

The second important conclusion, which follows from the obtained results, is that the "representation power" of the semiempirical model is quite sufficient to approximate complicated nonlinear functional dependencies defined on a wide range of values of their arguments, provided that there is a training set with the required level of informativity.

The results of computational experiments show high accuracy of both the obtained ANN model of the longitudinal motion of the aircraft and the corresponding aerodynamic characteristics.

REFERENCES

[1] Egorchev MV, Kozlov DS, Tiumentsev YV. Neural network adaptive semi-empirical models for aircraft controlled motion. In: 29th Congress of the International Council of the Aeronautical Sciences; 2014.

[2] Egorchev MV, Tiumentsev YV. Learning of semi-empirical neural network model of aircraft three-axis rotational motion. Opt Memory Neural Netw (Inf Opt) 2015;24(3):210–7. https://doi.org/10.3103/S1060992X15030042.

[3] Egorchev MV, Kozlov DS, Tiumentsev YV, Chernyshev AV. Neural network based semi-empirical models for controlled dynamical systems. Her Comput Inf Technol 2013;(9):3–10 (in Russian).

[4] Egorchev MV, Kozlov DS, Tiumentsev YV. Aircraft aerodynamic model identification: A semi-empirical neural network based approach. Her Mosc Aviat Inst 2014;21(4):13–24 (in Russian).

[5] Egorchev MV, Tiumentsev YV. Neural network semi-empirical modeling of the longitudinal motion for maneuverable aircraft and identification of its aerodynamic characteristics. Studies in computational intelligence, vol. 736. Springer Nature; 2018. p. 65–71.

[6] Egorchev MV, Tiumentsev YV. Semi-empirical neural network based approach to modelling and simulation of controlled dynamical systems. Proc Comput Sci 2017;123:134–9.

[7] Klein V, Morelli EA. Aircraft system identification: Theory and practice. Reston, VA: AIAA; 2006.

[8] Tischler M, Remple RK. Aircraft and rotorcraft system identification: Engineering methods with flight-test examples. Reston, VA: AIAA; 2006.

[9] Jategaonkar RV. Flight vehicle system identification: A time domain methodology. Reston, VA: AIAA; 2006.

[10] Berestov LM, Poplavsky BK, Miroshnichenko LY. Frequency domain aircraft identification. Moscow: Mashinostroyeniye; 1985 (in Russian).

[11] Vasilchenko KK, Kochetkov YA, Leonov VA, Poplavsky BK. Structural identification of mathematical model of aircraft motion. Moscow: Mashinostroyeniye; 1993 (in Russian).

[12] Ovcharenko VN. Identification of aircraft aerodynamic characteristics from flight data. Moscow: The MAI Press; 2017 (in Russian).

[13] Etkin B, Reid LD. Dynamics of flight: Stability and control. 3rd ed. New York, NY: John Wiley & Sons, Inc.; 2003.

[14] Boiffier JL. The dynamics of flight: The equations. Chichester, England: John Wiley & Sons; 1998.

[15] Roskam J. Airplane flight dynamics and automatic flight control. Part I. Lawrence, KS: DAR Corporation; 1995.

[16] Roskam J. Airplane flight dynamics and automatic flight control. Part II. Lawrence, KS: DAR Corporation; 1998.
[17] Cook MV. Flight dynamics principles. Amsterdam: Elsevier; 2007.
[18] Hull DG. Fundamentals of airplane flight mechanics. Berlin: Springer; 2007.
[19] Stevens BL, Lewis FL, Johnson E. Aircraft control and simulation: Dynamics, control design, and autonomous systems. 3rd ed. Hoboken, New Jersey: John Wiley & Sons, Inc.; 2016.
[20] Nguyen LT, Ogburn ME, Gilbert WP, Kibler KS, Brown PW, Deal PL. Simulator study of stall/post-stall characteristics of a fighter airplane with relaxed longitudinal static stability. NASA TP-1538; Dec. 1979.
[21] Sonneveld L. Nonlinear F-16 model description. The Netherlands: Control & Simulation Division, Delft University of Technology; June 2006.
[22] Haykin S. Neural networks: A comprehensive foundation. 2nd ed. Upper Saddle River, NJ, USA: Prentice Hall; 1998.
[23] Hamel PG, Jategaonkar RV. Evolution of flight vehicle system identification. J Aircr 1996;33(1):9–28.
[24] Hamel PG, Kaletka J. Advances in rotorcraft system identification. Prog Aerosp Sci 1997;33(3–4):259–84.
[25] Jategaonkar RV, Fischenberg D, von Gruenhagen W. Aerodynamic modeling and system identification from flight data — recent applications at DLR. J Aircr 2004;41(4):681–91.
[26] Klein V. Estimation of aircraft aerodynamic parameters from flight data. Prog Aerosp Sci 1989;26(1):1–77.
[27] Iliff KW. Parameter estimation for flight vehicles. J Guid Control Dyn 1989;12(5):609–22.
[28] Morelli EA, Klein V. Application of system identification to aircraft at NASA Langley Research Center. J Aircr 2005;42(1):12–25.
[29] Wang KC, Iliff KW. Application of system identification to aircraft at NASA Langley Research Center. J Aircr 2004;41(4):752–64.
[30] Dietterich TG. Machine-learning research: Four current directions. AI Mag 1997;18(7):97–136.
[31] Joshi P, Kulkarni P. Incremental learning: Areas and methods — a survey, vol. 2. Int J Data Min Knowl Manag Process 2012;2(5):43–51.
[32] Niewald PW, Parker SL. Flight-test techniques employed to successfully verify F/A-18E in-flight lift and drag. J Aircr 2000;37(2):194–200.
[33] Mulder JA, van Sliedregt JM. Estimation of drag and thrust of jet-propelled aircraft by non-steady flight-test maneuvers. Delft Univ. of Technology, Memorandum M-255, Dec. 1976.
[34] Cybenko G. Approximation by superposition of a sigmoidal function. Math Control Signals Syst 1989;2(4):303–14.
[35] Hornik K, Stinchcombe M, White H. Multilayer feedforward networks are universal approximators. Neural Netw 1989;2(5):359–66.
[36] Gorban AN. Generalized approximation theorem and computational capabilities of neural networks. Sib J Numer Math 1998;1(1):11–24 (in Russian).
[37] Muja M, Lowe DG. Scalable nearest neighbor algorithms for high dimensional data. IEEE Trans Pattern Anal Mach Intell 2014;36(11):2227–40.
[38] Egorchev MV, Tiumentsev YV. Neural network based semi-empirical approach to the modeling of longitudinal motion and identification of aerodynamic characteristics for maneuverable aircraft. Tr MAI 2017;(94):1–16 (in Russian).

APPENDIX

A

Results of Computational Experiments With Adaptive Systems

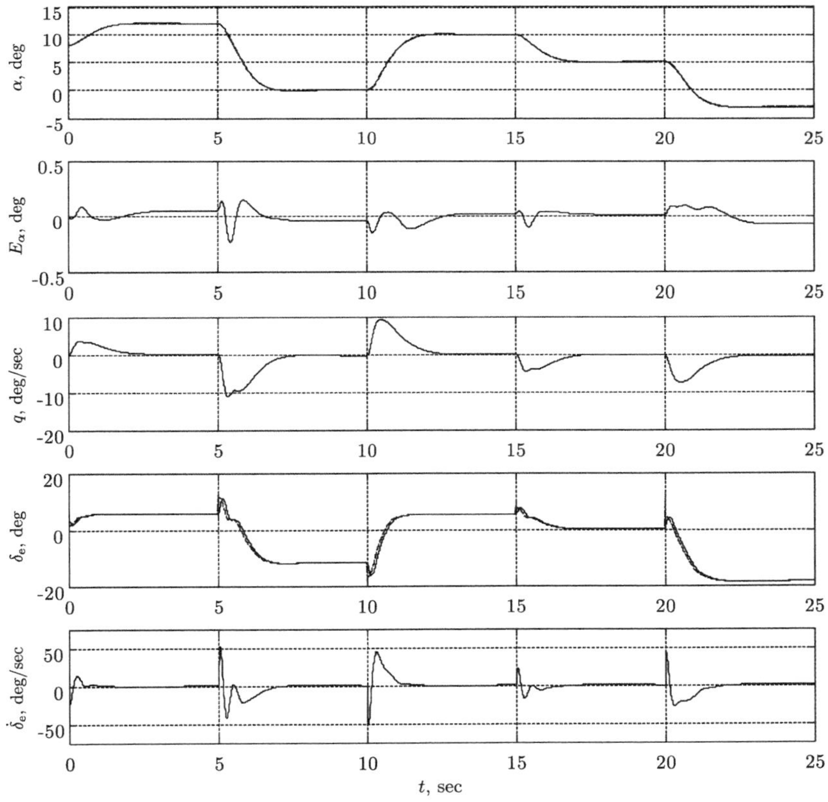

FIGURE A.1 The results of a computational experiment for an MRAC-type control system applied to the hypersonic research vehicle X-43 for estimating the influence of the natural frequency ω_{rm} of the reference model ($\omega_{rm} = 2$; stepwise reference signal on the angle of attack; flight mode $M = 6$, $H = 30$ km). Here α is angle of attack, deg (dotted line for reference model, solid line for plant); E_α is tracking error for the given angle of attack, deg; q is the pitch angular velocity, deg/sec; δ_e is command signal of the actuator (dotted line) and angle of deflection of elevons (solid line), deg; $d\delta_e/dt$ is the angular velocity of the deflection of the elevons, deg/sec; t is time, sec.

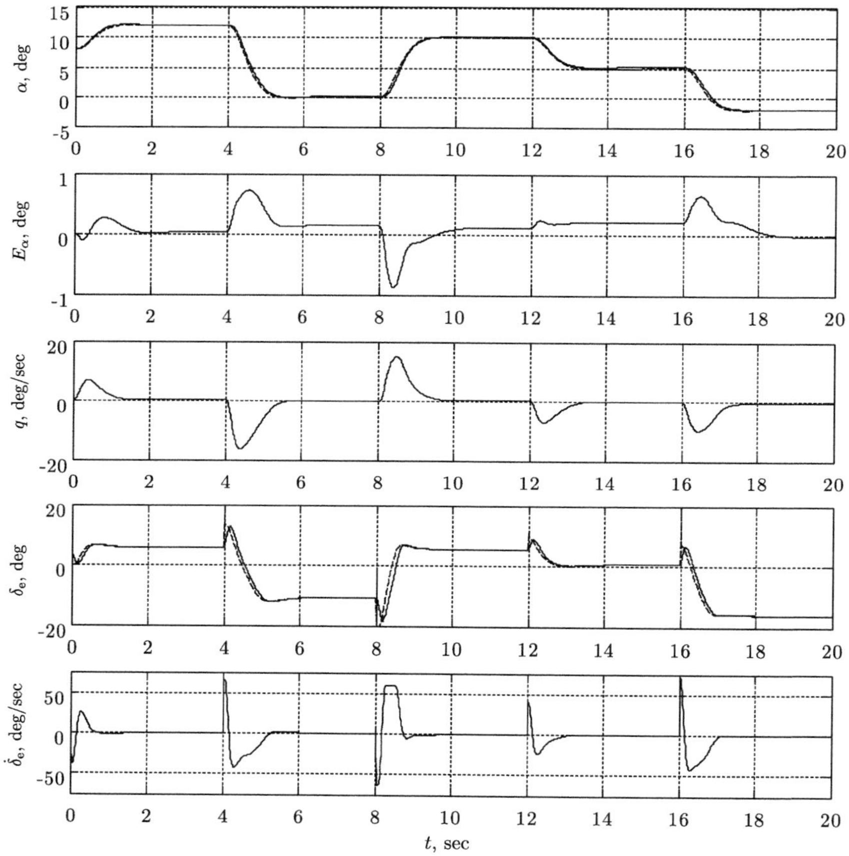

FIGURE A.2 The results of a computational experiment for an MRAC-type control system applied to the hypersonic research vehicle X-43 for estimating the influence of the natural frequency ω_{rm} of the reference model ($\omega_{rm} = 3$; stepwise reference signal on the angle of attack; flight mode $M = 6$, $H = 30$ km). Here α is angle of attack, deg (dotted line for reference model, solid line for plant); E_α is tracking error for the given angle of attack, deg; q is the pitch angular velocity, deg/sec; δ_e is command signal of the actuator (dotted line) and angle of deflection of elevons (solid line), deg; $d\delta_e/dt$ is the angular velocity of the deflection of the elevons, deg/sec; t is time, sec.

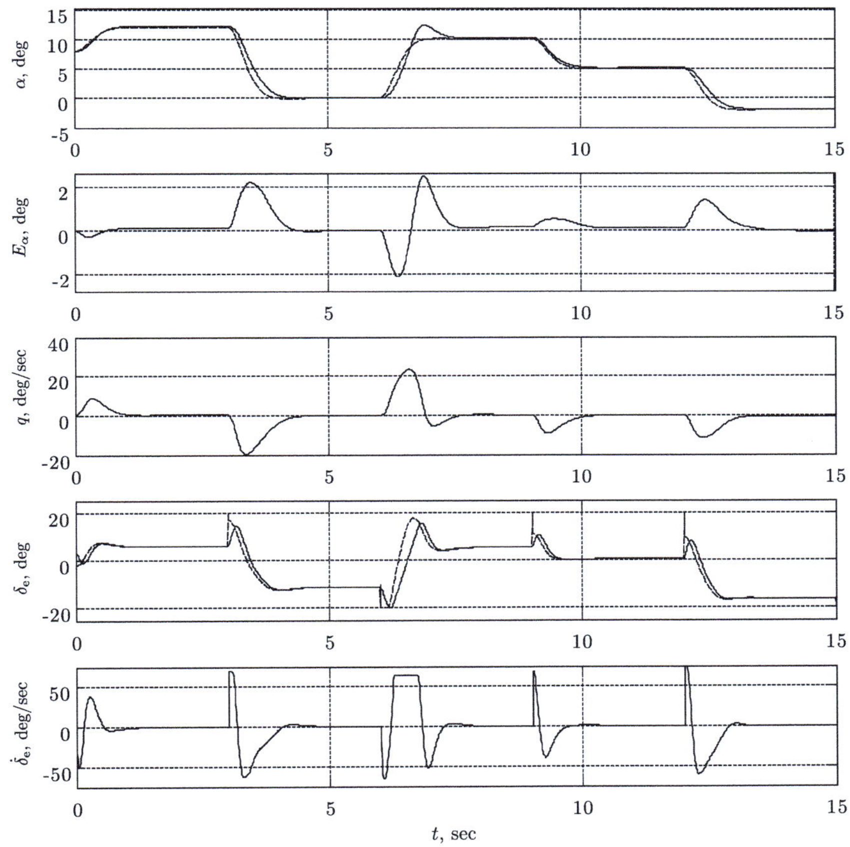

FIGURE A.3 The results of a computational experiment for an MRAC-type control system applied to the hypersonic research vehicle X-43 for estimating the influence of the natural frequency ω_{rm} of the reference model ($\omega_{rm} = 4$; stepwise reference signal on the angle of attack; flight mode $M = 6$, $H = 30$ km). Here α is angle of attack, deg (dotted line for reference model, solid line for plant); E_α is tracking error for the given angle of attack, deg; q is the pitch angular velocity, deg/sec; δ_e is command signal of the actuator (dotted line) and angle of deflection of elevons (solid line), deg; $d\delta_e/dt$ is the angular velocity of the deflection of the elevons, deg/sec; t is time, sec.

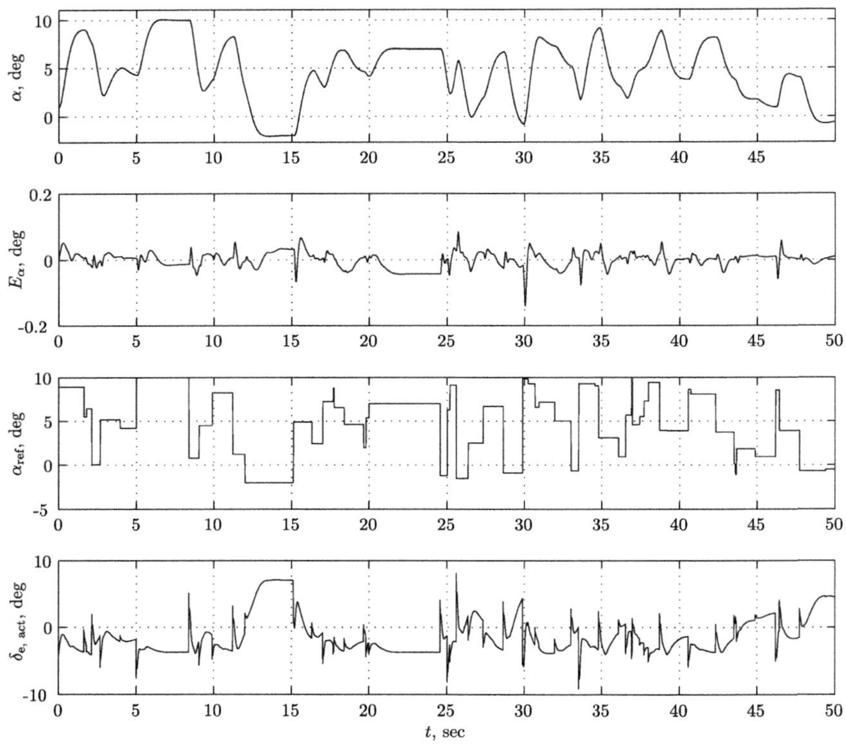

FIGURE A.4 The results of a computational experiment for the MRAC-type control system with the compensator (hypersonic research vehicle X-43, flight mode $M = 6$). Here α is angle of attack, deg (dotted line for reference model, solid line for plant); E_α is tracking error for the given angle of attack, deg; α_{ref} is reference signal on the angle of attack, deg; $\delta_{e,act}$ is command signal for the elevator actuator, deg; t is time, sec.

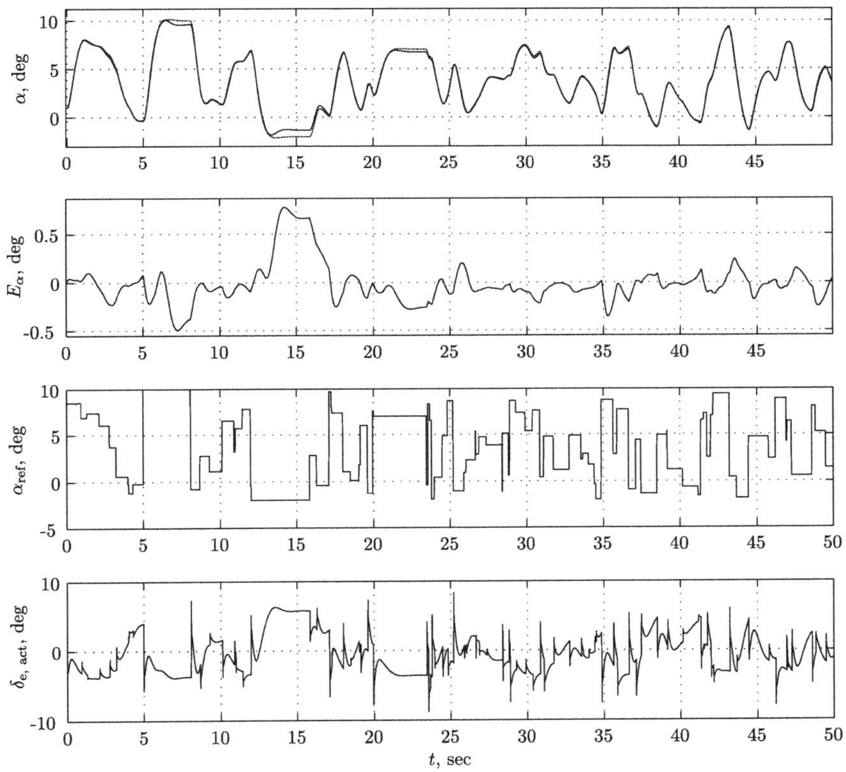

FIGURE A.5 The results of a computational experiment for the MRAC-type control system without the compensator (hypersonic research vehicle X-43, flight mode $M = 6$). Here α is angle of attack, deg (dotted line for reference model, solid line for plant); E_α is tracking error for the given angle of attack, deg; α_{ref} is specifying signal on the angle of attack, deg; $\delta_{e,\,act}$ is command signal for the elevator actuator, deg; t is time, sec.

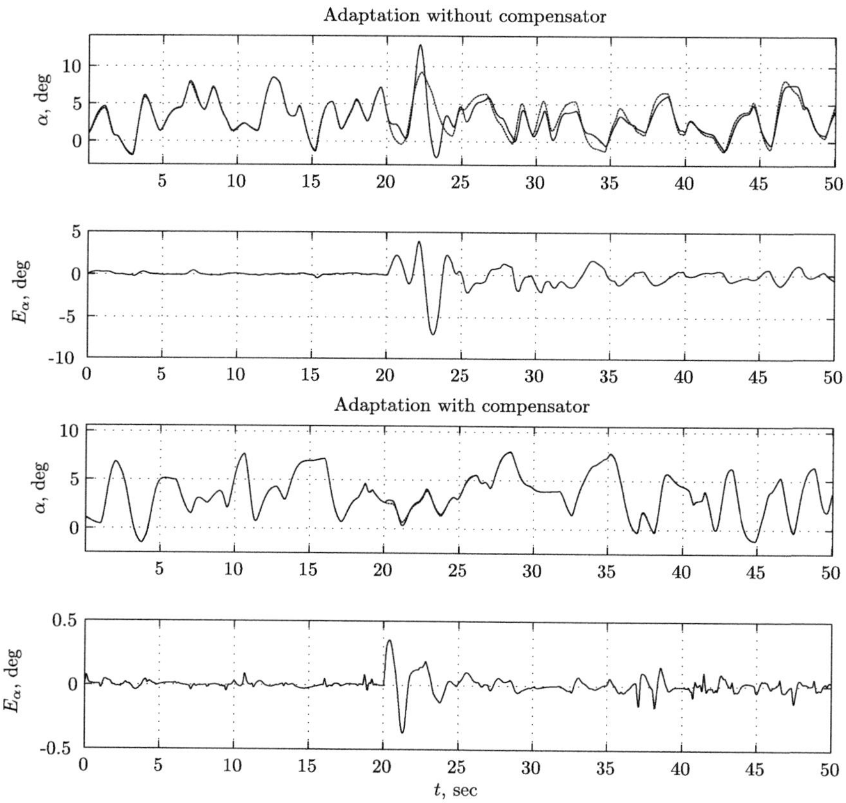

FIGURE A.6 The results of a computational experiment for the MRAC-type control system without the compensator (top pair of graphs) and with the compensator (bottom pair of graphs) for the hypersonic research vehicle X-43, flight mode $M = 6$, in the event of a failure situation at time $t = 20$ sec (backwards shift of the aircraft center of gravity by 5%). Here α is angle of attack, deg (dotted line for reference model, solid line for plant); E_α is tracking error for the given angle of attack, deg; t is time, sec (see α_{ref} and $\delta_{e,\,act}$ signals related to these cases in Fig. A.7).

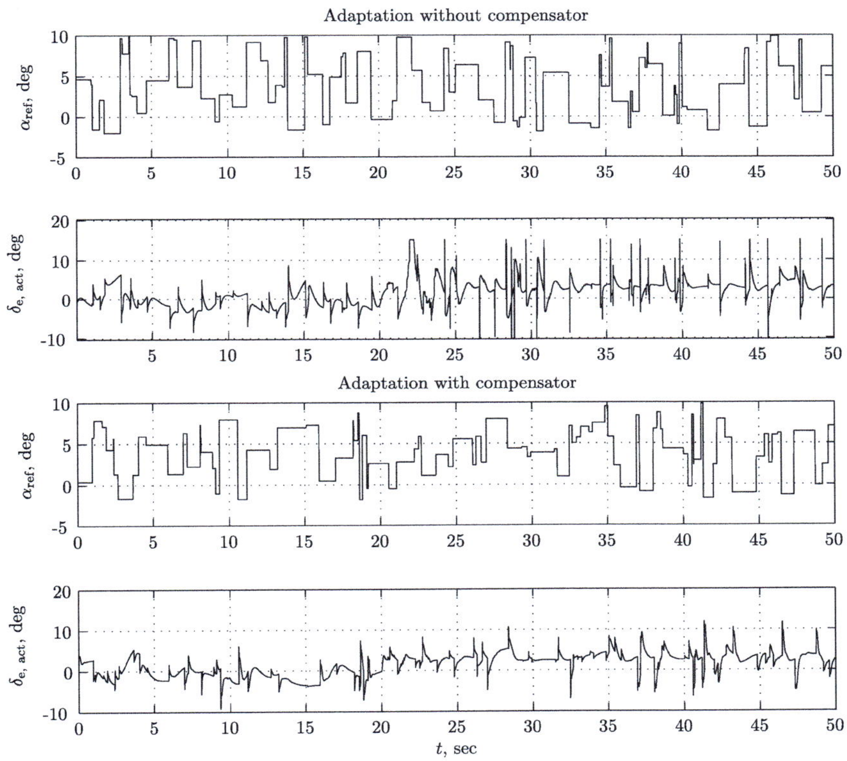

FIGURE A.7 The results of a computational experiment for the MRAC-type control system without the compensator (top pair of graphs) and with the compensator (bottom pair of graphs) for the hypersonic research vehicle X-43, flight mode $M = 6$, in the event of a failure situation at time $t = 20$ sec (aircraft center of gravity shift by 5% back). Here α_{ref} is specifying signal on the angle of attack, deg,; $\delta_{e,\,act}$ is command signal for the elevator actuator, deg; t is time, sec.

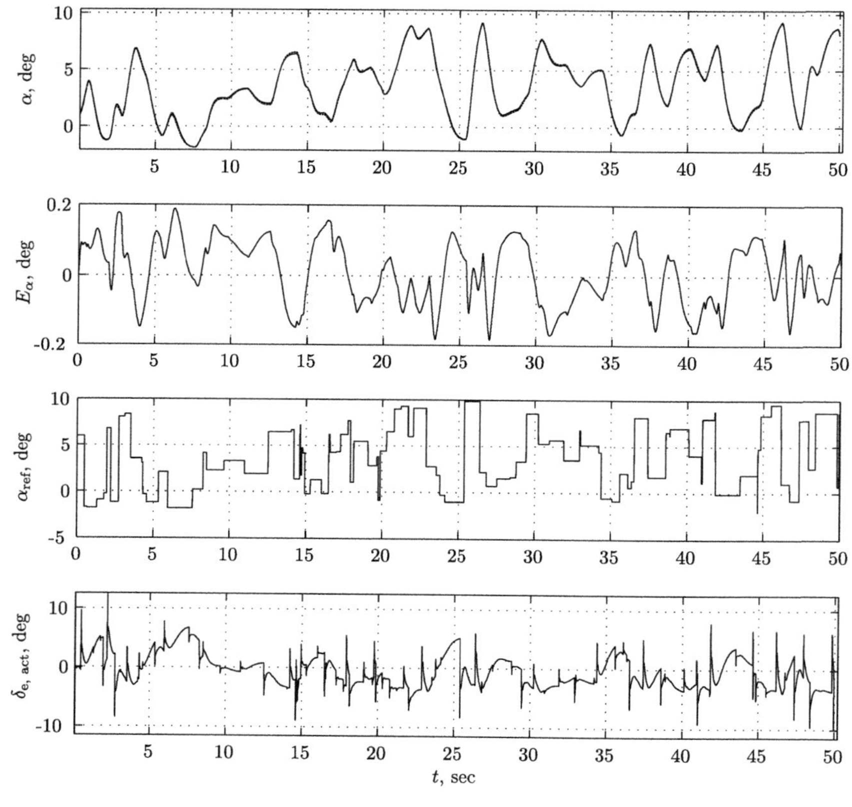

FIGURE A.8 The results of a computational experiment for a model reference control system with a compensator for an insufficiently accurate ANN model of the controlled object (hypersonic research vehicle X-43, flight regime $M = 6$). Here α is angle of attack, deg (dotted line for reference model, solid line for plant); E_α is the angle of attack tracking error, deg; α_{ref} is the angle of attack reference signal, deg; $\delta_{e,\,act}$ is command signal for the elevator actuator, deg; t is time, sec.

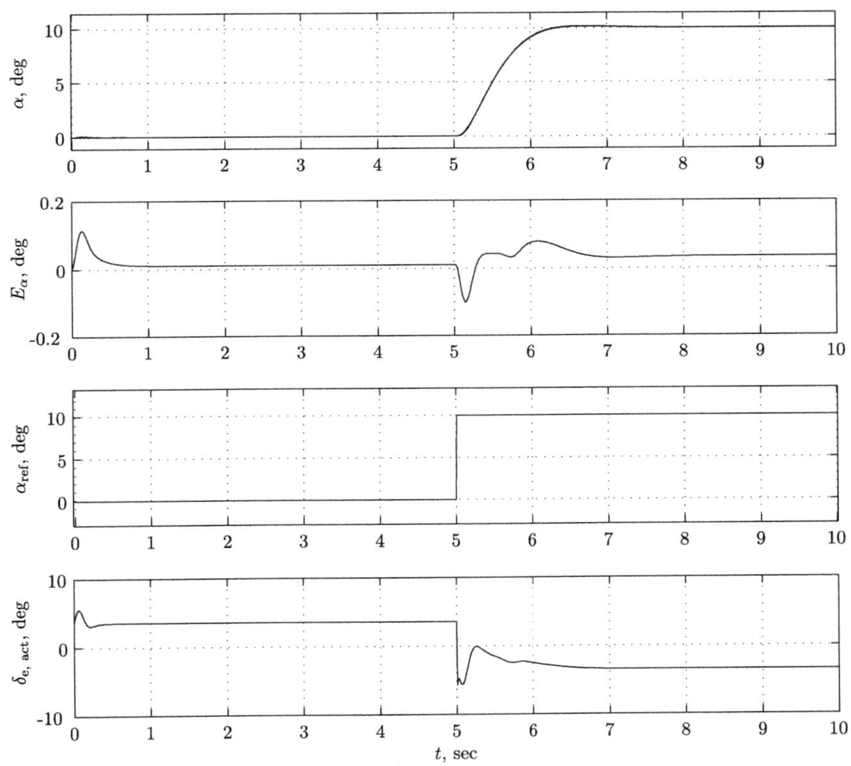

FIGURE A.9 The results of a computational experiment for a model reference control system with a compensator for the stepwise reference signal (hypersonic research vehicle X-43, flight regime $M = 6$). Here α is angle of attack, deg (dotted line for reference model, solid line for plant); E_α is the angle of attack tracking error, deg; α_{ref} is the angle of attack reference signal, deg; $\delta_{e,\,act}$ is command signal for the elevator actuator, deg; t is time, sec.

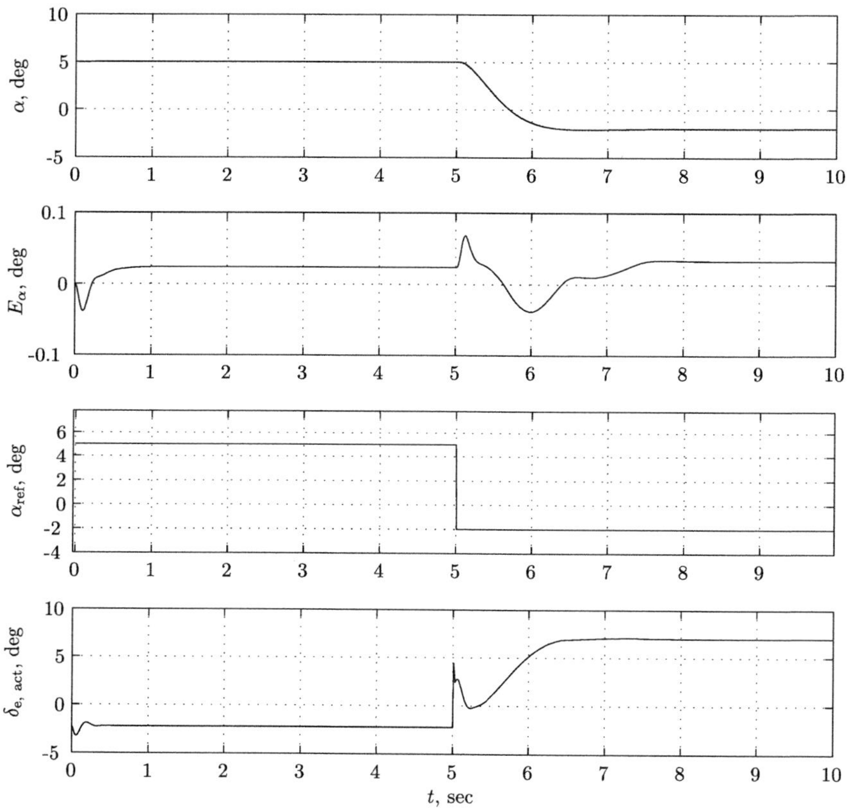

FIGURE A.10 The results of a computational experiment for a model reference control system with a compensator for the stepwise reference signal (hypersonic research vehicle X-43, flight regime $M = 6$, variant No. 2). Here α is angle of attack, deg (dotted line for reference model, solid line for plant); E_α is the angle of attack tracking error, deg; α_{ref} is the angle of attack reference signal, deg; $\delta_{e,\, act}$ is command signal for the elevator actuator, deg; t is time, sec.

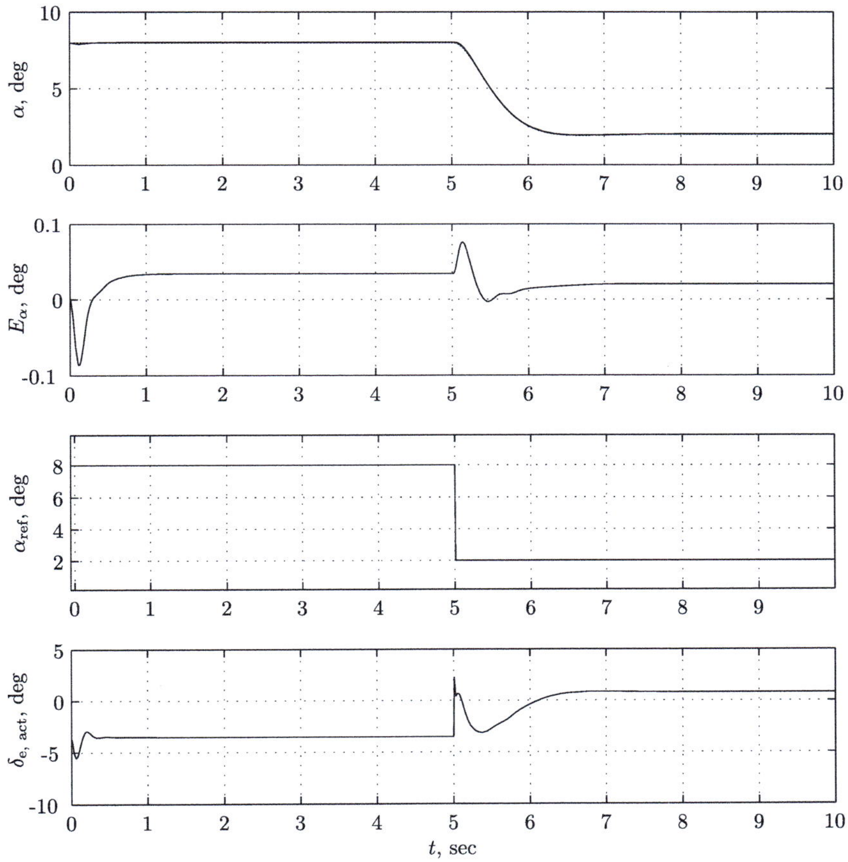

FIGURE A.11 The results of a computational experiment for a model reference control system with a compensator for the stepwise reference signal (hypersonic research vehicle X-43, flight regime $M = 6$, variant No. 3). Here α is angle of attack, deg (dotted line for reference model, solid line for plant); E_α is the angle of attack tracking error, deg; α_{ref} is the angle of attack reference signal, deg; $\delta_{e,\,act}$ is command signal for the elevator actuator, deg; t is time, sec.

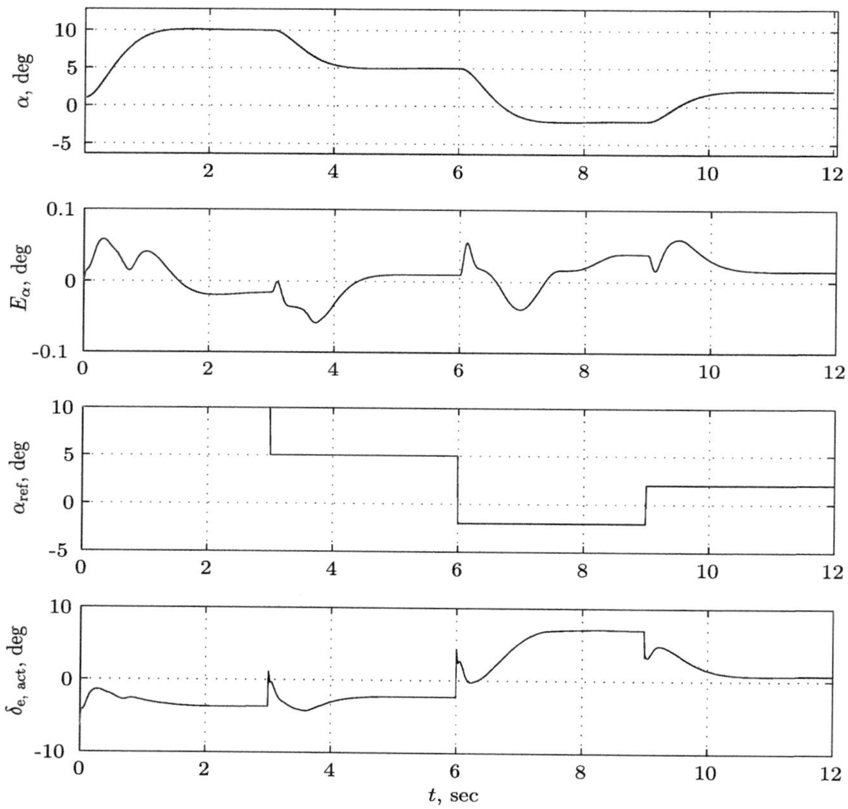

FIGURE A.12 The results of a computational experiment for a model reference control system with a compensator for the sequence of stepwise reference signals (hypersonic aircraft X-43, flight regime $M = 6$, variant No. 3). Here α is angle of attack, deg (dotted line for reference model, solid line for plant); E_α is the angle of attack tracking error, deg; α_{ref} is the angle of attack reference signal, deg; $\delta_{e,\,act}$ is command signal for the elevator actuator, deg; t is time, sec.

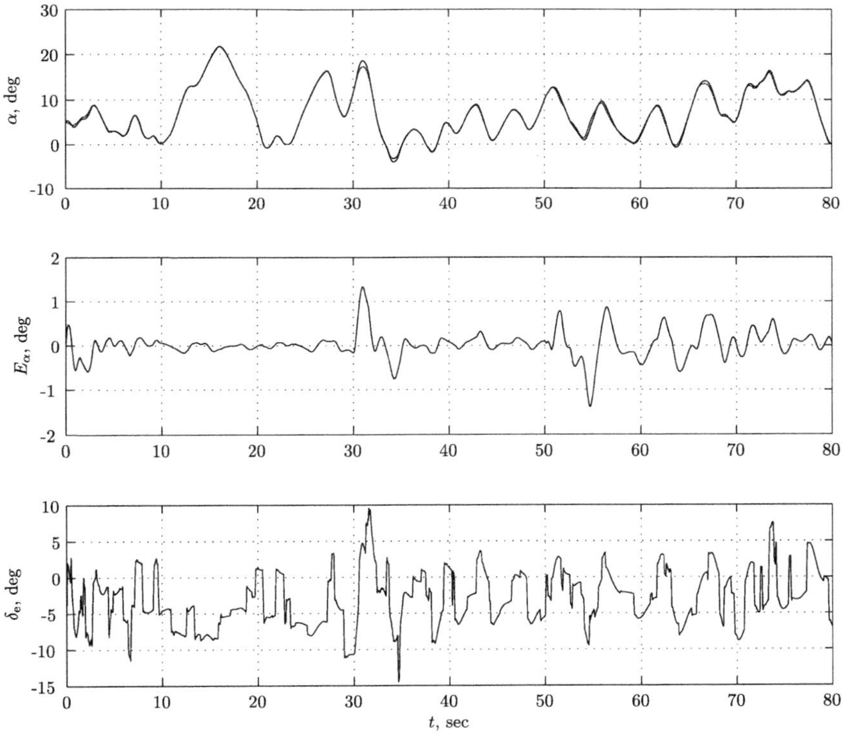

FIGURE A.13 The results of a computational experiment for the MRAC-type control system with compensator (F-16 aircraft, flight mode with a test speed $V_{ind} = 300$ km/h). Adaptation to the change in the dynamics of the controlled object: backwards shift of the center by 10% ($t = 30$ sec); 50% decrease in the effectiveness of the control ($t = 50$ sec). Here α is angle of attack, deg (dotted line for reference model, solid line for plant); E_α is tracking error for the reference angle of attack, deg; δ_e is the deflection angle of the stabilizer, deg; t is time, sec.

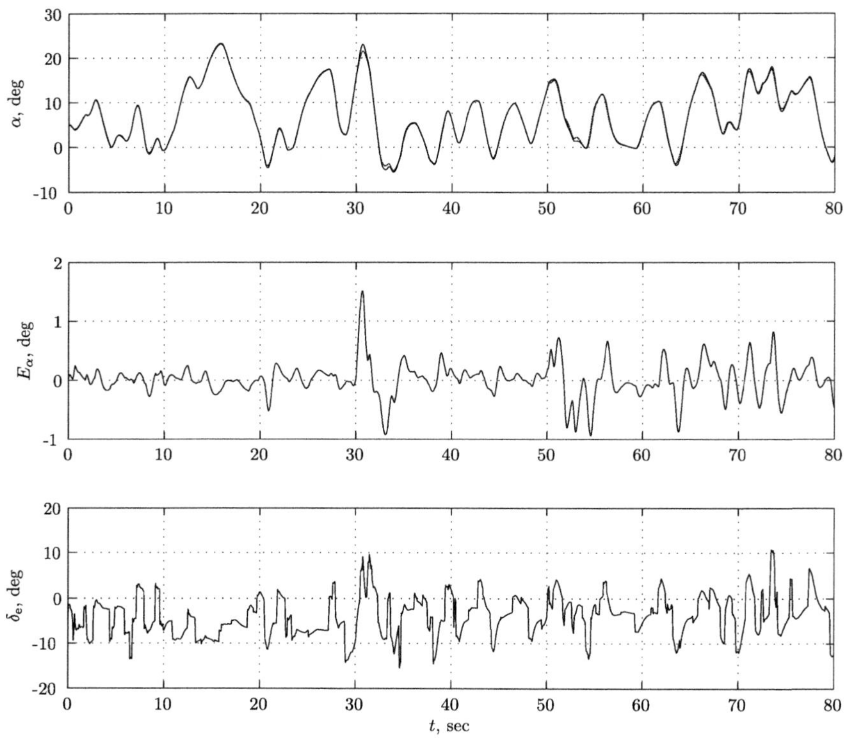

FIGURE A.14 The results of a computational experiment for the MRAC-type control system with compensator (F-16 aircraft, flight mode with a test speed $V_{ind} = 400$ km/h). Adaptation to the change in the dynamics of the controlled object: backwards shift of the center by 5% ($t = 30$ sec); 50% decrease in the effectiveness of the control ($t = 50$ sec). Here α is angle of attack, deg (dotted line for reference model, solid line for plant); E_α is tracking error for the reference angle of attack, deg; δ_e is the deflection angle of the stabilizer, deg; t is time, sec.

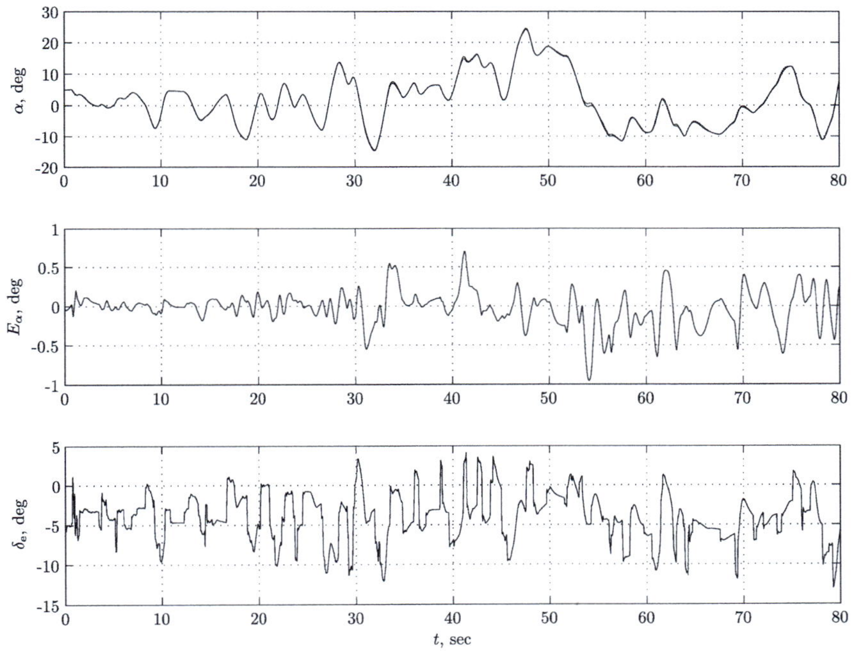

FIGURE A.15 The results of a computational experiment for the MRAC-type control system with compensator (F-16 aircraft, flight mode with a test speed $V_{ind} = 500$ km/h). Adaptation to the change in the dynamics of the controlled object: backwards shift of the center by 5% ($t = 30$ sec); 50% decrease in the effectiveness of the control ($t = 50$ sec). Here α is angle of attack, deg (dotted line for reference model, solid line for plant); E_α is tracking error for the reference angle of attack, deg; δ_e is the deflection angle of the stabilizer, deg; t is time, sec.

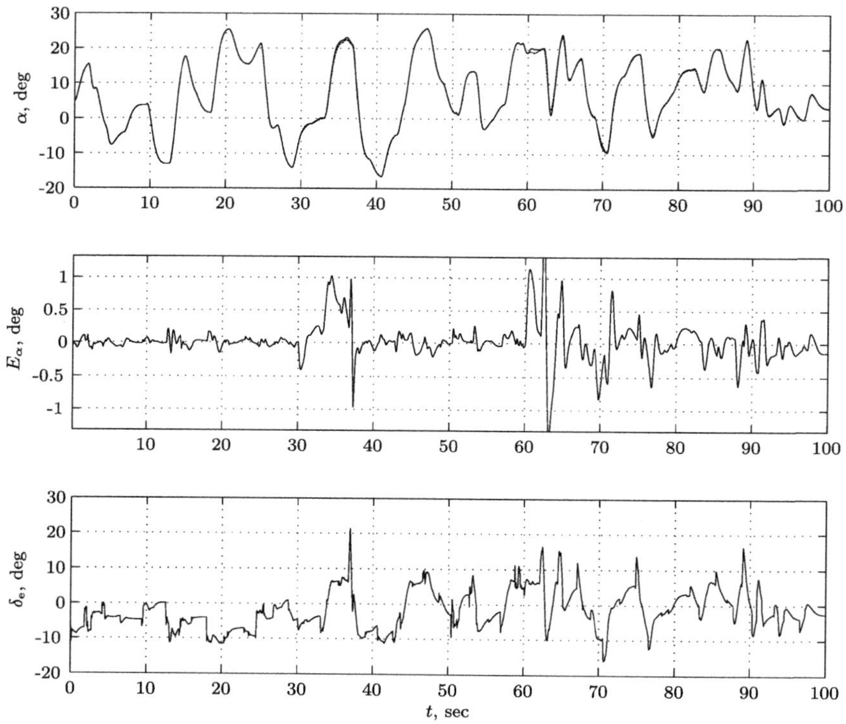

FIGURE A.16 The results of a computational experiment for the MRAC-type control system with compensator (F-16 aircraft, flight mode with a test speed $V_{ind} = 700$ km/h). Adaptation to the change in the dynamics of the controlled object: backwards shift of the center by 10% ($t = 30$ sec); 50% decrease in the effectiveness of the control ($t = 50$ sec). Here α is angle of attack, deg (dotted line for reference model, solid line for plant); E_α is tracking error for the reference angle of attack, deg; δ_e is the deflection angle of the stabilizer, deg; t is time, sec.

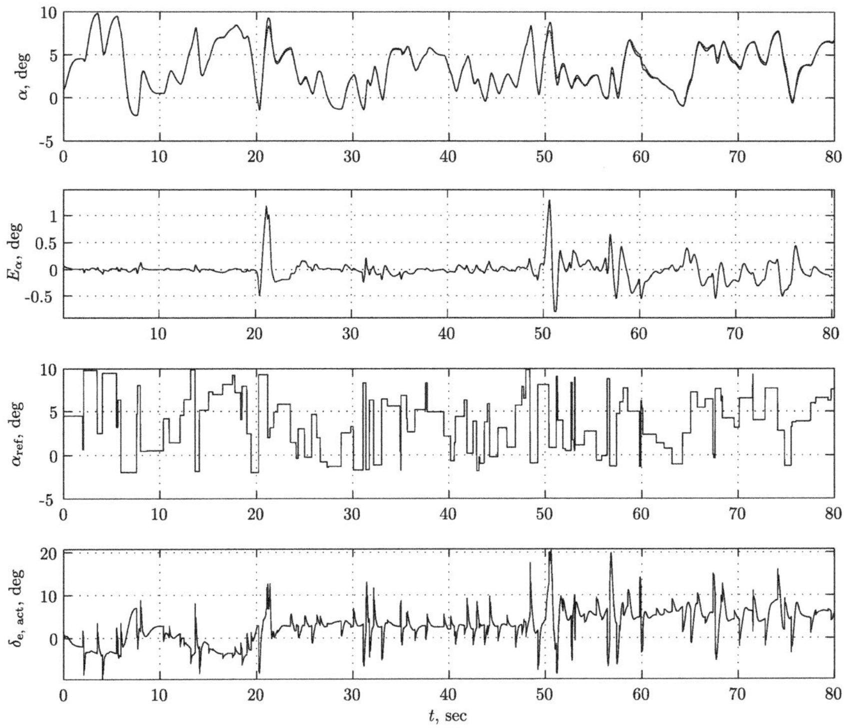

FIGURE A.17 The results of a computational experiment for the MRAC-type control system with compensator (hypersonic research vehicle X-43, flight mode $M = 6$). Adaptation to the change in the dynamics of the controlled object: backwards shift of the center by 5% ($t = 20$ sec); 50% decrease in the effectiveness of the control ($t = 50$ sec). Here α is angle of attack, deg (dotted line for reference model, solid line for plant); E_α is tracking error for the given angle of attack, deg; α_{ref} is reference signal on the angle of attack, deg; $\delta_{e,\,act}$ is command signal for the elevator actuator, deg; t is time, sec.

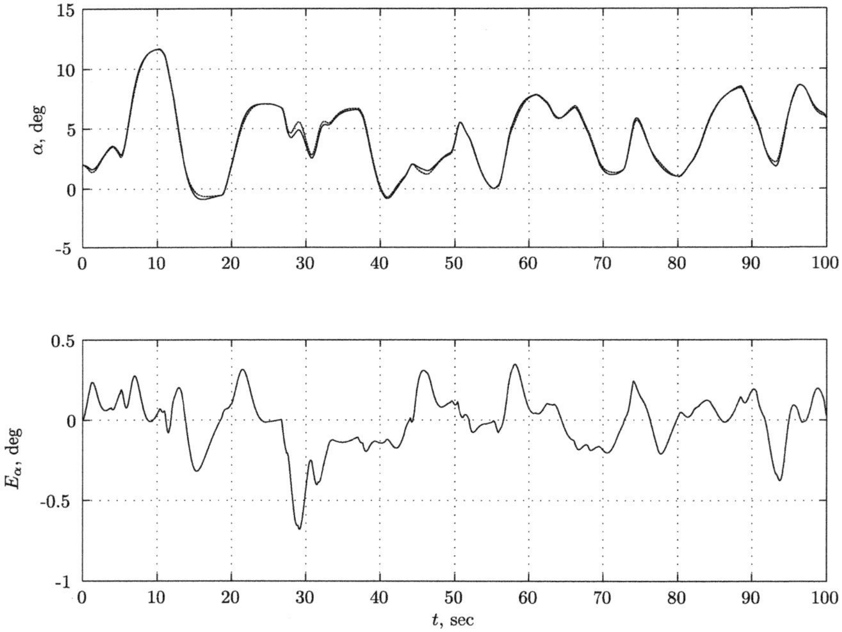

FIGURE A.18 Efficiency verification of the neurocontroller on the test set (for the hypersonic research vehicle NASP, flight mode $M = 6$). Here α is angle of attack, deg (dotted line for reference model, solid line for plant); E_α is tracking error for the given angle of attack, deg; t is time, sec.

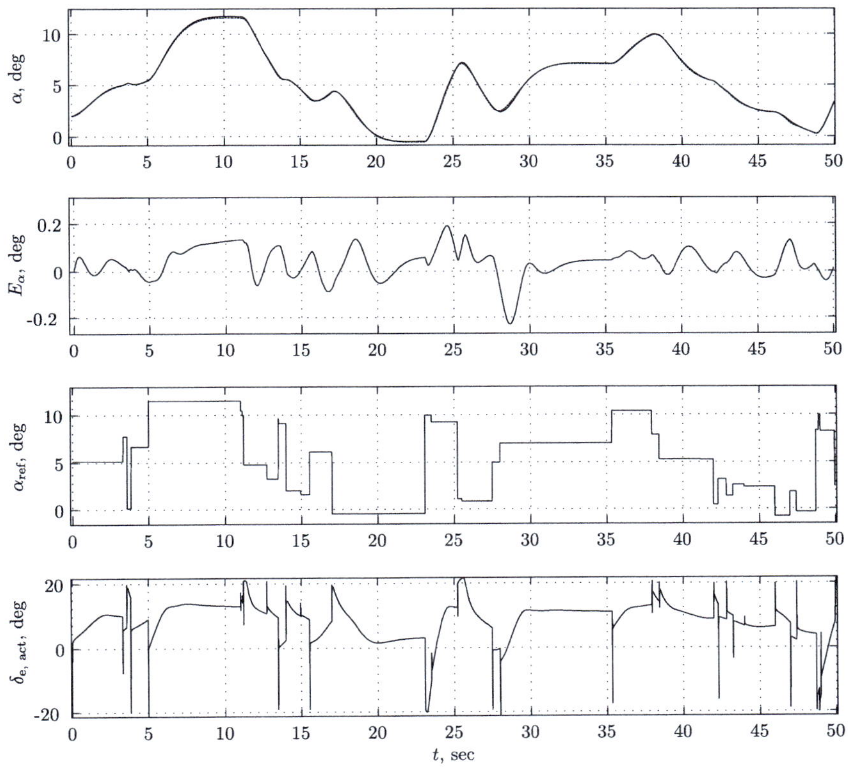

FIGURE A.19 The results of a computational experiment for the MRAC-type control system with the compensator (for the hypersonic research vehicle NASP, flight mode $M = 6$). Here α is angle of attack, deg (dotted line for reference model, solid line for plant); α_{ref} is reference signal on the angle of attack, deg; $\delta_{e,\,act}$ is command signal for the elevator actuator, deg; t is time, sec.

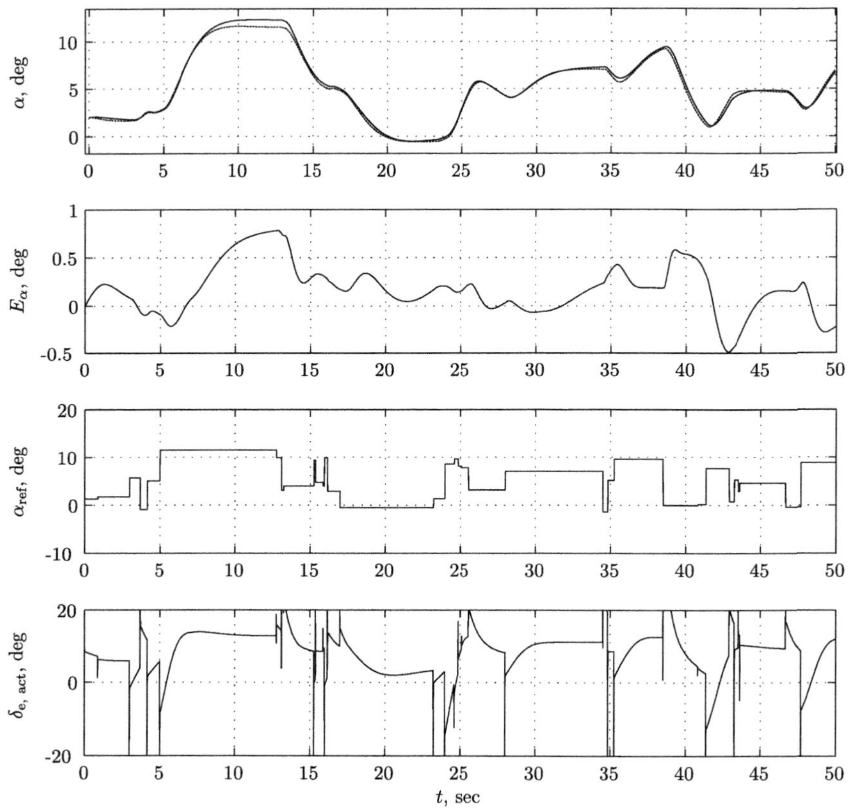

FIGURE A.20 The results of a computational experiment for the MRAC-type control system without the compensator (for the hypersonic research vehicle NASP, flight mode $M = 6$). Here α is angle of attack, deg (dotted line for reference model, solid line for plant); α_{ref} is reference signal on the angle of attack, deg; $\delta_{e,\,act}$ is command signal for the elevator actuator, deg; t is time, sec.

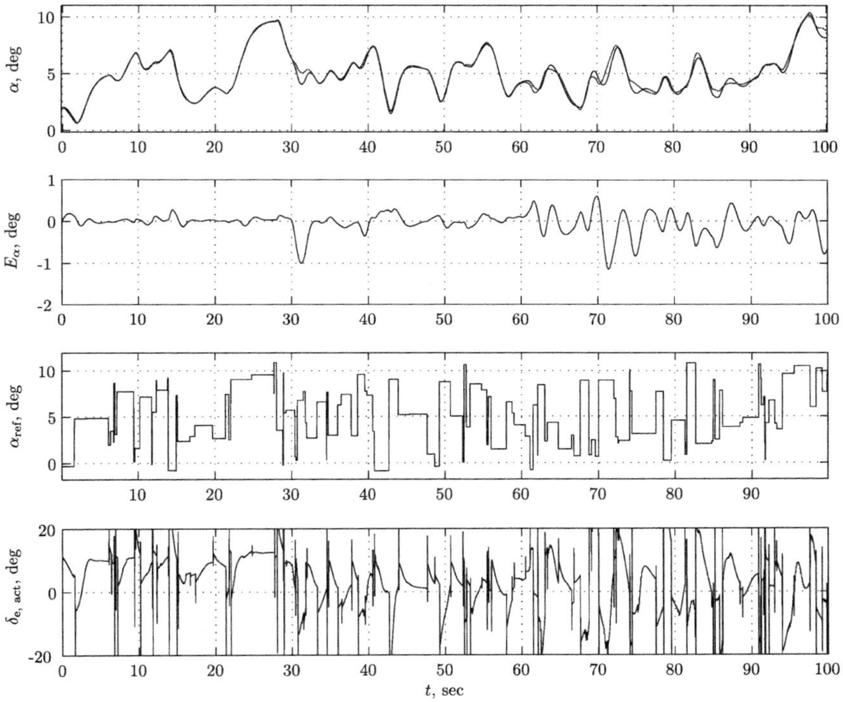

FIGURE A.21 The results of a computational experiment for the MRAC-type control system with the compensator (for the hypersonic research vehicle NASP, flight mode $M = 6$). Adaptation to the change in the dynamics of the controlled object: backwards shift of the center by 5% ($t = 30$ sec); 50% decrease in the effectiveness of the control ($t = 60$ sec). Here α is angle of attack, deg (dotted line for reference model, solid line for plant); e_α is tracking error for the given angle of attack, deg; α_{ref} is reference signal on the angle of attack, deg; $\delta_{e,\,act}$ is command signal for the elevator actuator, deg; t is time, sec.

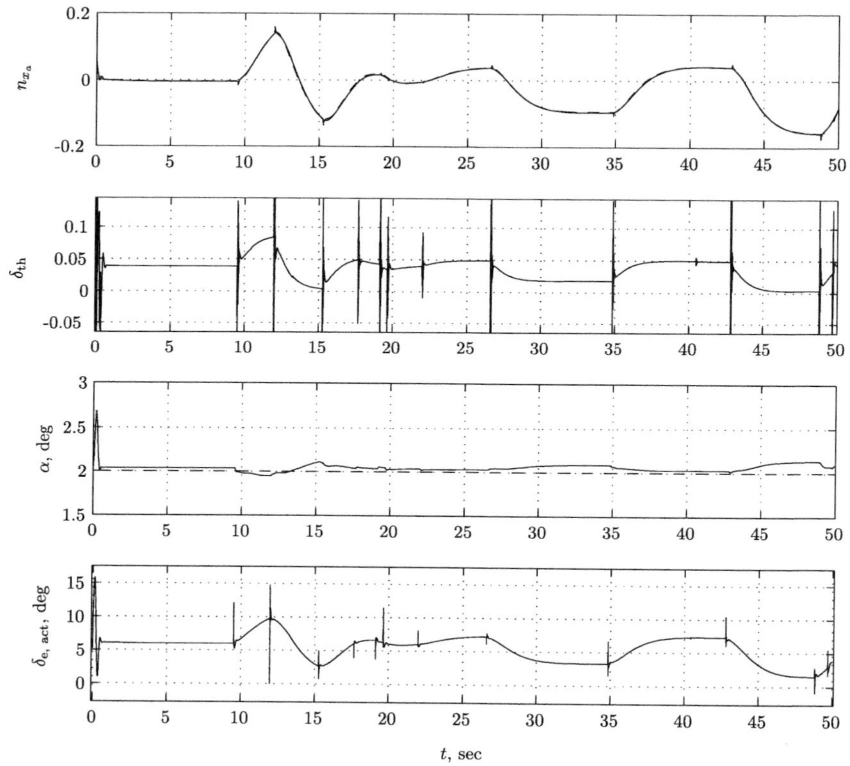

FIGURE A.22 The results of a computational experiment for the MRAC-type control system (hypersonic research vehicle X-43, flight mode $M = 6$). Constant load factor $n_{x_a} = 0$, the reference angle of attack is random. Compensators in both channels. I: Behavior of the object and the reference model. Here n_{x_a} is the tangential load factor; δ_{th} is engine control command signal; α is angle of attack, deg (dotted line for reference model, solid line for plant); $\delta_{e,\,act}$ is command signal for the elevator actuator, deg; t is time, sec.

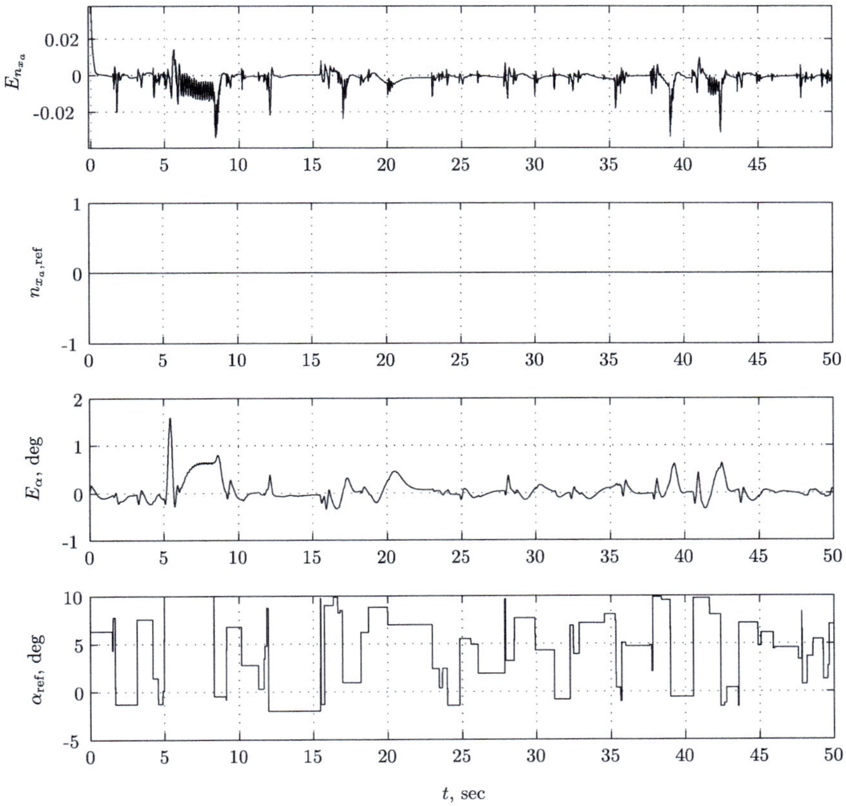

FIGURE A.23 The results of a computational experiment for the MRAC-type control system (hypersonic research vehicle X-43, flight mode $M = 6$). Constant load factor $n_{x_a} = 0$, the reference angle of attack is random. Compensators in both channels. II: The reference signals and tracking errors. Here $n_{x_a,ref}$ is the reference signal on tangential load factor; $E_{n_{xa}}$ is the tracking error for the specified value of the tangential load factor; E_α is the tracking error for the specified angle of attack value, deg; α_{ref} is reference signal on the angle of attack, deg; t is time, sec.

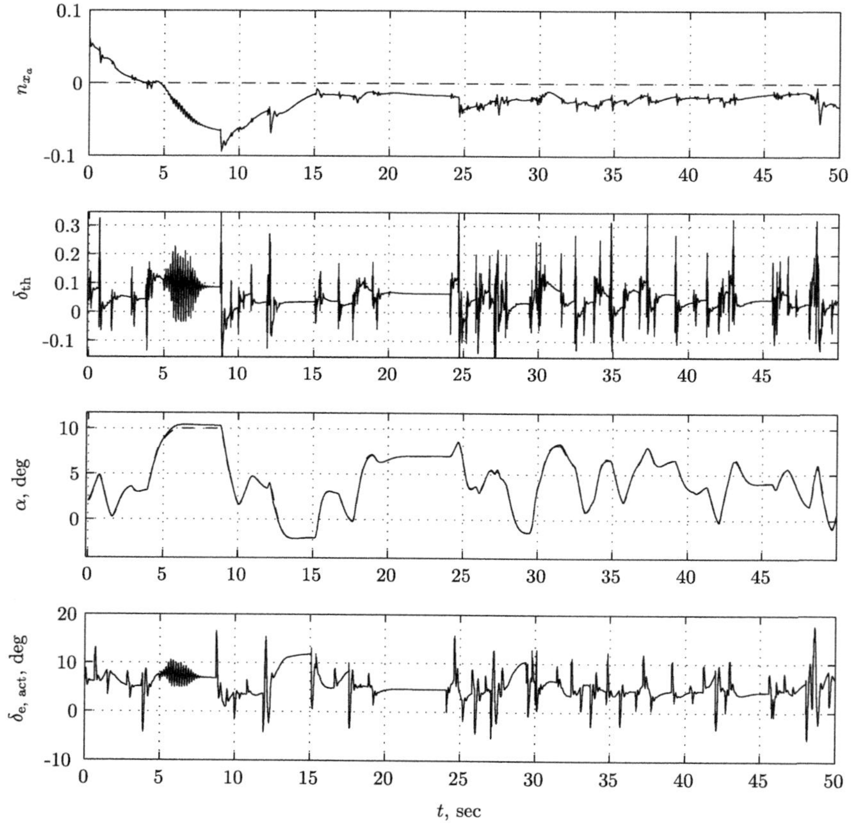

FIGURE A.24 The results of a computational experiment for the MRAC-type control system (hypersonic research vehicle X-43, flight mode $M = 6$). Constant load factor $n_{x_a} = 0$, the reference angle of attack is random. There is no compensator in the load factor channel. I: Behavior of the object and the reference model. Here n_{x_a} is the tangential load factor; δ_{th} is engine control command signal; α is angle of attack, deg (dotted line for reference model, solid line for plant); $\delta_{e,\,act}$ is command signal for the elevator actuator, deg; t is time, sec.

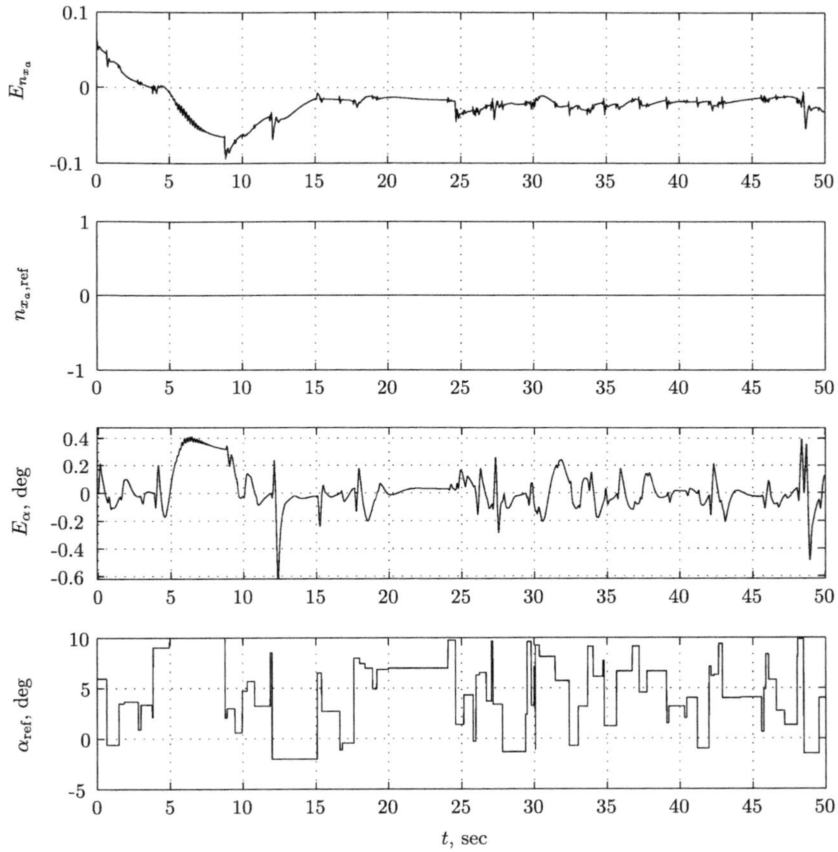

FIGURE A.25 The results of a computational experiment for the MRAC-type control system (hypersonic research vehicle X-43, flight mode $M = 6$). Constant load factor $n_{x_a} = 0$, the reference angle of attack is random. There is no compensator in the overload channel. II: The reference signals and tracking errors. Here $n_{x_a,ref}$ is the reference signal on tangential load factor; $E_{n_{xa}}$ is the tracking error for the specified value of the tangential load factor; E_α is the tracking error for the specified angle of attack value, deg; α_{ref} is reference signal on the angle of attack, deg; t is time, sec.

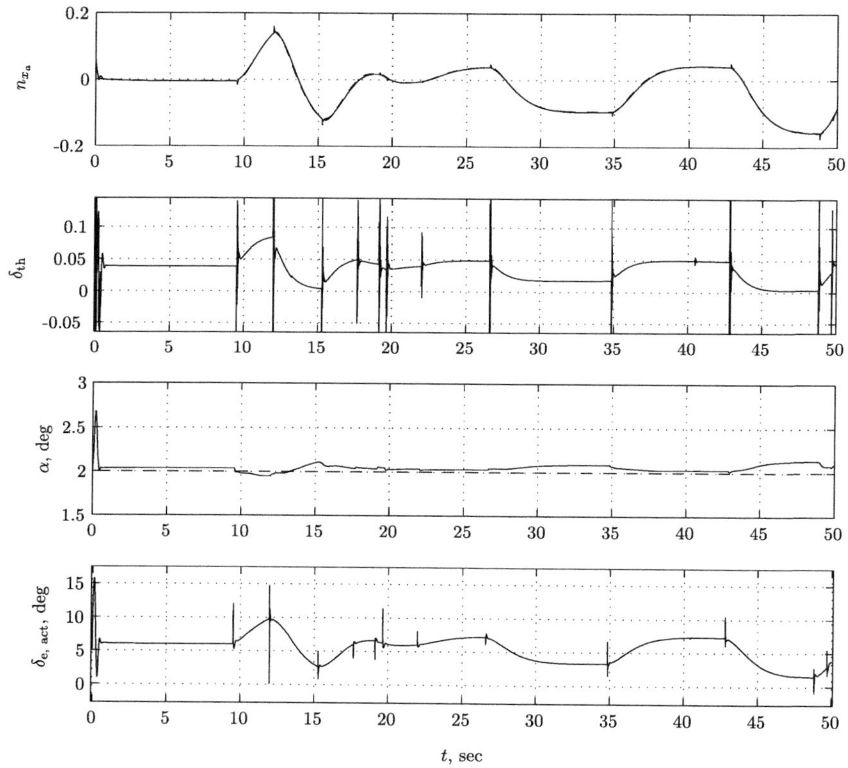

FIGURE A.26 The results of a computational experiment for the MRAC-type control system (hypersonic research vehicle X-43, flight mode $M = 6$). Constant predetermined angle of attack $\alpha = 2$ deg, random reference load factor. Compensators in both channels. I: Behavior of the object and the reference model. Here n_{x_a} is the tangential load factor; δ_{th} is engine control command signal; α is angle of attack, deg (dotted line for reference model, solid line for plant); $\delta_{e,\,act}$ is command signal for the elevator actuator, deg; t is time, sec.

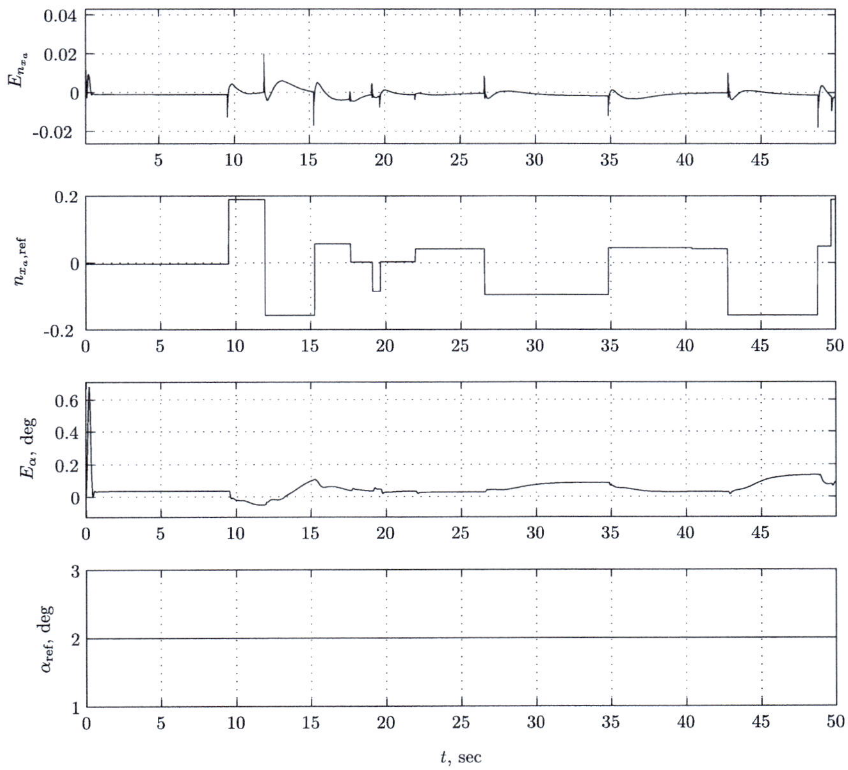

FIGURE A.27 The results of a computational experiment for the MRAC-type control system (hypersonic research vehicle X-43, flight mode $M = 6$). Constant predetermined angle of attack $\alpha = 2$ deg, random reference load factor. Compensators in both channels. II: Reference signals and tracking errors. Here $E_{n_{xa}}$ is tangential load factor error; $n_{x_a,ref}$ is the reference signal on tangential load factor; E_α is tracking error for the reference angle of attack, deg; α_{ref} is the reference signal on the angle of attack, deg; t is time, sec.

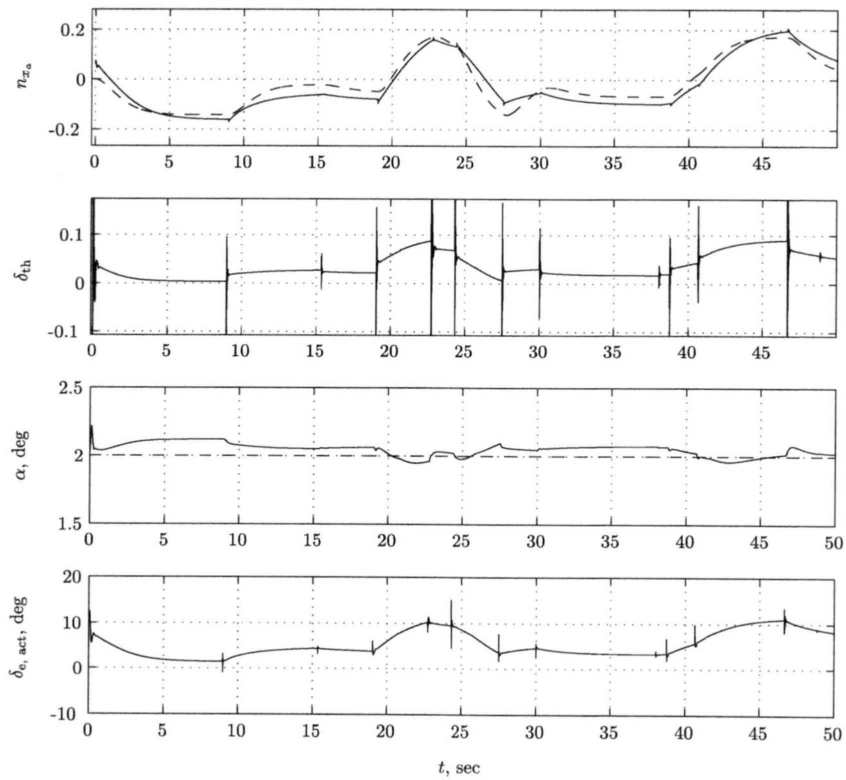

FIGURE A.28 The results of a computational experiment for the MRAC-type control system (hypersonic research vehicle X-43, flight mode $M = 6$). Constant predetermined angle of attack $\alpha = 2$ deg, random reference overload. There is no compensator in the overload channel. I: Behavior of the object and the reference model. Here n_{x_a} is the tangential load factor; δ_{th} is engine control command signal; α is angle of attack, deg (dotted line for reference model, solid line for plant); $\delta_{e,act}$ is command signal for the elevator actuator, deg; t is time, sec.

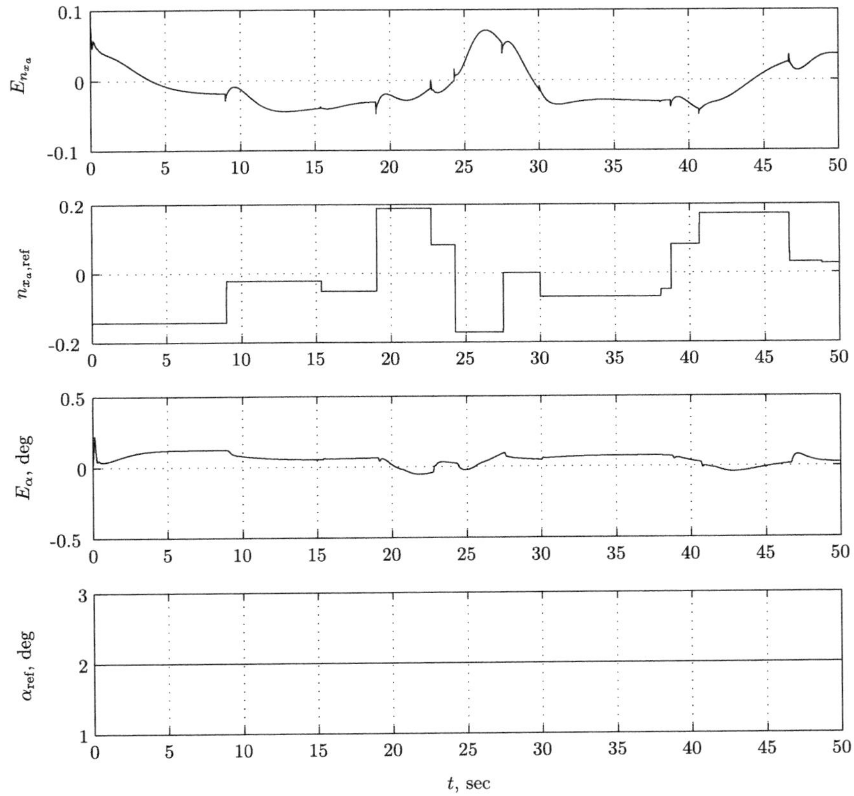

FIGURE A.29 The results of a computational experiment for the MRAC-type control system (hypersonic research vehicle X-43, flight mode $M = 6$). Constant predetermined angle of attack $\alpha = 2$ deg, random reference overload. There is no compensator in the overload channel. II: Reference signals and tracking errors. Here $E_{n_{xa}}$ is tangential load factor error; $n_{x_a,ref}$ is the reference signal on tangential load factor; E_α is tracking error for the reference angle of attack, deg; α_{ref} is the reference signal on the angle of attack, deg; t is time, sec.

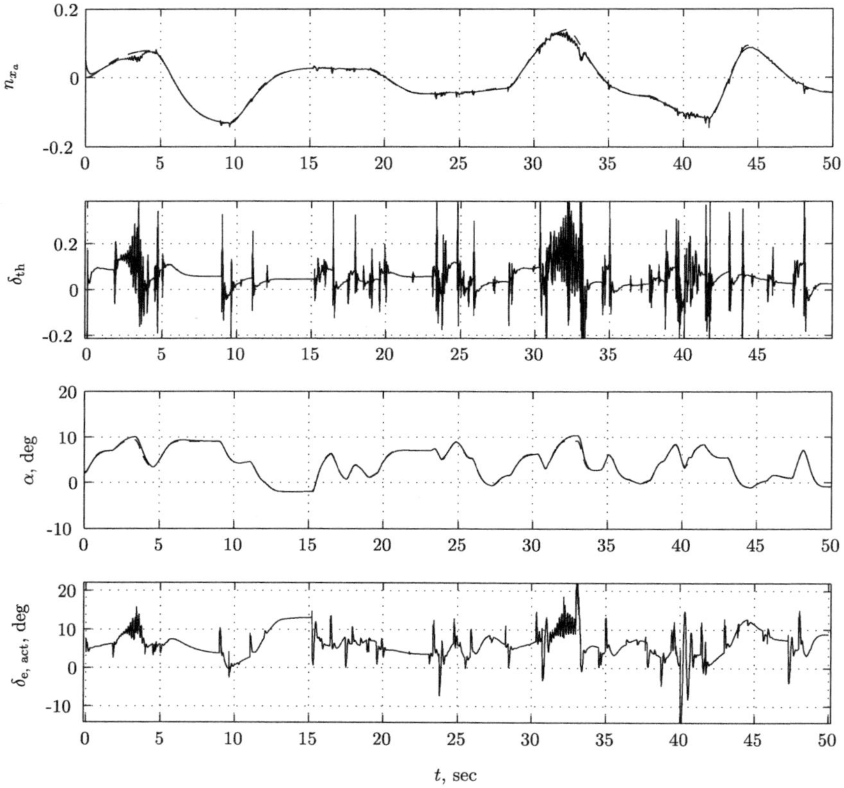

FIGURE A.30 The results of a computational experiment for the MRAC-type control system (hypersonic research vehicle X-43, flight mode $M = 6$). Both reference signals are random. Compensators in both channels. I: Behavior of the object and the reference model. Here n_{x_a} is the tangential load factor; δ_{th} is engine control command signal; α is angle of attack, deg (dotted line for reference model, solid line for plant); $\delta_{e,\,act}$ is command signal for the elevator actuator, deg; t is time, sec.

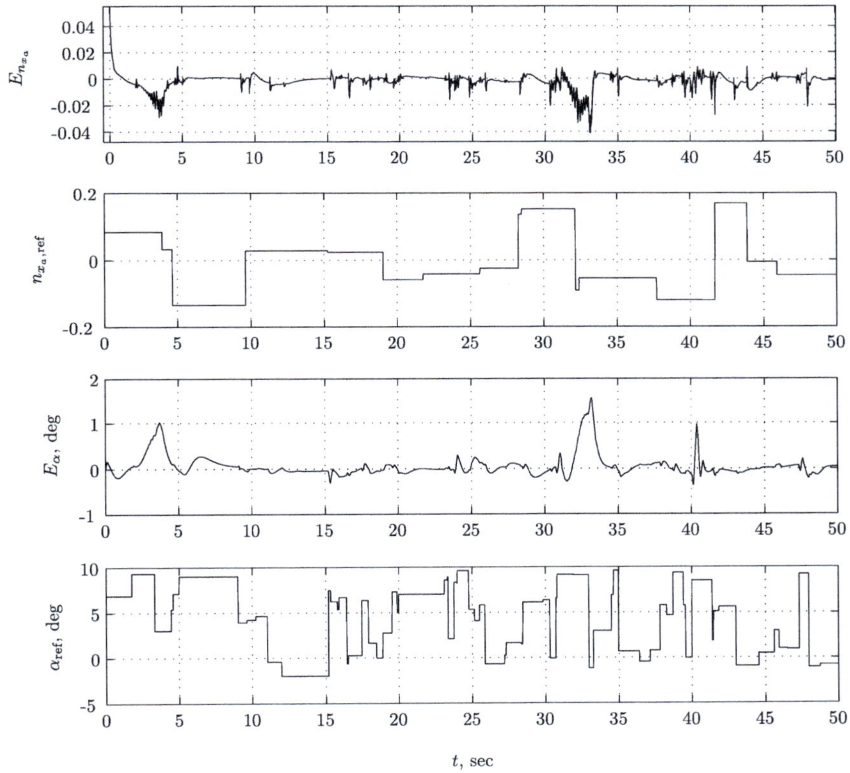

FIGURE A.31 The results of a computational experiment for the MRAC-type control system (hypersonic research vehicle X-43, flight mode $M = 6$). Both reference signals are random. Compensators in both channels. II: Reference signals and tracking errors. Here $E_{n_{xa}}$ is tangential load factor error; $n_{x_a,ref}$ is the reference signal on tangential load factor; E_α is tracking error for the reference angle of attack, deg; α_{ref} is the reference signal on the angle of attack, deg; t is time, sec.

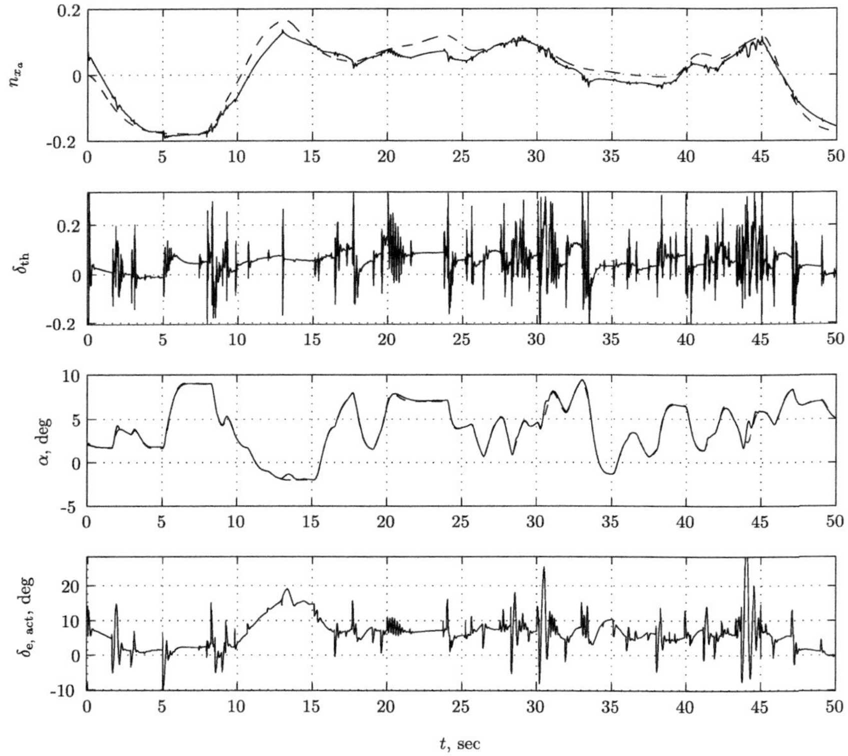

FIGURE A.32 The results of a computational experiment for the MRAC-type control system (hypersonic research vehicle X-43, flight mode $M = 6$). Both reference signals are random stepwise. There is no compensator in the overload channel. I: Behavior of the object and the reference model. Here n_{x_a} is the tangential load factor; δ_{th} is engine control command signal; α is angle of attack, deg (dotted line for reference model, solid line for plant); $\delta_{e,\,act}$ is command signal for the elevator actuator, deg; t is time, sec.

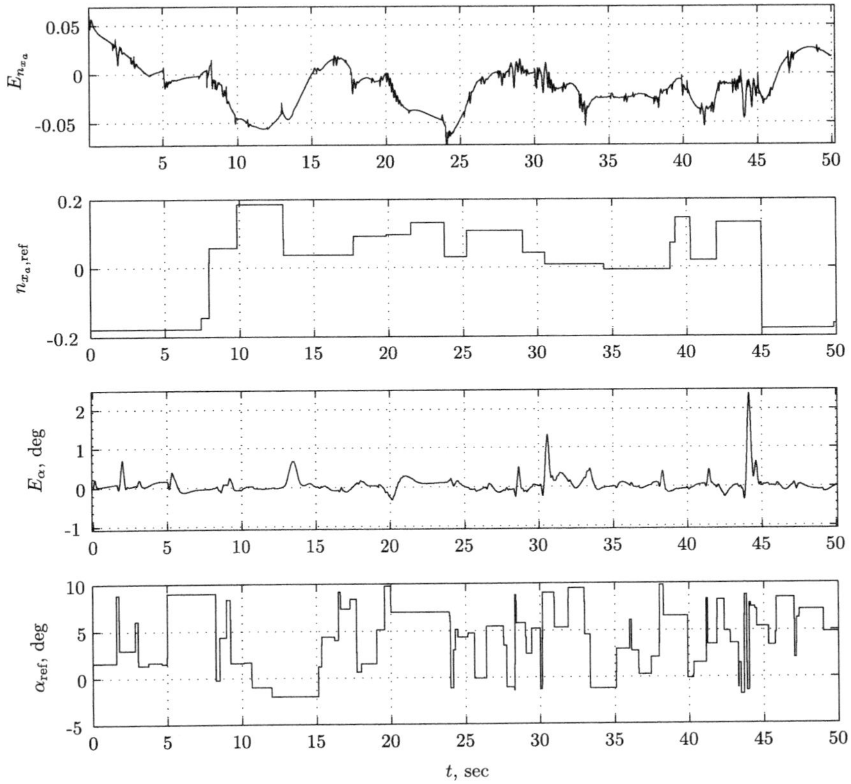

FIGURE A.33 The results of a computational experiment for the MRAC-type control system (hypersonic research vehicle X-43, flight mode $M = 6$). Both reference signals are random stepwise. There is no compensator in the overload channel. II: Reference signals and tracking errors. Here $E_{n_{xa}}$ is tangential load factor error; $n_{x_a,ref}$ is the reference signal on tangential load factor; E_α is tracking error for the reference angle of attack, deg; α_{ref} is the reference signal on the angle of attack, deg; t is time, sec.

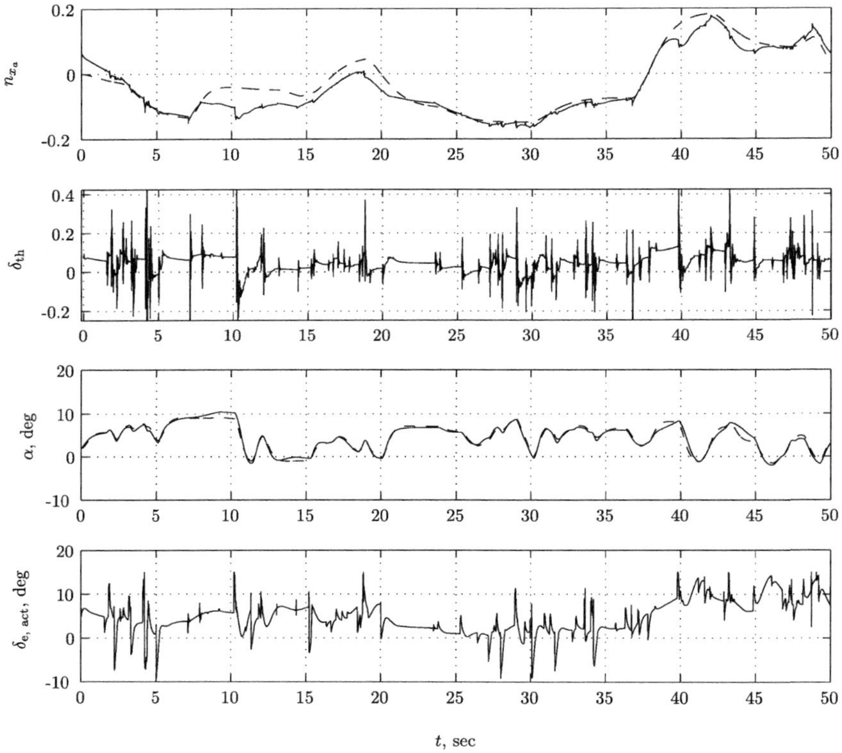

FIGURE A.34 The results of a computational experiment for the MRAC-type control system (hypersonic research vehicle X-43, flight mode $M = 6$). Both reference signals are random stepwise. There is no compensator in both channels. I: Behavior of the object and the reference model. Here n_{x_a} is the tangential load factor; δ_{th} is engine control command signal; α is angle of attack, deg (dotted line for reference model, solid line for plant); $\delta_{e,\,act}$ is command signal for the elevator actuator, deg; t is time, sec.

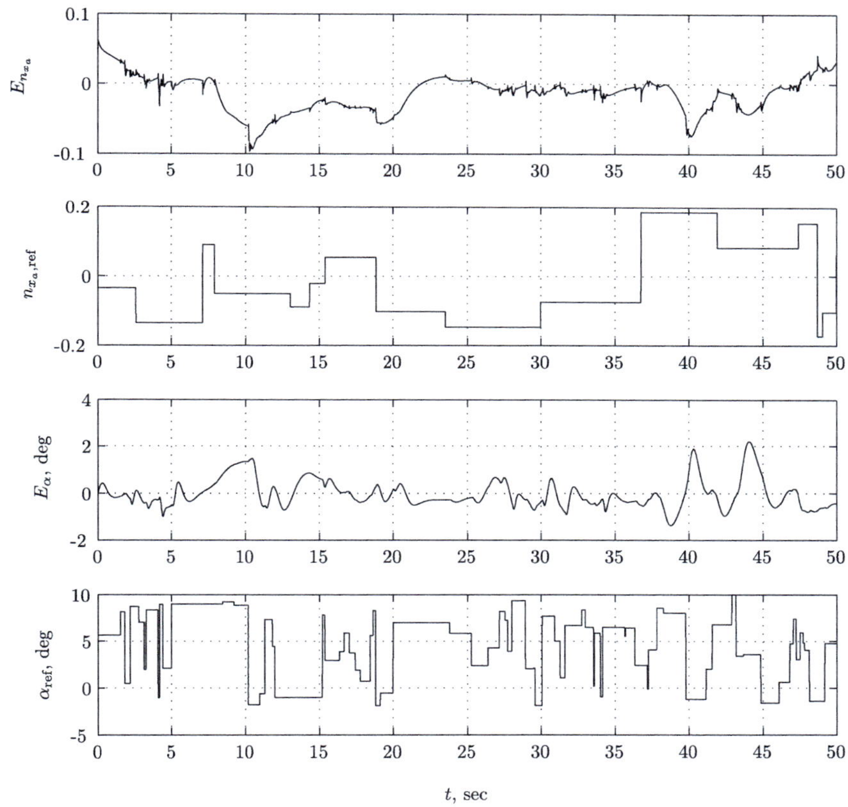

FIGURE A.35 The results of a computational experiment for the MRAC-type control system (hypersonic research vehicle X-43, flight mode $M = 6$). Both reference signals are random stepwise. There is no compensator in both channels. II: Reference signals and tracking errors. Here $E_{n_{xa}}$ is tangential load factor error; $n_{xa,ref}$ is the reference signal on tangential load factor; E_α is tracking error for the reference angle of attack, deg; α_{ref} is the reference signal on the angle of attack, deg; t is time, sec.

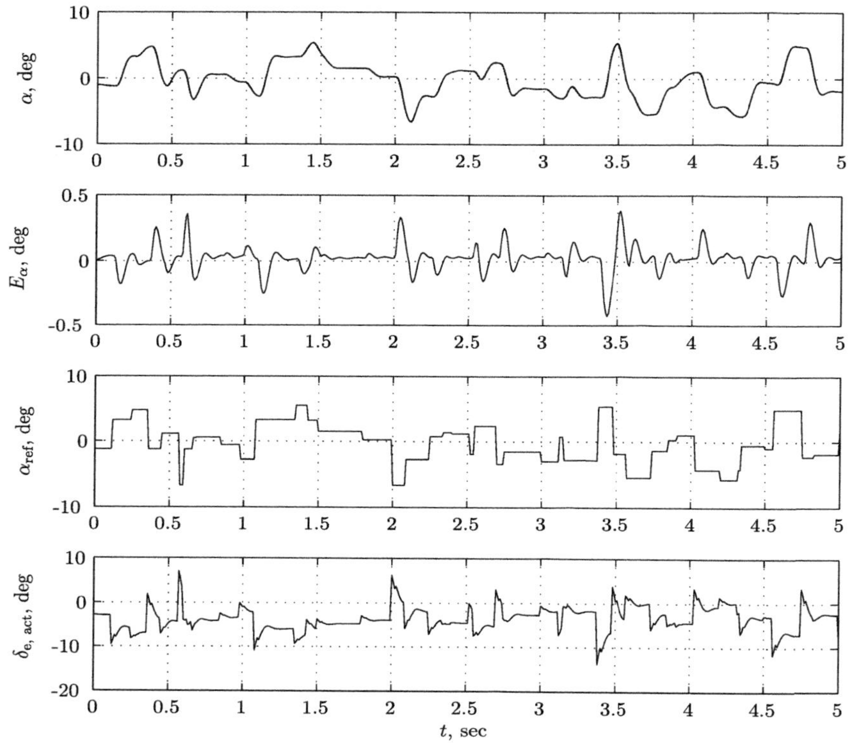

FIGURE A.36 The results of a computational experiment for the MRAC-type control system (micro-UAV "003"), a flight with airspeed $V_{ind} = 30$ km/h in normal operation conditions. Here α is angle of attack, deg (dotted line for reference model, solid line for plant); E_α is the tracking error for the given angle of attack, deg; α_{ref} is the reference signal on the angle of attack, deg; $\delta_{e,\,act}$ is the command signal for the elevator actuator, deg; t is time, sec.

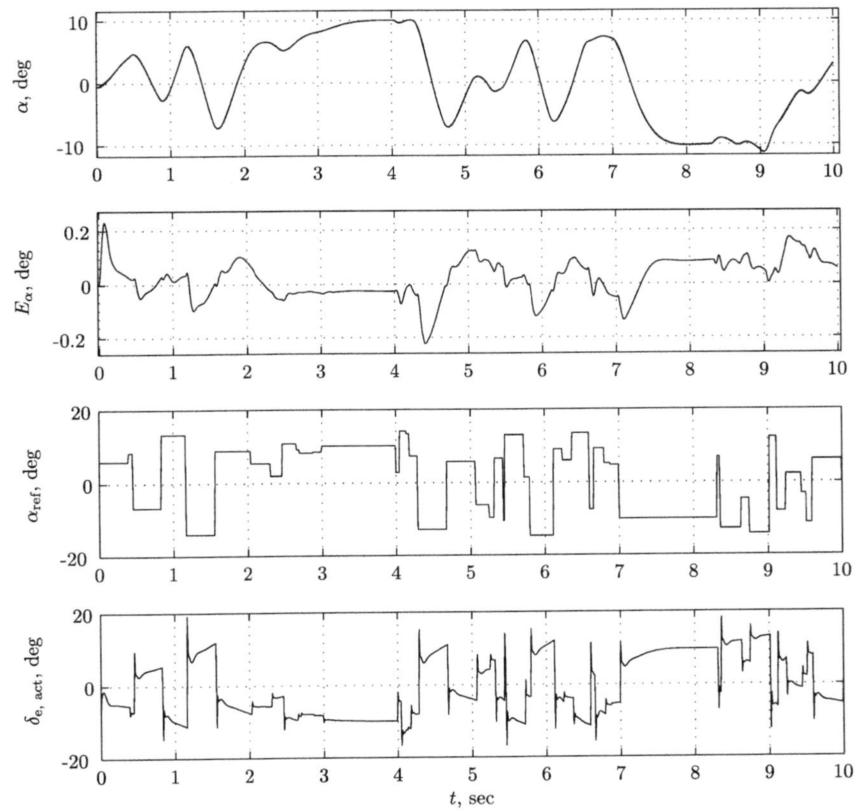

FIGURE A.37 The results of a computational experiment for the MRAC-type control system (mini-UAV X-04), a flight with airspeed $V_{ind} = 70$ km/h in normal operation conditions. Here α is angle of attack, deg (dotted line for reference model, solid line for plant); E_α is the tracking error for the given angle of attack, deg; α_{ref} is the reference signal on the angle of attack, deg; $\delta_{e,\,act}$ is the command signal for the elevator actuator, deg; t is time, sec.

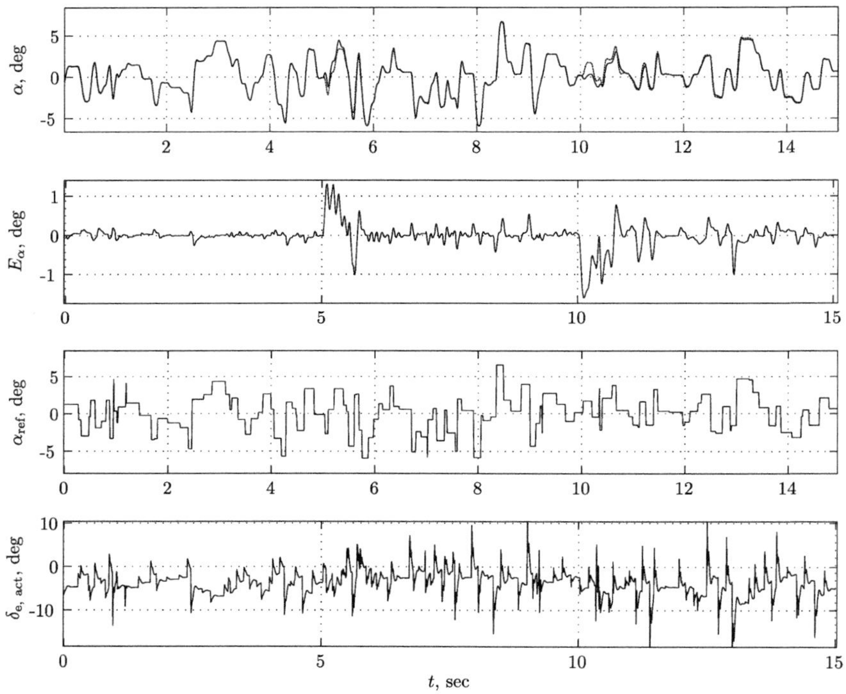

FIGURE A.38 The results of a computational experiment for the MRAC-type control system (micro-UAV "003"), a flight with airspeed $V_{ind} = 30$ km/h in the event of a failure situation at time $t = 5$ sec (backwards shift of the center of gravity by 10%) and then at the time $t = 10$ sec (50% decrease in the effectiveness of the longitudinal control organ). Here α is angle of attack, deg (dotted line for reference model, solid line for plant); E_α is the tracking error for the given angle of attack, deg; α_{ref} is the reference signal on the angle of attack, deg; $\delta_{e,\,act}$ is the command signal for the elevator actuator, deg; t is time, sec.

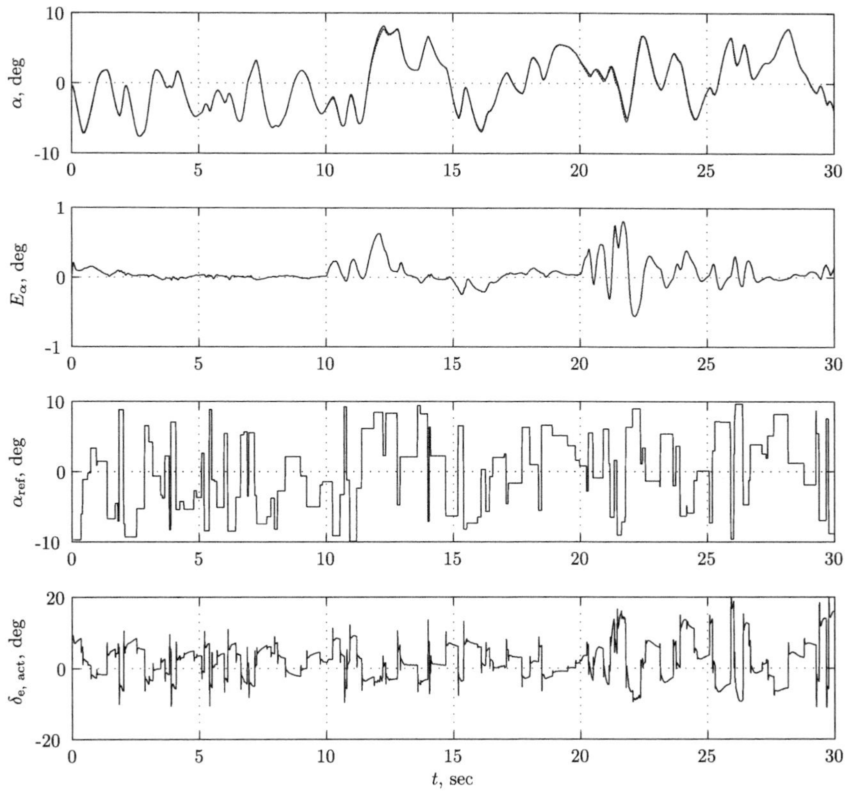

FIGURE A.39 The results of a computational experiment for the MRAC-type control system (mini-UAV X-04), a flight with airspeed $V_{ind} = 70$ km/h in the event of a failure situation at time $t = 10$ sec (backwards shift of the center of gravity by 10%) and then at the time $t = 20$ sec (50% decrease in the effectiveness of the longitudinal control organ). Here α is angle of attack, deg (dotted line for reference model, solid line for plant); E_α is the tracking error for the given angle of attack, deg; α_{ref} is the reference signal on the angle of attack, deg; $\delta_{e,act}$ is the command signal for the elevator actuator, deg; t is time, sec.

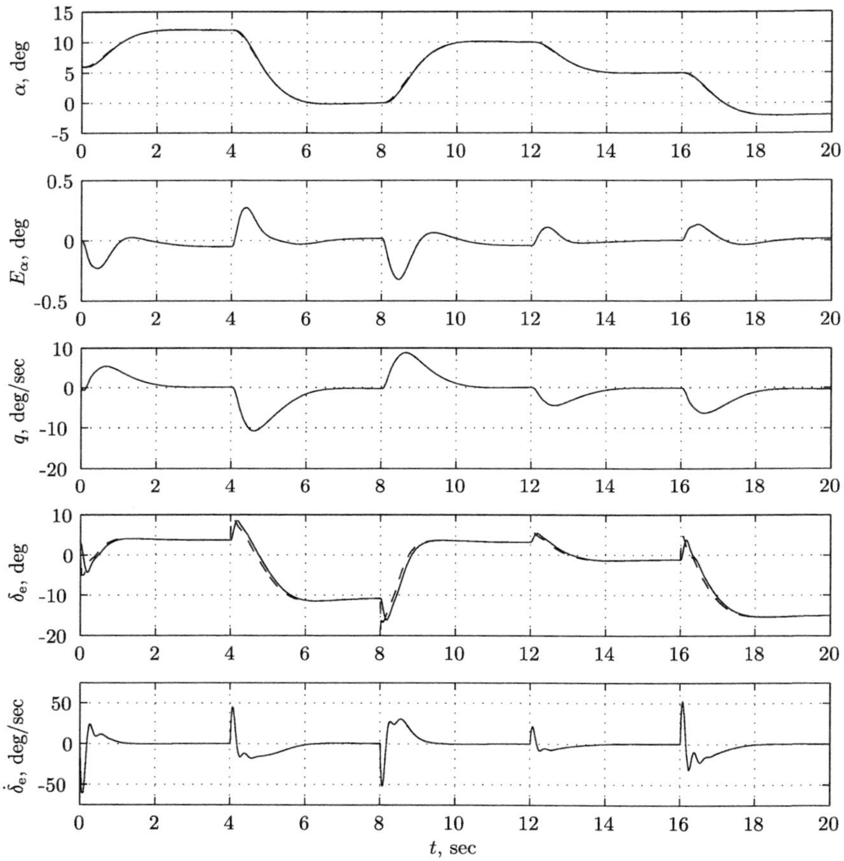

FIGURE A.40 The results of a computational experiment for the MPC-type control system applied to the hypersonic research vehicle with a stepwise reference signal on the angle of attack (flight mode $M = 5$, $H = 28$ km). Here α is angle of attack, deg (dotted line for reference model, solid line for plant); E_α is tracking error for the given angle of attack, deg; q is the pitch angular velocity, deg/sec; δ_e is the command signal of the actuator (dotted line) and the deflection angle of the elevons (solid line), deg; $d\delta_e/dt$ is the angular velocity of the deflection of the elevons, deg/sec; t is time, sec.

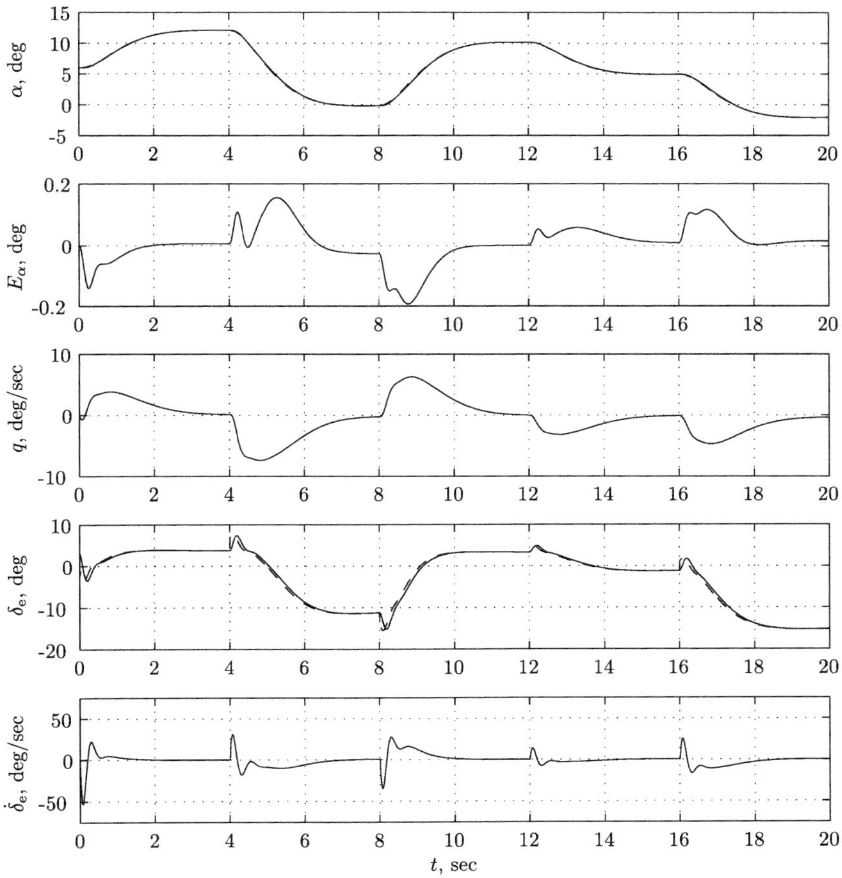

FIGURE A.41 The results of a computational experiment for the MPC-type control system applied to the hypersonic research vehicle with a stepwise reference signal on the angle of attack (flight mode $M = 5$, $H = 30$ km). Here α is angle of attack, deg (dotted line for reference model, solid line for plant); E_α is tracking error for the given angle of attack, deg; q is the pitch angular velocity, deg/sec; δ_e is the command signal of the actuator (dotted line) and the deflection angle of the elevons (solid line), deg; $d\delta_e/dt$ is the angular velocity of the deflection of the elevons, deg/sec; t is time, sec.

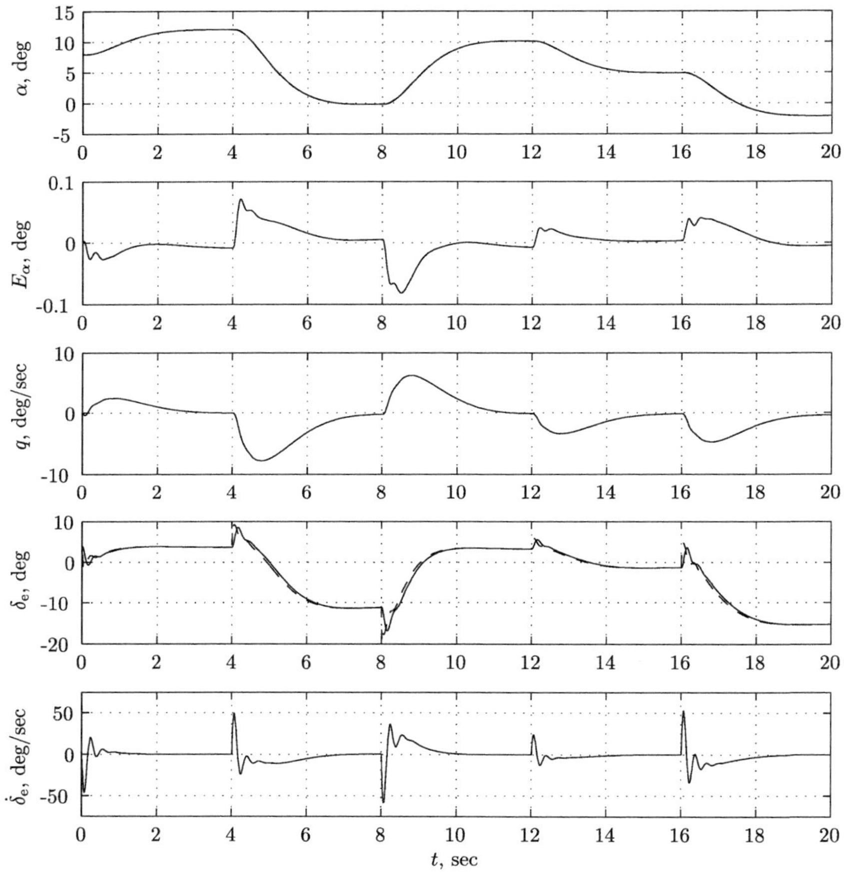

FIGURE A.42 The results of a computational experiment for the MPC-type control system applied to the hypersonic research vehicle with a stepwise reference signal on the angle of attack (flight mode $M = 5$, $H = 32$ km). Here α is angle of attack, deg (dotted line for reference model, solid line for plant); E_α is tracking error for the given angle of attack, deg; q is the pitch angular velocity, deg/sec; δ_e is the command signal of the actuator (dotted line) and the deflection angle of the elevons (solid line), deg; $d\delta_e/dt$ is the angular velocity of the deflection of the elevons, deg/sec; t is time, sec.

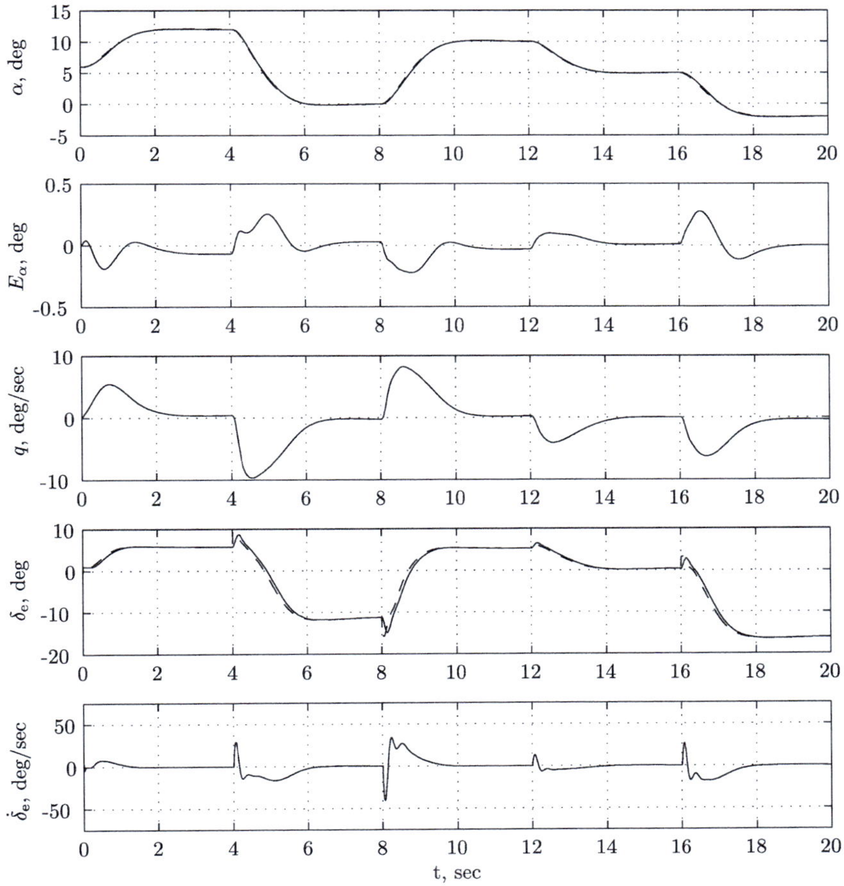

FIGURE A.43 The results of a computational experiment for the MPC-type control system applied to the hypersonic research vehicle with a stepwise reference signal on the angle of attack (flight mode $M = 6$, $H = 28$ km). Here α is angle of attack, deg (dotted line for reference model, solid line for plant); E_α is tracking error for the given angle of attack, deg; q is the pitch angular velocity, deg/sec; δ_e is the command signal of the actuator (dotted line) and the deflection angle of the elevons (solid line), deg; $d\delta_e/dt$ is the angular velocity of the deflection of the elevons, deg/sec; t is time, sec.

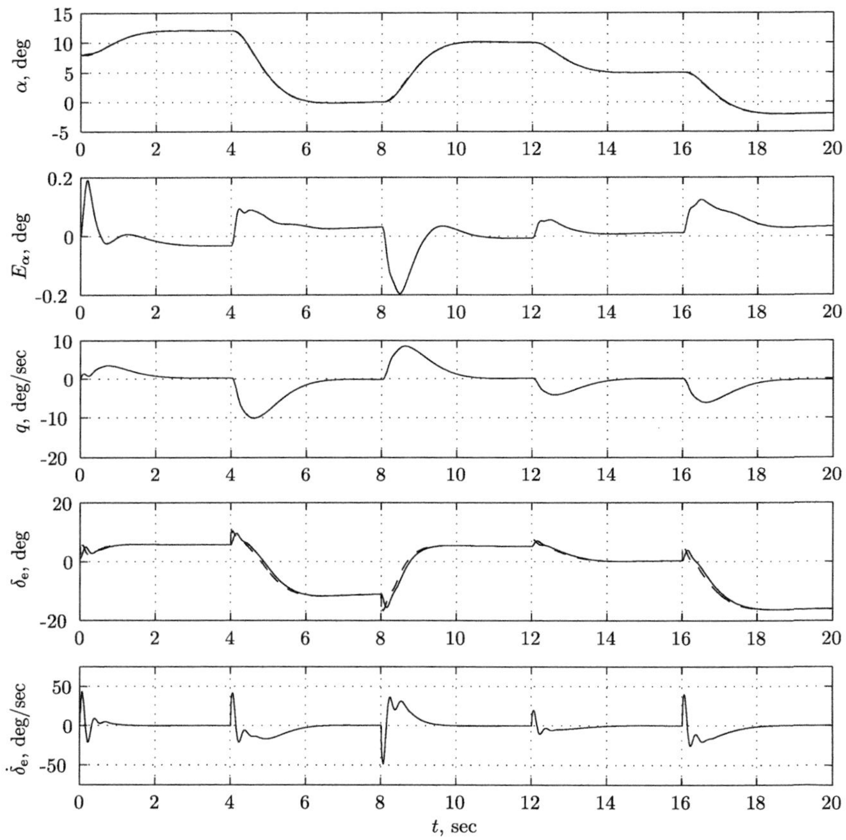

FIGURE A.44 The results of a computational experiment for the MPC-type control system applied to the hypersonic research vehicle with a stepwise reference signal on the angle of attack (flight mode $M = 6$, $H = 30$ km). Here α is angle of attack, deg (dotted line for reference model, solid line for plant); E_α is tracking error for the given angle of attack, deg; q is the pitch angular velocity, deg/sec; δ_e is the command signal of the actuator (dotted line) and the deflection angle of the elevons (solid line), deg; $d\delta_e/dt$ is the angular velocity of the deflection of the elevons, deg/sec; t is time, sec.

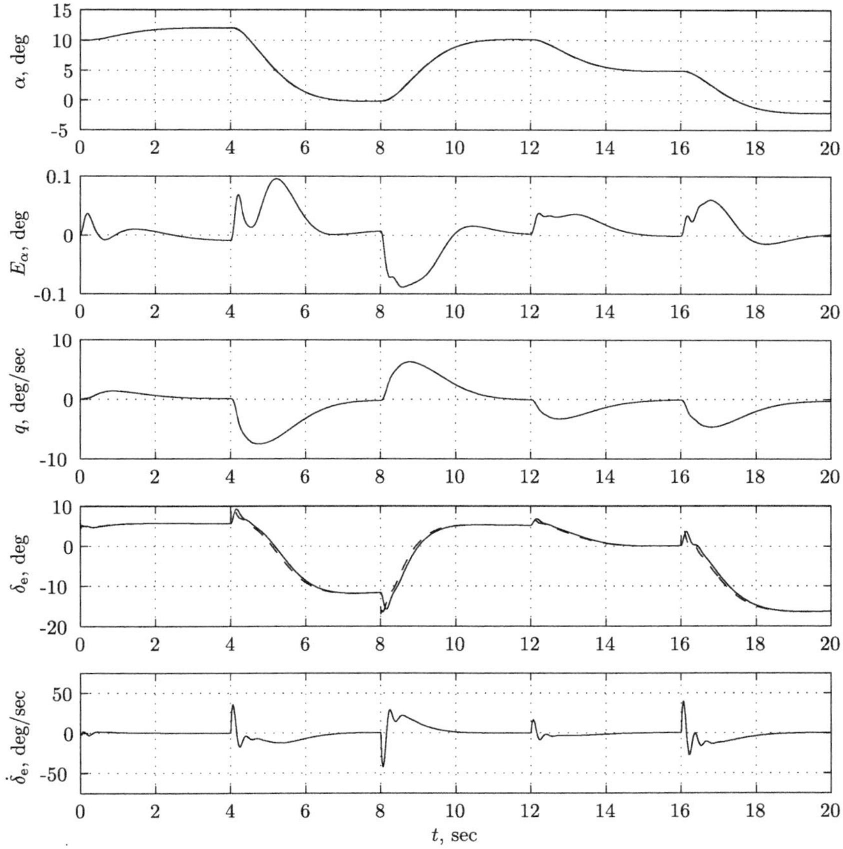

FIGURE A.45 The results of a computational experiment for the MPC-type control system applied to the hypersonic research vehicle with a stepwise reference signal on the angle of attack (flight mode $M = 6$, $H = 32$ km). Here α is angle of attack, deg (dotted line for reference model, solid line for plant); E_α is tracking error for the given angle of attack, deg; q is the pitch angular velocity, deg/sec; δ_e is the command signal of the actuator (dotted line) and the deflection angle of the elevons (solid line), deg; $d\delta_e/dt$ is the angular velocity of the deflection of the elevons, deg/sec; t is time, sec.

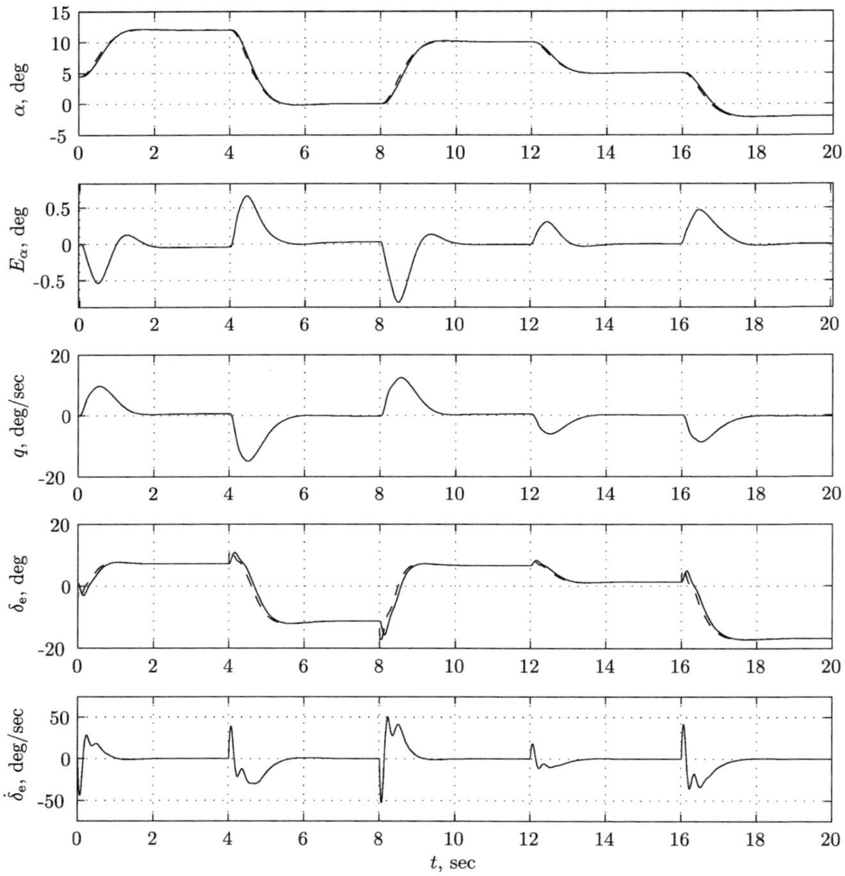

FIGURE A.46 The results of a computational experiment for the MPC-type control system applied to the hypersonic research vehicle with a stepwise reference signal on the angle of attack (flight mode $M = 7$, $H = 28$ km). Here α is angle of attack, deg (dotted line for reference model, solid line for plant); E_α is tracking error for the given angle of attack, deg; q is the pitch angular velocity, deg/sec; δ_e is the command signal of the actuator (dotted line) and the deflection angle of the elevons (solid line), deg; $d\delta_e/dt$ is the angular velocity of the deflection of the elevons, deg/sec; t is time, sec.

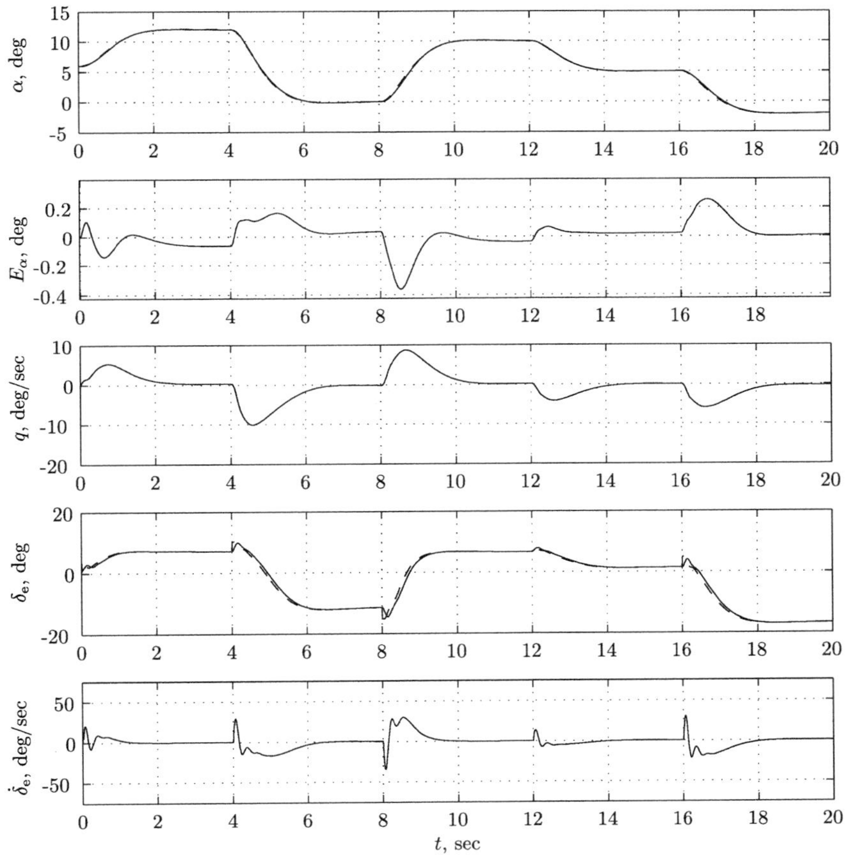

FIGURE A.47 The results of a computational experiment for the MPC-type control system applied to the hypersonic research vehicle with a stepwise reference signal on the angle of attack (flight mode $M = 7$, $H = 30$ km). Here α is angle of attack, deg (dotted line for reference model, solid line for plant); E_α is tracking error for the given angle of attack, deg; q is the pitch angular velocity, deg/sec; δ_e is the command signal of the actuator (dotted line) and the deflection angle of the elevons (solid line), deg; $d\delta_e/dt$ is the angular velocity of the deflection of the elevons, deg/sec; t is time, sec.

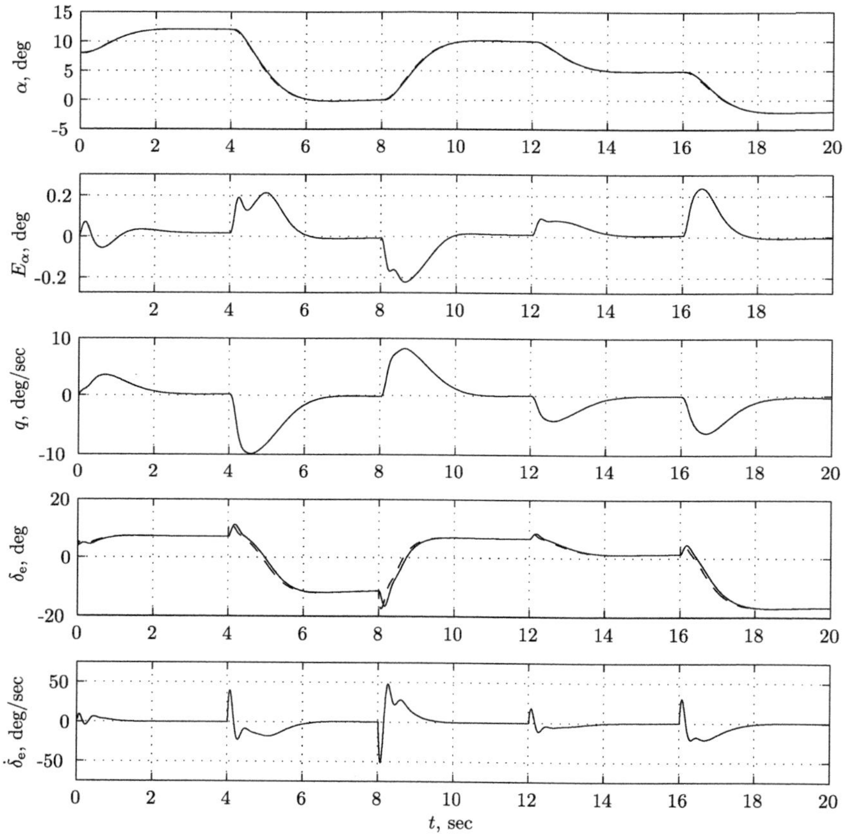

FIGURE A.48 The results of a computational experiment for the MPC-type control system applied to the hypersonic research vehicle with a stepwise reference signal on the angle of attack (flight mode $M = 7$, $H = 32$ km). Here α is angle of attack, deg (dotted line for reference model, solid line for plant); E_α is tracking error for the given angle of attack, deg; q is the pitch angular velocity, deg/sec; δ_e is the command signal of the actuator (dotted line) and the deflection angle of the elevons (solid line), deg; $d\delta_e/dt$ is the angular velocity of the deflection of the elevons, deg/sec; t is time, sec.

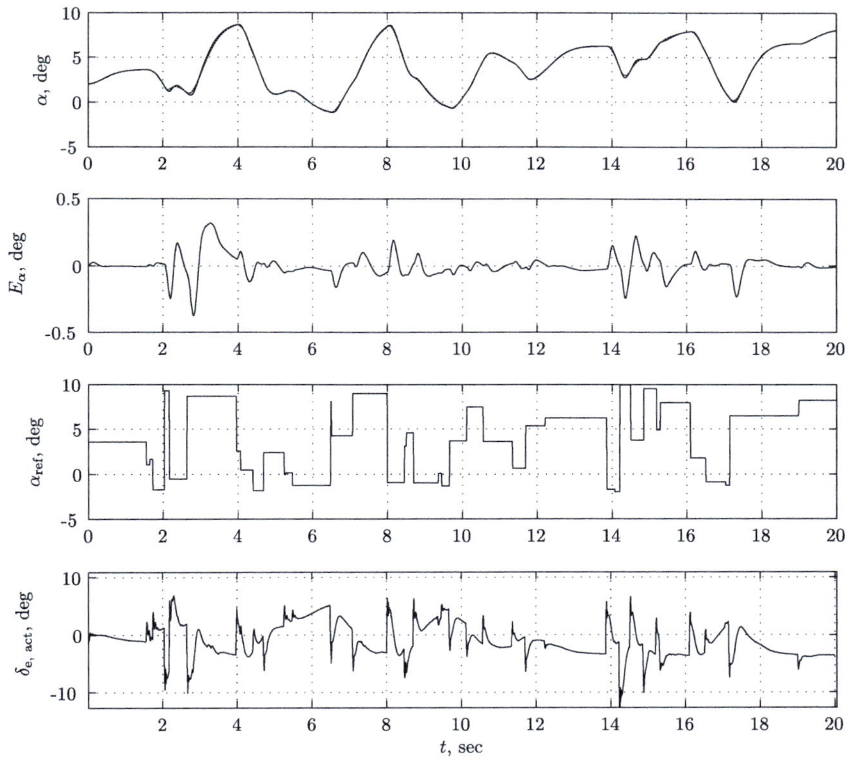

FIGURE A.49 The results of a computational experiment for the MPC-type control system with compensator (hypersonic research vehicle X-43; flight mode $M = 6$). Here α is angle of attack, deg (dotted line for reference model, solid line for plant); E_α is tracking error for the given angle of attack, deg; α_{ref} is reference signal on the angle of attack, deg; $\delta_{e,\,act}$ is command signal for the elevator actuator, deg; t is time, sec.

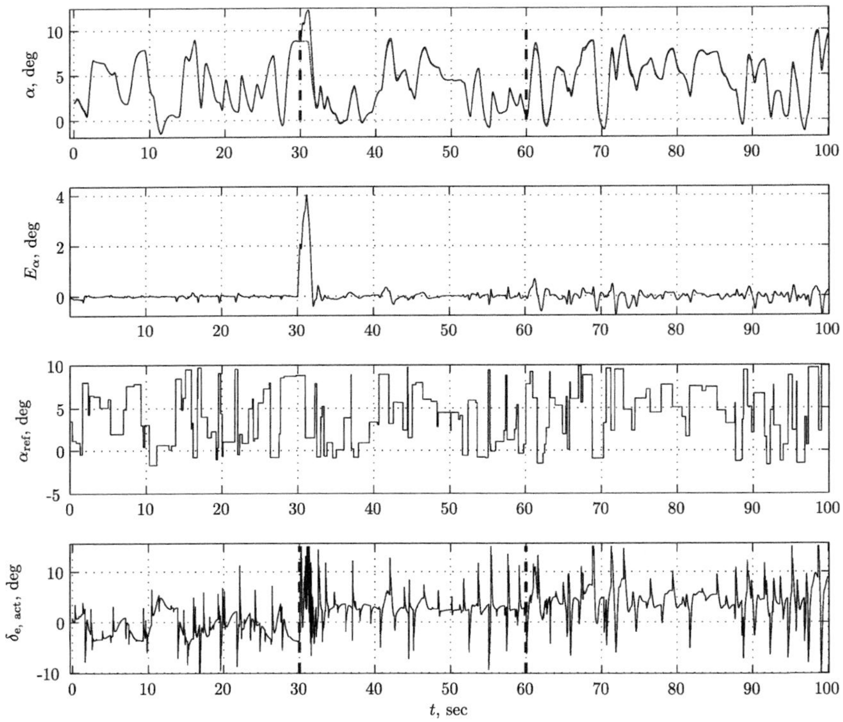

FIGURE A.50 The results of a computational experiment for the MPC-type control system with compensator (hypersonic research vehicle X-43; flight mode $M = 6$). Adaptation to the change in the dynamics of the control object due to failures: backwards shift of the center by 5% ($t = 30$ sec), 30% decrease in the effectiveness of the control ($t = 60$ sec). Here α is angle of attack, deg (dotted line for reference model, solid line for plant); E_α is tracking error for the given angle of attack, deg; α_{ref} is reference signal on the angle of attack, deg; $\delta_{e,\,act}$ is command signal for the elevator actuator, deg; t is time, sec.

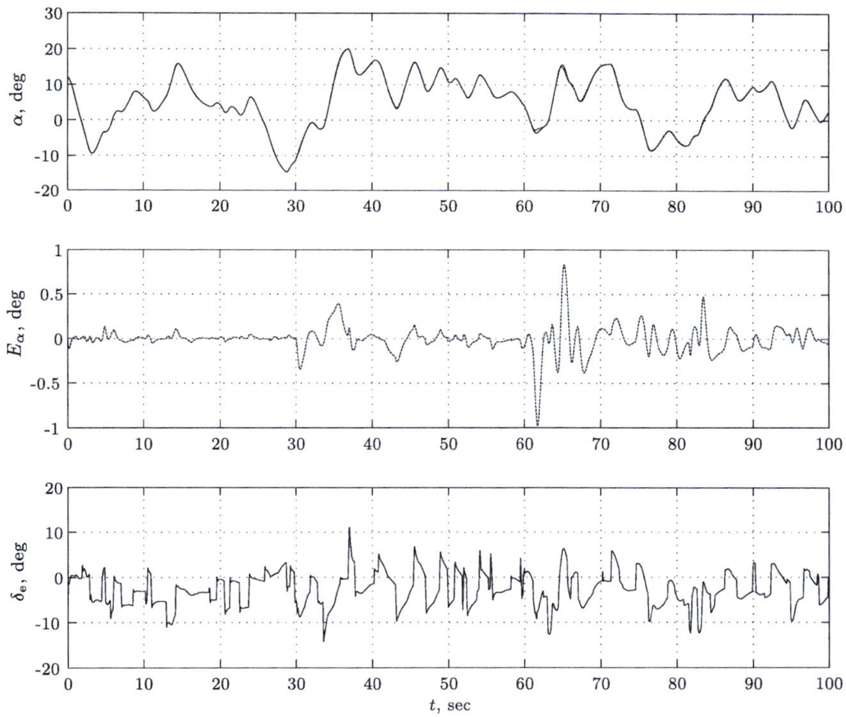

FIGURE A.51 The results of a computational experiment for the MPC-type control system with compensator (F-16 aircraft, flight mode with airspeed $V_{ind} = 300$ km/h). Adaptation to the change in the dynamics of the control object due to failures: backwards shift of the center by 5% ($t = 30$ sec); 50% decrease in the effectiveness of the control ($t = 60$ sec). Here α is angle of attack, deg (dotted line for reference model, solid line for plant); E_α is tracking error for the given angle of attack, deg; α_{ref} is reference signal on the angle of attack, deg; $\delta_{e,act}$ is command signal for the elevator actuator, deg; t is time, sec.

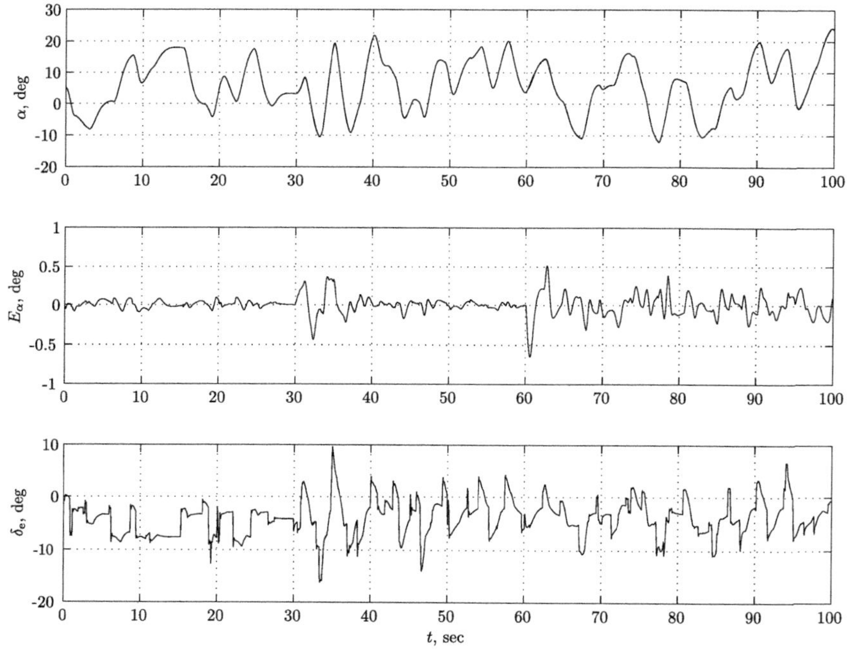

FIGURE A.52 The results of a computational experiment for the MPC-type control system with compensator (F-16 aircraft, flight mode with airspeed $V_{ind} = 500$ km/h). Adaptation to the change in the dynamics of the control object due to failures: backwards shift of the center by 5% ($t = 30$ sec); 50% decrease in the effectiveness of the control ($t = 60$ sec). Here α is angle of attack, deg (dotted line for reference model, solid line for plant); E_α is tracking error for the given angle of attack, deg; α_{ref} is reference signal on the angle of attack, deg; $\delta_{e,\,act}$ is command signal for the elevator actuator, deg; t is time, sec.

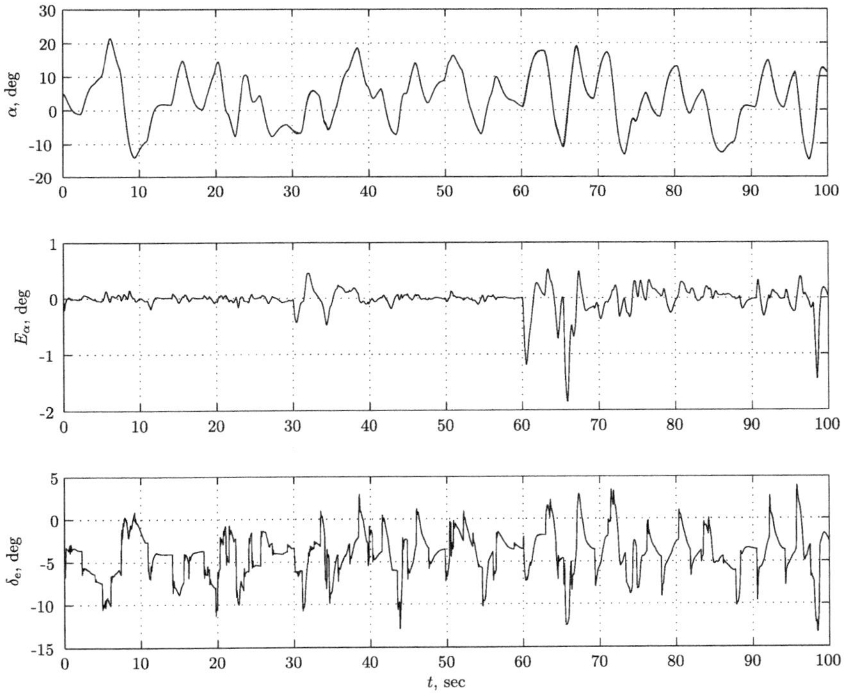

FIGURE A.53 The results of a computational experiment for the MPC-type control system with compensator (F-16 aircraft, flight mode with airspeed $V_{ind} = 700$ km/h). Adaptation to the change in the dynamics of the control object due to failures: backwards shift of the center by 5% ($t = 30$ sec); 50% decrease in the effectiveness of the control ($t = 60$ sec). Here α is angle of attack, deg (dotted line for reference model, solid line for plant); E_α is tracking error for the given angle of attack, deg; α_{ref} is reference signal on the angle of attack, deg; $\delta_{e,act}$ is command signal for the elevator actuator, deg; t is time, sec.

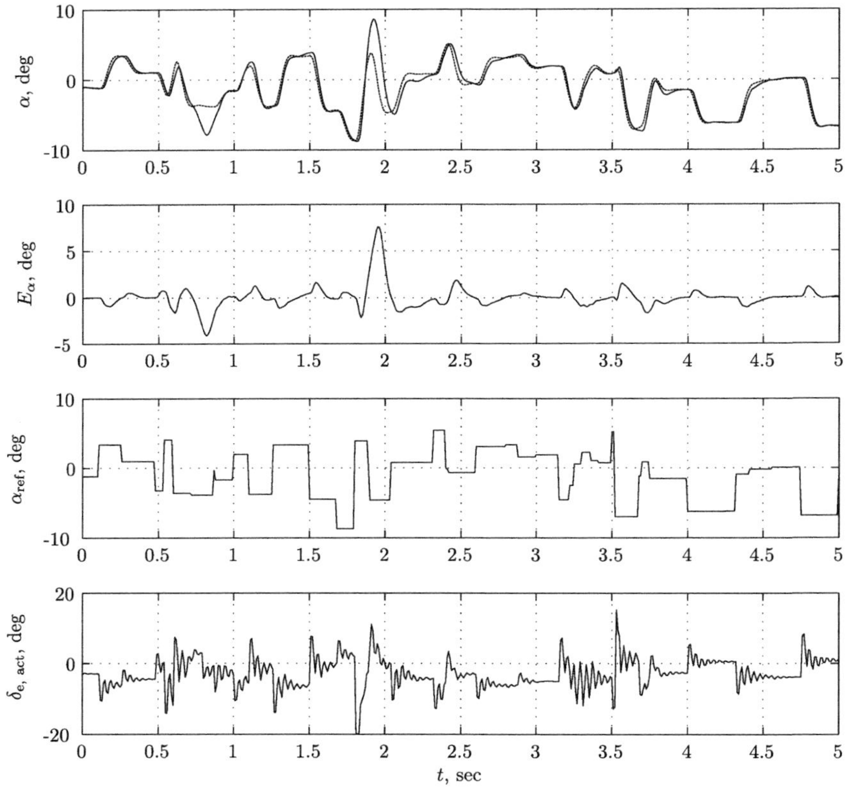

FIGURE A.54 The results of a computational experiment for the MPC-type control system with compensator (micro-UAV "003"), a flight with airspeed $V_{ind} = 30$ km/h in normal operation conditions. Here α is angle of attack, deg (dotted line for reference model, solid line for plant); E_α is tracking error for the given angle of attack, deg; α_{ref} is reference signal on the angle of attack, deg; $\delta_{e,\,act}$ is command signal for the elevator actuator, deg; t is time, sec.

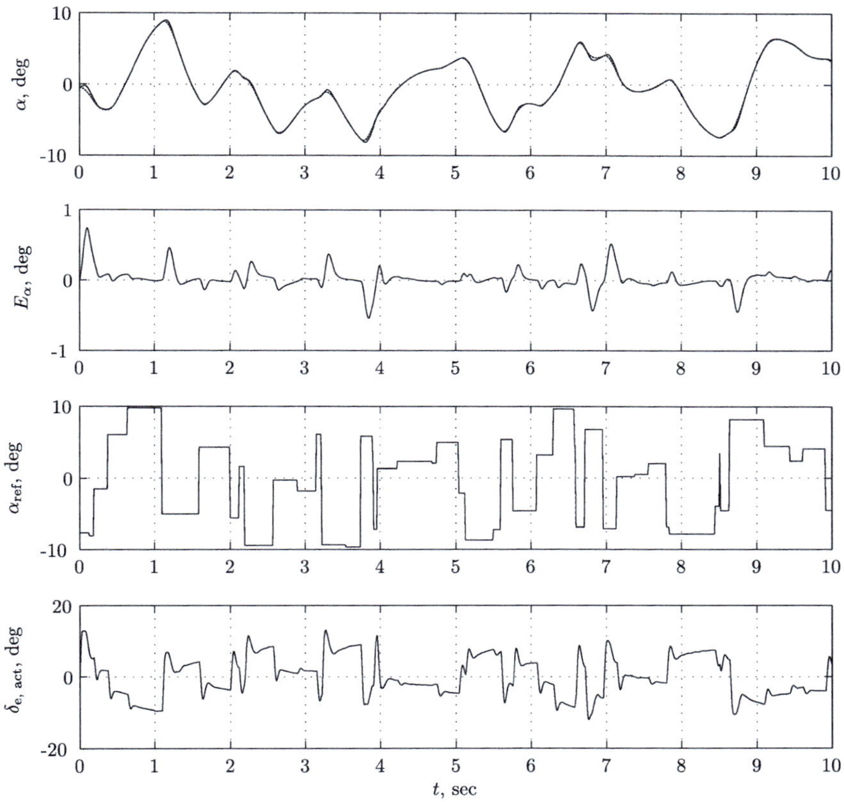

FIGURE A.55 The results of a computational experiment for the MPC-type control system with compensator (mini-UAV X-04), a flight with airspeed $V_{ind} = 70$ km/h in normal operation conditions. Here α is angle of attack, deg (dotted line for reference model, solid line for plant); E_α is tracking error for the given angle of attack, deg; α_{ref} is reference signal on the angle of attack, deg; $\delta_{e,\,act}$ is command signal for the elevator actuator, deg; t is time, sec.

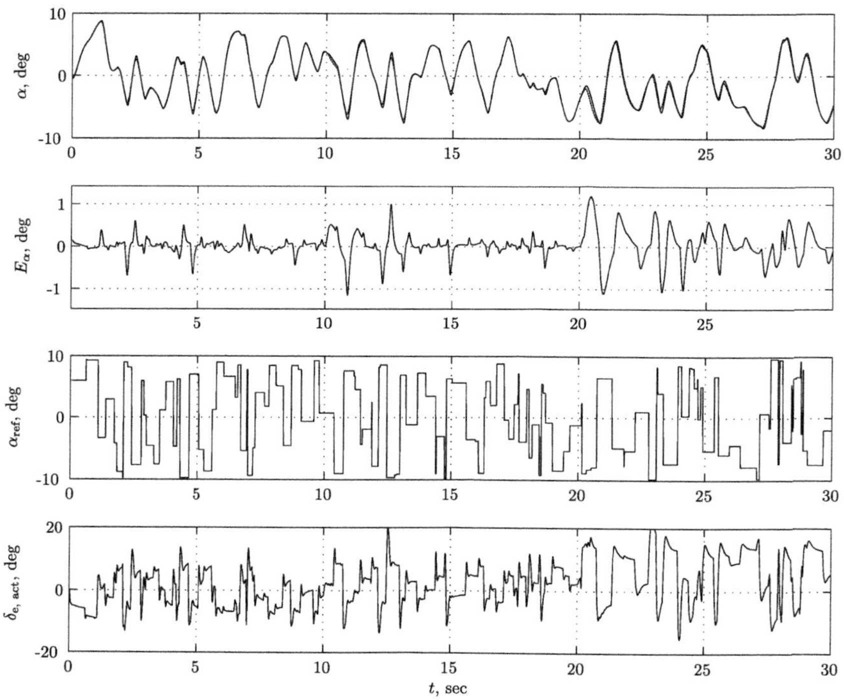

FIGURE A.56 The results of a computational experiment for the MPC-type control system (mini-UAV X-04), a flight with airspeed $V_{ind} = 70$ km/h when fault conditions occur at the time instant $t = 10$ sec (backwards shift of the center by 10%) and then at the time instant $t = 20$ sec (50% decrease in the effectiveness of the longitudinal control). Here α is angle of attack, deg (dotted line for reference model, solid line for plant); E_α is tracking error for the given angle of attack, deg; α_{ref} is reference signal on the angle of attack, deg; $\delta_{e,\,act}$ is command signal for the elevator actuator, deg; t is time, sec.

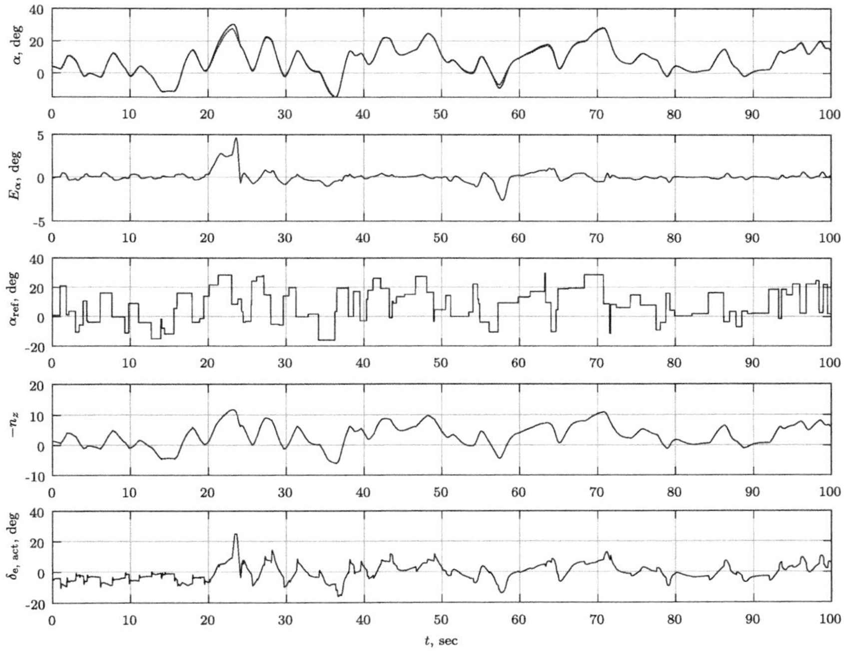

FIGURE A.57 Results of a computational experiment on the evaluation of the nature of the effect on the controlled system of atmospheric turbulence: the MRAC-type control system with PD compensator. No perturbing effect. Flight mode H = 100 m, V = 600 km/h. Here α is angle of attack, deg (dotted line for reference model, solid line for plant); α_{rm} is the angle of attack at the output of the reference model, deg; E_α is tracking error for the given angle of attack, deg; α_{rm} is reference signal on angle of attack, deg; n_z is the normal G-load; $\delta_{e,\,act}$ is command signal for actuator of the elevator, deg; t is time, sec.

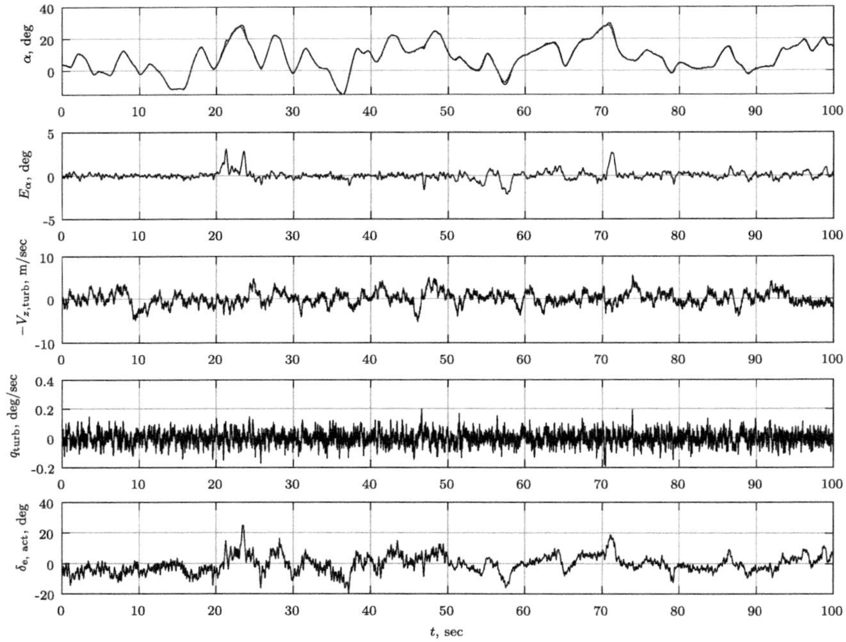

FIGURE A.58 Results of a computational experiment on the evaluation of the nature of the effect on the controlled system of atmospheric turbulence: the MRAC-type control system with PD compensator. Perturbing effect: $V_{y,turb}$ in the range ± 10 m/sec; Δq_{turb} in the range ± 2 deg/sec. Flight mode $H = 100$ m, $V = 600$ km/h. Here α is angle of attack, deg (dotted line for reference model, solid line for plant); α_{rm} is the angle of attack at the output of the reference model, deg; E_α is tracking error for the given angle of attack, deg; α_{rm} is reference signal on angle of attack, deg; n_z is the normal G-load; $\delta_{e,\,act}$ is command signal for actuator of the elevator, deg; t is time, sec.

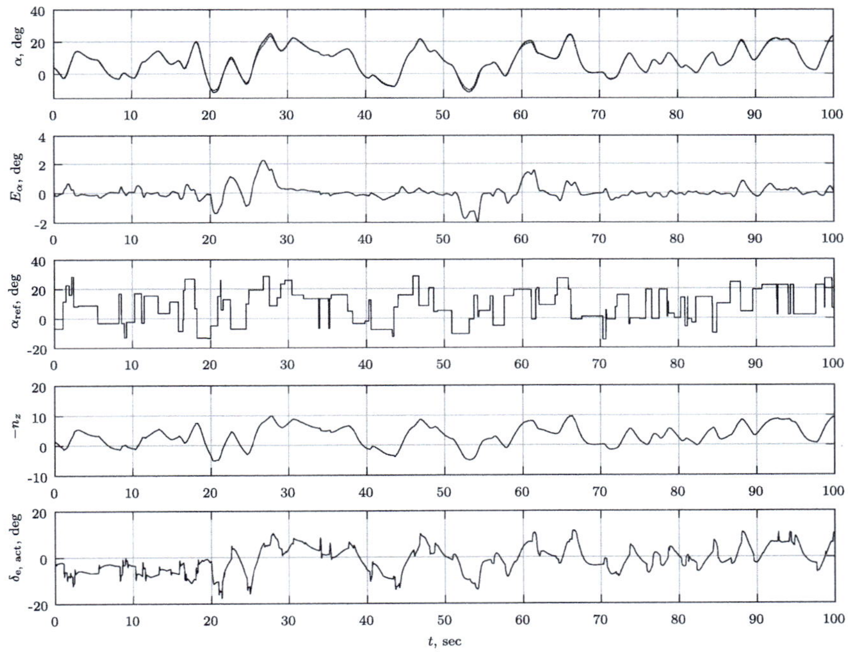

FIGURE A.59 Results of a computational experiment on the evaluation of the nature of the effect on the controlled system of atmospheric turbulence: the MRAC-type control system with PD compensator. No perturbing effect. Flight mode $H = 100$ m, $V = 600$ km/h. Here α is angle of attack, deg (dotted line for reference model, solid line for plant); α_{rm} is the angle of attack at the output of the reference model, deg; E_α is tracking error for the given angle of attack, deg; α_{rm} is reference signal on angle of attack, deg; n_z is the normal G-load; $\delta_{e,\,act}$ is command signal for actuator of the elevator, deg; t is time, sec.

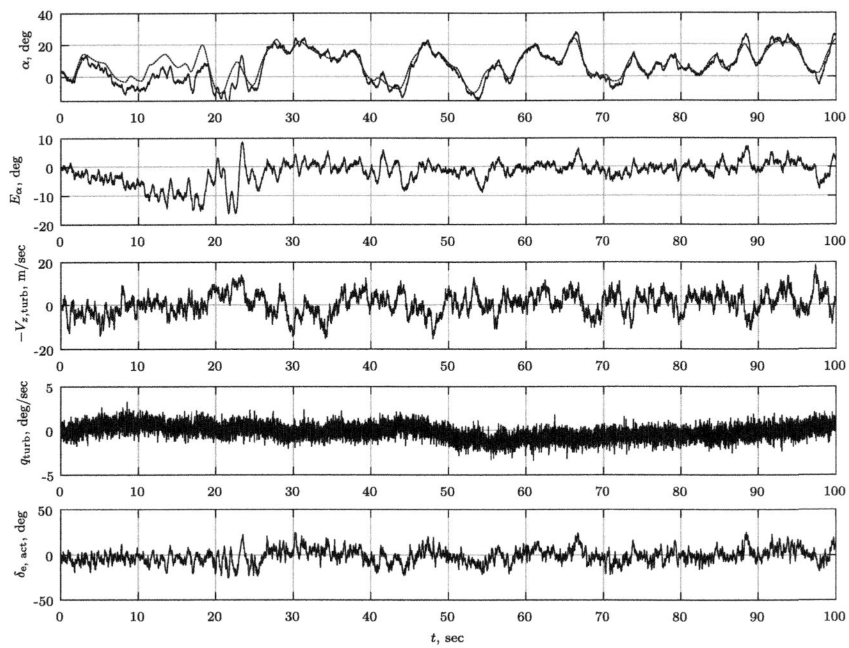

FIGURE A.60 Results of a computational experiment on the evaluation of the nature of the effect on the controlled system of atmospheric turbulence: the MRAC-type control system with PD compensator. Perturbing effect: $V_{y,turb}$ in the range ± 20 m/sec; Δq_{turb} in the range ± 2 deg/sec. Flight mode $H = 100$ m, $V = 600$ km/h. Here α is angle of attack, deg (dotted line for reference model, solid line for plant); α_{rm} is the angle of attack at the output of the reference model, deg; E_α is tracking error for the given angle of attack, deg; α_{rm} is reference signal on angle of attack, deg; n_z is the normal G-load; $\delta_{e,\,act}$ is command signal for actuator of the elevator, deg; t is time, sec.

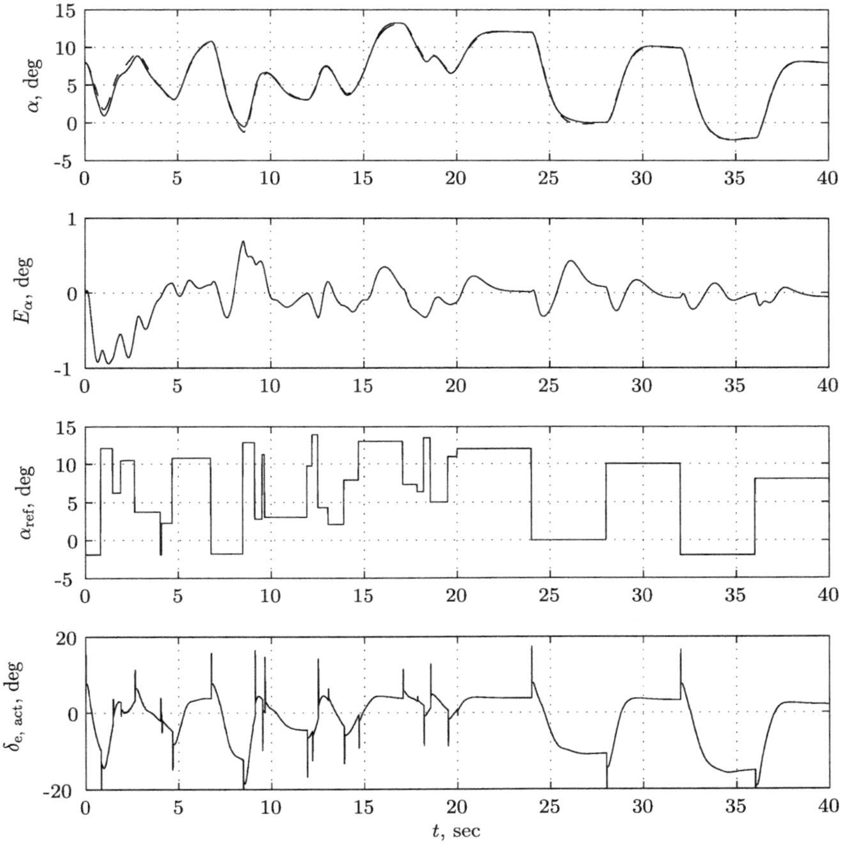

FIGURE A.61 The results of a computational experiment on adaptation to uncertainty in the source data for an MRAC control system applied to the hypersonic research vehicle X-43 (flight mode $M = 5$, $H = 28$ km, $\omega_{rm} = 2$). Here α is angle of attack, deg (dotted line for reference model, solid line for plant); E_α is tracking error for the given angle of attack, deg; α_{ref} is the reference signal on the angle of attack, deg; $\delta_{e,\,act}$ is the command signal for the elevator actuator, deg; t is time, sec.

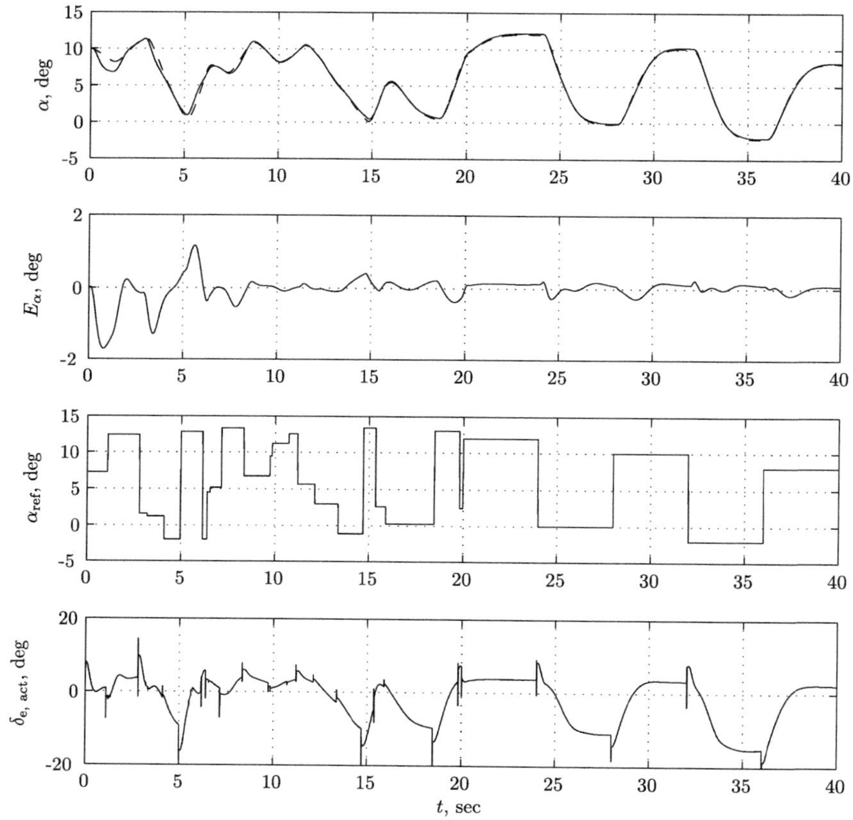

FIGURE A.62 The results of a computational experiment on adaptation to uncertainty in the source data for an MRAC control system applied to the hypersonic research vehicle X-43 (flight mode $M = 5$, $H = 30$ km, $\omega_{rm} = 1.5$). Here α is angle of attack, deg (dotted line for reference model, solid line for plant); E_α is tracking error for the given angle of attack, deg; α_{ref} is the reference signal on the angle of attack, deg; $\delta_{e,\,act}$ is the command signal for the elevator actuator, deg; t is time, sec.

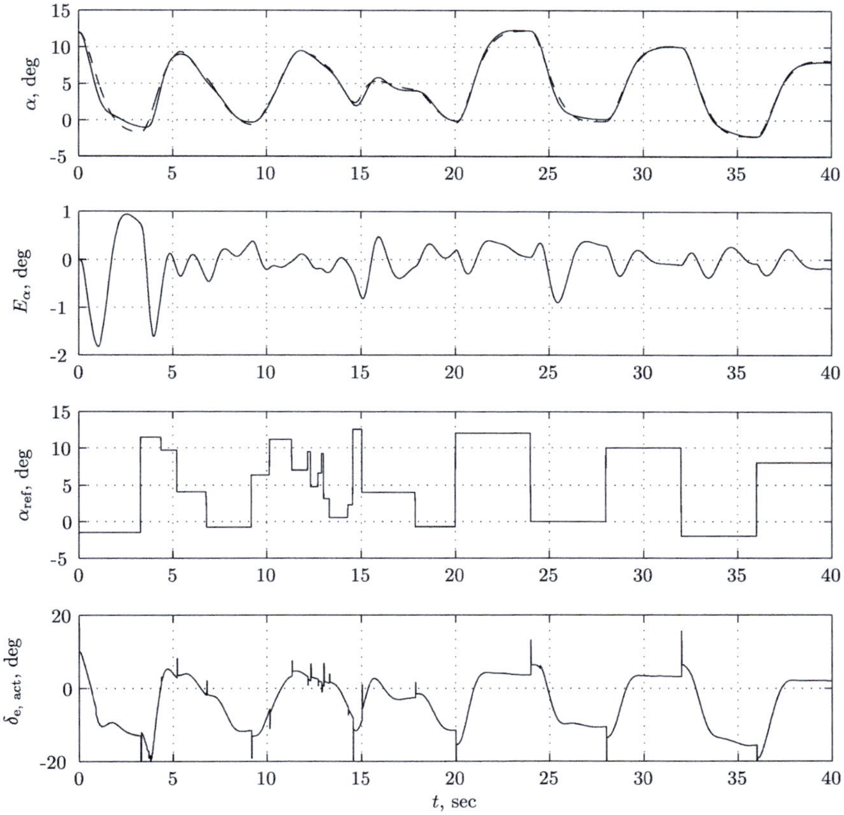

FIGURE A.63 The results of a computational experiment on adaptation to uncertainty in the source data for an MRAC control system applied to the hypersonic research vehicle X-43 (flight mode $M = 5$, $H = 32$ km, $\omega_{rm} = 1.5$). Here α is angle of attack, deg (dotted line for reference model, solid line for plant); E_α is tracking error for the given angle of attack, deg; α_{ref} is the reference signal on the angle of attack, deg; $\delta_{e,\,act}$ is the command signal for the elevator actuator, deg; t is time, sec.

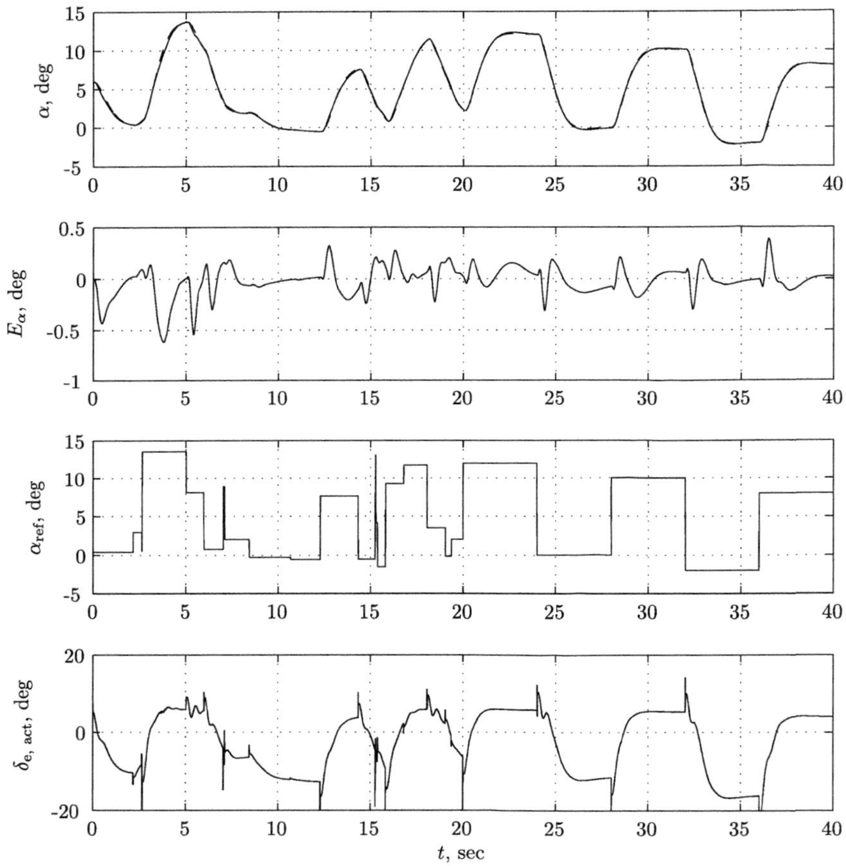

FIGURE A.64 The results of a computational experiment on adaptation to uncertainty in the source data for an MRAC control system applied to the hypersonic research vehicle X-43 (flight mode $M = 6$, $H = 28$ km, $\omega_{rm} = 2$). Here α is angle of attack, deg (dotted line for reference model, solid line for plant); E_α is tracking error for the given angle of attack, deg; α_{ref} is the reference signal on the angle of attack, deg; $\delta_{e,\,act}$ is the command signal for the elevator actuator, deg; t is time, sec.

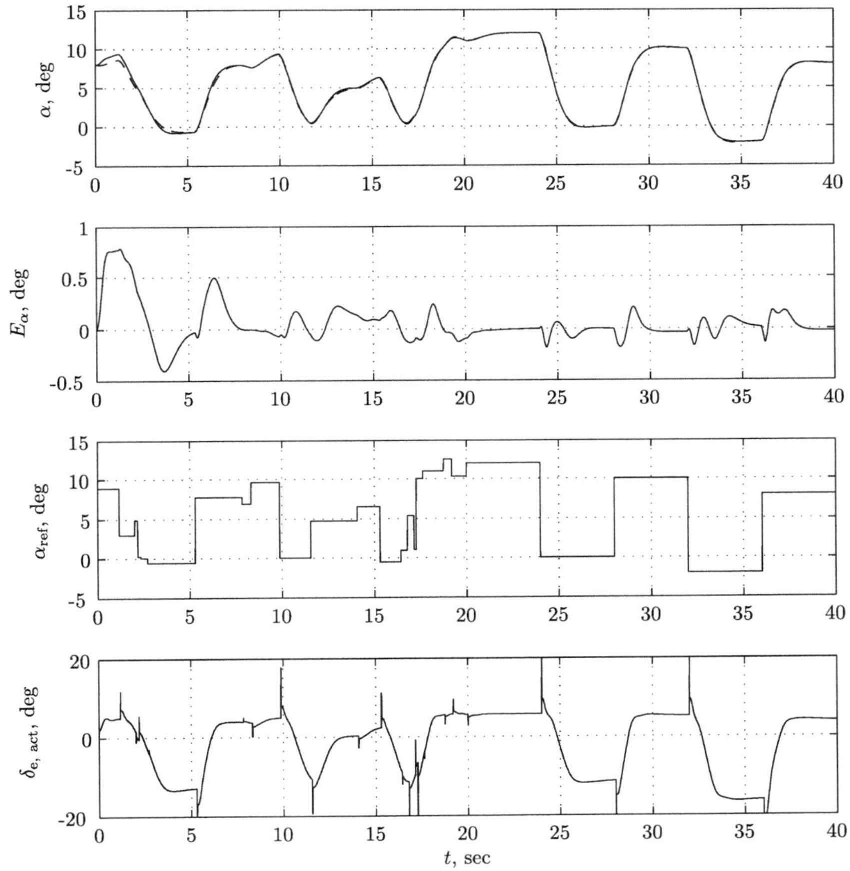

FIGURE A.65 The results of a computational experiment on adaptation to uncertainty in the source data for an MRAC control system applied to the hypersonic research vehicle X-43 (flight mode $M = 6$, $H = 30$ km, $\omega_{rm} = 2$). Here α is angle of attack, deg (dotted line for reference model, solid line for plant); E_α is tracking error for the given angle of attack, deg; α_{ref} is the reference signal on the angle of attack, deg; $\delta_{e,\,act}$ is the command signal for the elevator actuator, deg; t is time, sec.

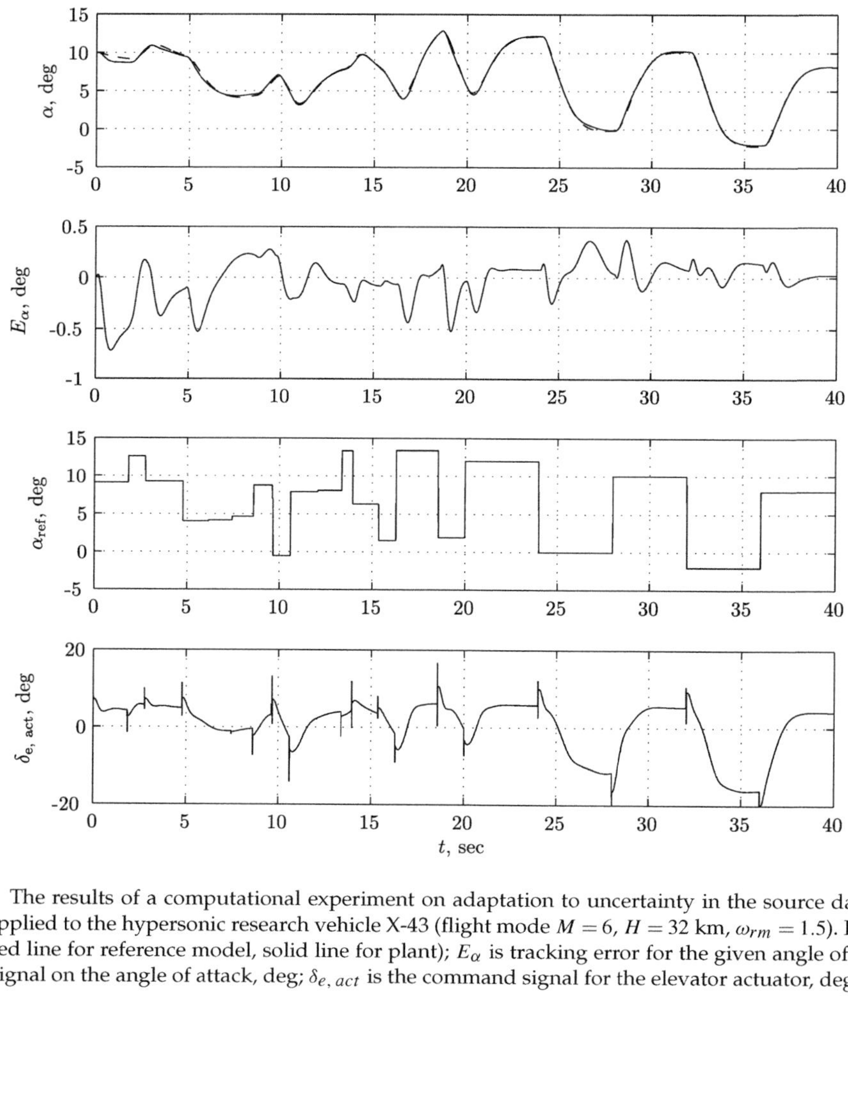

FIGURE A.66 The results of a computational experiment on adaptation to uncertainty in the source data for an MRAC control system applied to the hypersonic research vehicle X-43 (flight mode $M = 6$, $H = 32$ km, $\omega_{rm} = 1.5$). Here α is angle of attack, deg (dotted line for reference model, solid line for plant); E_α is tracking error for the given angle of attack, deg; α_{ref} is the reference signal on the angle of attack, deg; $\delta_{e,\,act}$ is the command signal for the elevator actuator, deg; t is time, sec.

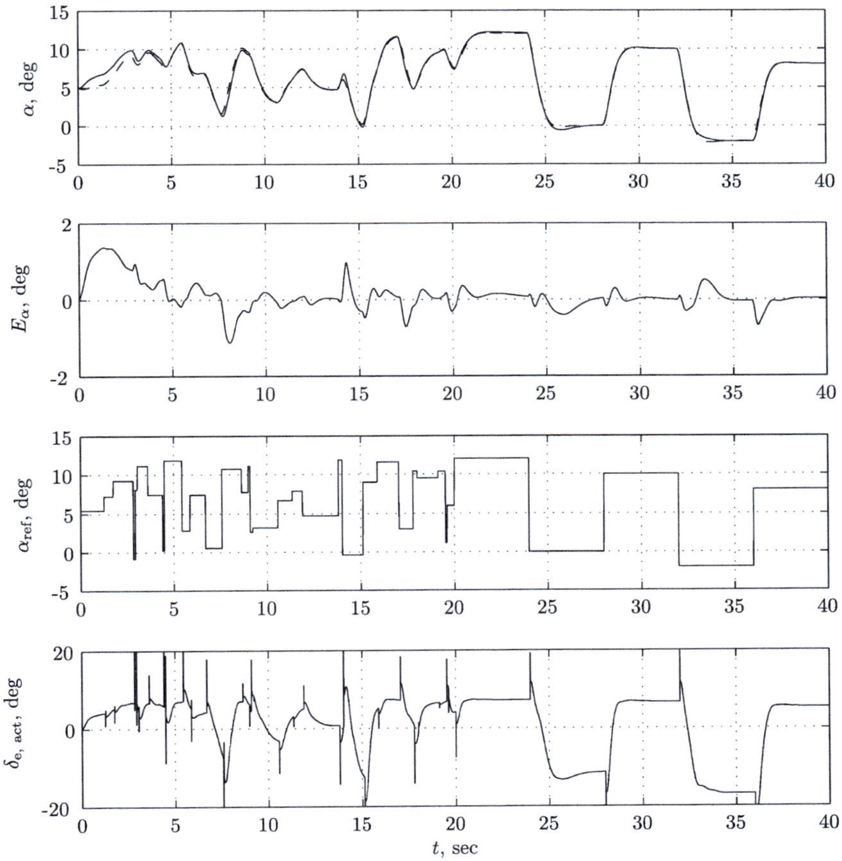

FIGURE A.67 The results of a computational experiment on adaptation to uncertainty in the source data for an MRAC control system applied to the hypersonic research vehicle X-43 (flight mode $M = 7$, $H = 28$ km, $\omega_{rm} = 3$). Here α is angle of attack, deg (dotted line for reference model, solid line for plant); E_α is tracking error for the given angle of attack, deg; α_{ref} is the reference signal on the angle of attack, deg; $\delta_{e,\,act}$ is the command signal for the elevator actuator, deg; t is time, sec.

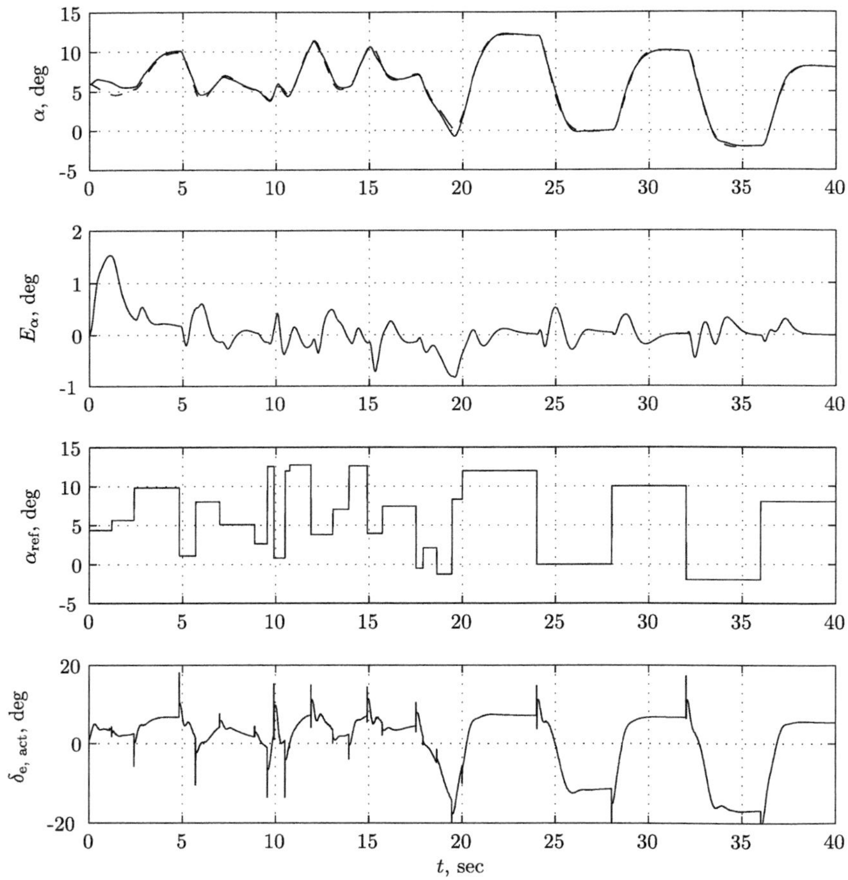

FIGURE A.68 The results of a computational experiment on adaptation to uncertainty in the source data for an MRAC control system applied to the hypersonic research vehicle X-43 (flight mode $M = 7$, $H = 30$ km, $\omega_{rm} = 2$). Here α is angle of attack, deg (dotted line for reference model, solid line for plant); E_α is tracking error for the given angle of attack, deg; α_{ref} is the reference signal on the angle of attack, deg; $\delta_{e,\,act}$ is the command signal for the elevator actuator, deg; t is time, sec.

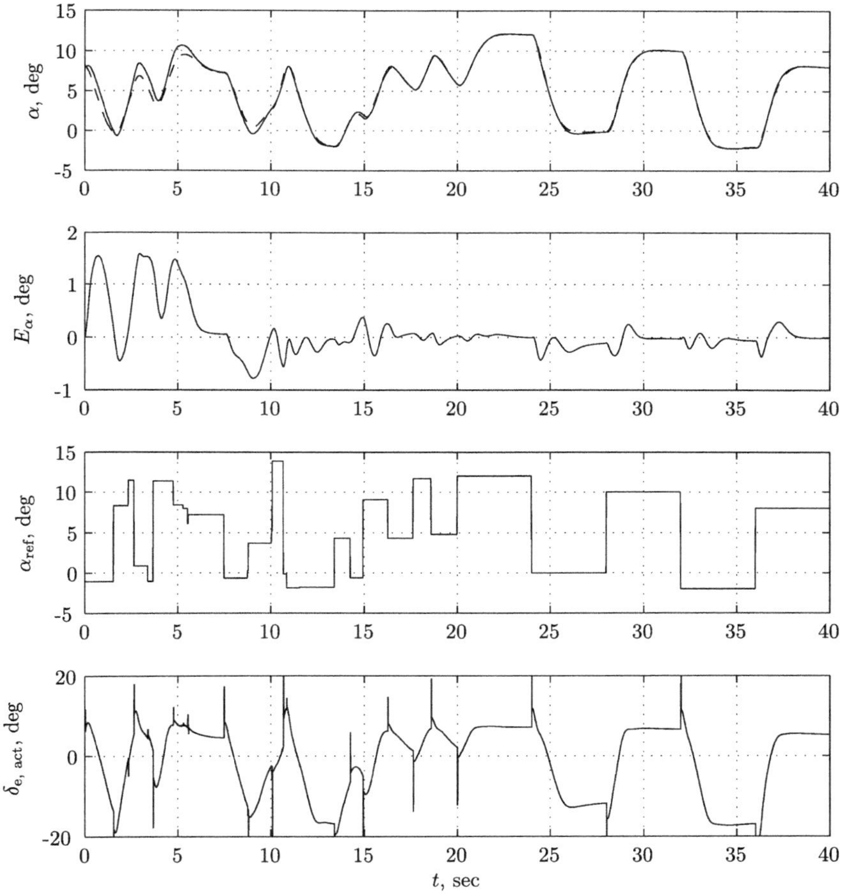

FIGURE A.69 The results of a computational experiment on adaptation to uncertainty in the source data for an MRAC control system applied to the hypersonic research vehicle X-43 (flight mode $M = 7$, $H = 32$ km, $\omega_{rm} = 2$). Here α is angle of attack, deg (dotted line for reference model, solid line for plant); E_α is tracking error for the given angle of attack, deg; α_{ref} is the reference signal on the angle of attack, deg; $\delta_{e,\,act}$ is the command signal for the elevator actuator, deg; t is time, sec.

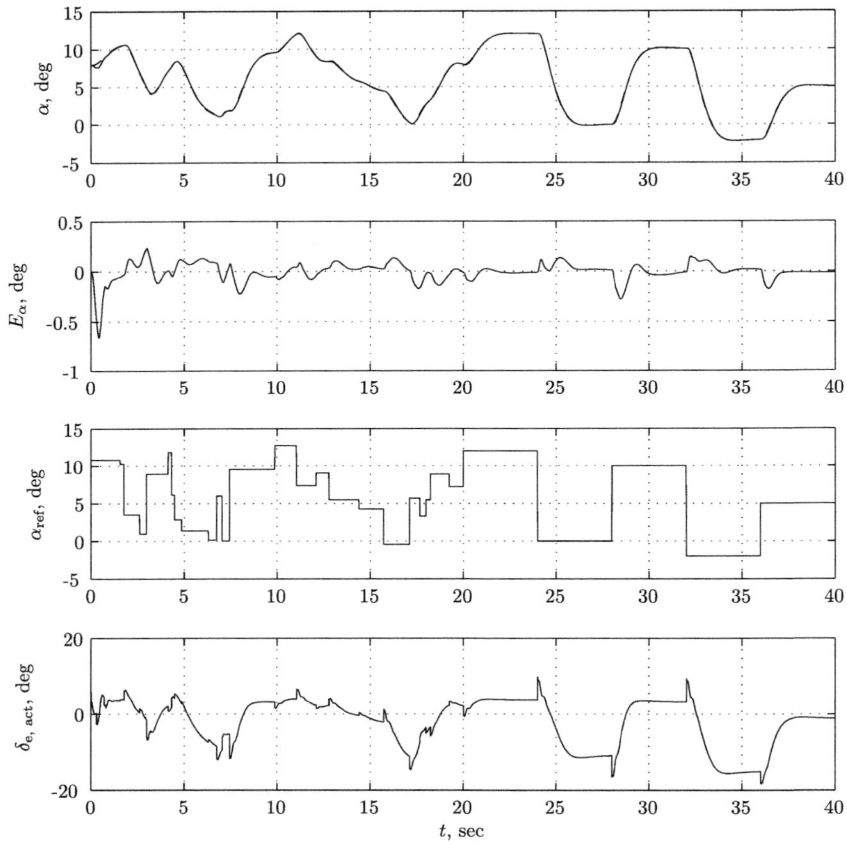

FIGURE A.70 The results of a computational experiment on adaptation to uncertainty in the source data for an MPC control system applied to the hypersonic research vehicle X-43 (flight mode $M = 5$, $H = 28$ km, $\omega_{rm} = 2$). Here α is angle of attack, deg (dotted line for reference model, solid line for plant); E_α is tracking error for the given angle of attack, deg; α_{ref} is the reference signal on the angle of attack, deg; $\delta_{e,\,act}$ is the command signal for the elevator actuator, deg; t is time, sec.

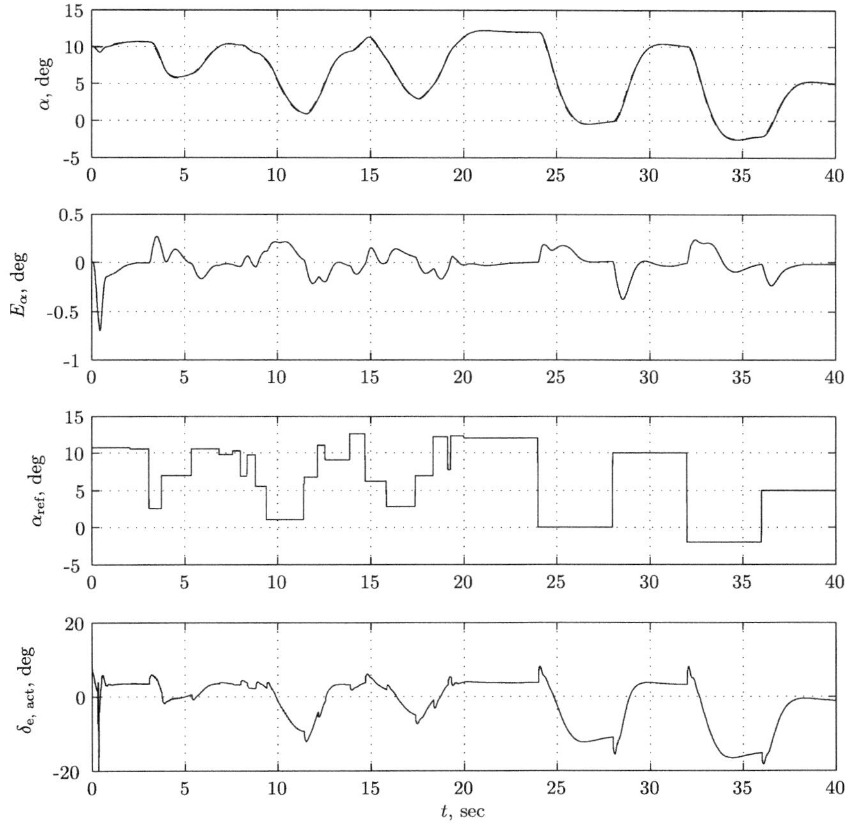

FIGURE A.71 The results of a computational experiment on adaptation to uncertainty in the source data for an MPC control system applied to the hypersonic research vehicle X-43 (flight mode $M = 5$, $H = 30$ km, $\omega_{rm} = 1.5$). Here α is angle of attack, deg (dotted line for reference model, solid line for plant); E_α is tracking error for the given angle of attack, deg; α_{ref} is the reference signal on the angle of attack, deg; $\delta_{e,\,act}$ is the command signal for the elevator actuator, deg; t is time, sec.

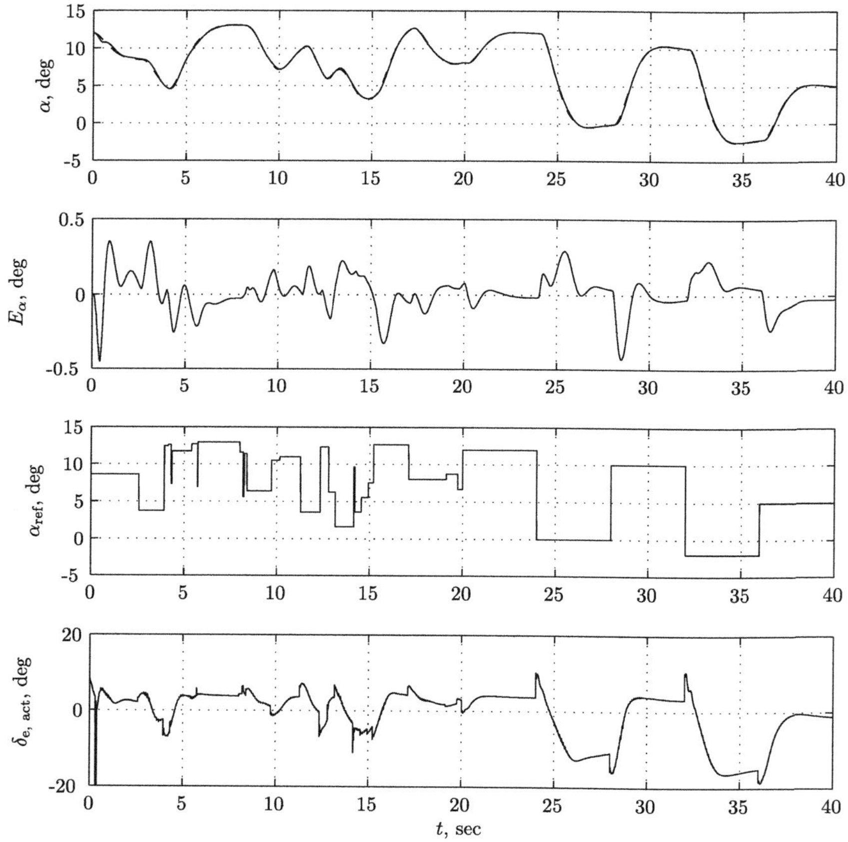

FIGURE A.72 The results of a computational experiment on adaptation to uncertainty in the source data for an MPC control system applied to the hypersonic research vehicle X-43 (flight mode $M = 5$, $H = 32$ km, $\omega_{rm} = 1.5$). Here α is angle of attack, deg (dotted line for reference model, solid line for plant); E_α is tracking error for the given angle of attack, deg; α_{ref} is the reference signal on the angle of attack, deg; $\delta_{e,\,act}$ is the command signal for the elevator actuator, deg; t is time, sec.

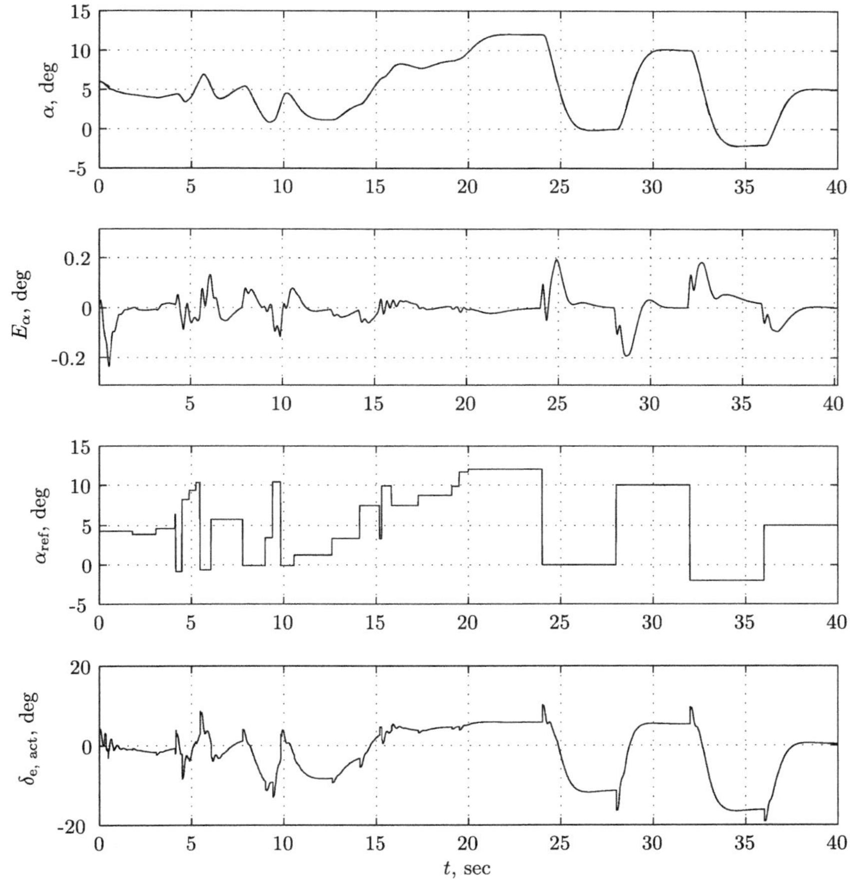

FIGURE A.73 The results of a computational experiment on adaptation to uncertainty in the source data for an MPC control system applied to the hypersonic research vehicle X-43 (flight mode $M = 6$, $H = 28$ km, $\omega_{rm} = 2$). Here α is angle of attack, deg (dotted line for reference model, solid line for plant); E_α is tracking error for the given angle of attack, deg; α_{ref} is the reference signal on the angle of attack, deg; $\delta_{e,\,act}$ is the command signal for the elevator actuator, deg; t is time, sec.

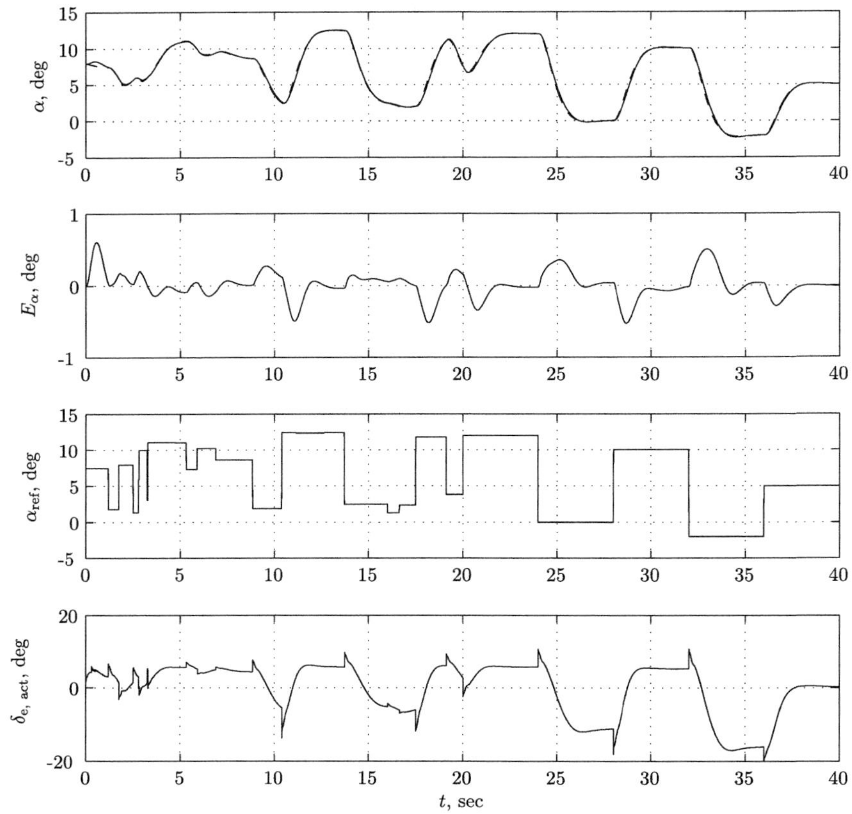

FIGURE A.74 The results of a computational experiment on adaptation to uncertainty in the source data for an MPC control system applied to the hypersonic research vehicle X-43 (flight mode $M = 6$, $H = 30$ km, $\omega_{rm} = 2$). Here α is angle of attack, deg (dotted line for reference model, solid line for plant); E_α is tracking error for the given angle of attack, deg; α_{ref} is the reference signal on the angle of attack, deg; $\delta_{e,\,act}$ is the command signal for the elevator actuator, deg; t is time, sec.

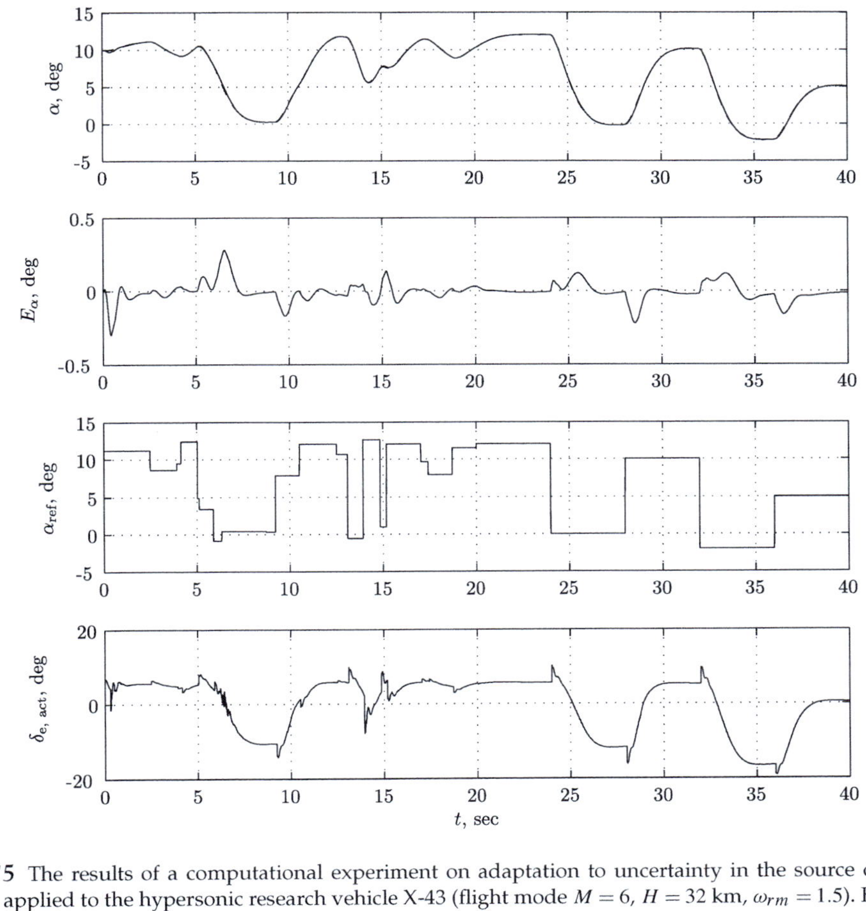

FIGURE A.75 The results of a computational experiment on adaptation to uncertainty in the source data for an MPC control system applied to the hypersonic research vehicle X-43 (flight mode $M = 6$, $H = 32$ km, $\omega_{rm} = 1.5$). Here α is angle of attack, deg (dotted line for reference model, solid line for plant); E_α is tracking error for the given angle of attack, deg; α_{ref} is the reference signal on the angle of attack, deg; $\delta_{e,\,act}$ is the command signal for the elevator actuator, deg; t is time, sec.

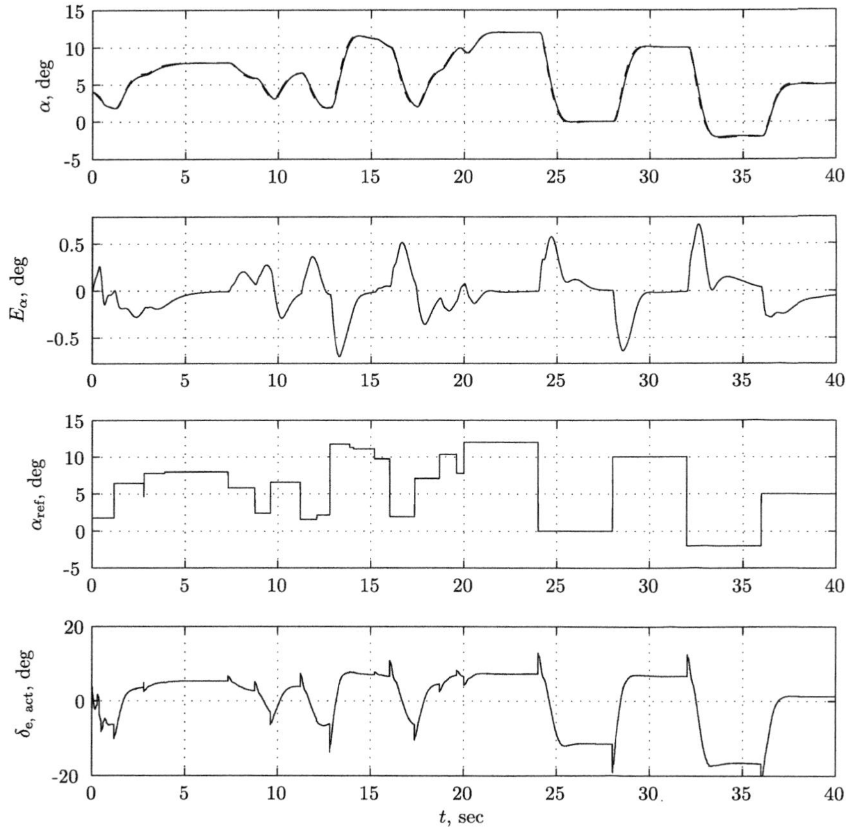

FIGURE A.76 The results of a computational experiment on adaptation to uncertainty in the source data for an MPC control system applied to the hypersonic research vehicle X-43 (flight mode $M = 7$, $H = 28$ km, $\omega_{rm} = 3$). Here α is angle of attack, deg (dotted line for reference model, solid line for plant); E_α is tracking error for the given angle of attack, deg; α_{ref} is the reference signal on the angle of attack, deg; $\delta_{e,\,act}$ is the command signal for the elevator actuator, deg; t is time, sec.

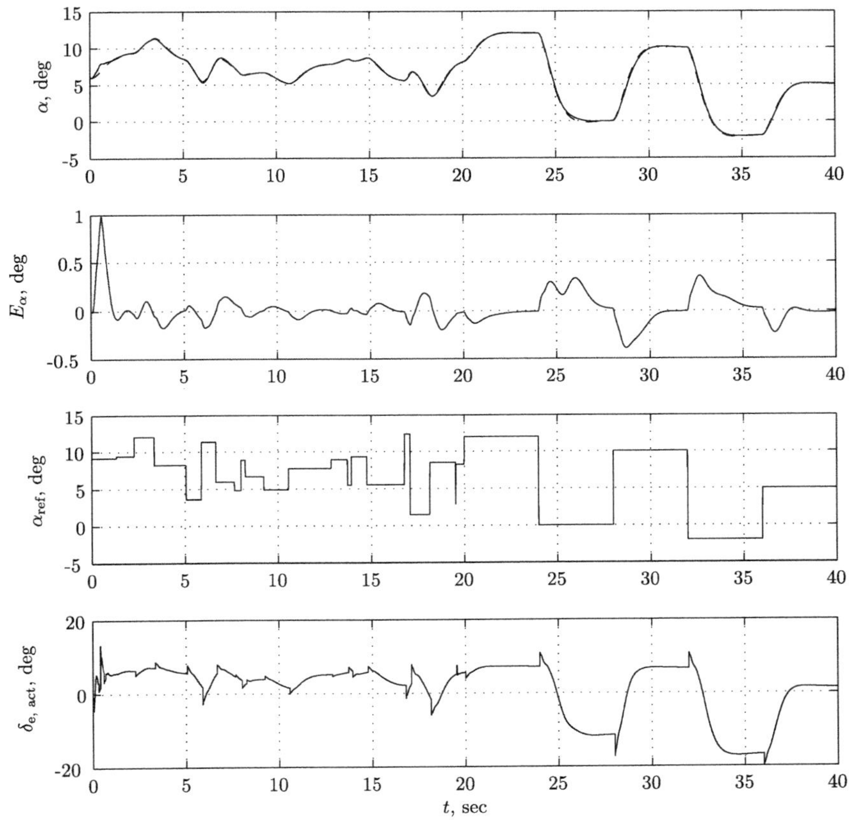

FIGURE A.77 The results of a computational experiment on adaptation to uncertainty in the source data for an MPC control system applied to the hypersonic research vehicle X-43 (flight mode $M = 7$, $H = 30$ km, $\omega_{rm} = 2$). Here α is angle of attack, deg (dotted line for reference model, solid line for plant); E_α is tracking error for the given angle of attack, deg; α_{ref} is the reference signal on the angle of attack, deg; $\delta_{e,\,act}$ is the command signal for the elevator actuator, deg; t is time, sec.

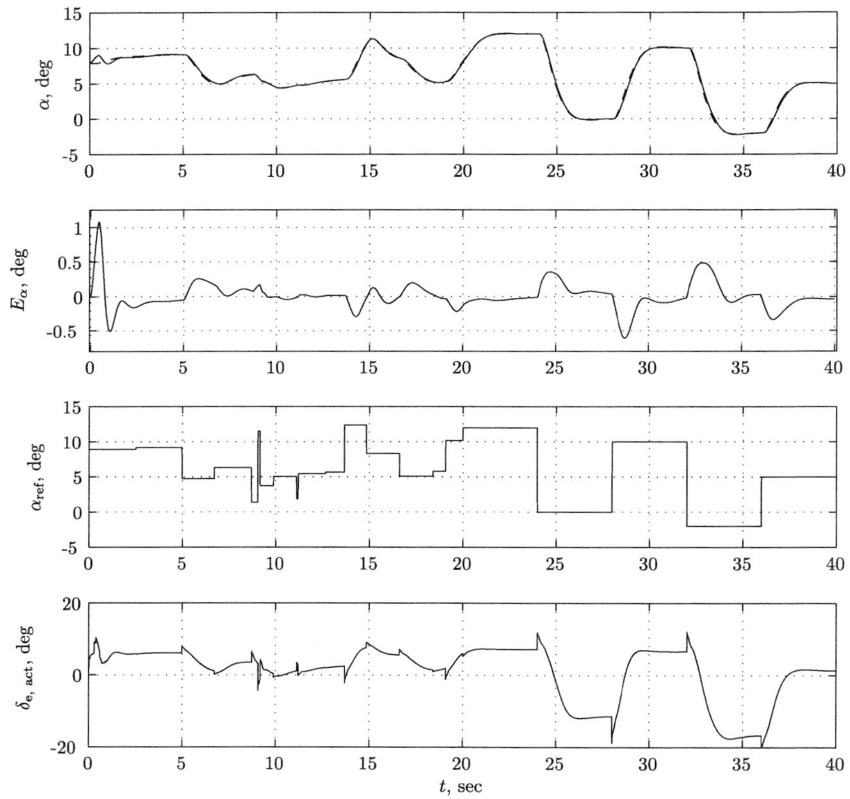

FIGURE A.78 The results of a computational experiment on adaptation to uncertainty in the source data for an MPC control system applied to the hypersonic research vehicle X-43 (flight mode $M = 7$, $H = 32$ km, $\omega_{rm} = 2$). Here α is angle of attack, deg (dotted line for reference model, solid line for plant); E_α is tracking error for the given angle of attack, deg; α_{ref} is the reference signal on the angle of attack, deg; $\delta_{e,\,act}$ is the command signal for the elevator actuator, deg; t is time, sec.

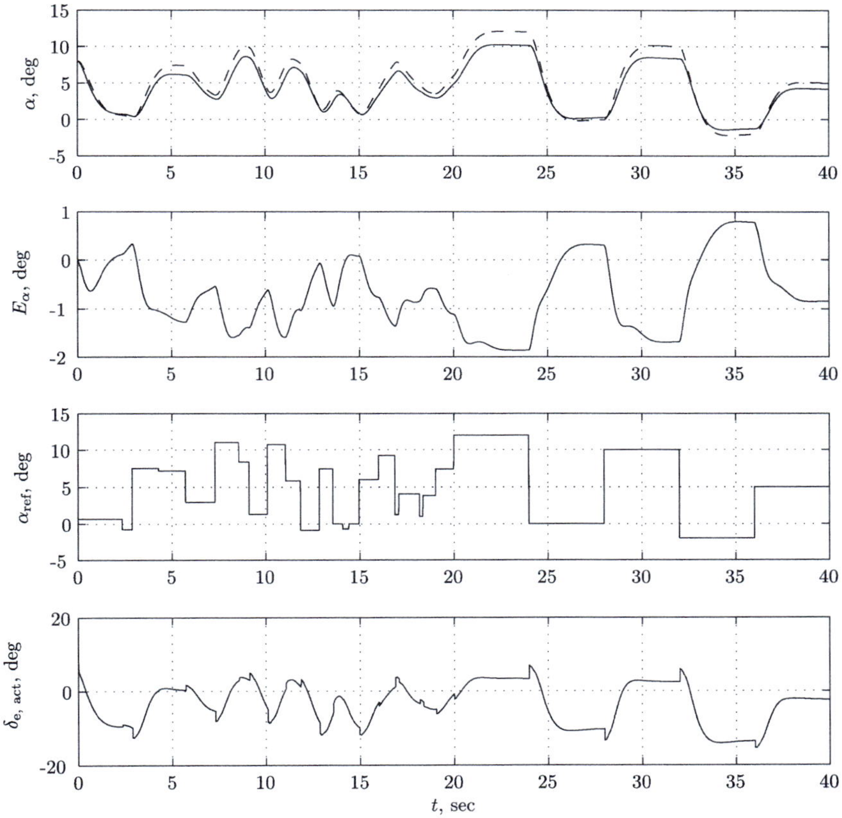

FIGURE A.79 The results of a computational experiment to evaluate the significance of the adaptation mechanism in the MPC control system for the hypersonic research vehicle X-43 (flight mode $M = 5$, $H = 28$ km, $\omega_{rm} = 2$). Here α is angle of attack, deg (dotted line for reference model, solid line for plant); E_α is tracking error for the given angle of attack, deg; α_{ref} is the reference signal on the angle of attack, deg; $\delta_{e,\,act}$ is the command signal for the elevator actuator, deg; t is time, sec.

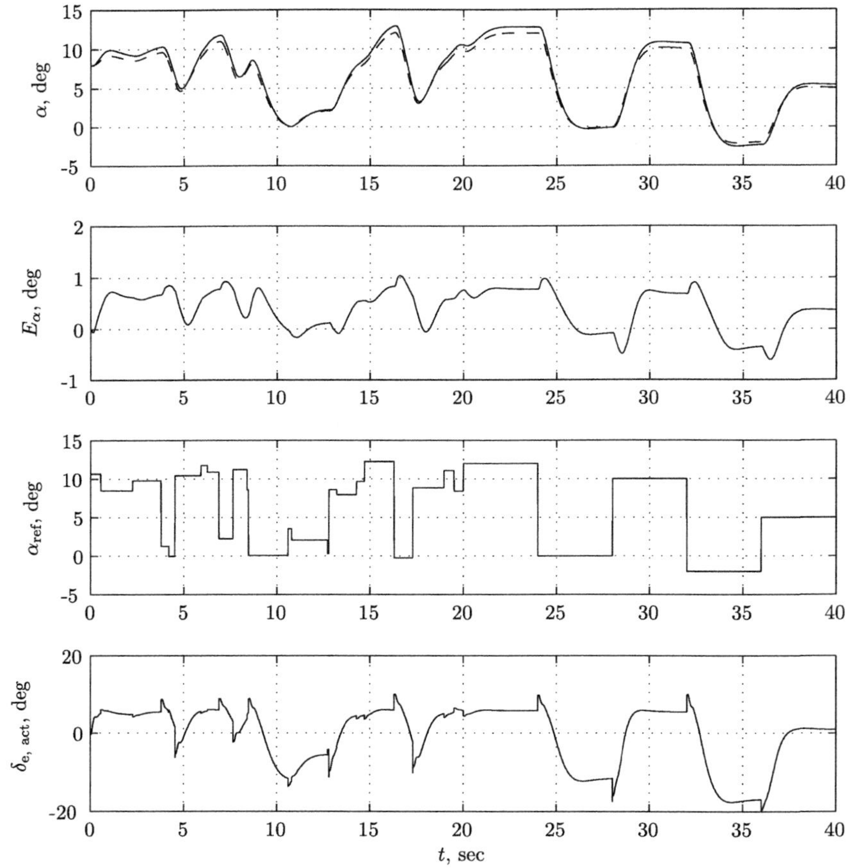

FIGURE A.80 The results of a computational experiment to evaluate the significance of the adaptation mechanism in the MPC control system for the hypersonic research vehicle X-43 (flight mode $M = 6$, $H = 30$ km, $\omega_{rm} = 2$). Here α is angle of attack, deg (dotted line for reference model, solid line for plant); E_α is tracking error for the given angle of attack, deg; α_{ref} is the reference signal on the angle of attack, deg; $\delta_{e,\,act}$ is the command signal for the elevator actuator, deg; t is time, sec.

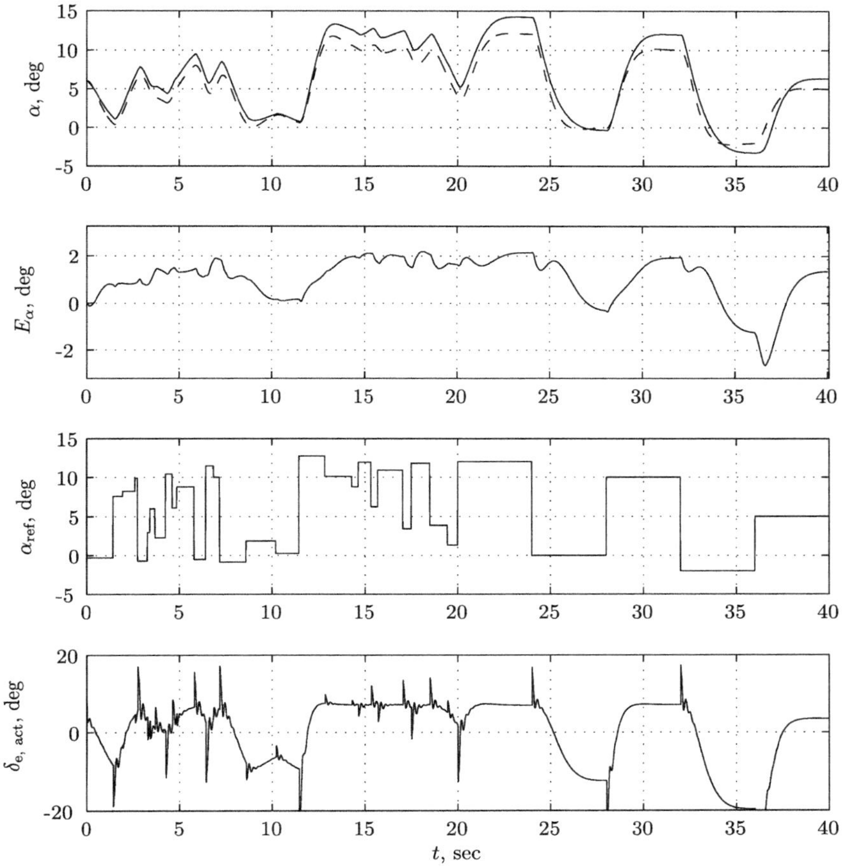

FIGURE A.81 The results of a computational experiment to evaluate the significance of the adaptation mechanism in the MPC control system for the hypersonic research vehicle X-43 (flight mode $M = 7$, $H = 30$ km, $\omega_{rm} = 2$). Here α is angle of attack, deg (dotted line for reference model, solid line for plant); E_α is tracking error for the given angle of attack, deg; α_{ref} is the reference signal on the angle of attack, deg; $\delta_{e,\,act}$ is the command signal for the elevator actuator, deg; t is time, sec.

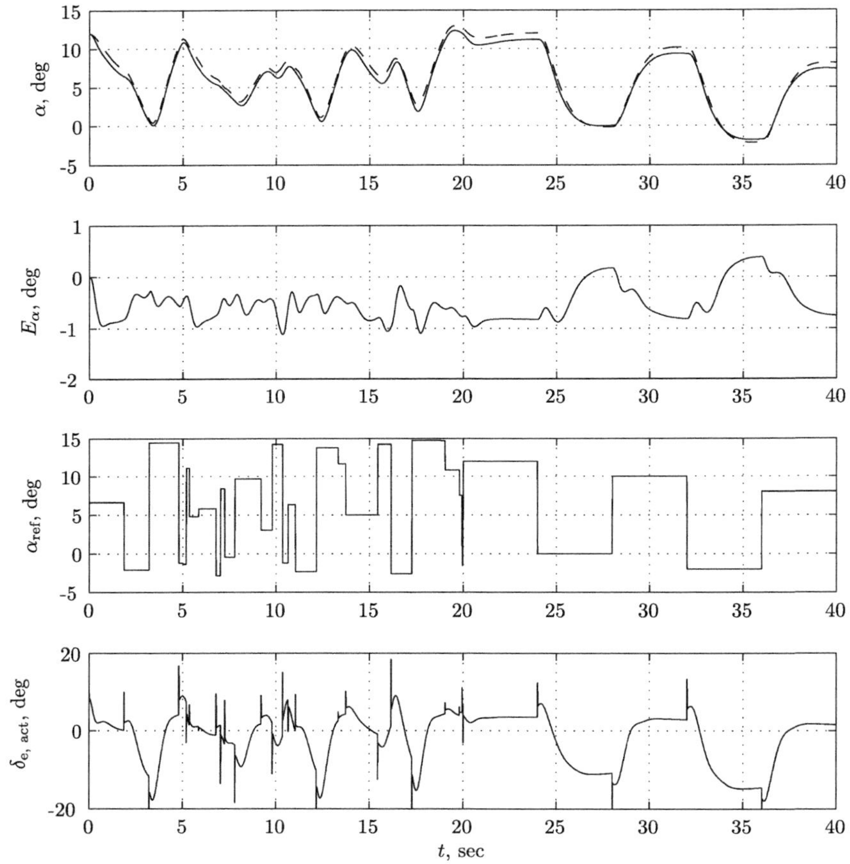

FIGURE A.82 The results of a computational experiment to evaluate the significance of the adaptation mechanism in the MRAC control system for the hypersonic research vehicle X-43 (flight mode $M = 5$, $H = 32$ km, $\omega_{rm} = 1.5$). Here α is angle of attack, deg (dotted line for reference model, solid line for plant); E_α is tracking error for the given angle of attack, deg; α_{ref} is the reference signal on the angle of attack, deg; $\delta_{e,\,act}$ is the command signal for the elevator actuator, deg; t is time, sec.

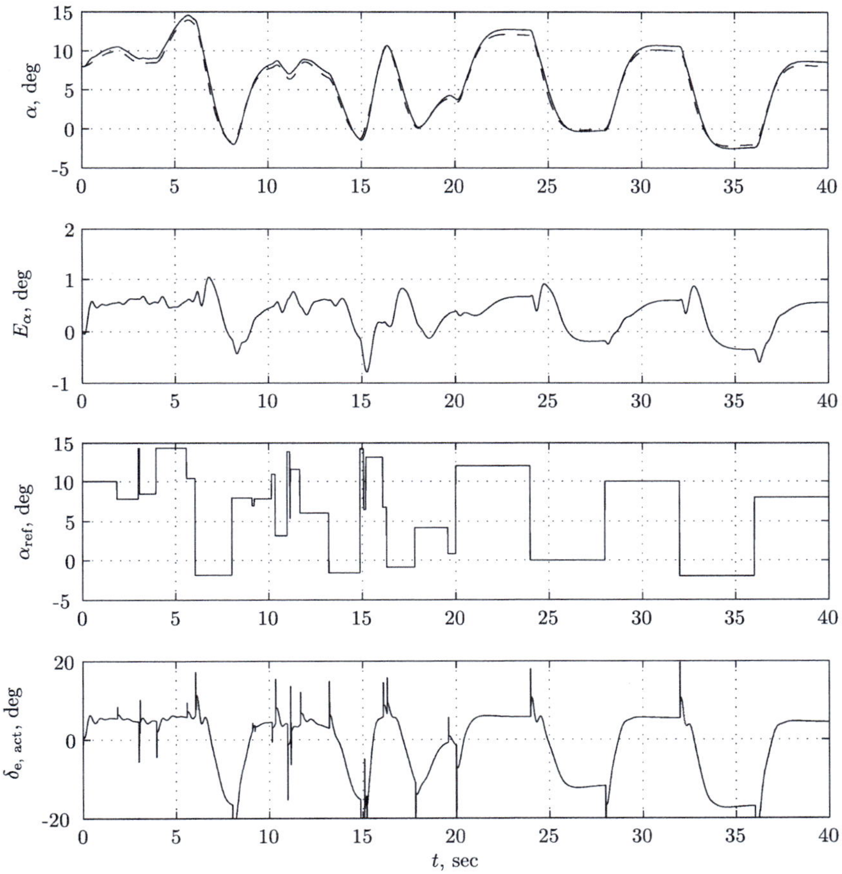

FIGURE A.83 The results of a computational experiment to evaluate the significance of the adaptation mechanism in the MRAC control system for the hypersonic research vehicle X-43 (flight mode $M = 6$, $H = 30$ km, $\omega_{rm} = 2$). Here α is angle of attack, deg (dotted line for reference model, solid line for plant); E_α is tracking error for the given angle of attack, deg; α_{ref} is the reference signal on the angle of attack, deg; $\delta_{e,\,act}$ is the command signal for the elevator actuator, deg; t is time, sec.

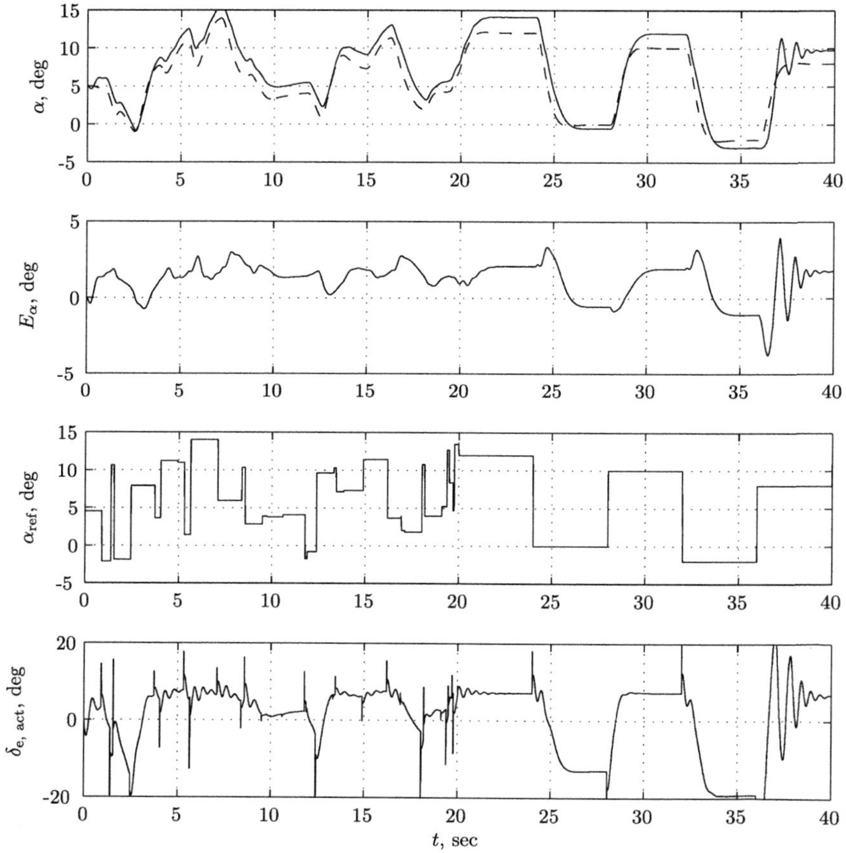

FIGURE A.84 The results of a computational experiment to evaluate the significance of the adaptation mechanism in the MRAC control system for the hypersonic research vehicle X-43 (flight mode $M = 7$, $H = 28$ km, $\omega_{rm} = 3$). Here α is angle of attack, deg (dotted line for reference model, solid line for plant); E_α is tracking error for the given angle of attack, deg; α_{ref} is the reference signal on the angle of attack, deg; $\delta_{e,\,act}$ is the command signal for the elevator actuator, deg; t is time, sec.

Index

A

Actuator, 18, 78, 108, 109, 134, 148, 149, 159, 201, 209, 210
- command signal, 151
- model, 109
- system, 116, 117

Adaptability, 3, 4, 21, 29, 94, 112
Adaptation
- algorithms, 153, 154, 161
- law, 26, 140
- mechanisms, 6, 11, 25, 157, 158, 160–162
- object, 22
- scheme, 25, 151, 154
- time, 24, 159

Adaptive
- adjustment, 159, 161
- control, 5, 24–27, 134, 135, 140, 149, 151, 157, 158
  - laws, 140
  - problem, 23
  - schemes, 4, 26, 140, 147, 149, 153, 155, 160
  - systems, 26, 140, 154, 158
- neural network control system, 154, 157
- systems, 6, 13, 21, 24, 27, 29, 139

Adjusting
- controller, 104, 106, 107
- effects, 25

Aerodynamic characteristics, 3, 4, 12, 19, 71, 82, 208, 210, 213, 215–217, 225
Aerodynamic forces, 4, 10, 12, 17, 199, 200, 208–210, 212, 216
Aircraft
- aerodynamic characteristics, 199, 221
- characteristics, 10
- control, 11
  - algorithms, 10
  - problems, 24, 102
  - surfaces, 12
  - system, 21, 29, 139
- dynamics, 112, 118, 158
- engine thrust, 216
- flight dynamics, 201, 209, 217
- motion, 4, 5, 72, 101–103, 108, 117, 131, 133, 135, 139, 152, 154, 158, 168, 199, 200, 208
  - ANN model, 131, 134
  - lateral, 8
  - longitudinal, 158
- wing, 210, 218

Angle pitch, 18, 105, 134, 201, 209, 217
ANN
- control, 118
- learning algorithm, 51, 217
- model, 39, 154
  - for aircraft motion, 103
  - for nonlinear dynamical systems, 101
- modeling, 3, 4, 35, 76, 80, 94, 133, 208
- modeling methods, 216
- modules, 207, 209, 216, 217, 223–225
- outputs, 108
- semiempirical, 4, 199–201, 208–210, 212, 217, 221, 224, 225

Antiaircraft missiles, 15, 23
Artificial neural network (ANN), 3, 27, 35, 51, 102, 106, 113, 216
Attack angle, 105, 109, 134, 135, 142, 151–154, 157, 159–161, 200, 201, 204, 206, 208, 209, 218, 225

B

Behavior
- aircraft, 11
- control, 2, 21
- network, 51

Black box, 3, 5, 94, 95, 101, 165, 167, 170, 200, 208
Black box neural network, 4, 5, 94

C

Command control, 104
Command control signal, 135
Command signal, 105–107, 134, 135, 147, 150, 161, 201, 209, 210, 213, 217
- actuator, 151
- input, 157

Compensating loop, 27, 145, 150, 160, 161
Computational experiments, 5, 6, 107–109, 135, 147, 149, 152, 153, 157–160, 171, 192, 201, 204, 210, 217, 225
Continuous derivatives, 183–186
Continuous time derivatives, 183
Control
- actions, 2, 17, 24, 25, 27, 68, 73, 107, 154, 159, 161, 210, 219
- adaptive, 5, 24–27, 134, 135, 140, 149, 151, 157, 158
- aircraft, 11, 149
- algorithms, 2, 4, 24, 25, 139
- ANN, 118
- flight, 132
- goals adaptation, 22, 23
- horizon, 154, 155
- inputs, 63, 85
- law, 13, 21, 25, 117, 118, 120, 121, 125, 139, 157–161

neural network, 103, 110
object, 24, 106
performance, 145, 147, 149
problem, 5, 114, 124
quality, 13, 117, 154, 158
scheme, 106, 141, 153
signal, 80, 101, 116, 131, 154, 155, 171, 183, 193, 194, 201, 203, 212, 219
signals, 80
surfaces, 149, 159, 208, 213
synthesis, 19, 94
theory, 102, 112
variables, 17, 72, 73, 75, 77, 78, 80, 82, 96, 101, 103, 105, 108, 173, 202, 210, 217
vector, 17, 18, 115, 156
Controllability, 11, 139, 208
Controllable
influences, 18
Controllable system, 12, 13, 15
Controller, 24–27, 102, 104, 116–119, 124, 140, 144, 147–149, 156, 159
adjusting, 104, 106, 107
network, 140
PI, 139
Controlling, 6, 9, 25, 93, 117, 160
Correcting controller (CC), 75

D
Deflection angle, 18, 78, 105, 108, 109, 134, 157, 200, 201, 208, 209, 217, 225
Derivatives
error function, 53, 59, 60, 64, 178
Designing control laws
for control systems, 112
for multimode objects, 5
Discrete time, 19, 20, 85, 171
instants, 63, 182
state space, 170
Disturbed motion, 86, 115–117, 160, 208

E
Elevator actuator, 134, 150, 157
ENC optimality criterion, 121, 123
Engine thrust, 134, 153, 212, 218
Engine thrust aircraft, 216
Engine thrust control, 134, 152
Error
covariance, 67, 68
covariance estimation, 67, 68
function, 52–59, 62, 63, 65, 67, 156, 178, 183, 187, 189–192, 195
derivatives, 53, 59, 60, 64, 178
Hessian, 182, 191
landscape, 187
value, 182
gradient, 66
signal, 131
values, 150, 152–154
EXogeneous inputs, 46, 96, 98
Extended Kalman filter (EKF), 67

F
Feedforward network, 39–41, 43, 65, 96, 97, 99, 101
Feedforward neural network, 41, 43, 96, 99, 100, 117, 166, 189, 192
Flight, 12, 15, 17, 18, 25, 105, 109, 112, 149, 153, 157, 158, 201, 204, 206, 217
aircraft, 10
altitude, 160, 210, 218
conditions, 102, 160
control, 132
mode, 112, 139, 158
path angle, 18, 217
regimes, 12, 21, 118, 139, 153, 157, 158, 160, 200
Functional basis (FB), 35

G
Gain Scheduling (GS), 25
Gradient descent (GD), 53

H
Hessian, 55, 57, 59–61, 63, 65, 178, 180, 181, 183, 190, 191
computation, 55
error function, 182, 191
matrix, 57, 62
nonsingular, 54
Hidden layer neurons, 46, 60, 61, 132, 133, 176
Hidden layers, 43, 46, 108, 117, 118, 121, 125, 224
Hidden layers outputs, 101
Hybrid ANN models, 101
Hypersonic aircraft, 135, 145

I
Identification problem, 4, 5, 140, 159, 208, 213, 216, 217, 225
Incremental learning, 52, 212
Incremental learning process, 72
Indirect adaptive control, 25, 26
Indirect approach, 73, 80, 103–106
Inputs, 39, 41–43, 48, 50, 51, 59, 61, 62, 66, 70, 85, 95–98, 101, 107, 224
ANN, 108
control, 63, 85
network, 43, 45, 60–62
neurons, 45, 50
Instantaneous error function, 64, 65, 178
Intelligent control, 102
International Standard Atmosphere (ISA), 218
Interneurons, 69, 70
Interpolation error, 185
Intersubnets, 70
Inverse Hessian, 55, 56
Inverse Hessian approximation, 55, 56
Inverse Hessian matrix, 194

K
Kalman filter (KF), 68, 133

L
Layered ANN model, 39, 132
Layered feedforward networks, 42, 101
Layered Feedforward Neural Network (LFNN), 45, 53, 58, 60, 64–66, 166, 168, 175, 176, 181
Learning
algorithm, 67, 77

ANN, 144
neural network, 203
process, 4, 69, 77, 132, 140, 144
rate, 53
reinforcement, 51, 52
strategy, 211
supervised, 51, 52, 132
unsupervised, 51
Linear ANN model, 43

M

Maneuverable aircraft, 5, 6, 78, 134, 139, 200, 208, 210, 216, 217, 219, 225
Manned aircraft, 1, 2
Model predictive control (MPC), 5, 139, 154, 156, 157
Model reference adaptive control (MRAC), 5, 27, 139
Motion
aircraft, 4, 5, 72, 101–103, 108, 117, 131, 133, 135, 139, 152, 154, 158, 168, 199, 200, 208
model, 108, 109, 118, 134, 153, 199
plant, 103
MRAC system, 145, 150, 151, 154, 157–159, 161
Multilayer neural network, 5, 131, 134, 156
Multimode dynamical system (MDS), 5, 117

N

NARX model, 132, 133, 176, 206
neural network, 132
Network
controller, 140
inputs, 43, 45, 60–62
operation, 42, 52
outputs, 45, 46, 60, 64
outputs derivatives, 61, 66
Network models (NM), 38
Neural controllers, 5, 102, 110, 112, 119, 120, 124, 125, 144
Neural network
black box, 4, 94
control, 103, 110
control system, 160
learning, 51, 203
motion model, 111
NARX model, 132
outputs, 58
stabilizer actuator, 109
supervised learning, 52
training, 57, 58, 62
Neurocontroller
configuration, 144
ensemble, 120
mutually agreed, 120
mutually coordinated, 124
synthesis, 5, 114, 141, 156
Neurocorrector, 118, 125
Neurons
inputs, 45, 50
outputs, 50
Nonlinear
controlled systems, 100
dynamical systems, 4, 5
dynamical systems control, 101, 139
ODEs, 217

O

Object
adaptation, 22
control, 24, 106
model, 154
moving, 16
Onboard control systems, 94
Online adaptation, 63, 192
Operation
modes, 72, 95, 116, 118, 120, 123
network, 42, 52
Optimal
control law, 121
control problem, 193
synthesis, 5
tracking control problem, 135
Optimality criterion, 118, 120–124, 126, 193
Ordinary differential equations (ODEs), 3, 4, 19, 28, 29, 165, 168–170, 172, 177, 180–183, 191, 199, 200, 209
nonlinear, 217
Outputs, 39, 41, 42, 48–50, 59, 66, 93–99, 101, 106–108, 117, 131, 132, 140, 144
ANN, 108
network, 45, 46, 60, 64
neural network, 58
neurons, 50
target, 107
true, 58, 63

P

Parametric adaptation, 22
Partial derivatives, 61, 62, 200, 208
Partial differential equations (PDE), 3, 28, 165
Phase angles, 85, 87
Pitch, 18, 105, 106, 108, 134, 158, 200, 201, 208–210, 217
angle, 18, 77, 78, 105, 134, 201, 209, 217
angular velocity, 76, 78
channel, 208
moment, 105, 152, 200, 216–218
moment coefficient, 18, 134, 201
Pitching
moment coefficient, 78
Pitching moment, 77
Plant, 75, 93, 102–104, 106, 108, 116, 125
model, 107, 108
motion, 103
state space, 103
Polyharmonic control signal, 212
Prediction error, 63, 67, 97, 175–178

R

Recurrent neural network architectures, 172
Recurrent neural network training, 65, 189
Recurrent neural networks (RNN), 43, 47, 65, 66, 97, 99, 101, 170, 172, 173, 187, 192
Reference model (RM), 25
Roll angle, 18
Rotational motion, 221

S

Semiempirical

ANN, 4, 199–201, 208–210, 212, 217, 221, 224, 225
ANN model, 170, 204, 212
ANN model learning, 222
approach, 5, 166, 170, 200, 208
model, 5, 166–168, 173, 176, 177, 179–181, 183, 184, 193, 204, 206–208, 212, 216, 219, 225
modeling, 166, 200, 208, 216
neural network, 208
Simulation error, 31, 96, 97
Stabilizer actuator, 108, 109, 217, 219
neural network, 109
State space, 4, 5, 46, 63, 67, 81, 82, 94, 95, 98, 101, 135, 151, 159, 162, 168–170, 172, 183, 184, 193, 214, 219
discrete time, 170
plant, 103
Stochastic gradient descent (SGD), 57
Synthesis, 2–4, 13, 21, 31, 103, 110–112, 116, 121, 125, 134, 135, 139, 158
control, 19, 94
neurocontroller, 5, 114, 141, 156
optimal, 5
problem, 29
processes, 118
System time, 11, 161

T

TDL elements, 43, 132, 140
Thrust, 10, 17, 18, 105, 109, 134, 135, 152, 154, 217, 218
Thrust control, 134, 152
Thrust control channel, 134
Thrust level, 135
Time Delay Neural Network (TDNN), 43
Translational motion, 210, 212
Truncation error, 184, 185, 191

U

UAV behavior, 2
UAV behavior control system, 1
UAV behavior intelligent control, 2
UAV motion control, 2
Uncertainty conditions, 2, 5, 19, 29, 139, 158, 162
Uncertainty factors, 10, 13, 14, 16, 21, 24, 208
Uncontrollability, 12
Uncontrollable, 12, 16
disturbances, 13
effects, 30
influences, 17, 18
Uncontrolled
aircraft, 11
disturbances, 10, 19, 29, 30, 93
missile, 16
rocket, 12
Unmanned aerial vehicles (UAV), 1, 9

Y

Yaw angle, 18

NPH

CPI Antony Rowe
Eastbourne, UK
June 18, 2019